FLOOD GEOMORPHOLOGY

Flood Geomorphology

Editor:

Navodita Bhatnagar

Delve Publishing

www.delvepublishing.com

Flood Geomorphology

Editor: **Navodita Bhatnagar**

708 3rd Avenue, 6th Floor
New York, NY 10017
United States of America
www.delvepublishing.com

ISBN: 978-1-68095-824-9

Library of Congress Control Number: 2016951566

About the Editor

Navodita Bhatnagar

Navodita Bhatnagar finished her masters in technology and currently pursuing her PhD in environmental sciences from Institute of Technology, Carlow, Ireland. She's a keen enthusiast of environment protection and management. An alumunus of Indo-German centre for sustainibility which is essentially a collaborative interface to gather scientists in the field of environment sciences to come together and work for the common cause which is environment protection.

Contents

List of Contributors

B. Merz
GFZ German Research Centre for Geosciences, 14473 Potsdam, Germany

J. Aerts
Institute for Environmental Studies, VU University Amsterdam, 1081HV Amsterdam, the Netherlands

K. Arnbjerg-Nielsen
Dept. of Env. Eng., Technical University of Denmark, Lyngby, Denmark

M. Baldi
CNR-Institute of Biometeorology, 00185 Rome, Italy

A. Becker
Deutscher Wetterdienst, Offenbach am Main, Germany

A. Bichet
University of Toronto, Physics Department, Toronto, Canada

G. Blöschl
Institute of Hydraulic Engineering and Water Resources Management, Vienna University of Technology, Vienna, Austria

L. M. Bouwer
Deltares, Delft, the Netherlands

A. Brauer
GFZ German Research Centre for Geosciences, 14473 Potsdam, Germany

F. Cioffi
DICEA, Department of Civil, Constructional & Environmental Engineering, University of Rome, La Sapienza, Rome, Italy

J. M. Delgado
University of Potsdam, Institute of Earth and Environmental Science, Potsdam, Germany

M. Gocht
Technische Universität Braunschweig, 38106 Braunschweig, Germany

F. Guzzetti
CNR IRPI, 06128 Perugia, Italy

S. Harrigan
ICARUS, Dept. of Geography, National University of Ireland, Maynooth, Ireland

K. Hirschboeck
Laboratory of Tree-Ring Research, University of Arizona, Tucson, Arizona, USA

C. Kilsby
School of Civil Engineering and Geosciences, Newcastle University, Newcastle, UK

W. Kron
Munich Re, 80802 Munich, Germany

H.-H. Kwon
Department of Civil Engineering, Chonbuk National University, Jeonju, Republic of Korea

U. Lall
Dept. of Earth & Env. Eng., Columbia University, New York, USA

R. Merz
Helmholtz Centre for Environmental Research, Leipzig, Germany

K. Nissen
Institute for Meteorology, Freie Universität Berlin, 12165 Berlin, Germany

P. Salvatti
CNR IRPI, 06128 Perugia, Italy

T. Swierczynski
GFZ German Research Centre for Geosciences, 14473 Potsdam, Germany

U. Ulbrich
Institute for Meteorology, Freie Universität Berlin, 12165 Berlin, Germany

A. Viglione
Institute of Hydraulic Engineering and Water Resources Management, Vienna University of Technology, Vienna, Austria

P. J. Ward
Institute for Environmental Studies, VU University Amsterdam, 1081HV Amsterdam, the Netherlands

M. Weiler
Chair of Hydrology, Faculty of Environment and Natural Resources, University of Freiburg, Freiburg, Germany

B. Wilhelm
EDYTEM, Université de Savoie, 73376 Le Bourget-du-Lac, France

M. Nied
GFZ German Research Centre for Geosciences, 14473 Potsdam, Germany

Ewa Bogdanowicz
Institute of Meteorology and Water Management, Podleśna 61, 01-673 Warsaw, Poland

Witold G. Strupczewski
Institute of Geophysics, Polish Academy of Sciences, Księcia Janusza 64, 01-452 Warsaw, Poland

Krzysztof Kochanek
Institute of Geophysics, Polish Academy of Sciences, Księcia Janusza 64, 01-452 Warsaw, Poland

Iwona Markiewicz
Institute of Geophysics, Polish Academy of Sciences, Księcia Janusza 64, 01-452 Warsaw, Poland

Subhankar Karmakar
Centre for Environmental Science and Engineering, Indian Institute of Technology Bombay, Mumbai, India

Slobodan P. Simonovic
Department of Civil and Environmental Engineering, The University of Western Ontario, London, Canada

Angela Peck
Department of Civil and Environmental Engineering, The University of Western Ontario, London, Canada

Jordan Black
Department of Civil and Environmental Engineering, The University of Western Ontario, London, Canada

Nilay Kanti Barman
Department of Geography, Hijli College, Kharagpur-721306, India

Soumendu Chetterjee
Department of Geography, Presidency University, Kolkata-700073, India

Ansar Khan
Department of Geography and Environment Management, Vidyasagar University, Midnapore-721102, India
Mariana Madruga de Brito
Department of Geography, University of Bonn, Bonn, Germany

Mariele Evers
Department of Geography, University of Bonn, Bonn, Germany

U. C. Nkwunonwo
Department of Geoinformatics and Surveying, University of Nigeria, Enugu Campus, Nigeria
School of Earth and Environmental Sciences, University of Portsmouth, Portsmouth, UK

M. Whitworth
School of Earth and Environmental Sciences, University of Portsmouth, Portsmouth, UK

B. Baily
Department of Geography, University of Portsmouth, Portsmouth, UK

Audrey C. Rule
Department of Curriculum and Instruction, University of Northern Iowa, Cedar Falls, USA

Jean Suchsland Schneider
Department of Curriculum and Instruction, University of Northern Iowa, Cedar Falls, USA

Denise A. Tallakson
Department of Curriculum and Instruction, University of Northern Iowa, Cedar Falls, USA

Diane Highnam
Area Education Agency 267, Cedar Falls, USA

Karl Erich Lindenschmidt
Manitoba Water Stewardship, Surface Water Management, Winnipeg, Manitoba, Canada

Robert Harrison
Manitoba Water Stewardship, Surface Water Management, Winnipeg, Manitoba, Canada

Martina Baborowski
Department of River Ecology, UFZ-Centre for Environmental Research, Magdeburg, Germany.

Lorenzo Alfieri
Institute for Environment and Sustainability, European Commission – Joint Research Centre, 21027 Ispra, Italy

Luc Feyen
Institute for Environment and Sustainability, European Commission – Joint Research Centre, 21027 Ispra, Italy

Peter Salamon
Institute for Environment and Sustainability, European Commission – Joint Research Centre, 21027 Ispra, Italy

Jutta Thielen
Institute for Environment and Sustainability, European Commission – Joint Research Centre, 21027 Ispra, Italy

Alessandra Bianchi
Institute for Environment and Sustainability, European Commission – Joint Research Centre, 21027 Ispra, Italy

Francesco Dottori
Institute for Environment and Sustainability, European Commission – Joint Research Centre, 21027 Ispra, Italy

Peter Burek
International Institute for Applied Systems Analysis (IIASA), 2361 Laxenburg, Austria

Raffaele Albano
School of Engineering, University of Basilicata, Potenza, Italy

Aurelia Sole
School of Engineering, University of Basilicata, Potenza, Italy

Francesco Sdao
School of Engineering, University of Basilicata, Potenza, Italy

Luciana Giosa
School of Engineering, University of Basilicata, Potenza, Italy

Andrea Cantisani
School of Engineering, University of Basilicata, Potenza, Italy

Stefania Pascale
School of Engineering, University of Basilicata, Potenza, Italy

R. Albano
School of Engineering, University of Basilicata, Potenza, Italy

A. Sole
School of Engineering, University of Basilicata, Potenza, Italy

J. Adamowski
Bioresource Department, McGill University, Montreal, Canada

L. Mancusi
Sustainable Development and Energy Resources Department, Research on Energy Systems SpA, Milano, Italy

K. Papagiannaki
Institute of Environmental Research and Sustainable Development, National Observatory of Athens, Athens, Greece

K. Lagouvardos
Institute of Environmental Research and Sustainable Development, National Observatory of Athens, Athens, Greece

V. Kotroni
Institute of Environmental Research and Sustainable Development, National Observatory of Athens, Athens, Greece

A. Bezes
Institute of Environmental Research and Sustainable Development, National Observatory of Athens, Athens, Greece

G. Bambara
Irstea, 3275 Route de Cézanne, CS40061, 13182 Aix-en-Provence CEDEX 5, France

L. Peyras
Irstea, 3275 Route de Cézanne, CS40061, 13182 Aix-en-Provence CEDEX 5, France

H. Felix
Irstea, 3275 Route de Cézanne, CS40061, 13182 Aix-en-Provence CEDEX 5, France

D. Serre
RESCUE Solutions, 7 avenue Jeanne d'Arc, 38100 Grenoble, Switzerland
Urban Water, 38 rue Croix des petits Champs, 75001 Paris, France
HDR associé à l'UMR LIED, Université Paris-Diderot, Sorbonne Paris Cité, France

Nkwunonwo UC
Department of Geoinformatics and Surveying, University of Nigeria, Enugu Campus, Nigeria
School of Earth and Environmental Sciences, University of Portsmouth, UK

Whitworth Malcolm
School of Earth and Environmental Sciences, University of Portsmouth, UK

Baily Brian
Department of Geography, University of Portsmouth, UK
D. J. Wagenaar
Deltares, Delft, the Netherlands

K. M. de Bruijn
Deltares, Delft, the Netherlands

L. M. Bouwer
Deltares, Delft, the Netherlands

H. de Moel
Institute for Environmental Studies, Faculty of Earth and Life Sciences, VU University Amsterdam, Amsterdam, the Netherlands

A. J. Stevens
Faculty of Engineering and Environment, University of Southampton, University Road, Highfield, Southampton, SO17 1BJ, UK

D. Clarke
Faculty of Engineering and Environment, University of Southampton, University Road, Highfield, Southampton, SO17 1BJ, UK
Tyndall Centre for Climate Change Research, Norwich, UK

R. J. Nicholls
Faculty of Engineering and Environment, University of Southampton, University Road, Highfield, Southampton, SO17 1BJ, UK
Tyndall Centre for Climate Change Research, Norwich, UK

M. P. Wadey
Ocean and Earth Science, National Oceanography Centre, University of Southampton Waterfront Campus, European Way, Southampton, SO14 3ZH, UK

María Carolina Rogelis
UNESCO IHE, P.O. Box 3015, 2601DA Delft, the Netherlands

Micha Werner
UNESCO-IHE, P.O. Box 3015, 2601DA Delft, the Netherlands
Deltares, P.O. Box 177, 2600MH Delft, the Netherlands

Nelson Obregón
Universidad Javeriana, KR 7 No 40-62, Bogotá, Colombia

Nigel Wright
School of Civil Engineering, University of Leeds, Leeds, LS2 9JT, UK

Preface

Floods are essentially a natural phenomenon that affects the geomorphology of surrounding landform and may result in events like erosion, sedimentation or deposition. Certain areas are more prone to floods than the others. It is important to study the concepts behind their occurrence, frequency, vulnerability of an area to floods and the geomorphological changes post floods owing to their damaging effects on the environment.

Floods arise due to many reasons ranging from geomorphological setting, climate of a particular area, nature of the coastline, or anthropological factors like man made dams. Flood damages continue to rise as there is increased global warming and glaciers melting leading to rise in sea levels combined with increased industrialization and urbanization. A very high percentage of population is living and working in flood prone areas and continues to do so.

There are economic and social implications when in post flood scenario. In this book, articles have been included to provide an insight into the flood risk assessment, developing models to study the impact on affected areas, socio-economic aspects of flood damage, and use of Geographic information systems for risk evaluation. I hope the book will suffice its purpose.

Editor
Navodita Bhatnagar

CHAPTER 1

FLOODS AND CLIMATE: EMERGING PERSPECTIVES FOR FLOOD RISK ASSESSMENT AND MANAGEMENT

B. Merz[1], J. Aerts[2], K. Arnbjerg-Nielsen[3], M. Baldi[4], A. Becker[5], A. Bichet[6] , G. Blöschl[7] , L. M. Bouwer[8] , A. Brauer[1] , F. Cioffi[9] , J. M. Delgado[10], M. Gocht[11], F. Guzzetti[12], S. Harrigan[13], K. Hirschboeck[14], C. Kilsby[15], W. Kron[16] , H.-H. Kwon[17], U. Lall[18], R. Merz[19], K. Nissen[20], P. Salvatti[12], T. Swierczynski[1] , U. Ulbrich[20], A. Viglione[7] , P. J. Ward[2] , M. Weiler[21], B. Wilhelm[22], and M. Nied[1]

[1]GFZ German Research Centre for Geosciences, 14473 Potsdam, Germany

[2]Institute for Environmental Studies, VU University Amsterdam, 1081HV Amsterdam, the Netherlands

[3]Dept. of Env. Eng., Technical University of Denmark, Lyngby, Denmark

[4]CNR-Institute of Biometeorology, 00185 Rome, Italy

[5]Deutscher Wetterdienst, Offenbach am Main, Germany

[6]University of Toronto, Physics Department, Toronto, Canada

[7]Institute of Hydraulic Engineering and Water Resources Management, Vienna University of Technology, Vienna, Austria

[8]Deltares, Delft, the Netherlands

[9]DICEA, Department of Civil, Constructional & Environmental Engineering, University of Rome, La Sapienza, Rome, Italy

[10]University of Potsdam, Institute of Earth and Environmental Science, Potsdam, Germany

[11]Technische Universität Braunschweig, 38106 Braunschweig, Germany

[12]CNR IRPI, 06128 Perugia, Italy

[13]ICARUS, Dept. of Geography, National University of Ireland, Maynooth, Ireland

[14]Laboratory of Tree-Ring Research, University of Arizona, Tucson, Arizona, USA

[15]School of Civil Engineering and Geosciences, Newcastle University, Newcastle, UK

[16]Munich Re, 80802 Munich, Germany

[17]Department of Civil Engineering, Chonbuk National University, Jeonju, Republic of Korea

[18]Dept. of Earth & Env. Eng., Columbia University, New York, USA

[19]Helmholtz Centre for Environmental Research, Leipzig, Germany

[20]Institute for Meteorology, Freie Universität Berlin, 12165 Berlin, Germany

[21]Chair of Hydrology, Faculty of Environment and Natural Resources, University of Freiburg, Freiburg, Germany

[22]EDYTEM, Université de Savoie, 73376 Le Bourget-du-Lac, France

ABSTRACT

Flood estimation and flood management have traditionally been the domain of hydrologists, water resources engineers and statisticians, and disciplinary approaches abound. Dominant views have been shaped; one example is the catchment perspective: floods are formed and in- fluenced by the interaction of local, catchment-specific characteristics, such as meteorology, topography and geology. These traditional views have been beneficial, but they have a narrow framing. In this paper we contrast traditional views with broader perspectives that are emerging from an improved understanding of the climatic context of floods. We come to the following conclusions: (1) extending the traditional system boundaries (local catchment, recent decades, hydrological/hydraulic processes) opens up exciting possibilities for better understanding and improved tools for flood risk assessment and management. (2) Statistical approaches in flood estimation need to be complemented by the search for the causal mechanisms and dominant processes in the atmosphere, catchment and river system that leave their fingerprints on flood characteristics. (3) Natural climate variability leads to time-varying flood characteristics, and this variation may be partially quantifiable and predictable, with the perspective of dynamic, climate-informed flood risk management. (4) Efforts are needed to fully account for factors that contribute to changes in all three risk components (hazard, exposure, vulnerability) and to better understand the interactions between society and floods. (5) Given the global scale and societal importance, we call for the organization of an international multidisciplinary collaboration and datasharing initiative to further understand the links between climate and flooding and to advance flood research.

INTRODUCTION

Flood estimation and flood management have a long history, over which dominant views have been shaped. One example is the river basin paradigm. In this, floods are seen from the catchment perspective; they are formed and influenced by the interaction of local, i.e. catchment-specific, characteristics of meteorology, topography, geology, vegetation, etc. Another traditional view is that floods are essentially random events: typically, for statistical convenience, extreme flood occurrence and magnitude are assumed to correspond to an independent, identically distributed (iid) random process represented by an appropriate probability density function

During the EGU topical meeting "Floods and Climate: Understanding and exploiting the link between floods and climate" on 4–5 October 2012 in Potsdam, the climate context of floods was discussed: how are floods related to climate and, in particular, to climate variability? To what extent can we take advantage of linkages between climate and floods? The workshop participants felt that, although the traditional views have been beneficial, they have a narrow framing, and that flood estimation and flood management could benefit from a broader perspective. For instance, the local catchment perspective needs to be complemented by the global view, since floods are embedded in a global climate context. The actual flood potential of a catchment and the characteristics of floods may be influenced by global climate mechanisms. Hirschboeck (1987) offered the hypothesis that unusually large floods in drainage basins of all sizes may be related to large-scale atmospheric circulation anomalies. Nakamura et al. (2013) show that 20 major flood events over the last 100 years in the Ohio River basin have nearly identical storm tracks, moisture source and delivery patterns. Such persistent anomalies may in turn reflect modulation of the storm tracks by identifiable climate phenomena such as the El Niño–Southern Oscillation (ENSO) and Pacific Decadal Oscillation (PDO). Consequently, it seems necessary to complement the local catchment perspective with a global view in order to investigate whether floods are purely random and whether the widespread iid assumption holds, and to develop flood risk assessment methods which are capable of taking advantage of such extended perspectives. New perspectives are emerging even more clearly in the field of flood risk management. Traditionally, flood management strategies have focused on reducing the flood hazard, i.e. the probability of flooding. Hence, flood research has centred on the physical processes within catchments and river systems. Over the past two decades there has been a shift from the hazard-focused view to the broader riskbased perspective, including societal processes and implications (Merz et al., 2010a). Flood disaster risk results from the interaction of hazard, exposure and vulnerability. Hazard represents the probability and intensity of flooding; exposure describes the elements at risk – for example, the people and their assets that may be affected by flooding; and vulnerability describes the susceptibility or propensity of elements at risk to be adversely affected. This concept is at the heart of the Global Assessment Reports of the United Nations Of- fice for Disaster Risk Reduction (UNISDR, 2011, 2013), and the IPCC (2012) SREX report adopted it for analysing risks related to climate change.

It is taken for granted that changes in climate or human interventions in catchments and river systems may change flood hazard and, as a consequence, flood risk. Within this view, floods are evaluated from a hazard perspective, focusing on hydrologic/hydraulic parameters such as discharge, water level or inundation extent. Societal processes are often neglected, which implicitly means they are assumed to be constant or, if random, a stationary process. However, some socio-economic processes, like population growth and economic development, may change at a faster pace than long-term physical changes (for example, the impacts of climate change on discharge), and exposure and vulnerability to floods can be highly dynamic. Against this background, societal processes need to be addressed within a risk-based approach, where next to the hazard, societal exposure and vulnerability play a decisive role. A particularly interesting question is how space–time variations in flood hazard that may be related to climate variability and change intersect with the changing nature of the flood exposure and vulnerability. As an example, globalization

Citation: Merz, B., Aerts, J., Arnbjerg-Nielsen, K., Baldi, M., Becker, A., & Bichet, A. et al. (2014). Floods and climate: emerging perspectives for flood risk assessment and management. Nat. Hazards Earth Syst. Sci., 14(7), 1921-1942.

leads to potential pathways of supply chain disruption as seen with the long-duration floods in Thailand (2011); Queensland, Australia (2010, 2011); the Indus River (2010); and the Mississippi River (2012), which led to significant local and global economic impacts.

Table 1 contrasts the traditional narrow framing of floods with the broader perspective that is emerging from an improved understanding of the climatic context of flood generation. The different aspects of Table 1 are closely linked. This becomes obvious in Sect. 2, which focuses on understanding climate–flood linkages. In Sect. 3, we discuss how this broader framing may be used for improved flood estimation and risk management. In particular, we explore how an improved process-based understanding of climate–flood linkages at multiple scales can be exploited to address critical management and societal issues in a rapidly changing world.

Table 1. Contrasting traditional views with emerging perspectives on flood hazard and risk.

Aspect	Traditional view	Emerging perspective
	Understanding climate–flood linkages	
Randomness	*Random*: floods are random events with flood magnitude quantified by a probability distribution.	*Causal*: flood occurrence and magnitude depend on a causal network of processes in atmosphere, catchment and river systems. A fraction of the flood variability is described by deterministic processes: for example, by using climate information as co-variates in flood probability distributions.
Spatial perspective	*Local*: floods are events that can be described fully by processes on a catchment scale.	*Global*: floods occur within the spatial framework of large-scale circulation patterns and global climate mechanisms.
Natural variability and floods	*Stationary*: flood characteristics are stationary and represent the long-term natural variability of the climate–catchment system.	*Time-varying*: flood characteristics change in time due to climate variability at different timescales.
Temporal perspective	*Recent*: flood characteristics result from current catchment characteristics and are derived from recent observations.	*Long-term*: flood characteristics result from the long-term interplay of climate, geology, topography, vegetation (biology), and humans. To fully understand floods, this long-term interplay has to be disentangled.
	Exploiting climate–flood linkages	
Flood estimation	*Process-neutral*: flood estimation does not differentiate between different flood event types and processes. Flood frequency analysis is based on iid assumption.	*Process-based*: flood events of different types occur in a given catchment. Knowledge on flood generation processes provides information on flood probability estimation.
Flood projections under climate change	*Model chain*: flood scenarios are the result of model chains, from emission scenarios to climate models to flood frequency estimation.	*Model-chain-augmented*: in addition to model chains, a range of approaches for assessing climate-related flood changes are used, such as assessing historical climate variability, or using ocean source–atmospheric moisture transport–flood linkages.
Flood risk management	*Hazard-focused, static*: flood management focuses on flood hazard reduction within a static framework, principally using structural or zoning flood proofing or financial instruments (insurance).	*Risk-oriented, dynamic*: risk management takes into account changing hazard, exposure and vulnerability, and the combined application of financial, structural and non-structural measures. The best way to mitigate floods depends on how well changes in flood risk can be predicted at short and long timescales.

This paper is based on the EGU topical meeting "Floods and Climate: Understanding and exploiting the link between floods and climate" in October 2012 in Potsdam. This interdisciplinary meeting was dedicated to discuss innovative approaches and to develop new perspectives in the broad field of how floods are embedded in the climate context. This paper is the attempt to organize the different aspects in this field within one paper, putting emphasis on how different ideas have been emerging during the last years and how they can be connected.

Understanding climate–flood linkages

Climate–flood linkages operate across multiple spatial and temporal scales determined by (i) the nature of the meteorological and climatic processes that drive the atmospheric components of the hydrologic cycle and (ii) the nature of the hydrologic land-surface and subsurface processes within which a flood develops. As implied in Table 1, the traditional approach for linking atmospheric components to flood analysis has been primarily hydro-meteorological, i.e. framed by local weather events of recent decades interacting with current catchment characteristics. The emerging view of climate–flood linkages is process driven and seeks to understand and analyse flood events in the context of their long-term history of variation – in magnitude, frequency, and seasonality – and within the climatic framework of the global and regional atmospheric circulation patterns and processes that drive changing combinations of meteorological elements at the catchment scale. In the following, we discuss these contrasting perspectives.

RANDOMNESS OR CAUSALITY?

Floods are typically seen as events that are generated by the random superposition of processes in the atmosphere, catchment and river system. The idea of very strong or even complete randomness prevails in the history of flood prediction. Much effort has been spent on describing flood occurrence and magnitude by probability distributions, a large share of the flood research has been focused on statistical aspects, and the role of randomness has been emphasized. This avenue has been shaped by the pioneers of flood frequency analysis; for example, in 1941 Emil Julius Gumbel mentioned the flood estimation problem: "The author believes it is possible to give exact solutions, exactitude being interpreted from the standpoint of the calculus of probabilities." (Gumbel, 1941, p. 163).

Figure 1 provides examples of the atmospheric moisture transport averaged across the top 10 floods over a 60- year period in 4 regions of North America. It is clear that there is a distinct large-scale circulation pattern with different moisture sources and storm tracks associated with these extreme events. Such determinism is not restricted to the mid-latitudes. The 2011 persistent flooding in Thailand was marked by five typhoon landings over a period of 100 days, and the typhoon tracks were very similar over this period with a similar area of origin and path. The prevailing La Niña conditions in the western Pacific (the source region of the storms) and phase of the Madden–Julian Oscillation (20– 40-day) period are thought to have interacted to produce the convergence of moisture into Thailand per event. To be sure of this, causality may also be found in other aspects of the climate–catchment system. For example, Nied et al. (2013) show that floods in the Elbe catchment in central Europe are preferentially initiated by specific large-scale soil moisture patterns, and Sivapalan et al. (2005) demonstrate that the flood peaks increase notably if seasonality of rainfall and antecedent soil moisture are in phase.

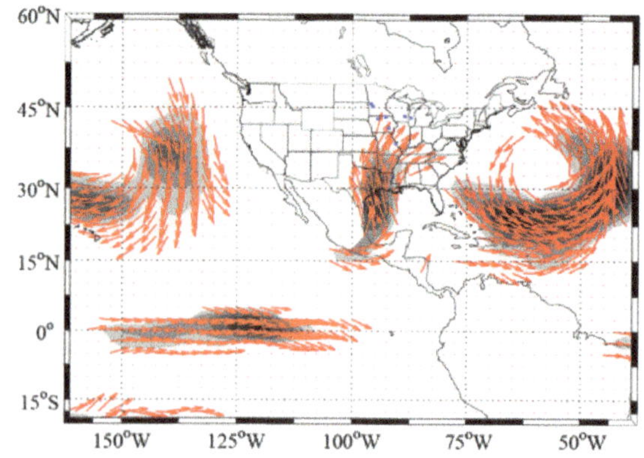

Figure 1. Vertically integrated moisture flux averaged over the dates of the 10 largest floods in the Ohio River basin in North America from 1948 to 2005. The flux is computed daily over 1000–850 mb, using the NOAA-NCEP reanalysis data set. The arrows show the wind direction, and the shading shows the intensity of the moisture flux. Only pixels whose moisture flux magnitude is in the top 5 % of the visible map are shown. Note that a large-scale meridional lowlevel flow into the region is active averaged across these 10 events.

If one takes a causal approach, arguing that flood occurrence and magnitude contain a significant fraction of determinism, then identifying the signatures of major floods in terms of the associated climatic and catchment mechanisms becomes important. A fruitful direction of research could be to identify the dominant mechanisms associated with regional floods, and to see how the occurrence and intensity of those mechanisms lead to the space and time expression of floods. The analysis would move from annual maximum or peak over threshold floods to looking at extreme floods in terms of discharge, volume and duration and their space and time structure over perhaps a season, a decade or longer as could be informed by the nature of the underlying mechanisms. The volume and duration of flooding are directly tied to inundation characteristics and hence represent how the strength and recurrence characteristics of the large-scale atmospheric fluxes that bring moisture into the region interact with catchment hydrologic processes. These in turn determine how the static and dynamic aspects of topography, soils, channel networks and flood control infrastructure process the atmospheric inputs. Given the uncertainties attendant to all these processes, the integrated causal view could provide a basis for better probabilistic modelling of floods from a physical as well as statistical perspective.

Local or global?

Frequently, floods have been understood as local phenomena, driven by the perspective of flood management at the local level, e.g. at the municipal or county scale. Within the last two decades there have been many attempts to reconcile the spatial frame of this societal lens with the traditional hydrological view. Here, floods are considered through the local catchment lens, and are shaped by catchment meteorology, hydrology and river processes. The emerging view extends this spatial framework to the continental and global scale to accommodate the interactions between local flooding and global climate mechanisms.

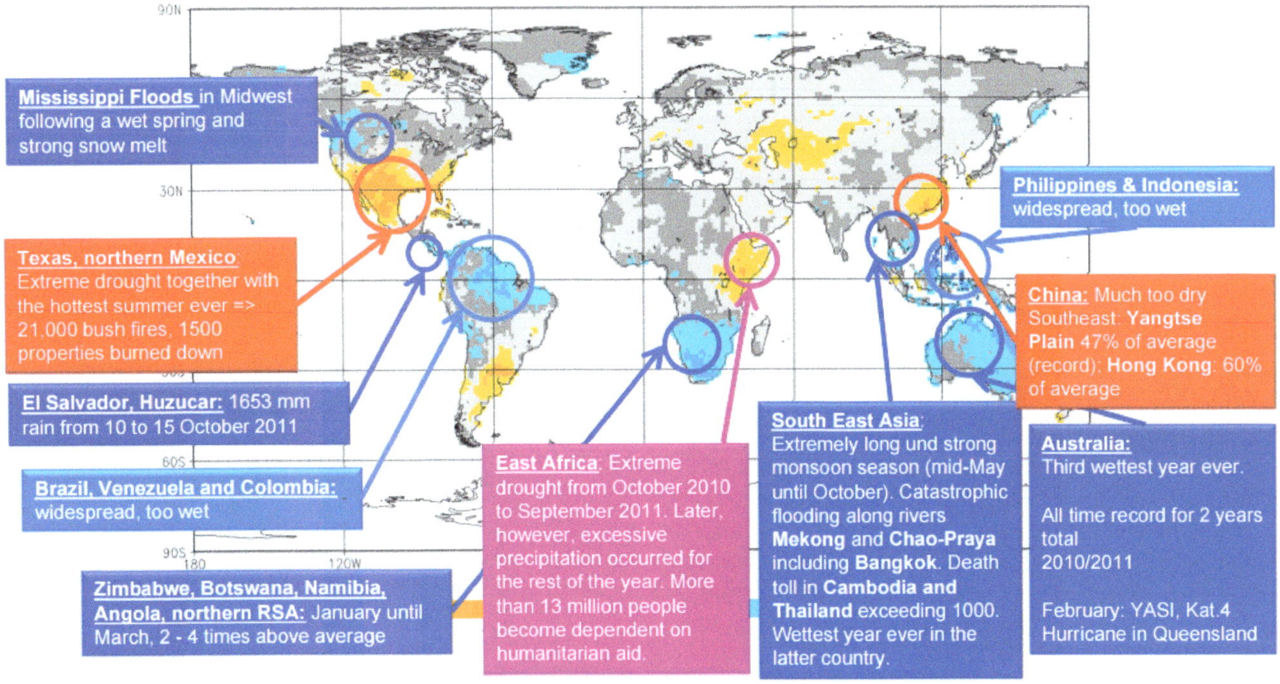

Figure 2. Centennial climatology of ENSO sensitive regions with regard to the global land-surface precipitation from 1901 to 2010. Blue (red) areas have enhanced (reduced) precipitation during La Niña periods. In addition, the circles and boxes point to extreme hydro-meteorological events (including devastating floods) which occurred during the strong La Niña year 2011.

Probably the most famous and well-studied global linkage between large-scale climate oscillations and hydrometeorological variables is ENSO, with variations on a 2– 7-year timescale. For example, linkages can clearly be seen between ENSO indices and gridded precipitation data sets. Such an analysis has been carried out by Becker et al. (2013) using global land-surface precipitation of the past 110 years at a spatial resolution of 0.5° (Schneider et al., 2011). At this scale, the data are capable of resolving global and synopticscale precipitation patterns. Over this longer than centennial time period, the correlation of each grid cell against the negative Southern Oscillation Index (i.e. the occurrence of La Niña events) reveals that years with strong ENSO phases (such as the La Niña years 2010 and 2011) feature precipitation anomaly patterns in the same overall ENSO sensitive regions. Therefore, these predictable anomaly patterns point to regions where extreme precipitation events (Thailand; Queensland, Australia; El Salvador) or droughts (northern Mexico, southern US, southeastern Africa, Hong Kong) have also occurred (Fig. 2) during a specific year or period.

ENSO has been linked to floods in Peru (Waylen and Caviedes, 1986), the USA (Cayan et al., 1999; Jain and Lall, 2000, 2001; Pizaro and Lall, 2002; Sankarasubramanian and Lall, 2003), China (Lin et al., 2005; Zhang et al., 2007), Australia (Kiem et al., 2003) and generally to the extra tropics (Ward et al., 2010), as well as to precipitation anomalies in Europe (Bichet et al., 2014). PDO and the North Atlantic Oscillation (NAO) have been shown to influence precipitation and flood regimes and to lead to flood episodes of varying intensity in various regions (e.g. Cayan, 1996; Jain and Lall, 2000; Bouwer et al., 2006; Kingston et al., 2006; Maraun et al., 2010; Bichet et al., 2011; Rana et al., 2011; Gregersen et al., 2013). Moreover, Ward et al. (2010, 2014b) examine correlations between observed peak discharge and the Southern Oscillation index (SOI) at the global scale and show that peak discharges have a strong sensitivity to ENSO in catchments over large parts of the globe.

In Mediterranean Europe, most of the weather systems causing intense and/or persistent rainfall originate over the Atlantic Ocean and are carried towards the Mediterranean region by westerly winds. These systems can interact with regional topography and produce localized extreme precipitation (Rudari et al., 2005; Pinto et al., 2013), inducing severe river floods. In the upper troposphere over the Euro-Atlantic region, the westerly flow is characterized by two jets: the Atlantic jet, which crosses the ocean with a northeasterly trajectory, and the African jet, which flows above the coast of North Africa. The cross-jet circulation of the Atlantic jet favours storm activity in its exit region, while the cross-jet circulation of the African jet suppresses this kind of activity in its entrance region. It follows that the rainfall distribution downstream of these cross-jet circulations is strongly influenced by their relative positions (Gaetani et al., 2011; Toreti et al., 2010). Specifically, in autumn, rainfall is coplous and more persistent in the western Mediterranean Basin when the Atlantic jet is relatively strong but its northeasterly tilt is small, and the African jet is in its easternmost position. In winter, rainfall is abundant in the eastern Mediterranean Basin; this is when the Atlantic jet reaches the Scandinavian Peninsula and the African jet is in its westernmost position. In spring, when the two jets weaken, the Atlantic jet retreats over the ocean, but the African jet stays in its winter position, and rainfall is more persistent and abundant in the Alpine region and in the Balkans. Therefore, the origin for many floods is not necessarily local, but is related to large-scale circulation patterns like ENSO or jet stream tracks. These examples show that the investigation of global climate processes might help to understand and predict local floods. However, the potential benefit goes further. Linkages across different hazards and across large distances are possible, since the flood and drought potential for far catchments may be related. Specifically, one can hypothesize that persistent spatial patterns of surface temperature in the tropical oceans lead to enhanced probabilities of regions where strong convection takes place and provides a moisture source that could lead to floods in certain areas. The same anomalous surface temperature patterns could also lead to suppression of convection in other tropical areas, as well as to changes in the position and strength of the mid-latitude jet stream and the eddies that are coupled to it. The net result could be that persistent centres of low and high pressure are set up in the mid-latitudes, such that tropical moisture is funnelled repeatedly into the persistent low-pressure regions leading to floods, and droughts persist associated with the atmospheric blocks associated with the high-pressure centres. The net result may be synchronous flooding and drought patterns at seasonal to annual scales across the world. The potential dynamics of such a system at a global scale under a changing climate are discussed in Karamperidou et al. (2012, 2013).

Stationary or time-Varying Random Variables?

Traditional flood frequency analysis assumes stationarity. This assumption implies that flood characteristics fluctuate around a constant value, and that flood quantiles can be determined from data collected over relatively short time periods, typically a few decades. It is assumed that the floodgenerating processes remain constant in time, and that flood probability represents the long-term natural variability of the climate–catchment system. Changes in flood characteristics are expected to result from anthropogenic interventions in this system, such as human-induced climate change, land-use change or river training.

Stationarity implies that each flood record has the same probability of occurrence at any time. However, Sects. 2.1 and 2.2 suggest that there may be periods with distinct flood peak distributions for catchments with strong climate–flood links. Analyses of long-observation data and historical flood records give evidence that floods tend to cluster in time in many regions, and hence that the traditional assumption of time-invariant probability distribution does not hold. Periods with higher or more frequent floods can be differentiated from periods with minor flood activity (e.g. Jacobeit et al., 2003; Schmocker-Fackel and Naef, 2010; for a compilation of European studies, see Hall et al., 2014). This is attributed to low-frequency climate variability, as climate has wellorganized modes of interannual, interdecadal and lowerfrequency variability modulating atmospheric moisture uptake, transport and deposition (precipitation) processes, and hence flood conditions (Hirschboeck, 1988). Knox (2000) studied the effect of climate variability on flood probability at the Mississippi River. He divided the 120-year-long observation period into four decadal-scale sub-periods according to their prevailing global circulation patterns and calculated flood quantiles for each sub-period separately. The 50-year flood was approximately 50 % larger during episodes with more frequent meridional circulation patterns than in periods with more common zonal circulation patterns. This approach of stratifying the flood sample into sub-samples with different climate modes has been repeated, for example, by Kiem et al. (2003) for 40 flood gauges in New South Wales, Australia. The stratification of a regional flood index according to ENSO and the Interdecadal Pacific Oscillation (IPO) leads to very marked differences in flood quantiles. In flood frequency estimation, such variations will show up as a bias as well as an underestimation of the uncertainty of projecting future flood hazards because the interdecadal and longer oscillations are ignored.

Consequently, even in the absence of human-induced changes, the statistical properties of flood time series may vary significantly over time. Flood characteristics may not be fixed, but may float dynamically in time. These regimelike patterns with quasi-cyclic behaviour of flood-rich and flood-poor periods need to be understood, and those regions and conditions under which marked regimes exist need to be identified.

An interesting research question is whether flood regimes occur in accordance across large distances. If the flood behaviour of large regions responds in a similar way to largescale or global climate regimes, this should be seen in concurrent flood-rich clusters and flood-poor periods. For instance, Schmocker-Fackel and Naef (2010) find that large floods tend to synchronously occur in different regions of Europe since 1500 (e.g. in Switzerland and the Czech Republic). Willems (2013) identifies north–south patterns of multidecadal oscillations in rainfall extremes across Europe for the past 100 years and explains them by persistence in atmospheric circulation patterns over the North Atlantic during periods of 10 to 15 years. In the late 19th century, central Europe was particularly affected by recurring large and destructive floods (e.g. Pfister, 1999; Mudelsee et al., 2006; Brazdil et al., 2005; Schmocker-Fackel and Naef, 2010). Using an atmospheric general circulation model (GCM), Bichet et al. (2014) found that this 15-year period of destructive floods in central Europe was induced by a change in the atmospheric circulation (positive PNA-like – Pacific–North American – pattern and negative summer NAO pattern), itself driven by changes in the tropical SSTs (eastern tropical Pacific and Indian Ocean sea surface temperatures). Although the impact of tropical SSTs on European precipitation has previously been documented in the literature (e.g. Bronnimann, 2007), there is still no clear understanding of how and under which circumstances the SST signal transmitted from the tropics to Europe can lead to precipitation extremes associated with flooding. Although other parts of the world were also affected by above-average precipitation in the late 19th century (Gergis and Ashcroft, 2012), the reasons why this period was so exceptional in Europe are still unclear.

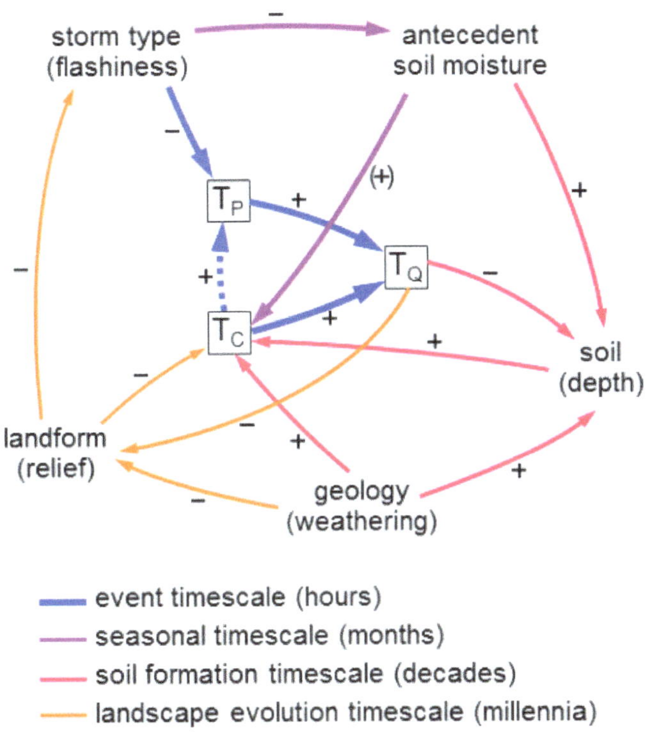

Figure 3. Schematic of the coupling of process controls on the flood timescale. Plus and minus signs indicate whether coupling is positive or negative, i.e. whether an increase in a variable increases or decreases another variable. $_{TP}$: precipitation timescale, characteristic duration of rainfall producing a maximum annual flood; T_C: catchment response timescale, time delay of catchment routing; T_Q: flood timescale, characteristic flood duration (from Gaál et al., 2012).

Recent or Long Term?

Traditionally, understanding of flood characteristics and flood frequency analysis have been focused on data from the most recent decades, based on the idea that flood characteristics result from current catchment characteristics. From a long-term perspective, it must be recognized that flood characteristics are the result of an interplay of climate, geology, topography, vegetation (biology) and humans. As an example, consider the way precipitation and catchment timescales are related to flood timescales. Figure 3 (from Gaál et al., 2012) illustrates some of the factors that influence these three timescales. Note that Fig. 3 could be extended by feedbacks in the climate system and the role of humans as strong floods may lead people to build flood control systems that modify hydrology (Di Baldassarre et al., 2013). The timescales shown in Fig. 3 are not independent, and the interplay amongst them can be interpreted differently at different timescales, from hours to millennia. The events that produce the maximum annual floods are those for which the storm duration is close to the concentration time of the catchment, because the catchment-response timescales filter the distribution of all storms to produce the distribution of floodproducing storms (Viglione and Blöschl, 2009). This is the reasoning behind the rational method for flood estimation and it applies at the event scale (e.g. Pilgrim and Cordery, 1993, p. 9.13). At the seasonal timescale, flood characteristics tend to be closely related to the seasonal water balance, and, conversely, runoff event types affect the seasonal water balance through rainfall and snowmelt (Sivapalan et al., 2005). At the timescale of decades, however, the flow paths as well as soil moisture affect erosion during floods and soil evolution (modulated by differences in geology), while soil depth and permeability affect flow paths and therefore the flood response at the event scale. At the landscape evolution timescale there are further interactions (Abrahams and Ponczynski, 1984; Tucker and Bras, 2000). To fully understand floods, this long-term interplay has to be disentangled. Because of the coupling between different processes across many spatial and temporal scales, the interplay between climate and catchments needs to be seen as a complex system (Rihani, 2002; Raupach, 2005; Kumar, 2007; Blöschl and Merz, 2010), i.e. systems with a large number of strongly interdependent variables at multiple space- and timescales.

One promising avenue to better understand this long-term interplay and the role of climate for flood characteristics is the use of historical and geological archives (e.g. Kochel and Baker, 1982; Baker, 2008; Glaser et al., 2010; Gilli et al., 2012). Historical archives provide exact dates of catastrophic floods, often together with detailed descriptions of the hydro-meteorological conditions and the resulting damage. However, these archives are limited to regions that were populated over a long period and only reach back a couple of centuries. Millennial-scale variations from potentially all regions worldwide can be revealed from river and lake sediment records (e.g. Benito et al., 2003a, b; Swierczynski et al., 2012, 2013; Wilhelm et al., 2012, 2013). While river sediments typically reflect periods of both accumulation and erosion and thus mainly reflect highest-amplitude floods, lakes represent ideal natural traps which continuously collect suspended detrital sediments transported by floods. In the case of annually laminated sediments, even the seasons of palaeofloods can be precisely determined by the microstratigraphic position of a detrital layer within the annual succession of lake deposition (Mangili et al., 2005; Czymzik et al., 2010).

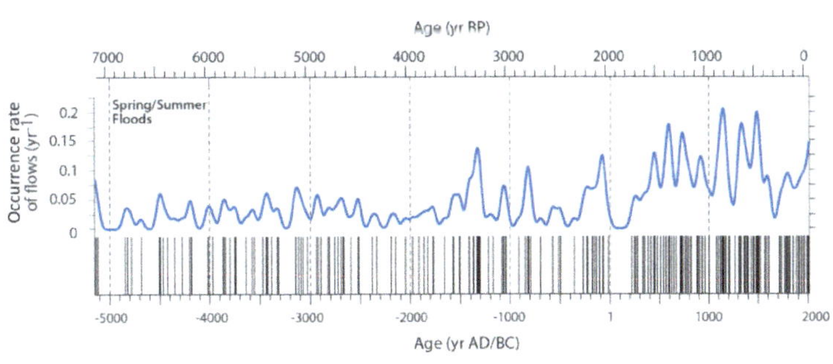

Figure 4. A 7000-year reconstruction of flood events from Mondsee, Upper Austria. Lower panel: occurrence of flood events. Upper panel: occurrence rate calculated by using a Gaussian kernel regression with a 30-year bandwidth (data from Swierczynski et al., 2013).

Figure 4 shows exemplarily the reconstruction of spring and summer floods from laminated sediments of Mondsee, Austria. The amount of temporal variability in the occurrence of floods is striking. There is a period of more than 200 years (21 BC–AD 216) without any flood documented, whereas the average frequency is 0.04 floods year^{-1} , yielding nine floods for such a time interval. Similar fluctuations are seen in other palaeoflood records (e.g. Ely et al., 1993; Knox, 2000; Benito et al., 2003a, b).

Combining flood reconstructions with climate reconstructions provides a great potential to better understand climate– flood linkages. However, interpreting sedimentary palaeo- flood records requires the precise timing of detrital flood layer deposition and a detailed understanding of the mechanisms from the sediment source to the sink. The entire chain of processes needs to be captured, from the hydrometeorological event and the resulting erosion in the catchment to sediment transport into the lake and distribution of suspended material within the lake basin. This in turn requires both detailed investigation of the lacustrine sediment formation including the influence of land-use and vegetation changes on catchment erosion and sedimentation, and calibration of flood layer deposits with meteorological and hydrological observations. Such a detailed understanding allows for an assessment of the completeness of the palaeo- flood record and whether certain flood types are missing. Where very long flood time series extending over millennia can be reconstructed with sufficient reliability, they provide a data-based approach to separate the impact of anthropogenic climate change on floods from natural climate variability. Given the large uncertainty of today's models to simulate floods under changing climate, reliable palaeoflood reconstructions have an exciting potential to determine the natural variability of the flood activity, to better constrain flood projections under climate change, and to better assess flood recurrence times and flood–climate relations for various climate states and timescales.

This long-term view has been successfully used for flood management in dam safety considerations by defining upper limits of flooding through palaeoflood and documentary studies (Enzel et al., 1993; Levish, 2002). However, the systematic integration of palaeo- and historical records into flood frequency analysis has been limited. An argument that is often raised is that if climate is now changing, the long-term record is of little benefit because the past is no longer the key to the present. However, even with a changing climate, from a meteorological/mechanistic perspective, the laws of physics which result in rain, snow and floods are time-invariant. Non-stationarity is produced by changes of these processes in their frequency, magnitude, location, persistence, intensity and clustering. These are driven by the interplay described earlier and are partially deterministic (Sect. 2.1). Hence the extreme events of the past are indeed important indicators of what the atmosphere–catchment system is capable of, given the right interplay of factors. They have left evidence in the landscape of the occurrence of a real event (not something emerging from modelling). Floods – by their very nature – are rare events; hence there is a need to identify them in all temporal contexts (palaeo, historical and systematically gauged) and explain them from a mechanistic understanding of causal drivers.

EXPLOITING CLIMATE–FLOOD LINKAGES TO BETTER MANAGE FLOOD RISK

How can a better understanding of climate–flood linkages be used for improved flood risk management? In the following we contrast the traditional approaches for flood estimation, flood projections under climate change, and for flood management with emerging perspectives based on this improved understanding of floods in their climatic context.

Flood Estimation: Process-Neutral Or Process-Based?

Flood estimation does not traditionally differentiate between flood event types and flood generation processes and can be termed process-neutral. The main approach to estimate flood probabilities is based on extreme value statistics, i.e. fitting distribution functions to ordered sequences of observed flood peaks, ignoring the underlying processes and extrapolating the tails of the distribution to low exceedance probabilities. It is typically assumed that the flood data used are independent and identically distributed. This has been popular largely because most common statistical methods for the analysis of extreme values make this assumption. Consequently, even in locations where a mix of mechanisms such as frontal storms, localized thunderstorms and rain-on-snow events, which have very different space and time attributes, leads to floods, most analyses focus on finding a single best distribution to fit them instead of identifying and modelling the mixtures. Flood heterogeneity has been shown to invalidate the iid assumption in different catchments (e.g. Murphy, 2001; Todhunter, 2012). Further, flood clustering may introduce serial correlation in the time series and may invalidate the independence assumption.

Even more disturbing is the possibility that climate-related fluctuations may significantly bias flood quantile estimates. Franks and Kuczera (2002) point to the need to distinguish between conditional and unconditional flood probability. The unconditional, long-term probability averages across different climate states, whereas the conditional probability depends on the current climate state. Traditional flood frequency analysis results in sound estimates of unconditional flood probability only if the observed data accurately represent the different climate states. Moreover, it may seriously over- or underestimate flood probability in a given year. In a comparison of unconditional and conditional flood probability, Kwon et al. (2008) found that the conditional 100-year flood fell outside the confidence interval of the unconditional estimate. Jain and Lall (2001) explore how the long-term, ENSO-related climate variability impacts the exceedances of flood thresholds. From their bootstrap analysis they conclude that there is a much higher chance of obtaining substantially different numbers of exceedances of the design flood derived from the iid assumption. Hence, low-frequency fluctuations may lead to fat tails of flood distributions and may increase the potential for surprise.

The contrasting, process-based perspective is founded on the premise that flood events of different types occur in a given catchment, and that flood characteristics are linked to the hydro-meteorological processes driving the water input into catchments, to the runoff generation processes, and to the flood-routing processes. The expectation is that processbased approaches can provide information on flood probability estimation and allow for more reliable and catchmentspecific characterization of the flood hazard.

The analysis of flood events in terms of their meteorological causes is an important starting point for understanding how flood–climate linkages can influence flood probability distribution functions (Hirschboeck, 1988). Different storm types can trigger completely different flood processes in the same catchment. For example, in the southwestern United States (Fig. 5), intense summer convective thunderstorm cells interact selectively with a watershed's drainage network and topography, typically generating flashy flood hydrographs. In contrast, synoptic-scale winter storms deliver snow and rain to large areas and tend to generate flood hydrographs characterized by sustained and cumulative flows. Catchments with flood peaks generated by a mix of climatic mechanisms may exhibit complex heterogeneous frequency distributions, but stratifying the flood record according to climatic cause (Fig. 5) reveals useful information about the individual processes that shape the catchment's overall flood frequency distribution. If marked climate–flood linkages can be identified, it allows for flood estimation and design to be improved. Kwon et al. (2008) stress the reduction in uncertainty of the 100-year estimate when they include climate information. This reduction can be used to decrease the costs of over- or underdesign. However, in order to benefit, there is a need for more statistical and theoretical analysis of temporal variability in extremes in order to understand how temporal clustering or existence of a range of stochastic processes or mixture of distributions affect the tails of flood distributions.

The reasoning that better process understanding improves flood estimation can be extended further along the chain of flood processes. Runoff able to generate floods is produced in three main ways: (1) at the surface by infiltration excess (Hortonian overland flow) due to low or no infiltration intensities of the soil or surface, (2) by saturation excess (saturation overland flow) due to saturating the porous soil matrix, or (3) by fast subsurface flow, usually triggered by a variety of mechanism and conditions (Bachmair and Weiler, 2011). Based on the specific setting of a given catchment (geology, soils, topography, vegetation cover, land use, etc.), the three processes interact differently; therefore in conjunction with the hydro-meteorological conditions, different runoff generation processes can dominate the flood generation (Weiler et al., 2009; Steinbrich and Weiler, 2012). Rogger et al. (2012) give the following example, which illustrates the benefits of understanding the dominant controls on flood probability. Empirical distribution functions of flood peaks in small catchments sometimes show abrupt changes in the slope; that is, the largest flood peaks are significantly larger than the rest of the record. For Austrian Alpine catchments, Rogger et al. (2012) suggest that these abrupt changes are the consequence of a threshold of storage capacity being exceeded, which causes fast surface runoff in large parts of the catchments. This effect is not captured by flood frequency statistics, which tend to underestimate floods in these catchments.

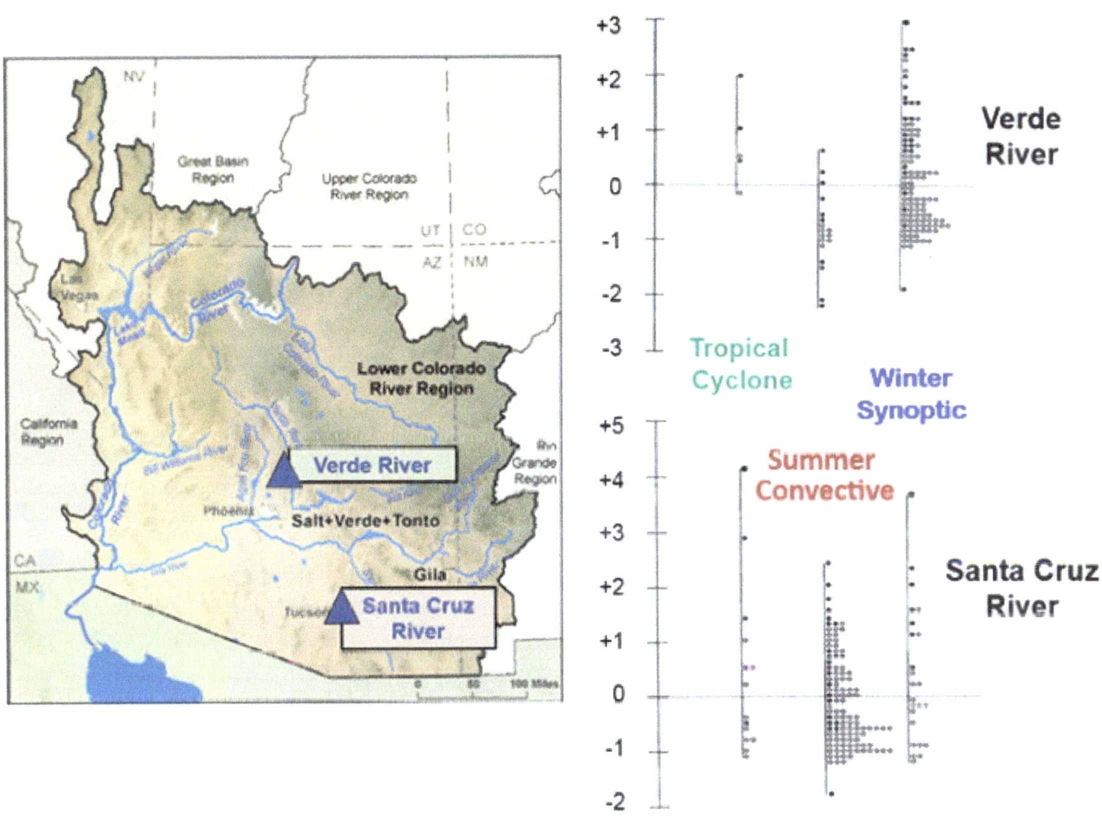

Figure 5. Stratification of flood events according to climatic mechanism. Climate-based flood heterogeneity is evident in two Arizona watersheds when observed flood peak distributions are stratified according to the three dominant flood-producing mechanisms in the southwestern United States. Shown are histograms of instantaneous peaks above base, plotted as standardized dimensionless z scores for comparison. The Santa Cruz River at Tucson floods most frequently in response to summer convective thunderstorm rainfall, but the two largest peaks of record were produced by tropical-cyclone-enhanced rainfall and precipitation from winter synoptic storms, respectively. In contrast, in the Verde River below Tangle Creek, some 300 km to the north, the most frequent and largest floods are produced by winter precipitation events (based on House and Hirschboeck, 1997).

Another example of the effect of runoff processes on the flood distribution function is shown in Fig. 6. The left graphs show the flood frequency curve and the associated event runoff coefficients for the Weißach catchment at Zwing located in western Austria (Merz and Blöschl, 2008a). Owing to orographic enhancement of northwesterly airflows, rainfall is high and persistent and runoff coefficients tend to be constantly high. As rainfall becomes more extreme, runoff coefficients only increase moderately as they are already close to unity, and the flood frequency curve of the catchment shows a downward curvature. Assuming that the rainfall regime remains similar, and on the basis of the runoff coefficients, one would expect that the trend continues as one extrapolates to higher return periods. In contrast, runoff coefficients of Wulka at Schützen in the dryer eastern part of Austria are much lower and increase with increasing return periods. The flood frequency curve shows an upward curvature, and, on the basis of the analysis of the runoff coeffi- cients, one would expect that this trend continues, assuming the rainfall regime remains similar. Hence, the information on runoff coefficients gives more confidence in extrapolating to higher return periods.

The idea that the flood frequency curve is a fingerprint of important processes can finally be extended to flood routing and river–floodplain interactions. Changes in the slope of the flood frequency curve may also result from threshold processes in the river system: for example, when the river banks are inundated and large additional storage capacity is activated. Apel et al. (2009) show for the Lower Rhine in Germany that dike breaches lead to significant retention effects by pruning the flood waves, which in turn modify the flood frequency curve downstream of breach locations. This effect only occurs beyond a certain (extreme) threshold, and it is typically not contained in observed time series. Hence, a process-neutral, purely statistical approach, i.e. not including this specific process in the extrapolation to extreme floods, results in significant overestimation of large flood quantiles.

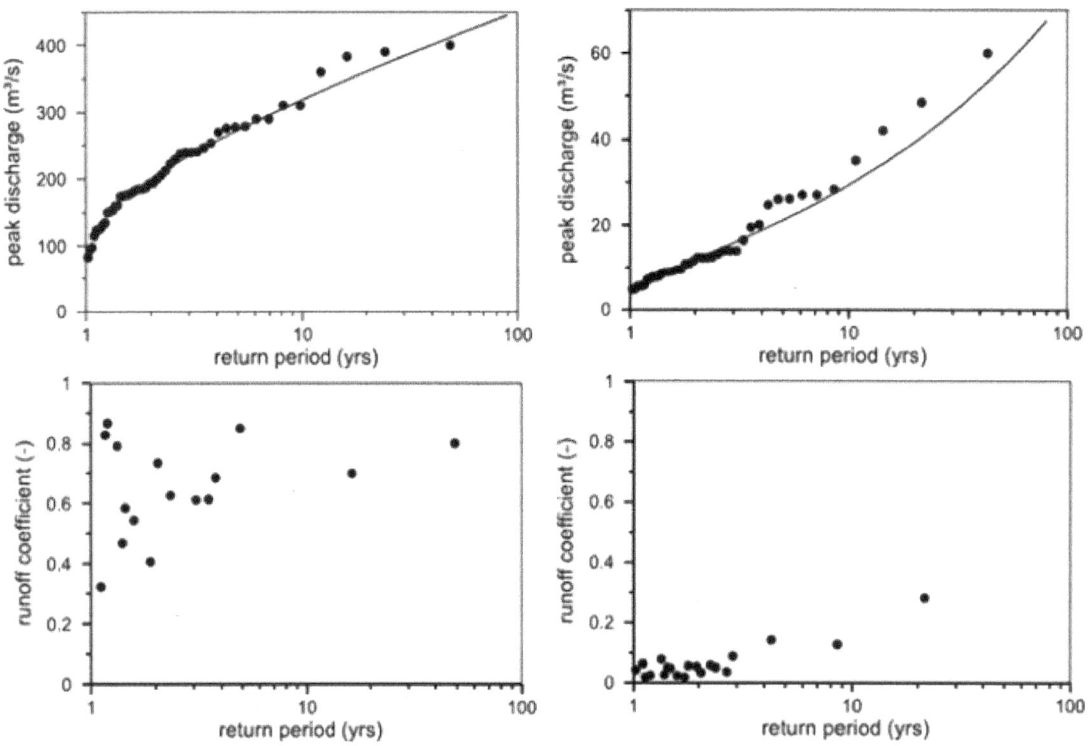

Figure 6. Flood frequency plots (top panels) and runoff coefficients (bottom panels) of the associated flood events for Weißach at Zwing (199 km²) (left panels) and Wulka at Schützen (383 km²) (right panels) (from Merz and Blöschl, 2008a).

We support the reasoning of Merz and Blöschl (2008a, b), who argue that much better use should be made of the wealth of knowledge on hydrological processes in estimating flood probabilities, and have introduced the concept of flood frequency hydrology. It is based on a systematic expansion of information on flood processes beyond the flood sample at the site of interest in order to derive more informed estimates of flood frequencies. By combining local flood data statistics with additional temporal information on historic floods, spatial information on floods in neighbouring catchments and causal information on the flood processes, the subtleties of the flood characteristics at the site of interest can be more fully captured and the uncertainty of the flood quantile estimation reduced. It is important to note that the complete spectrum of processes – in the atmosphere, catchment, river system and floodplains – should be included in a process-based flood estimation. This is a very broad definition of flood frequency hydrology and subsumes, for example, elements of the flood hydro-climatology concept of Hirschboeck (1988).

These process-based approaches also have their downsides. For example, they require a detailed understanding of the flood-generating processes in a catchment. Furthermore, the stratification of flood events into different flood types increases the sampling uncertainty of the frequency analysis. Nevertheless, the small sample of floods produced in a catchment by an infrequently occurring process provides some of the best information we have in the observed record on what future flooding from that same process might be like. Hence, we conclude that the past dominance of processneutral, purely statistical approaches to the flood estimation problem should be augmented with a broader approach: the development of process-based methods to complement standard approaches to flood estimation, in particular for assessing extreme floods. There is a need for innovative statistical extreme value methods (e.g. as discussed in Serinaldi and Kilsby, 2014) and process-based approaches, and for merging them into tools that can be used by practitioners.

Climate Change: Model Chain Versus Model-Chainaugmented?

Anthropogenic climate change is perceived as particularly relevant by authorities responsible for flood design and management. Therefore, in this section the benefit of understanding climate–flood linkages for projections of future floods under climate change is elaborated.

The typical approach for deriving future flood hazard scenarios under climate change is to implement model chains consisting of the following elements: "emission scenario→general circulation model (GCM)→downscaling, possibly including bias correction→hydrological catchment model→flood frequency analysis" (e.g. Dankers and Feyen, 2009; Kay et al., 2009; Ott et al., 2013), with the emission scenarios stemming from scenarios of future economic and social development (Nakicenovic and Swart, 2000). The IPCC has changed the climate change impact modelling concept slightly with the introduction of the concept of representative concentration pathways (Moss et al., 2008), but this will not affect the above-mentioned modelling chain except for replacing the initial element from "emission scenario" to "representative concentration pathways" (Moss et al., 2010).

In either case the natural variability in the general circulation models is produced without a link to the observed timing of variability mode stages. The "pure" climate change signal can thus be analysed in ensembles of model simulations. Multi-model ensembles can reduce the uncertainties associated with deviations of individual model climatologies. However, uncertainty is still large in projected flood changes, and the IPCC SREX report (2012, p. 178) states that "overall there is low confidence in projections of changes in fluvial floods". One of the main uncertainty sources is the problem all GCMs have in simulating regional and local rainfall. Causes for these shortcomings are influences of local orography and land surface that are not resolved and smallscale processes which cannot be represented by a GCM. Under such conditions, it is not recommendable to infer future changes in flooding directly from the GCM-produced rainfall at individual grid points. This constraint will ease as climate models are applied at higher resolutions – for example, in the work by Kendon et al. (2014) using a regional climate model at up to 1.5 km resolution to resolve convective rainfall in the UK.

Statistical downscaling can be used to estimate the local rainfall from large-scale parameters. It can, for example, be based on the relationship between extreme rainfall and large-scale atmospheric circulation patterns (ArnbjergNielsen et al., 2013). Circulation patterns or weather types, either subjectively (e.g. Hess and Brezowsky, 1952) or objectively (e.g. Philipp et al., 2007) derived, can take different atmospheric variables into account (e.g. direction of flow, atmospheric humidity, temperature). Climate change signals in the relevant weather types, in particular in the frequency of occurrence and persistence, can then be identified for the individual climate change simulations. Instead of (or in addition to) using weather types, it is also possible to identify cyclone tracks which are associated with extreme precipitation and enhanced flood risk in specific regions. A wellknown example of a pathway associated with major flooding in central Europe during summer is van Bebber's Vb track (e.g. Kundzewicz et al., 2005).

Where a strong climate–flood link can be identified for a given catchment, this link can be used to directly project future flood changes. For river gauges in the Mekong Basin, Delgado et al. (2012) identify a statistical relationship between flood variance and the western Pacific monsoon variance. A non-stationary flood frequency model is developed in which the monsoon variance drives temporal changes in flood quantiles (Delgado et al., 2014). For climate change scenarios, the variance of the monsoon is extracted from GCM runs. The simplicity of the monsoon–flood link allows for the derivation of large ensembles of flood projections under climate change almost immediately when GCM output is available. Such approaches are promising – and complement the above-mentioned climate change model chains – because they bypass precipitation, which is poorly simulated by global and regional climate models. Rather than using precipitation, they are based on circulation-linked atmospheric variables, such as large-scale pressure fields and wind velocities, which are frequently much better represented by climate models. Problems with this approach, as in all statistical downscaling procedures, occur when there is non-stationarity in the link between the scales and parameters, a lack of data for estimating them, or a need for extrapolation to situations far from the area for which the statistical link has been derived. However, we see a great potential in such low-dimensionality models for understanding flood changes because they force the modeller to identify the dominant processes and they offer the possibility for establishing direct causality links for observed and projected flood changes (Hall et al., 2014).

Notwithstanding the approach that is selected, one has to be aware of the current limitations of global and regional climate models, which go beyond errors in simulating precipitation. For example, SSTs are one of the most important drivers for precipitation on decadal timescales (e.g. Bichet et al., 2011). The spatial patterns of SSTs are particularly relevant in driving regional precipitation. This is an issue of concern, considering that coupled atmosphere–ocean GCMs from the Coupled Model Intercomparison Project Phase 3 (CMIP3) and 5 (CMIP5) globally predict warming of SSTs that is too uniform (e.g. Shin and Sardeshmukh, 2011).

Another approach with great potential beyond the typical use of model chains is the development of processbased scenarios of flood response to realistic shifts in climate. For instance, in the southwestern US, where different flood-generating storm types dominate in different seasons (Fig. 5), the circulation features associated with each storm type are distinctly different. This information can be used to design alternative scenarios for a catchment's flood response to climate change. A poleward shift in the latitude of the current winter storm track, an increase or decrease in the magnitude or frequency of eastern North Pacific tropical cyclones, or an expansion of the low-latitude tropical moisture belt that enhances the summer thunderstorm season would each have a profound impact on the overall probability distribution of flood peaks in southwestern US watersheds. Catchmentbased future flooding situations constructed by sampling observed floods within each of these flood types could be used to develop alternative planning scenarios for different regional atmospheric circulation responses to changing global climate drivers based on GCM output or projected circulation regime shifts (Fig. 7)

Such scenarios could be derived via simulation or via data analysis: for example, relying on historical periods with analogous conditions in the same catchment. Alternatively, one could derive possible future flood changes by investigating catchments which have a climatic regime that is similar to the regime we expect in the future in a specific catchment. Such process-based scenario approaches do not need to be based on the relation between floods and atmospheric moisture transport or storm tracks. They can be applied more directly to surface hydro-meteorological variables. For example, in several European regions, human-induced climate change is associated with an increase in winter precipitation and a higher snow line. Although we may not be very con- fident about the actual future precipitation, we can be confi- dent about the general trend of increasing winter precipitation and decreasing fraction of snowfall. Indeed watersheds in mountainous areas of the western United States that rely on seasonal snowpacks are already vulnerable to shifts in the timing of runoff and increases in early spring flooding because warmer temperatures promote an earlier snowmelt season and more precipitation falling as rain instead of snow.

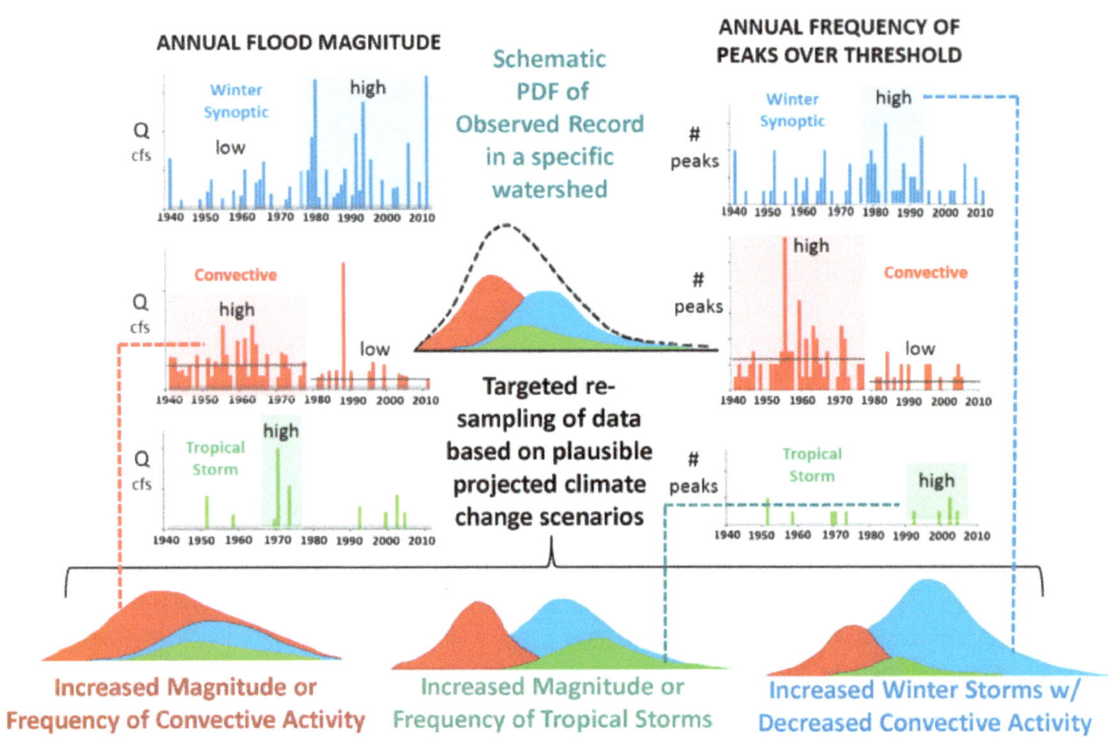

Figure 7. Schematic illustration of a proposed method for the development of process-based flooding scenarios by re-sampling distinctive segments of the observed flood record. Extreme flooding variability often occurs as clusters or episodes of high or low magnitudes of annual flood maxima (left-hand side of figure), and/or high or low frequencies of flood peaks that exceed a defined threshold (right-hand side of figure). To use these modes of variability in scenario development, the flood record can be partitioned into separate time series for each atmospheric-circulation-linked flood-generating process (i.e. winter synoptic, convective, or tropical storm) and then analysed using subjective and objective methods (e.g. change-point analysis) to identify statistically significant and climatically unique episodes of extreme flooding behaviour (shaded segments of flood time series). The magnitude and frequency attributes of the flood peaks in each unique episode can then be iteratively sampled and recombined to simulate new modes of flood-generating behaviour during plausible projected climatic shifts (e.g. increased magnitude and frequency of winter synoptic storms combined with decreased frequency of convective activity).

Such process-based scenarios might provide a space possible future flood characteristics, conditioned on our current knowledge. This possibility space of flooding characteristics in a given catchment could be compared to the vulnerability of the elements at risk, separating catchments which require special attention from those which are less vulnerable to the conceivable changes in climate. Prudhomme et al. (2010) introduced the closely related "scenario-neutral approach to climate change impact studies", which is based on sensitivity analyses of catchment response to a plausible range of climate changes. An advantage of this approach and related ideas, such as what-if analysis or cause-and-effect analysis, is that it is much easier to understand and interpret changes in flood behaviour. However, the complexity of the typical climate model chains impedes the understanding of how changes in climate propagate to changes in flood characteristics. Hence, in parallel to improving climate model chains and their reliability for flood projections under climate change, a better understanding of climate–flood linkages has a large potential for flooding under climate change.

Currently, human-induced climate change and natural climate variability are not operationally incorporated into flood management practice, in part because floodplain managers are constrained by the absence of accepted alternatives to standard stochastic methods. To overcome this impasse, there is a need for methods that can incorporate realistic, processbased climate change scenarios into the statistics of gauge records in ways that demonstrate an improvement on the traditional flood frequency analysis approach. Probabilistic methods now being used with palaeoflood data present one approach, but they are not directed toward future climate change and are limited by the availability of palaeoflood indicators in the landscape. Methods are needed that can augment existing flood peak records in ways that reflect how the climatic drivers of local hydro-meteorological processes have changed and will change in the past, present and future.

Table 2. Average economic flood losses in Australia stratified according to ENSO phase. Analysis based on data from EM-DAT (CRED, Belgium) for 1971–2010, and insured loss data normalized for changes in wealth from the Insurance Council of Australia for 1967–2010 (for method, see Crompton et al., 2008). The t test and Mann–Whitney U test (MWU test) were applied indicating the significance of the differences in average losses.

Data set	N	Average damage		t test	MWU test
		La Niña	Non-La Niña		
EM-DAT [USD]	57	1070	142	$p=0.026$	$p=0.014$
Insurance Council of Australia [AUD]	50	369	61	$p=0.014$	$p=0.024$

Flood Risk Management

What is the benefit of an improved understanding of climate–flood linkages for flood risk management? In the last two decades, flood change research has been dominated by studies looking at changes in flood hazard, for instance due to human-induced climate change (e.g. Feyen et al., 2012; Ott et al., 2013, Ward et al., 2014a), land-use change or river training (e.g. Bronstert et al., 2007). Today it is recognized that all risk elements are dynamic through time, not only hazard (IPCC, 2012; Jongman et al., 2012).

Bubeck et al. (2012) show that societal responses are critical in understanding how vulnerability to floods changes over time. This study surveyed 752 households along the Rhine in Germany that experienced floods, with two major floods in 1993 and 1995. The results indicate that flood damage mitigation measures were implemented by households gradually over time, with major flood events being important triggers for accelerated implementation. Especially in the aftermath of the severe flood in 1993, a remarkable increase in the number of measures undertaken was observed. Kuhlicke et al. (2011) analyse the social vulnerability of households to floods for three European case studies. Using the definition of Blaikie et al. (1994), who understand vulnerability as "... the characteristics of a person or group in terms of their capacity to anticipate, cope with, resist, and recover from the impact of a natural hazard", they find that social vulnerability to floods is not a static characteristic. It is highly dynamic and may change even in the course of one single flood event. A household may be vulnerable in certain event phases – anticipation, resistance and coping, recovery and reconstruction – and not vulnerable in others.

Further, a single driver may influence different risk components. For example, the societal perception of flood risk may be strongly influenced by a damaging flood, which may trigger investments in structural flood defence, such as flood retention basins, and may change flood hazard. In parallel, it may change exposure by flood-affected companies and private households migrating out of heavily flooded areas, and vulnerability by triggering private precaution in the flooded areas (Petrow et al., 2006; Kreibich et al., 2011). In addition, spatio-temporal interdependencies between vulnerability, exposure and hazard have to be expected. Current risk assessments, if they include dynamics at all, often examine dynamics in one of the components, whereas the interdependencies could be crucial (Di Baldassarre et al., 2013). For example, flood protection by dikes aimed at reducing the flood hazard can lead to increased development behind dikes, thus increasing exposure, the so-called levee effect (Tobin, 1995). In areas with high protection standards by dikes, individuals and societies may have low risk perceptions, and thus be less prepared for floods; in other words, they may have high vulnerability (Bubeck et al., 2012; Zaalberg and Midden, 2013). Spatio-temporal changes in all of these factors influence the overall flood risk.

Another factor that has received little attention in flood risk studies is the role of climate variability. As discussed earlier, many studies have found linkages between climate variability and extreme streamflow. A number of studies are now taking the step of examining whether such linkages can be detected between climate variability and flood losses. A preliminary analysis of recorded economic flood costs in Australia shows these to be sensitive to climate variability. Table 2 illustrates that average flood damages in Australia are significantly higher in La Niña years than in non-La Niña years. This signal appears to be robust to the loss data used, with significant differences for both loss data sets (EM-DAT, CRED, Belgium, and data on insured losses from the Insurance Council of Australia).

The work of Ward et al. (2010, 2014b) on relationships between ENSO and river flooding at the global scale is currently extended to assess the impacts of ENSO on risk (rather than discharge). Using the global flood risk modelling framework described in Ward et al. (2013), preliminary results presented at the EGU topical meeting in Potsdam show that there are clear and significant regional patterns of ENSO in- fluence on disaster risk, in terms of both affected population and economic damage. Hence, socio-economic risks may to some extent be linked to climate regimes, which could have major implications for the insurance and re-insurance industry. For example, clustering of flood events increases the likelihood of a high number of damage claims within a short period of time – an issue which is aggravated if clustering is spatially coherent and damaging events occur at different locations within a short time interval.

Figure 8 illustrates the concept of dynamic flood risk and dynamic risk management, based on the climate-informed risk management approach of Pizarro et al. (2009). Since all components of risk vary in time, flood risk itself is dynamic. Some aspects of flood risk dynamics may be predictable, many others not. For example, changing climatic boundary conditions at interannual to decadal to century scales may be due to structured natural variability or anthropogenic factors or their interaction. Some of these changes may be smooth and potentially predictable over a certain timescale, while others (e.g. a reversal of the Atlantic Gulf Stream) may be abrupt and can be anticipated but not predicted. Given the fact that ENSO events are predictable to some extent across seasons (Cheng et al., 2011), it appears worth testing the predictive skill for seasonal flood forecasts in those regions where Ward et al. (2014b) have shown a strong sensitivity of peak discharges to ENSO. Drivers other than climate may be predictable to some extent as well. For example, exposition in the tourism and agricultural sector may vary regularly in the course of the year.

The potential predictability of flood hazard and risk varying at seasonal to decadal timescales opens the door to a dynamic risk management paradigm. Assume that one can quantify the climate mechanisms that result in changing flood characteristics over a certain future time period. Then one could use the projected statistics and their uncertainty distribution to provide information on risk reduction measures. For example, if one knows that a flood-rich period is active in a given region, and that one has to expect higher probabilities for damaging floods during the next few years, this knowledge can be used to optimize disaster management (optimized strategies in Fig. 8). Relief stocks and emergency capabilities could be temporarily increased, reservoir water levels could be temporarily lowered to allow for more flood retention volume, temporary levees could be built or floodplain zoning rules for different activities could be temporarily imposed. The predictability of the climate–flood link would allow for the portfolio of risk reduction measures to be optimized.

On the other hand, many future flood risk changes are expected to be severe but are not predictable. In such cases robustness is an important criterion for designing risk management strategies. Robustness describes how well a measure performs under different possible but initially uncertain future developments. Flood-proofing strategies – such as elevated configuration of buildings, sealing of buildings to prevent water entrance, or the use of building materials in such a Figure 8. Drivers of flood risk change, dynamic risk and dynamic flood risk management. way that the impact of inundation is minimized – are robust in the sense that these measures will lower the damage in the case of flooding, regardless of the exact future development of the flood hazard (Merz et al., 2010b). Another example is the enhancement of risk awareness and self-protecting behaviour of people at risk. If residents are aware of their flood risk and of their possibilities to undertake effective precautionary, adverse flood impacts will be reduced under different possible but initially uncertain future developments. Risk reduction measures that are designed to be robust differ from measures that are the result of an optimization for the most likely future development. They represent trade-offs and are associated with real or opportunity costs (Heltbert et al., 2009). Hence, the degree of predictability of future changes in flood hazard and risk influences the role that different criteria play in designing risk management measures (Blöschl et al., 2013).

As an example, consider seasonal flood forecasts in the insurance and reinsurance industry. Companies try to take short-term changes in loss probabilities into account in managing their lines of business. If, for instance, a wet year with above-average flood losses is expected in a certain region (such as in northeastern Australia during La Niña), less insurance may be offered to reduce accumulation risks. Apart from underwriting-related reactions on seasonal forecasts, preparation on high/low losses in terms of management of reserves may be done; short-term liquidity for flood losses can be raised or reduced. Also, reinsurers and insurers "like" to provide services to their clients in the form of recommendations based on the characteristics of the upcoming flood season, not least to show their expertise. The advantage the insurance sector derives from seasonal forecasts results from the fact that it does not deal with a specific, localized issue. For an emergency planner in a city, a shift in event probability of a few percent is almost meaningless; in the case of a (re-)insurance company, this shift concerns a whole portfolio with a large number of exposed items, and hence even a small shift in probability may be traced in the total amount of losses.

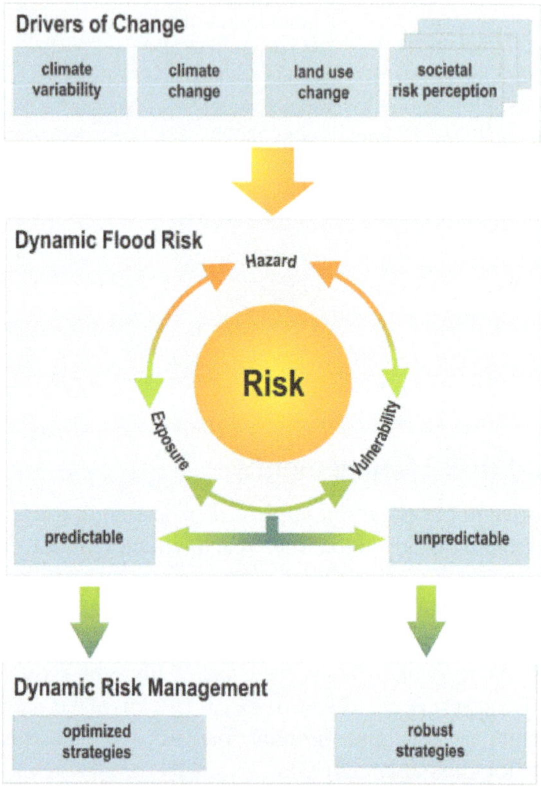

Figure 8. Drivers of flood risk change, dynamic risk and dynamic flood risk management.

Although the dynamic flood risk management concept which considers climate–flood linkages is theoretically appealing, it is necessary to explore whether and in which regions these linkages constitute a sufficiently large part of the variability to be exploited in risk management strategies. Furthermore, how dynamic risk management could be implemented also needs to be explored. It would raise new questions, such as "what is the necessary change in flood probability for the coming season, and what is the necessary reliability with which this change can be quantified in order to impose floodplain zoning rules for certain activities?". Such questions need to be addressed against the given economic, legal and social environment.

CONCLUSIONS: BROADER FRAMING AND GLOBAL INITIATIVES

Traditionally, flood estimation and flood management have mainly been the domain of hydrologists, water resources engineers and statisticians, and disciplinary approaches abound. Embedding floods in the climate context, as laid out in this paper, requires the extension of this narrow framing. It is timely to bring together the different aspects of floods at the interfaces of the climate–catchment–society system. The discussions during the EGU topical meeting "Floods and Climate" converged to a common understanding that extending the perspectives in flood research opens up exciting possibilities for better understanding and for providing improved tools for flood risk assessment and management. The conclusions can be summarized as follows.

Extend the System Boundaries

We argue for extending the traditional system boundaries (local catchment, recent decades, hydrological/hydraulic processes) in flood estimation and management. Extension in space goes beyond the inclusion of regional information which has proven beneficial in regional flood frequency analysis. Floods are embedded in global climate mechanisms, and deciphering these relations assures improvement in flood estimation and management, such as better understanding the tail behaviour of flood distributions or improved flood forecasting. However, it has been found that catchments respond rather differently to global climate variability (e.g. Glaser et al., 2010, for 12 major rivers in Europe for 1500–today). Hence, there exists a large variety of climate– flood linkages, and the existence of a marked climate–flood relation within one catchment does not automatically allow for it to be transferred to other catchments in the same region. This extension of perspective concerns the temporal system boundaries as well. For example, new technological developments allow for reconstruction of millennial-scale flood time series differentiating between floods in different seasons. The integration of systematic measurements with historical and palaeodata offers exciting prospects for quantifying flood variability and extremes at very different timescales, from the decadal to the millennial timescale. Finally, it is noted that a richer understanding of floods requires considering the interplay of climate, geology, topography, vegetation (biology) and humans. Experiences, methods and data from a variety of scientific disciplines (e.g. meteorology, climatology, palaeohydrology, geography, economics) and from practitioners (e.g. insurance, disaster management and water authorities) need to be integrated.

Understand the Dominant Processes

We argue that different societal needs, such as flood estimation and forecasting, benefit from a better understanding of the processes that dominate the tail behaviour of flood distributions. Although statistical approaches have played and will play an important role, they have to be complemented by the search for the causal mechanisms and dominant processes in the atmosphere, catchment and river system that leave their fingerprints on flood characteristics. Understanding the dominant processes is expected to increase the reliability of extrapolation to rare floods and to reduce the potential for surprise. It is also an avenue for flood projections under climate change – complementing the widespread model chain approach. Understanding the different flood event types that occur in a given catchment and their dominant processes may provide an opportunity to hypothesize from a causal, mechanistic perspective about possible future flood changes. For example, spatial shifts in storm tracks or changes in the snowfall fraction may impact flood characteristics in certain ways. There is a need to develop processbased methods which can be applied by practitioners. One push in this direction are the guidelines on the estimation of flood probabilities, recently published by the German Water Association (DWA, 2012), which recommend the enhancement of extreme value statistics by incorporating hydrological understanding and additional data beyond the observed flood time series.

Quantify the Role of Natural Variability

One of the core concepts of this paper is that natural variability has an important role to play in flood estimation and flood management besides delivering the random component. We argue that flood characteristics vary in time that this variation may be partially quantifiable and predictable, and that knowledge about this structured variability may be exploited in flood estimation and management. It is necessary to understand the ratio of structured variability to random variability, and how this ratio varies in space, with flood types, and with flood magnitude. For example, the structured variability for most regions in Europe is significantly lower than in the US. One task is to identify regions for which predictable climate– flood links play a significant role and regions where randomness dominates. Another mainly open question is how the type and strength of climate–flood linkages vary with flood magnitude. Floods of different magnitudes may correspond to different climate modes (for an illustrative example, see Jain and Lall, 2000), and these variations need to be understood. Maybe the most important aspect is the persistence of climate–flood linkages: how stable are linkages between climate and floods? Results are mixed: for instance, floodfavouring circulation patterns have been found in many studies, but it has been found as well that different circulation patterns have dominated during different flood-rich periods of a given river (Schmocker-Faeckel and Naef, 2010). This is part of the more general discussion on the evidence of cycles in climate phenomena: as Burroughs (2003, p. 5) states in his comprehensive treatise on such cycles: ". . . a fundamental weakness of many apparently convincing examples of 'weather cycles': they come and go in a most tantalizing manner." The central question for a dynamic flood risk management concept which considers climate–flood linkages is whether these linkages constitute a sufficiently large part of the observed variability to be exploited in risk management strategies. Although there are some convincing examples, the potential of climate-informed risk management still needs to be explored.

Integrate Hazard with Societal Aspects

Although the focus of this paper is on the role of climate, we stress the tremendous role of society. Ideally, risk assessment and management consider changes in hazard and societal aspects, and vary according to the predictability with which short- and long-term changes in flood risk can be quantified. Increasing efforts are needed to fully account for factors that contribute to changes in risk, which implies that changes in vulnerability (e.g. early warning, building codes) and exposure (e.g. population growth, spatial planning) need to be considered when interpreting records of observed flood losses. However, the understanding of the interplay between changes in hazard, exposure and vulnerability is hampered by data scarcity. The knowledge on disaster costs is incomplete and fragmented, with many cost types unaccounted for in current analyses, and a heavy focus on direct losses and less complete understanding of longer-term impacts and indirect costs (Merz et al., 2010b). While changes in the number of exposed assets and people can be reasonably well accounted for, temporal changes in vulnerability that are critically important are poorly understood and quantified. The interactions and feedbacks between society and floods are little understood. For instance, is the societal response to floods different for clustered events compared to random events? Here, we support the call for socio-hydrology which explicitly studies the co-evolution of humans and water (Sivapalan et al., 2012).

Organize Data and Knowledge Exchange Globally

The advancement of flood research in the discussed broader framework is fundamentally dependant on (1) access to and exchange of fit-for-purpose data on a global scale and (2) increased international and national collaboration between different disciplines, in particular hydrologists and meteorologists/climatologists. On a global level the identification of climate-driven changes in river flow is hindered due to a lack of long-term, high-quality data from rivers limited from artificial human disturbances (Whitfield et al., 2012). Therefore, in the past 15 years, several countries have established reference hydrologic networks (RHN) with the sole purpose of assessing hydrological responses to fluctuations in climate (Hannaford and Marsh, 2008; Whitfield et al., 2012; Murphy et al., 2013). Despite these efforts, the majority of regions across the globe have yet to establish such databases. Data users have a responsibility to promote the need of such data for their analyses in order to increase the global distribution of RHN-like networks. However, significant economic and institutional barriers to the open access to hydrometeorological data remain (Viglione et al., 2010; Hannah et al., 2011) and need to first be overcome. Hydrologists could learn from the climate community that actively advocate open data access and support data exchange (Hannah et al., 2011). Given the global scale and societal importance, we call for the organization of an international multidisciplinary collaboration and data-sharing initiative to understand further the links between climate and flooding. This would require the integration of meteorological, climatological and hydrological expertise in conjunction with increased access to coupled RHN and meteorological databases. While this is ambitious, discussions need to take place now for such an initiative to ever be realized even within medium-term flood risk planning horizons.

There have been some initiatives for studying floods in their global context. One of the earliest was the Natural Disaster Hotspots study of the World Bank (Arnold et al., 2005), which prepared a preliminary map of global flood hazard frequency, and examined mortality and economic risk. Recently, modelling efforts have been undertaken with the aim of assessing global flood hazard and risk (UNISDR, 2011; Hirabayashi et al., 2013; Ward et al., 2013; Winsemius et al., 2013). A different and very interesting path has been initiated by the earthquake risk community by launching the Global Earthquake Model (GEM, www.globalquakemodel. org). GEM is a collaborative effort and involves scientists and stakeholders from around the globe. Data, models, opensource software and knowledge are shared with the final aim of expanding the science of earthquake risk, informing decision making and increasing earthquake resilience worldwide. We feel that the flood research community should foster global approaches and should study physical and societal mechanisms of flooding within their global interconnectedness. This global perspective would not only improve our understanding and exploitation of linkages between climate and flooding but would also help in understanding and mitigating far-distance impacts of flooding, such as the disruption of global supply chains.

ACKNOWLEDGEMENTS

This paper is a result of the EGU topical meeting "Floods and Climate: Understanding and exploiting the link between floods and climate" held on 4–5 October 2012, at GFZ German Research Centre for Geosciences in Potsdam. Financial support for this meeting by the European Geosciences Union, AXA Research Fund, Free University Berlin and geo.X is gratefully acknowledged. We would like to acknowledge the support of the ERC advanced grant "FloodChange", project no. 291 152, and the Austrian Academy of Sciences (ÖAW) project "Mountain Floods".

REFERENCES

1. Abrahams, A. D. and Ponczynski, J. J.: Drainage density in relation to precipitation intensity in the USA, J. Hydrol., 75, 383–388, 1984.
2. Apel, H., Merz, B., and Thieken, A. H.: Influence of dike breaches on flood frequency estimation, Comput. Geosci., 35, 907–923, doi:10.1016/j.cageo.2007.11.003, 2009.
3. Arnbjerg-Nielsen, K., Willems, P., Olsson, J., Beecham, S., Pathirana, A., Gregersen, I. B., Madsen, H., and Nguyen, V.-T.- V.: Impacts of climate change on rainfall extremes and urban drainage systems: a review, Water Sci. Technol., 68, 16–28, doi:10.2166/wst.2013.251, 2013.
4. Arnold, M., Dilley, M., Deichmann, U., Chen, R. S., and LernerLam, A. L.: Natural Disaster Hotspots: A Global Risk Analysis, World Bank, Washington, D.C., 2005.
5. Bachmair, S. and Weiler, M.: New dimensions of hillslope hydrology, in: Forest Hydrology and Biogeochemistry: Synthesis of Research and Future Directions, edited by: Levia, D., CarlyleMoses, D., and Tanaka, T., Springer Science + Business Media B.V., 455–481, 2011.
6. Baker, V. R.: Paleoflood hydrology: Origin, progress, prospects, Geomorphology, 101, 1–13, doi:10.1016/j. geomorph.2008.05.016, 2008.
7. Becker, A., Finger, P., Meyer-Christoffer, A., Rudolf, B., Schamm, K., Schneider, U., and Ziese, M.: A description of the global land-surface precipitation data products of the Global Precipitation Climatology Centre with sample applications including centennial (trend) analysis from 1901–present, Earth Syst. Sci. Data, 5, 71–99, doi:10.5194/essd-5-71-2013, 2013.
8. Benito, G., Sopena, A., Sanchez-Moya, Y., Machado,

M. J., and Perez-Gonzalez, A.: Palaeoflood record of the Tagus River (Central Spain) during the Late Pleistocene and Holocene, Quaternary Sci. Rev., 22, 1737–1756, 2003a.

9. Benito, G., Díez-Herrero, A., and Fernández de Villalta, M.: Magnitude and frequency of flooding in the Tagus Basin (Central Spain) over the last millennium, Climatic Change, 58, 171–192, 2003b.

10. Bichet, A., Wild, M., Folini, D., and Schär, C.: Global precipitation response to changing forcings since 1870, Atmos. Chem. Phys., 11, 9961–9970, doi:10.5194/acp-11-9961-2011, 2011.

11. Bichet, A., Folini, D., Wild, M., and Schär, C.: Enhanced Central European summer precipitation in the late 19th century: a link to the Tropics, Q. J. Roy. Meteorol. Soc., 140, 111–123, doi:10.1002/qj.2111, 2014.

12. Blaikie, P. M. Cannon, T., Davis, I., and Wisner, B.: At Risk: Natural Hazards, People's Vulnerability and Disasters, Routledge, London, 1994.

13. Blöschl, G. and Merz, R.: Landform-hydrology feedbacks, in: Landform: Structure, Evolution, Process Control, edited by: Otto, J.-C. and Dikau, R., Springer, Wien, Heidleberg, 117–126, 2010.

14. Blöschl, G., Viglione, A., and Montanari, A.: Emerging approaches to hydrological risk management in a changing world, in: Climate Vulnerability, Chapter 5.05, edited by: Pielke, R. A., Sr., Elsevier Inc., Academic Press, 3–10, doi:10.1016/B978-0-12- 384703-4.00505-0, 2013.

15. Bouwer, L. M., Vermaat, J. E., and Aerts, J. C. J. H.: Winter atmospheric circulation and river discharge in northwest Europe, Geophys. Res. Lett., 33, L06403, doi:10.1029/2005GL025548, 2006.

16. Brazdil, R., Pfister, C., Wanner, H., Storch, H. V., and Luterbacher, J.: Historical climatology in Europe: the state of the art, Climatic Change, 70, 363–430, 2005.

17. Bronnimann, S.: Impact of El Nino – Southern Oscillation on European climate, Rev. Geophys., 45, RG3003, doi:10.1029/2006RG000199, 2007. Bronstert, A., Bárdossy, A., Bismuth, C., Buiteveld, H., Disse, M., Engel, H., Fritsch, U., Hundecha, Y., Lammersen, R., Niehoff, D., and Ritter, N.: Multi-scale modelling of land-use change and

18. river training effects on floods in the Rhine basin, River Res. Appl., 23, 1102–1125, 2007.

19. Bubeck, P., Botzen, W. J. W., and Aerts, J. C. J. H.: A review of risk perceptions and other factors that influence mitigation behaviour?, Risk Analysis, 32, 1481–1495, doi:10.1111/j.1539- 6924.2011.01783.x, 2012.

20. Burroughs, W. J.: Weather Cycles, Real or Imaginary?, 2nd Edn., Cambridge University Press, 317 pp., 2003.

21. Cayan, D. R.: Interannual climate variability and snowpack in the Western United States, J. Climate, 9, 928–948, 1996.

22. Crompton, R. P. and McAneney, K. J.: Normalised Australian insured losses from meteorological hazards: 1967–2006, Environ. Sci. Policy, 11, 371–378, 2008.

23. Czymzik, M., Dulski, P., Plessen, B., von Grafenstein, U., Naumann, R., and Brauer, A.: A 450 year record of spring/summer flood layers in annually laminated sediments from Lake Ammersee (southern Germany), Water Resour. Res., 46, W11528, doi:10.1029/2009WR008360, 2010.

24. Dankers, R. and Feyen, L.: Flood hazard in Europe in an ensemble of regional climate scenarios, J. Geophys. Res.-Atmos., 114, D16108, doi:10.1029/2008JD011523, 2009.

25. Delgado, J. M., Merz, B., and Apel, H.: A climate-flood link for the lower Mekong River, Hydrol. Earth Syst. Sci., 16, 1533–1541, doi:10.5194/hess-16-1533-2012, 2012.

26. Ely, L., Enzel, Y., Baker, V. R., and Cayan, D. R.: A 5000-year record of extreme floods and climate change in the southwestern United States, Science, 262, 410–412, 1993.

27. Enzel, Y., Ely, L., House, P. K., Baker, V. R., and Webb, R. H.: Paleoflood evidence for a natural upper bound to flood magnitudes in the Colorado River Basin, Water Resour. Res., 29, 2287–2297, doi:10.1029/93WR00411, 1993.

28. Feyen, L., Dankers, R., Bódis, K., Salamon, P., and Barredo, J.: Fluvial flood risk in Europe in present and future climates, Climatic Change, 112, 47–62, doi:10.1007/s10584-011-0339-7, 2012.

29. Gaál, L., Szolgay, J., Kohnová, S., Parajka, J., Merz, R., Viglione, A., and Blöschl, G.: Flood timescales: Understanding the interplay of climate and catchment processes through comparative hydrology, Water Resour. Res., 48, W04511, doi:10.1029/2011WR011509, 2012.

30. Gergis, J. and Ashcroft, L.: Rainfall variations in south-eastern Australia, part 2: a comparison of documentary, early instrumental and paleoclimate records, 1788–2008, Int. J. Climatol., 33, 2973–2987, doi:10.1002/joc.3639, 2012.

31. Glaser, R., Riemann, D., Schönbein, J., Barriendos, M., Brázdil, R., Bertolin, C., Camuffo, D., Deutsch, M., Dobrovolný, P., van Engelen, A., Enzi, S., Halícková, M., Koenig, S. J., Kotyza, O., Limanówka, D., Macková, J., Sghedoni, M., Martin, B., and Himmelsbach, I.: The variability of European floods since AD 1500, Climatic Change, 101, 235–256, 2010.

32. Gumbel, E. J.: The return period of flood flows, Ann. Math. Stat., 12, 163–190, 1941.

33. Hall, J., Arheimer, B., Borga, M., Brázdil, R., Claps, P., Kiss, A., Kjeldsen, T. R., Kriauciunien, J., Kundzewicz, Z. W., Lang, M., Llasat, M. C., Macdonald, N., McIntyre, N., Mediero, L., Merz, B., Merz, R., Molnar, P., Montanari, A., Neuhold, C., Parajka, J., Perdigão, R. A. P., Plavcová, L., Rogger, M., Salinas, J. L., Sauquet, E., Schär, C., Szolgay, J., Viglione, A., and Blöschl, G.: Understanding flood regime changes in Europe: a state-ofthe-art assessment, Hydrol. Earth Syst. Sci., 18, 2735–2772, doi:10.5194/hess-18-2735-2014, 2014..

34. Hannah, D. M., Demuth, S., van Lanen, H. A. J., Looser, U., Prudhomme, C., Rees, G., Stahl, K., and Tallaksen, L. M.: Large-scale river flow archives: importance, current status and future needs, Hydrol. Process., 25, 1191–1200, doi:10.1002/hyp.7794, 2011.

35. House, P. K. and Hirschboeck, K. K.: Hydroclimatological and paleohydrological context of extreme winter flooding in Arizona, 1993, in: Storm-Induced Geological Hazards: Case Histories from the 1992–1993 Winter Storm in Southern California and Arizona, vol. XI, edited by: Larson, R. A. and Slosson, J. E., Geological Society of America Reviews in Engineering Geology, Boulder, Colorado, 1–24, 1997.

36. IPCC: Managing the Risks of Extreme Events and Disasters to Advance Climate Change Adaptation, A Special Report of Working Groups I and II of the Intergovernmental Panel on Climate Change, Cambridge University Press, Cambridge, New York, 2012.

37. Jacobeit, J., Glaser, R., Luterbacher, J., and Wanner, H.: Links between flood events in central Europe since AD 1500 and largescale atmospheric circulation modes, Geophys. Res. Lett., 30, 1172, doi:10.1029/2002GL016433, 2003.

38. Jain, S. and Lall, U.: Magnitude and timing of annual maximum floods: Trends and large-scale climatic associations for the Blacksmith Fork River, Utah, Water Resour. Res., 36, 3641– 3651, 2000.

39. Jongman, B., Ward, P. J., and Aerts, J. C. J. H.: Global exposure to river and coastal flooding: Long term trends and changes, Global Environ. Change, 22, 823–835, doi:10.1016/j.gloenvcha.2012.07.004, 2012.

40. Karamperidou, C., Cioffi, F., and Lall, U.: Surface temperature gradients as diagnostic indicators of midlatitude circulation dynamics, J. Climate, 25, 4154–4171, 2012.

41. Karamperidou, C., Cane, M. A., Lall, U., and Wittenberg, A. T.: Intrinsic modulation of ENSO predictability viewed through a local Lyapunov lens, Clim. Dynam., 25, 4154–4171, 2013.

42. Kendon, E., Roberts, N., Fowler, H. J., Roberts, M., Chan, S. C., and Senior, C.: Heavier summer downpours with climate change revealed by weather forecast resolution model, Nat. Clim. Change, 4, 570–576, doi:10.1038/nclimate2258, 2014.

43. Levish, D. R.: Paleohydrologic bounds-nonexceedance information for flood hazard assessment, in: Ancient Floods, Modern Hazards: Principles and Applications of Paleoflood Hydrology, vol. 5, edited by: House, P. K., Webb, R. H., Baker, V. R., and Levish, D. R., American Geophysical Union Water Science and Application Series, American Geophysical Union, Washington, D.C., 175–190, 2002.

44. Lin, Z., Levy, J. K., Xu, X., Zhao, S., and Hartmann, J.: Weather and seasional climate prediction for flood planning in the Yangtze River Basin, Stoch. Environ. Res. Risk Assess., 19, 428–437, 2005.

45. Mangili, C., Brauer, A., Moscariello, A., and Naumann, R.: Microfacies of detrital event layers deposited in Quaternary varved lake sediments of the Piànico-Sèllere Basin (northern Italy), Sedimentology, 52, 927–943, 2005.

46. Maraun, D., Rust, H. W., and Osborn, T. J.: Synoptic airflow and UK daily precipitation extremes: Development and validation of a vector generalised linear model, Extremes, 13, 133–153, doi:10.1007/s10687-010-0102-x, 2010.

47. Moss, R. H., Babiker, M., Brinkman, S., Calvo, E., Carter, T., Edmonds, J., Elgizouli, I., Emori, S., Erda, L., Hibbard, K., Jones, R., Kainuma, M., Kelleher, J., Lamarque, J. F., Manning, M., Matthews, B., Meehl, J., Meyer, L., Mitchell, J., Nakicenovic, N., O'Neill, B., Pichs, R., Riahi, K., Rose, S., Runci, P., Stouffer, R., van Vuuren, D., Weyant, J., Wilbanks, T., van Ypersele, J. P., and Zurek, M.: Towards new scenarios for analysis of emissions, climate change, impacts and response strategies, Technical Summary, Intergovernmental Panel on Climate Change, Geneva, 25 pp., 2008

48. Nakamura, J., Lall, U., Kushnir, Y., Robertson, A. W., and Seager, R.: Dynamical structure of extreme floods in the U.S. Midwest and the United Kingdom, J. Hydrometeorol., 14, 485–504, doi:10.1175/JHM-D-12-059.1, 2013.

49. Nakicenovic, N. and Swart, R.: IPCC Special Report on Emissions Scenarios, Cambridge University Press, Cambridge, UK, 570 pp., 2000.

50. Ott, I., Duethmann, D., Liebert, J., Berg, P., Feldmann, H., Ihringer, J., Kunstmann, H., Merz, B., Schaedler, G., and Wagner, S.: High-Resolution Climate Change Impact Analysis on MediumSized River Catchments in Germany: An Ensemble Assessment, J. Hydrometeorol., 14, 1175–1193, doi:10.1175/JHM-D- 12-091.1, 2013.

51. Petrow, T., Thieken, A. H., Kreibich, H., Merz, B., and Bahlburg, C. H.: Improvements on flood alleviation in Germany: Lessons learned from the Elbe flood in August 2002, Environ. Manage., 38, 717–732, doi:10.1007/s00267-005-6291-4, 2006.

52. Rana, A., Uvo, C. B., Bengtsson, L., and Sarthi, P. P.: Trend analysis for rainfall in Delhi and Mumbai, India, Clim. Dynam., 38, 1-2, 45-56, doi:10.1007/s00382-011-1083-4, 2011.

53. Raupach, M. R.: Simplicity, complexity and scale in terrestrial biosphere modelling, in: Predictions in Ungauged Basins: International Perspectives on the State-of-the-Art and Pathways Forward, edited by: Franks, S.,Sivapalan, M., Takeuchi, K., and Tachikawa, Y., IAHS Publication 301, IAHS, Wallingford, 239– 274, 2005.

54. Rogger, M., Pirkl, H., Viglione, A., Komma, J., Kirnbauer, R., Merz, R., and Blöschl, G.: Step changes in the flood frequency curve – process controls, Water Resour. Res., 48, W05544, doi:10.1029/2011WR011187, 2012.

55. Schneider, U., Becker, A., Finger, P., Meyer-Christoffer, A., Rudolf, B., and Ziese, M.: GPCC Full Data Reanalysis Version 6.0 at 0.5∘ : Monthly Land-Surface Precipitation from Rain-Gauges built on GTS-based and Historic Data, Global Precipitation Climatology Centre, GPCC, http:// gpcc.dwd.de/ at Deutscher Wetterdienst, Offenbach, Germany, doi:10.5676/DWD_GPCC/FD_M_V6_050, 2011.

56. Sivapalan, M., Blöschl, G., Merz, R., and Gutknecht, D.: Linking flood frequency to long-term water balance: Incorporating effects of seasonality, Water Resour. Res., 41, W06012, doi:10.1029/2004WR003439, 2005.

57. Swierczynski, T., Brauer, A., Lauterbach, S., Martín-Puertas, C., Dulski, P., von Grafenstein, U., and Rohr, C.: A 1600 yr seasonally resolved record of decadal-scale flood variability from the Austrian Pre-Alps, Geology, 40, 1047–1050, doi:10.1130/G33493.1, 2012.

58. Toreti, A., Xoplaki, E., Maraun, D., Kuglitsch, F. G., Wanner, H., and Luterbacher, J.: Characterisation of extreme winter precipitation in Mediterranean coastal sites and associated anomalous atmospheric circulation patterns, Nat. Hazards Earth Syst. Sci., 10, 1037–1050, doi:10.5194/nhess-10-1037-2010, 2010.

59. Tucker, G. and Bras, R.: A stochastic approach to modeling the role of rainfall variability in drainage basin evolution, Water Resour. Res., 36, 1953–1964, 2000.

60. UNISDR: Global Assessment Report on Disaster Risk Reduction 2013, From Shared Risk to Shared Value: the Business Case for Disaster Risk Reduction, United Nations International Strategy for Disaster Reduction Secretariat, Geneva, 2013.

61. Viglione, A. and Blöschl, G.: On the role of storm duration in the mapping of rainfall to flood return periods, Hydrol. Earth Syst. Sci., 13, 205–216, doi:10.5194/hess-13-205-2009, 2009.

62. Viglione, A., Borga, M., Balabanis, P., and Blöschl, G.: Barriers to the exchange of hydrometeorological data in Europe: Results from a survey and implications for data policy, J. Hydrol., 394, 63–77, 2010.

63. Ward, P. J., Jongman, B., Sperna Weiland, F., Bouwman, A., Van Beek, R., Bierkens, M. F. P., Ligtvoet, W., and Winsemius, H. C.: Assessing flood risk at the global scale: model setup, results, and sensitivity, Environ. Res. Lett., 8, 044019, doi:10.1088/1748- 9326/8/4/044019, 2013.

64. Ward, P. J., Eisner, S., Flörke, M., Dettinger, M. D., and Kummu, M.: Annual flood sensitivities to El Niño–Southern Oscillation at the global scale, Hydrol. Earth Syst. Sci., 18, 47–66, doi:10.5194/hess-18-47-2014, 2014b.

65. Whitfield, P. H., Burn, D. H., Hannaford, J., Higgins, H., Hodgkins, G. A., Marsh, T., and Looser, U.: Reference hydrologic networks I. The status and potential future directions of national reference hydrologic networks for detecting trends, Hydrolog. Sci. J., 57, 1562–1579, 2012.

66. Winsemius, H. C., Van Beek, L. P. H., Jongman, B., Ward, P. J., and Bouwman, A.: A framework for global river flood risk assessments, Hydrol. Earth Syst. Sci., 17, 1871–1892, doi:10.5194/hess-17-1871-2013, 2013.

67. Zaalberg, R. and Midden, C. J. H.: Living behind dikes: mimicking flooding experiences, Risk Anal., 33, 866–876, doi:10.1111/j.1539-6924.2012.01868.x, 2013.

68. Zhang, Q., Xu, C.-Y., Jiang, T., and Wu, Y.: Possible influence of ENSO on annual maximum streamflow of the Yangtze River, China, J. Hydrol., 333, 265–274, 2007.

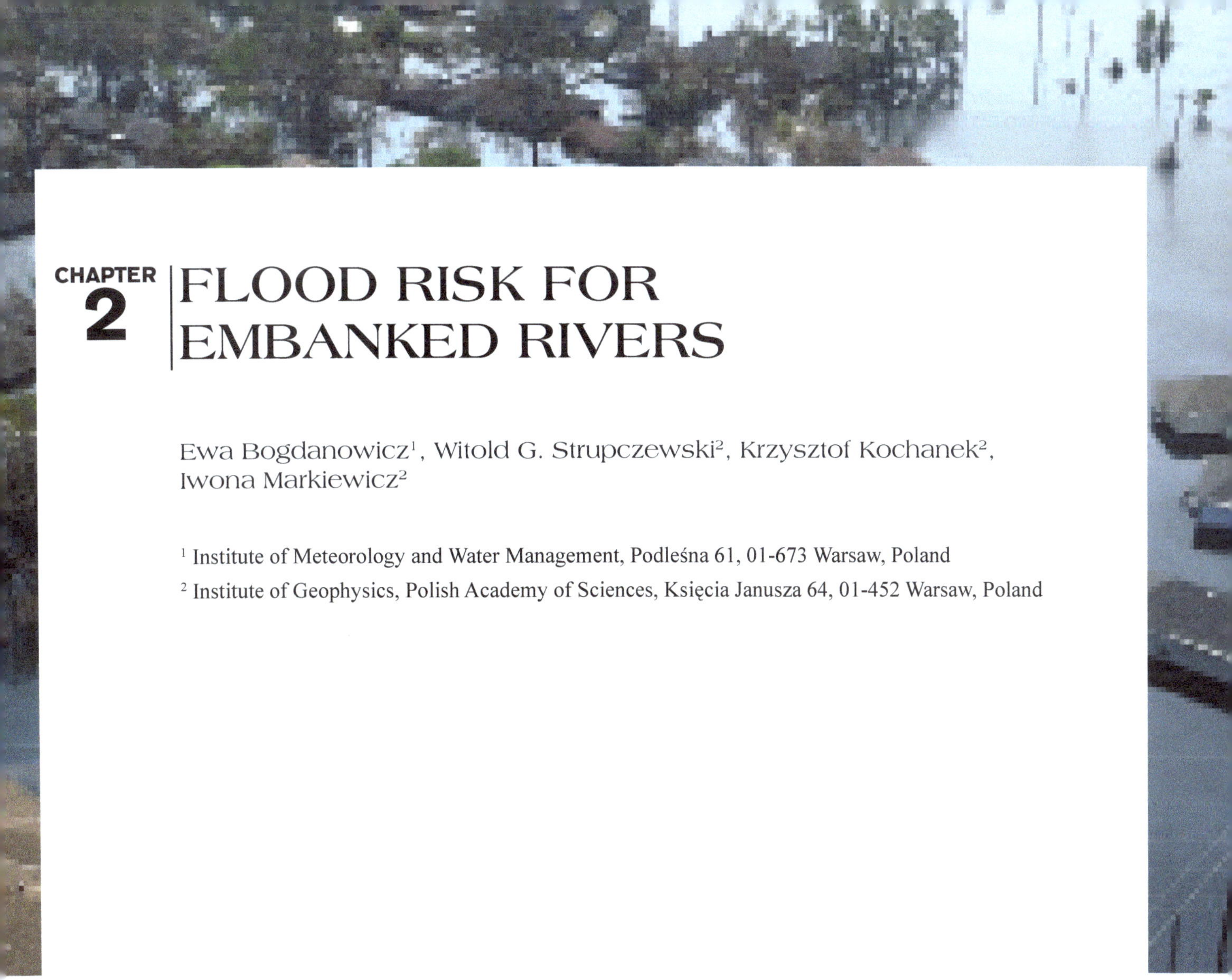

CHAPTER 2

FLOOD RISK FOR EMBANKED RIVERS

Ewa Bogdanowicz[1], Witold G. Strupczewski[2], Krzysztof Kochanek[2], Iwona Markiewicz[2]

[1] Institute of Meteorology and Water Management, Podleśna 61, 01-673 Warsaw, Poland

[2] Institute of Geophysics, Polish Academy of Sciences, Księcia Janusza 64, 01-452 Warsaw, Poland

ABSTRACT

Flood frequency analysis (FFA) concentrates on peak flows of flood hydrographs. However, floods that last years devastated large parts of Poland lead us to revision of the views on the assessment of flood risk in Poland. It turned out that it is the prolonged exposure to high water on levees that causes floods, not only the water overflowing the levee crest. This is because, the levees are weakened by water and their disruption occurs when it seems that the danger is over, i.e. after passing culmination. Two main causes of inundation of embanked rivers, namely over-crest flow and wash out of the levees, are combined to assess the total risk of inundation. Therefore the risk of inundation is the total of risk of exceeding embankment crest by flood peak and risk of washout of levees. Hence, while modeling the flood events in addition to the maximum flow one should consider also the duration of high water in a river channel, Analysis of the frequency of annual peak flows based on annual maxima and peaks over threshold is the subject of countless publications. Therefore we will here mainly modeling the duration of high water levels. In the paper the two-component model of flood hydrograph shape i.e. "duration of flooding-discharge- probability of nonexceedance" (DqF), with the methodology of its parameters estimation for sta- tionary case was developed as a completion to the classical FFA with possible extension to non stationary flood regime. The model combined with the technical evaluation of probability of levees breach due to the d-days duration of flow above alarm stage gives the annual probability of inundation caused by the embankment breaking. The results of theoretical research were supplemented by a practical example of the model application to the series for daily flow in the Vistula River in Szczucin. Regardless promising results, this method is still in its infancy despite its great cognitive potential and practical importance. Therefore, we would like to point to the usefulness and necessity of the DqF models to the one-dimensional analysis of the peak flood hydrographs and to flood risk analysis. This approach constitutes a new direction in FFA for embanked rivers.

Citation: Bogdanowicz, E. , Strupczewski, W. , Kochanek, K. and Markiewicz, I. (2014) Flood Risk for Embanked Rivers. Journal of Geoscience and Environment Protection, 2, 135-143. doi: 10.4236/gep.2014.23018.

INTRODUCTION

The most popular way of flood protection in Poland is the embankment of the rivers. In consequence of this passive way of protection, floods in Poland occur mostly due to the levee breach or very rarely to flow over the crest of dikes. Sense of security in floodplains of embanked rivers results from the belief that levees protect against the flood magnitude for which they were designed. So it creates the illusion that if the actual forecasted flood peak does not exceed the safety levels related to levee's designed value one can assume that the risk of water overtopping the dike crest is negligible and so is the risk of flooding in the protected area. The records of floods in Poland show that this is not true; more often the floods are the result of the prolonged exposure to high water on levees. The levees are weakened by water and their disruption occurs when it seems that the danger is over, i.e. after passing culmination. This is particularly dangerous because when the staff responsible for flood protection and local residents breathe sigh of relief the worst is yet to come.

Therefore, apart from the magnitude of the peak flows another important factor should be taken into consideration, i.e. the persistence of flood described by duration of high water levels. Long-lasting high stages may weaken the levees' structure (soaking) and cause dangerous leaks, blurs and breaks that threaten their destruction. That is why the classical flood frequency analysis (FFA) concerning only the frequency of the annual maximum (AM) flows (Bogdanowicz et al., 2011) is not suitable in this case and ought to be supplemented by the analysis of the duration of flows over the given threshold. The joint risk of inundation making allowance for the two main sources of vulnerability to flood hazard for areas protected by embankments, i.e. over-crest flow and levees failure, has been proposed and defined. In Poland, as in many other countries for each hydrological station two benchmark water levels, named the warning stage and the alarm stage, have been specified. Although warning and alarm stages are assigned to the places where water levels are observed, i.e. to the hydrological stations, their determination procedures as well as other inundation risk characteristics take into account, inter alia, the elevation of the embankment system for the whole adjacent river reach. So, the results of below analysis refer to the river reaches represented by data observed at hydrological stations. The frequency of annual maximum uninterrupted duration, D (in days), of flows over the flood alarm stage (Figure 1) can be used to assess the risk of flooding due to waning of the levees' strength. The aim of this study is to introduce formal aspects of the Duration-flow-Frequency (DqF) modelling in stationary and non-stationary conditions, to use it to assess the inundation risk due to the levees breach and to combine it with the AM flow model to get the cumulative probability of inundation.

To cater for the conventional FFA, the flow discharge (QA) corresponding to the alarm stage (HA) is used here, i.e. the upper limb of the rating curve is regarded as time invariant. The frequency of annual maximum uninterrupted duration of flows, D (in hours, days, etc.), over the flood alarm stage (HA) (or equivalently over the alarm flow (QA)) but excluding floods pouring over the embankment crest (which corresponds to flows exceeding the overtopping flow QB) serves to assess the inundation risk of flood spilling out of river channel caused by scouring the levees (Figure 2). Therefore, the dt = 0 in the [d] time-series denotes that the QA has not been exceeded during the t-thyear ($Q_{max}(t) < Q_A$) or that the peak flow has exceeded the overtopping flow ($Q_{max}(t) > Q_B$) where Q_{max} denotes the annual maximum discharge. Note that if more than one flood appears in a year, the D and the annual peak flow (Q_{max}) can correspond to different floods (Figure 1). In the presented statistical model, the duration (D) is considered as a random variable while the alarm flow discharge (QA) is the fixed value. The paper is built as follows: in the second section the concept of the inundation risk for embanked river is defined. Then the Duration-Flow discharge-Frequency (DqF) model is introduced and estimations of its parameter for stationary and non-stationary case are described and discussed. Taking into account the embankment resistance, the annual probability of inundation caused by levees breaching is introduced. To illustrate the proposed way of inundation risk assessment the case study for the Szczucin gauging station at the Vistula River (Southern Poland) is presented (Section 4). The probability of inundation due to levees breaching is compared and combined with the conventional probability of peak flow exceeding the levee crest and the cumulative probability of inundation are computed. The section 5 concludes the paper.

FLOOD RISK

Floods occur as a result of water spilling over the crest of embankment (Q > QB) or more often as a result of prolong existence of high water in the embanked river channel, i.e. when the peak flow discharge exceeds the alarm flow (QA) but is lower than the overtopping flow (QB, i.e. the discharge that overtops levee crests) (QAQ < Qmax < QB). One can also distinguish many other causes of floods, such as back water and ice-jams, etc., but they do not stem from the embankment failures and will not be considered in this study.

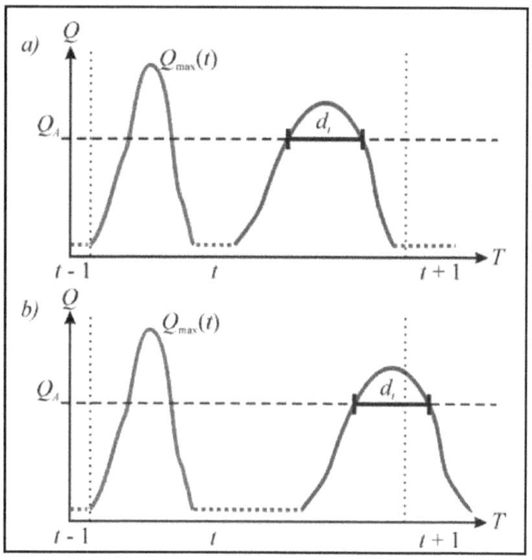

Figure 1. Definition of the threshold flow discharge and duration in DqF model: a) the flood wave of d_t duration entirely in the year t; b) the flood wave starts in the year t and continues in t + 1.

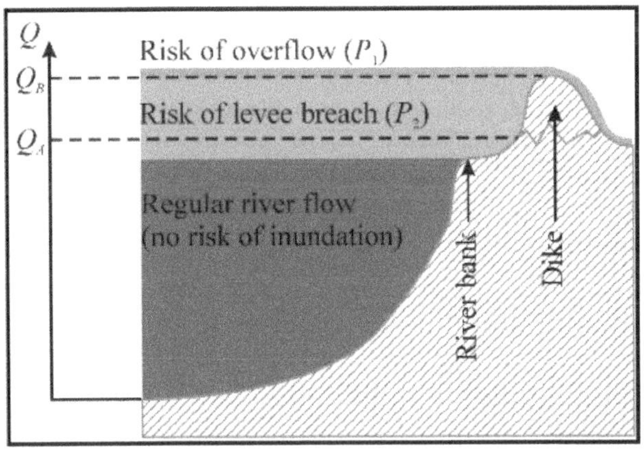

Figure 2. Two reasons of inundation—an illustration.

The annual probability of inundation for embanked river reach is expressed as the total of probability of the two exclusive events (Fl stands for "flood") (see Figure 2):

$$P(Fl) = P_1(Fl) + P_2(Fl) \tag{1}$$

where the first term comes from the conventional FFA

$$P_1(Fl) = p(Q_{max} > Q_B) \tag{2}$$

The second term of Equation (1) defines the probability of inundation caused by levees breaching which depends on both the flood persistency and levees resistance to high water stages which in turns depends on their design and technical condition. Therefore the P_2(Fl) is expressed as the integral of the product of the value of the hazard index p(Fl|d) which is defined as the probability of levee breaching caused by the d-days duration of flow over the flow level Q_A and of the pdf of the d—duration, i.e. f(d) for annual peak flows in the interval $Q_A < Q_{max}(t) < Q_B$.

$$P_2(Fl) = p\left(Fl \middle| (Q_A < Q_{max} \le Q_B)\right) = \int_{0^+}^{\infty} h(Fl|d) \cdot f(d) \cdot \mathrm{d}d \tag{3}$$

where

f(d)—pdf of the duration d of flows above the alarm stage;

h(Fl|d)—the hazard index being the probability of levee breaching caused by a high water of the duration d.

The value of the hazard index h(Fl|d) tends to 0 for d going to 0 and to 1 for d going to infinity. The hazard index h(Fl|d) is determined administratively for the river reach by the Regional Water Management Board based on the technical assessment of flood embankments.

Note that collating the annual maximum high flow duration data for analysis one puts dt = 0 both for $Q_{max} < Q_A$ and $Q_{max} >$ QB, i.e. the only one inundation yearly is considered and that caused by spilling over crest has the priority over one caused by prolonged high stages. Furthermore note that the weaker is the relationship between annual maximal values of peak flow and duration of flows above the alarm flow (QA) the more justified is the separate analysis of the both random variables.

The ratio of probabilities P_2 to P_1 and their total is helpful to determine the actions to reduce the risk of flooding, namely the strengthening or heighten the levees (or building parallel levees).

FORMAL ASPECTS OF THE DURATION-FLOW-FREQUENCY MODELLING

To address the flood risks arising from softening and washing out the river embankments, Bogdanowicz et al. (2011) proposed to take as the subject of analysis the frequency of annual maximum uninterrupted duration, D (in days), of flows over the flood alarm stage (Q_A), i.e. the duration (D) is considered as a random variable while the alarm flow discharge (Q_A) is the fixed value (Figure 2).

The time-series of annual maximum uninterrupted duration, D (in days), of flows over the flood alarm flow Q_A i.e. d = (d_1, d_2,..., d_t,..., d_T) is the subject of statistical modelling in stationary and non-stationary conditions. The d_t = 0, denotes that the Q_A has not been exceeded during the t-th year ($Q_{max}(t) < Q_A$) or that the peak flow has exceeded the overtopping flow ($Q_{max}(t) \ge Q_B$), which means that the priority of overtopping over breaching is given and we rule out the possibility of two inundation floods of the two different origins within one year. Note that the condition $Q_{max}(t) \ge Q_B$ is equivalent to the unconditional inundation, i.e. from Equation (2) $P_1(Fl|Q_{max}(t) \ge Q_B) = 1$, while $Q_B > Q(t) \ge Q_A$ points only possible inundation (see Equation (3)).

Frequency analyses of hydrological sample with zero events have received relatively little attention. Still there are several approaches for analysis of censored data, including probability plot regression, weighted-moment estimators, maximum likelihood estimators, and conditional probability analyses (Gilliom & Helsel, 1986; Hass & Scheff, 1990; Harlow, 1989; Helsel, 1990). A consistent approach to the frequency analysis of such data requires using discontinuous probability distribution functions. Jennings and Benson (1969), Interagency Advisory Committee on Water Data (1982), Woo and Wu (1989), Wang and Singh (1995) among others developed empirical three-parameter models for frequency analysis of hydrologic data containing zero values. When the available water stage records have been sampled in daily intervals, the d values are integer numbers and in reality correspond to the duration (d – 0.5, d + 0.5), while in particular for d = 0 to the interval (0, d + 0.5). If a flood starts at the end of a year and is continuing to the next year, the d value is derived for whole flood and ascribed to the year t when culmination occurred. To get an insight into flood persistence properties, the several threshold stages (Q_T) are considered but not only the alarm stage Q_A.

As far as the probability theory is concerned, the occurrence of zero events can be expressed by placing a non-zero probability mass on a zero value, i.e. P(D = 0) ≠ 0, where D is the random variable, and Pis the probability mass (e.g. Strupczewski et al., 2002, 2003; Węglarczyk et al., 2005). Therefore, the parent distribution functions of such hydrologic series would be discontinuous (with discontinuity at zero) and, using the theorem of total probability, their forms can be written as:

$$f(d) = \beta\delta(d) + (1-\beta) f^{\circ}(d;\mathbf{g}) \cdot 1(d) \tag{4}$$

where β denotes the probability of the zero event, i.e., β = P(D = 0), f(d; g) is the conditional probability density function (CPDF), i.e. f(d; g) ≡ f(d|D > 0), which is continuous in the range (0, +∞) with a lower bound of zero, and g is the vector of parameters (containing β or not), δ(d) is the Dirac's delta function and 1(d) is the unit step function. Assuming the infinite upper bound for D seems acceptable and facilitates modelling. Due to discretized duration d intervals, the probability of exceeding the Q_A flow during one day only equals to

$$P(d) = \int_{d-1/2}^{d+1/2} f(d) \cdot \mathrm{d}d \quad (5)$$

Hydrological samples with zero values are most frequently of exponential-like shape. Weglarczyk et al. (2005) model the continuous part of (4) by two-parameter distributions, namely by Generalized Pareto, Weibull and Gamma, estimating parameters by the maximum likelihood (ML) and the moments (MOM) methods.

ESTIMATION OF THE WEIGHT PARAMETER β

1) From the pdf of the duration d [Equation (4)] and the records $d = (d_1, d_2,\ldots, d_t,\ldots, d_T)$ for given alarm flow Q_A From Equation (4) one can write the likelihood function as:

$$L = \beta^{n_1} \cdot (1-\beta)^{n_2} \prod_{j=1}^{n_2} f^{\circ}(d_j; \mathbf{g}) \quad (6)$$

where n_1 and n_2 denote the number of zeros and non-zeros values, respectively.

If $\beta \notin g$, from ML-equations:

$$\frac{\partial \ln L}{\partial \beta} = \frac{n_1}{\beta} - \frac{n_2}{(1-\beta)} = 0 \quad (7)$$

one can easily find that the ML-estimate of β is

$$\hat{\beta} = \frac{n_1}{n_1 + n_2} \quad (8)$$

i.e. β and g are estimated by MLM independently.

2) From CDF of annual maximum floods obtained from FFA

The better estimate of the β parameter in the sense of definition (Equation (9)), not its standard error, can be obtained from the CDF of annual peaks providing the selected for Annual Maxima (AM) model fits well upper tail data. Note that the D = 0, denotes that the QA has not been exceeded during the t-th year ($Q_{max}(t) \leq Q_A$) or that the peak flow has exceeded the overtopping flow ($Q_{max}(t) > Q_B$) where Q_{max} denotes the annual maximum discharge, therefore, probability of zero value of D

$$\hat{P}(D=0) = \hat{P}(Q_{\max} < Q_A) + \hat{P}(Q_{\max} > Q_B) = \hat{\beta} \quad (9)$$

Should be estimated from CDF of annual peak flows got from FFA rather than from the (0, 1) time series of the d record. Having derived from FFA the CDF of the annual peaks $\hat{G}(Q_{\max}) \equiv \phi(Q_{max}, \hat{\mathbf{h}})$ where $\hat{h}$ is the vector of parameter estimates, one gets the estimate of β as

$$\hat{\beta} = \hat{G}(Q_{max} = Q_A) + \left(1 - \hat{G}(Q_{\max} = Q_B)\right) \quad (9a)$$

Note that if more than one flood appears in a year it may happen that the dt and the annual peak flow Q_{max}((t) correspond to different floods. Floods in excess of Q_B are unique in Polish rivers, but if they were they should be in the FFA treated as of unknown magnitude over the threshold Q_B, thus one deals with first order right censored sample.

ESTIMATION OF PARAMETERS OF THE CONTINUOUS PART OF EQUATION (4)

ML estimate of the parameters (g) of the continuous part of PDF (Equation 6), i.e. the conditional probability density function (CPDF). f (d; g) ≡ f(d|D > 0) of f(d; g), can be obtained by solving the ML system of equations:

$$\frac{\partial \ln L}{\partial \mathbf{g}} = \frac{\partial}{\partial \mathbf{g}} \sum_{j=1}^{n_2} \ln f^{\circ}(d_j; \mathbf{g}) = 0 \text{ for } \beta \notin \mathbf{g} \quad (10)$$

Since the samples with zero values are most frequently of exponential-like shape, the following distribution functions are recommended as candidates for f(dj; g) model: Exponential, Wei bull, Generalized Pareto, Generalized Exponential and Gamma. Note that Exponential distribution is a special case of all other mentioned above distributions. The detailed information on the models mentioned above with the methods of ML estimation, one can easily find in hydrological and statistical literature, e.g. in Rao (2000) and for GE in Gupta and Kundu (2000). Had the duration and peak flow were strongly linearly related to a legitimate adoption of a common distribution function truncated the d distribution function to the positive half i.e. to d > 0.

The basic assumption in the classical Flood Frequency Analysis and the Duration-Flood-Frequency modelling is that neither the adopted distribution function nor its parameters change in time it is possible to extend the described methodology of the inundation risk assessment for the case of non-stationarity of flood regime.

CASE STUDY—SZCZUCIN AT VISTULA RIVER (SOUTHERN POLAND)

To illustrate how the proposed approach works in practice the Szczucin gauge (southern Poland) at the Vistula River has been selected as the case study. The daily flows record covering the period 1951-2006 (n = 56 years) was used in this study. At first the daily records have been controlled and tested with regard to the sharp discontinuities and jumps in data—no particular irregularities have been detected (Figure 3).

The overtopping flow Q_B was assessed from the rating curve as 10,500 m3 /s which roughly corresponds to two-hundred-years return period of annual peak flow ($Q_{0.5\%}$), the base design value for the 1st class embankments. In fact, there are no annual peak flows exceeding this value in the record. Therefore the Q_B value does not affect the composition of the vector of observation values [d_t]. The alarm threshold for the Szczucin station QA = 1690 m^3 /s (which means flow of ca. 2-year return period, stage 660 cm), It gives n_2 = 23 flood hydrographs with Q_{max} above QA. The hazard index h(Fl|d) for QA = 1690 m^3 /s [Equation (3)] was assessed as:

$$h(Fl|d) = \begin{cases} 0.05 \cdot d & \text{for } d \leq 20 \text{ days} \\ 1 & \text{for } d > 20 \text{ days} \end{cases} \quad (11)$$

i.e. the embankments can not withstand the pressure of high waters of more than 20 days.

A visual inspection of flood hydrographs reveals the variety of their shapes and, as a consequence, the d distribution cannot be represented (or replaced rather) in FFA by Q_{max} distribution.

It implies the analysis of both d and Q_{max} by (perhaps) two different types of models. As a model for the parameters of the f function Generalised Exponential (GE) distribution has been chosen (e.g. Gupta & Kundu, 2000):

$$f^{\circ}(d;\alpha,\gamma)=\frac{\gamma}{\alpha}\exp(-d/\alpha)\left[1-\exp(-d/\alpha)\right]^{\gamma-1} \tag{12}$$

where α > 0 is the scale parameter and β > 0 is the shape parameter.

Among the set of distributions mentioned in the previous section the GE distribution performs relatively well in terms of the AIC value and shows stability of numerical ML solutions in estimation of f(d; g) parameters. An additional argument for choosing the GE distribution is that being relatively recently derived was not to our knowledge used in FFA. However following on the model adopted for AM distribution which is Generalised Extreme Value (GEV) distribution, the truncated at zero GEV distribution may be selected as the model of o f^0, in particular in case of a strong correlation between d(t) and $Q_{max}(t)$. The annual maxima are believed to be adequately described by the heavy-tailed distributions (e.g. Strupczewski et al., 2011), so to cater for the Flood Frequency Analysis (FFA) for extreme values (i.e. annual maxima) the values [Equation (8)] and $P_1(Fl)$ [Equation (2)] by means of Q_{max} series were calculated with the three-parameter GEV distribution:

$$G(q;\alpha,\gamma,\varepsilon)=\exp\left\{-\left[1-\frac{\gamma}{\alpha}(q-\varepsilon)\right]^{1/\gamma}\right\}=G_{GEV}^{\gamma}(q) \tag{13}$$

From the AM sample covering the period 1951–2006 (n=56 years) we got the ML estimates of GEV parameters equal: location ≡ $\hat{\varepsilon}$ = 1260.02 m^3/s, scale ≡ $\hat{\alpha}$ = 671.39 m^3/s and shape ≡ $\hat{\beta}$ = –0.33. Substituting for q into Equation (13) the Q_A value and then putting the corresponding probabilities to Equation (9a), one gets the estimates of the weighting parameter $\hat{\beta} = 0.577$. From Equation (8) we get quite similar result, i.e. $\hat{\beta}$ = = 33 56 0.559 and the confidence interval for proportion β includes the value estimated from AM distribution [Equation (9)]. ML parameter estimates of GE applied for o f arc $\hat{\alpha}$ − 3.2318 and $\gamma\hat{}$ = 0.8357.

ASSESSMENT OF PROBABILITY OF LEVEE BREACH ALONG SZCZUCIN REACH

Since the event of levee breach is conditioned by the peak flow being in the range of $[Q_A, Q_B]$, Equation (3) can be written as:

$$P_2(Fl)=(1-\beta)\int_{0^+}^{\infty} p(Fl|d)\cdot f^{\circ}(d)\cdot \mathrm{d}d \tag{14}$$

The pdf for d > 0 takes the form

$$\left(1-\hat{\beta}\right)\cdot f^{\circ}(d;\alpha=3.4238,\gamma=0.8357)=0.423\frac{0.2441\cdot\exp(-d/3.4238)}{\left[1-\exp(-d/3.4138)\right]^{0.1643}} \tag{15}$$

while the ML estimate of β was taken from Equation (9a) and it equals 0.577. Substituting it and the hazard index function defined by Equation (11) into Equation (3a) and integrating one gets the annual probability of levee breaching $P_2(Fl)$ = 0.064. Note that at the same time, and when the same GEV distribution is used the probability of flood caused by exceeding embankment crest by annual peak flow, i.e. $P_1(Fl)$ = P(Q_{max} > QB = 10,500 m^3/s) = 1 – G(QB) is equal to 0.005, i.e. it is almost insignificant (more than ten times smaller than P1), hence, the overall probability of flood along Szczucin reach P = P_1 + P_2 = 0.069. Variety of shapes of flood hydrographs can be evaluated by a measure of correlation strength between $Q_{max}(t)$ and d(t). Due to shape similarity of flood peak parts, a strong dependence between the peak flows (Q_{max}) and the duration above the alarm flow (d) can take place. If it is a case, the probability $P_2(Fl)$ can be assessed on the base of Q_{max} distribution $g(Q_{max})$. Assuming that d = $\psi(Q_{max})$ one can expressed in Eq.(15) the d variable by the Q_{max} getting

$$P_2(Fl)=p\left(Fl\middle|(Q_A<Q_{max}\le Q_B)\right)=\int_{Q_A}^{Q_B} h\left(Fl\middle|\psi(Q_{max})\right)\cdot g(Q_{max})\cdot \mathrm{d}Q_{max} \tag{16}$$

where per analogy to Euation (11) $h(Fl|\psi(Q_{max}))$ equals 0 and 1 for QA and QB, respectively. The Pearson's correlation coefficient $r(Q_{max}, d)$ for Szczucin equals to 0.83.

Of course, when estimating the risk of a levee breach except the time of high water residence, more technical parameters of levees should be analysed, such as the construction of the levee, the material used for its building, its age, susceptibility to softening, the regime of the river, wind-induced waving and so on. All in all, those who decided to build their houses in the river's proximity behind the levees, sooner or later do experience a catastrophe.

CONCLUSIONS

In the paper the new two-component model of flood waves, 'duration of flooding-discharge-probability of non-exceedance' (DqF), with the methodology of its parameters estimation was proposed as a completion to the classical FFA methods. Such model can estimate the duration (d) of stages (and flows) exceeding the assumed magnitude with a certain probability which is of key importance when the river's dikes are prone to the prolonged impact of high waters. The embankments may be weaken by the water, soak and eventually break—this is the most frequent cause of floods in Poland. However, in this study the two main causes of inundation of embanked rivers, namely over-crest flow and wash out of the levees, were combined to assess the total risk of inundation. The proposed DqF modeling approach can be generalized to the non-stationary conditions. Therefore, in addition to the maximum flow one should consider also the duration of high waters above the alarm flow Q_A in a river channel. The model combined with the technical evaluation of probability of levees breach expressed by the hazard index gives the annual probability of inundation caused by the embankment failure. The probability of inundation is the total of probabilities of exceeding embankment crest by flood peak and the probability of washout of levees.

The DqF modeling is the consequence of QdF approach developed by Javelle et al. (1999, 2000, 2002) and Bogdanowicz et al.(2008) but in the first model the gravity is put on the probability of the certain duration above alarming stage/discharge (Q_A) rather than on magnitude of flood itself

(Q_{max}) like in the latter case.

The DqF model in the form of Equation (4) consists of two terms: β·δ (d) deals with the zero event, i.e. D = 0, whereas the latter term (1 – β)·f(d;g)·1(d) stands for the events when the duration D > 0. In general both β and f° in non-stationary case may depend on time. The maximum likelihood method (MLM) was proposed for estimation of β and g parameters. In the non-stationary case it is convenient to describe the β (t; θ) by means of the logistic function (23). However, β and β (t; θ) can be also estimated by means of annual peak flows series, Q_{max}, using the routine flood frequency techniques (FF) with distribution functions commonly used in FFA (e.g. GEV) for stationary and non-stationary case, respectively. For f(d_j; g) model the exponential-like shaped distribution functions are recommended, such as: Exponential, Weibull, Pareto, Generalized Exponential, Gamma and similar.

The calculations for the Szczucin at the Vistula River case study have showed the similar results for the weighting factor estimated by ML method from the duration time-series and from annual peaks time-series. The peak flows that could overtop the embankments have not been detected in the Szczucin's record (1951-2005). According to the hazard function (11) the possibility of levees breaching increases almost tenfold the probability of inundation.

Work on the application of the model DqF provides insight with respect to the shape of the flood hydrographs. The shape of flood hydrograph varies according to a number of controlling factors in the drainage basin. For extreme floods the decisive factor is the rainfall/snowmelt time distribution while the catchment variable characteristics do not play a significant role. This results in shape similarity of peak parts of extreme flood waves of the same origin. For smaller floods the observed shapes are more diversified. This often leads to the rather strong, but not functional dependence between the main shape measures: duration d over a given threshold and the peak flow Q_{max}. This correlation is stronger when one type of floods (rainfall or thaw origin) dominates at the studied catchment closing profile. Then the questions arise if it is possible and profitable, in face of strong enough dependence of the two variables, to replace d in the persistence analysis P_2(Fl) by pdf of annual maxima and if both variables have to be of the same probability distribution. Although substitution of d by Q_{max} seems to be an effective way to simplify the calculations of the total risk of flooding, however this method reduce the observed variability of d values and so distorts the description of flood dynamics. Due to not exact dependence or not linear relation the distributions of d and Q_{max} can be different. It is difficult to point the lower limit of the coefficient of determination from which the above method could be used, but the weaker is the dependence, the more justified is to analyze the both random variables D and Q_{max} separately.

The DqF model proved to be the important completion to the traditional FFA concentrating on maximal seasonal or annual discharges. The DqF approach is especially useful in polish specific conditions where the flood protection infrastructure is dated and often does not survive confrontation with prolonged pressure of high waters.

Reliable data and information about floods are indispensable for better understanding the interactions between rivers and flood protection system: embankments, reservoirs and polders. Improvement of statistical models is essential for engineering design in general and in particular for implementation of flood risk mitigation procedures. Not only has the DqF modelling shown that actual flood risk is greater than the risk assessed by means of classical FFA but also provides quantitative measures which can be used in flood protection systems planning, exploitation and conservation. This measure in form of dependence of inundation risk on river flow (or water level) should be established for other hydrological stations on Polish rivers and their dimensionless versions compared. The geographic information systems technique (GIS) could be used to indicate locations prone to inundation, also the GIS can be a helpful tool to visualisation and testing trends in the structure of river network and to the regional analysis. These results can constitute the theoretical background to a number of practical decisions in water management issues.

Acknowledgements

This research project was partly financed by the grant of the Polish National Science Centre titled "Modern statistical models for analysis of flood frequency and features of flood waves", decision nr DEC-2012/05/B/ ST10/00482.

REFERENCES

1. Bogdanowicz, E., Strupczewski, W. G., & Kochanek, K. (2011). Persistence as a Factor of Flood Hazard for Embanked Rivers. EGU Leonardo Conference "Floods in 3D", Bratislava, 23-25 November 2011, Abstract in Proceedings.
2. Gilliom, R. J., & Helsel, D. R. (1986). Estimation of Distributed Parameters for Censored Trace Level Water Quality Data-1. Estimation techniques. Water Resources Research, 22, 1201-1206.
3. Gupta, R. D., & Kundu, D. (2000). Generalized Exponential Distribution: Different Method of Estimations. Journal of Statistical Computation and Simulation, 1-22.
4. Haas, C. N., & Scheff, P. A., (1990). Estimation of Averages in Truncated Samples. Environmental Science & Technology, 24, 912-919. http://dx.doi.org/10.1021/es00076a021
5. Harlow, D. G. (1989). Effect of Proof-Testing on the Weibull Distribution. Journal of Materials Science, 24, 1467-1473. http://dx.doi.org/10.1007/BF02397087
6. Helsel, D. R. (1990). Less Than Obvious: Statistical Treatment of Data Below Detection Limit. Environmental Science & Technology, 24, 1767-1774.
7. Interagency Advisory Committee on Water Data (1982). Guidelines for Determining Flood Flow Frequency. Bulletin 17B, U.S. Department of the Interior, Geological Survey, Office of Water Data, Reston, Va.
8. Rao, A. R., & Hamed, K. H. (2000). Flood Frequency Analysis. CRC Press.
9. Strupczewski, W. G., Singh, V. P., & Weglarczyk, S. (2002). Physically Based Model of Discontinuous Distribution for Hydrological Samples with Zero Values. In S. Al-Rashed, & A. A. Sherif (Eds.), Proceedings of the Int. Conf. On WaRMAR, Kuwait. Surface Water

Hydrology (pp. 523-537). Lisse: Balkema Publishers, Swets & Zeitlinger.

10. Strupczewski, W. G., Weglarczyk, S., & Singh, V. P. (2003). Impulse Response of the Kinematic Diffussion Model as a Probability Distribution of Hydrologic Samples with Zero Values. Journal of Hydrology, 270, 328-351. http://dx.doi.org/10.1016/S0022-1694(02)00309-8

11. Strupczewski, W. G., Kochanek, K., Markiewicz, I., Bogdanowicz, E., Weglarczyk, S., & Singh V. P. (2011). On the Tails of Distributions of Annual Peak Flow. Hydrology Research, 42, 171-192. http://dx.doi.org/10.2166/nh.2011.062

12. Wang, S. X., & Singh, V. P. (1995). Frequency Estimation for Hydrological Samples with Zero Values. Journal of Water Resources Planning and Management, ASCE, 121, 98-108. http://dx.doi.org/10.1061/(ASCE)0733-9496(1995)121:1(98)

13. Woo, M. K. & Wu, K. (1989). Fitting Annual Floods with Zero Flows. Canadian Water Resources Journal, 14, 10-16. http://dx.doi.org/10.4296/cwrj1402010

14. Weglarczyk, S., Strupczewski, W. G., & Singh, V. P. (2005). Three-Parameter Discontinuous Distributions for Hydrological Samples with Zero Values. Hydrologic Processes, 19, 2899-2914.

CHAPTER 3

AN INFORMATION SYSTEM FOR RISK-VULNERABILITY ASSESSMENT TO FLOOD

Subhankar Karmakar[1], Slobodan P. Simonovic[2], Angela Peck[2] , Jordan Black[2]

[1] Centre for Environmental Science and Engineering, Indian Institute of Technology Bombay, Mumbai, India

[2] Department of Civil and Environmental Engineering, The University of Western Ontario, London, Canada

ABSTRACT

An exhaustive knowledge of flood risk in different spatial locations is essential for developing an effective flood mitigation strategy for a watershed. In the present study, a risk-vulnerability analysis to flood is performed. Four components of vulnerability to flood: 1) physical, 2) economic, 3) infrastructure and 4) social; are evaluated individually using a Geographic Information System (GIS) environment. The proposed methodology estimates the impact on infrastructure vulnerability due to inundation of critical facilities, emergency service stations and bridges. The components of vulnerability are combined to determine an overall vulnerability to flood. The exposures of land use/land cover and soil type (permeability) to flood are also considered to include their effects on severity of flood. The values of probability of occurrence of flood, vulnerability to flood, and exposures of land use and soil type to flood are used to finally compute flood risk at different locations in a watershed. The proposed methodology is implemented for six major damage centers in the Upper Thames River watershed, located in the South-Western Ontario, Canada to assess the flood risk. An information system is developed for systematic presentation of the flood risk, probability of occurrence of flood, vulnerability to flood, and exposures of land use and soil type to flood by postal code regions or Forward Sortation Areas (FSAs). The flood information system is designed to provide support for different users, i.e., general public, decision-makers and water management professionals. An interactive analysis tool is developed within the information system to assist in evaluation of the flood risk in response to a change in land use pattern.

INTRODUCTION

Records of loss of life and damage caused by floods worldwide show that these have continued to rise steadily during recent years. Understandably, the response has been to call for increased efforts to protect life and property. The sustainable and effective management of floods demands a holistic approach—linking socio-economic development with the protection of natural ecosystems and appropriate management links between land and water uses. It is recognized that a watershed is a dynamic system in which there are many interactions between human population, land use and water bodies. Assessment and mapping of "flood risk" [1-7] and "vulnerability to flood", and dissemination of the appropriate

Citation: S. Karmakar, S. Simonovic, A. Peck and J. Black, "An Information System for Risk-Vulnerability Assessment to Flood," Journal of Geographic Information System, Vol. 2 No. 3, 2010, pp. 129-146. doi: 10.4236/jgis.2010.23020.

information to different stakeholders is a very important part of the flood management process. The general public may use the information in purchasing a house, or in selecting a site to start a business. Knowledge of flood risk could aid decision-makers in: developing land development plans and land use zoning; planning emergency response strategies; waste disposal site selections; preparing infrastructure budgetary decisions; developing guidelines for operation of existing infrastructure; and general policy development at all levels. Water management professionals can utilize the flood risk information in planning, design, construction, and operation & maintenance of flood protection infrastructure (e.g., reservoirs, dikes, drainage pipes, etc). Flood risk mapping has been performed extensively for effective flood management, starting with the pioneering work of Garrett [8]. The risk of flooding to towns and villages along 200 km of the River Thames and its tributaries are assessed using a mathematical model developed for Thames Water Rivers Division, UK. The River Thames Strategic Flood Defence Initiative examines the vulnerability of floodplain development along the river. The achievements of the Flood Risk Mapping Program, New Brunswick, Canada, are summarized by Burrell and Keefe [9]. The procedures used to produce flood risk maps are outlined very clearly along with an assessment of the accuracy achieved. Floyd [1] performs a flood risk assessment on the city of Bombay (Mumbai), India. The results provide an initial indication of the cost-effectiveness of different remedial measures. Morris and Flavin [10] present maps of England and Wales showing the built-up areas that would be at flood risk. Shrubsole [11] mentions government responsibilities in flood management of the Saguenay and Red River valley and provides alternative flood management strategies considering ecosystem management, partnerships and the role of science. Hall et al. [12] represents the processes of fluvial and coastal flooding over linear flood defence systems in sufficient detail to test alternative policy options for investment in flood management. Potential economic and social impacts of flooding are assessed using national databases of floodplain properties and demography. A case study of the river Parrett catchment and adjoining sea defences in Bridgwater Bay in England demonstrates the application of the method and presentation of results using Geographic Information System (GIS). Barredo et al. [13] aims to illustrate a framework for flood risk mapping at pan-European scale produced by the Weather-Driven Natural Hazards (WDNH). The threatening natural event is represented as the hazard component, and furthermore, exposure and vulnerability are considered as anthropogenic factors that contribute also to flood risk. The flood risk is considered on the light of exposure, vulnerability and hazard, and mathematically considered as product of hazard, exposure and vulnerability.

Vulnerability assessments have been undertaken to understand the "potential for loss", traditionally they focused on the nature of the hazard and who and what are exposed [14]. More recently, vulnerability assessments have explored the social, economic, and political conditions that are likely to affect the capacity of individuals or communities to cope with or adapt to hazards [15]. Bender [16] discusses the development and use of natural hazard vulnerability assessment techniques in the Americas. He emphasizes how and why a thorough vulnerability analysis is required for physical, economic and social planning in a watershed. There are numerous studies that have addressed contemporary vulnerability of different communities worldwide to flooding from the natural hazards perspective of understanding exposure and the number of people and structures affected [17,18] but few that explore the socio-economic aspects of flooding vulnerability [19-22]. Recently, the conceptualization on social vulnerability has gained prominence in the literature. It is related to characteristics that influence an individual's or group's ability or inability to anticipate, cope with, resist, and recover from or adapt to any external stress such as the impact of flooding [23-25]. Cutter et al. [26] present a method for assessing vulnerability in spatial terms using both biophysical and social indicators. Their results suggest that the most biophysically vulnerable places do not always spatially interact with the most vulnerable populations. Flax et al. [27] develop a risk-vulnerability assessment methodology named as Community Vulnerability Assessment Tool (CVAT), which assists emergency managers in their efforts to reduce vulnerabilities through hazard mitigation, comprehensive land use and development planning. Cutter et al. [28] list factors that have gained consensus among social scientists as contributing to social vulnerability to environmental hazards. Blong [29] introduces a new damage index for estimating the replacement cost of damaged buildings in vulnerability analysis. Carter [30] analyzes flood risk as a combination of threat, consequence, and vulnerability. He discusses the federal role in investment decisions for flood control infrastructure. Chakraborty et al. [31] develop two new quantitative indicators, i.e., a geophysical risk index, based on National Hurricane Center and National Flood Insurance Program data, and a social vulnerability index, based on census information. Rygel et al. [32] focus on constructing a social vulnerability index and its application to a case study of hurricane storm hazard. They demonstrate a method of aggregating vulnerability indices for different indicators using Pareto ranking that results in a composite index of vulnerability, which avoids the problems associated with assigning weights. Werritty et al. [33] discuss the social impacts of flood events in Scotland including attitude and behavior toward flooding events, warnings, evacuations and consequences. The study suggests ways for enhancing social resilience for sustainable flood management in Scotland.

GIS is considered as a key tool by many researchers [34-38] to map the spatial distribution of flood risk and vulnerability to flood. A GIS facilitates the input, storage, management, analysis, integration, and output of spatial data which can aid with real time decision making and strategic planning for effective risk management and hazard preparedness [39]. GIS can improve warning, evacuation, and emergency response systems by helping route emergency response vehicles and locating emergency response facilities [39-40]. Exposures of soil and geology to flood, urban infrastructure, and socioeconomic data, can be input and stored in a GIS and then analyzed to identify areas prone to flood, identify vulnerable populations, and forecast flood events, and aid in land use zoning decisions to improve flood mitigation and management [17,39].

The flood risk mapping and analysis on various flood prone watersheds have been performed by many researchers throughout the world. In a recent study on Romanian national strategy, Varga et al. [41] provide basic information for preliminary flood risk assessments and flood hazard mapping in all areas with a significant flood risk, according to the Flood Directive. The technical and scientific approach and the main steps in setting up the plan for flood prevention, protection and mitigation at the river basin level are presented. Apel et al. [7] perform flood risk analyses in Eilenburg, Germany, especially in urban areas and tested a number of combinations of models of different complexity both on the hazard (probability of occurrence) and on the vulnerability. Chang et al. [42] examine the anthropogenic and natural causes of flood risks in six representative cities in the Gangwon Province of Korea. Tran et al. [43] explore the impacts of floods on the economy, environment and society; and tries to clarify the rural community's coping mechanism to flood disasters in Central Viet Nam. They reveal that flooding is an essential element for a coastal population, whose livelihood depends on productive functions of cyclical floods. Forster et al. [44] assess flood risk for a rural detention area, alongside the Elbe River in Germany. They find that the losses in agricultural areas exhibit a strong seasonal pattern, and the flooding probability also has a seasonal variation. The flood risk is assessed for a planned detention area based on loss and probability estimation approaches of different time frames, namely a monthly and an annual approach.

The present research study is motivated by the Hotspots project [45,46] completed by the Center for Hazards and Risk Research (CHRR) at Columbia University and the World Bank's Disaster Management Facility [DMF), now the Hazard Management Unit (HMU). In the Hotspots project, the risk levels are estimated by combining hazard or probability of occurrence with historical vulnerability for two indicators of risk—population and Gross Domestic Product (GDP) per unit area—for six major natural hazards: earthquakes, volcanoes, landslides, floods, droughts, and cyclones. The relative risks for each grid cell, rather than country as a whole, is calculated at sub-national scales. Such information can inform a range of disaster prevention and preparedness measures, and development of long-term land-use plans and multihazard risk management strategies. Hotspots global analysis and case studies stimulate additional research, particularly at national and local levels. The present study develops an information system for risk-vulnerability analyses to flood and facilitates vulnerability mitigation by providing various flood information to different users. The information system is designed to provide selective access to information on the bases of user needs. This reduces the misuse of data and promotes data security. A set of suitable vulnerability indicators and the procedure for their integration into an overall vulnerability index with high spatial density represent the major analysis tool within the information system. The additional innovations of the information system include: 1) the computation of selected flood risk-vulnerability indicators organized into themes from four components of vulnerability to flood, i.e., physical, economic, infrastructure, and social vulnerability [15], 2) the spatial infrastructure vulnerability analysis to flood due to inundation of main communication routes and road bridges, 3) the spatial flood impacts due to inundation of critical facilities (schools, hospitals, and fire stations) and 4) quantification of exposures of land use/land cover and soil permeability to flood. The postal codes or Forward Sortation Areas (FSA) are considered for spatial discretization of the region and flood risk evaluation. An interactive analysis tool is also developed for calculation of flood risk as a function of change in land use. The proposed information system is implemented to six major damage centers in the Upper Thames River watershed, located in the South-Western Ontario, Canada. The study focuses only on floods which are caused by the overflow of river banks that are characteristics for the region of interest.

As a prerequisite, some relevant technical definitions are provided in the next section. The third section contains a detailed description of the study area—the Upper Thames River basin in South-Western Ontario, Canada. The fourth, fifth and sixth sections provide the details on determination of probability of occurrence, vulnerability and exposures of land use and soil permeability to flood, respectively; and summarize the results obtained. The seventh section describes the features of developed information system for risk-vulnerability analyses to flood. The eighth section summarizes the conclusions from the study.

GENERAL DEFINITIONS

The most common approach to define "flood risk" is the definition of risk as the product of "hazard", i.e., the physical and statistical aspects of the actual flooding (e.g., return period of the flood, extent and depth of inundation), and the "vulnerability", i.e., the exposure of people and assets to floods and the susceptibility of the elements at risk to suffer from flood damage [7,47,48]. This definition is adopted in the Flood Directive [49]. According to Forster et al. [44], flood risk is a combination of potential damage and probability of flooding. More precisely, risk is considered as the product of hazard and vulnerability of a region [47]. However, in this study flood risk is the product of probability of occurrence (p_e), vulnerability to flood (V), and exposures of land use (E^{Land}) and soil permeability (E^{Soil}) to flood, following the concept of Kron [50] and Barredo et al. [13], where flood risk is expressed as a function of the hazard, vulnerability and exposed values:

$$Flood\ Risk = (p_e) \times (V) \times (E^{Land}\ and\ E^{Soil}) \quad (1)$$

Hazard may be defined as a threatening event, or the "probability of occurrence (p_e)" of a potentially damaging phenomenon within a given time period and area [31, 47]. It frequently encompasses hydrological and hydraulic analyses and the mapping of flood lines on floodplain. Vulnerability to flood (V) is defined as a measure of a regions' or population susceptibility to flood damages [51-53]. Exposures of land use (E_{Land}) and soil permeability (E_{Soil}) to flood quantify their effect on the severity of flood. When the exposures of land use/land cover and soil permeability to flood are considered, these denote how land use and soil permeability affects the severity of flood. For example, urbanized land use pattern results in an impervious soil layer increasing the severity of flood and thus the exposure of land use pattern to flood in an urban area is high. The exposure of soil permeability to flood is also directly related to flood flow. The more permeable soil has more infiltration capacity and therefore reduces surface runoff, whereas less permeable soil has less infiltration capacity and is more prone to water logging [54].

In the present study, all the information on the above mentioned three components of flood risk are effectively presented and processed using GIS. The layout for collecting and integrating the data, along with the sequential procedural steps for data processing and representation are outlined in Figure 1. The vulnerability section in Figure 1, illustrates the concept of layering data using a GIS, as well as combining vulnerability components to assess the overall vulnerability to flood. The next section presents the detailed characteristics and geography of the study area.

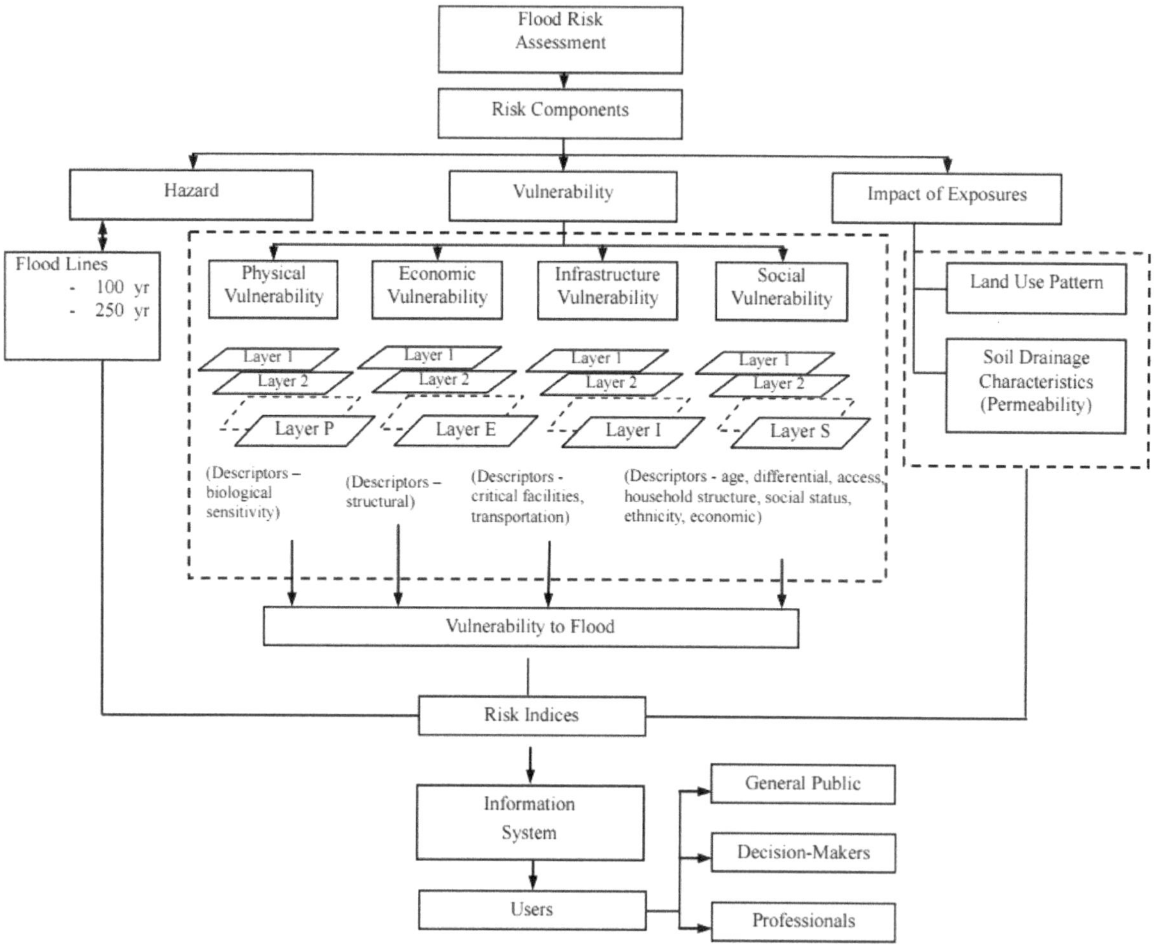

Figure 1. Organization of the flood information system.

STUDY AREA

The Upper Thames river basin lies in the middle of south western Ontario; drains an area of 3,500 km^2; and is populated by approximately 422,000 people. The basin is nested between the Great Lakes Huron and Erie. The basin has a well documented history of flooding events dating back to the 1700s. A detailed map of the Upper Thames River watershed in Ontario, Canada with a location map (inset) is shown in Figure 2. Two main tributaries of the Thames River, referred to as the North (1,750 km^2) and South (1,360 km^2) branches, join at a location in London known as "The Forks" (Figure 2). The Forks region has served as a historical landmark for London, and is characterized by both commercial and residential structures. Major flood damage centers in the watershed include communities of London, St. Marys, Ingersoll, Mitchell, Stratford and Woodstock. The Upper Thames river basin is an area of special importance for the sustainable socio-economic development of Ontario. This is a large and fertile area, and plays an important role in agriculture production from, fishing and aquaculture to perennial fruit trees. The flooding in the Upper Thames river watershed has the great effect on the fertility of soil and increase in the natural aquatic production. It is also the most dangerous natural disaster hazard affecting the socio-economic development and the life of the people in the area. Several studies have already been done to estimate the economic damage in the watershed due to flooding [55].

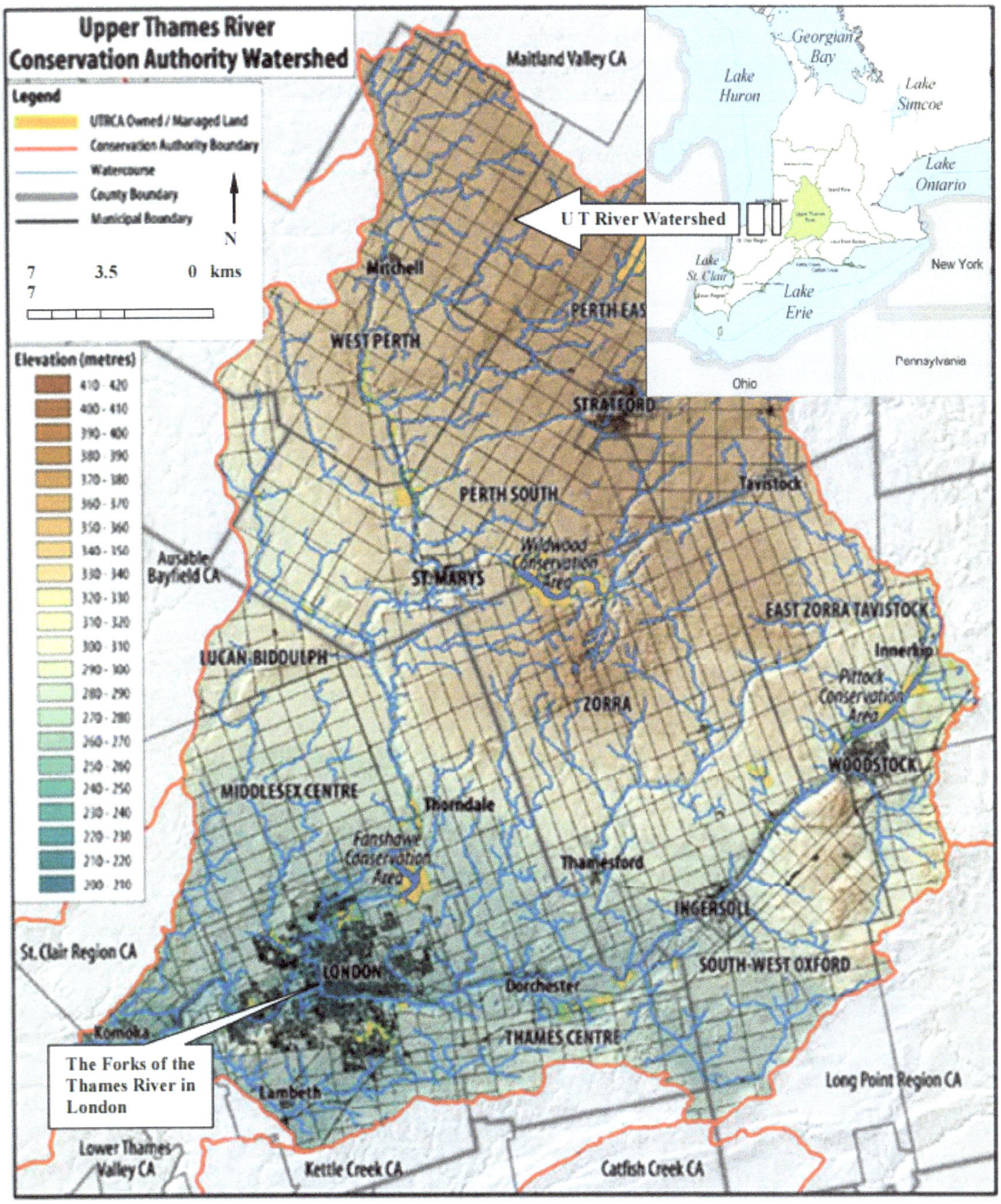

Figure 2. Detailed map and location map (inset) of the Upper Thames River watershed.

The study area consists of a number of major postal code regions, some of which extend beyond the watershed boundaries. The Forward Sortation Areas (FSAs) are distinguished by the first three characters of their postal code designation. The regions which historically experience more frequent flooding events are selected as areas of particular significance and the FSAs of these regions are selected for analysis. A total of 25 FSAs from these cities have been considered in the study, and are listed in Table 1. These FSAs are the smallest spatial geographic units considered in this study. The cities of important FSAs include London (17 FSAs), Woodstock (3 FSAs), Mitchell (1 FSA), St. Marys (1 FSA), Ingersoll (1 FSA) and Stratford (2 FSAs); with a particular emphasis on the city of London. Table 1 shows that the largest FSA by area in London is N6P; with an area of 103 km2 and in the entire study area is N0K; with an area of 1510.0 km^2. The smallest FSA by area in London is N6B; with an area of 3.2 km^2, and in the entire study area is N4Z; with an area of 0.5 km^2 . The average area of an FSA in London is 29.5 km^2 and the average area of all FSAs in the study area is 122.2 km^2 .

Numerical data necessary for the development of a flood information system has been collected from Statistics Canada, which provides updated national statistics consistently every five years following a Census of the population. Data is available for areas of various sizes, including FSAs as small census divisions which remain relatively stable over many years. Various layers and datasets compatible with the GIS software are collected from Statistics Canada, The Ontario Fundamental Dataset, Upper Thames River Conservation Authority (UTRCA), Surficial Geology of Southern Ontario (SGSO) dataset, and Route Logistics (RL). These datasets are available online or obtained from the Serge A. Sawyer map library and the Internet Data Library System (IDLS) at the University of Western Ontario (UWO), London, Canada. Table 2 presents the detailed information on different data layers with their formats and sources used in the present case study. All three components of the flood risk are assessed and represented in the information system separately in dissimilar ways. Each component is represented graphically, numerically, or using a combination of both. Next two sections describe the methodologies for determination of probability of occurrence and vulnerability to flood.

PROBABILITY OF OCCURRENCE

Probability of occurrence of flood (pe) [31] describes a physical threat from a flood occurring and a region becoming inundated. The consideration of probability of occurrence as a flood risk component is essential, since flood risk of a highly vulnerable population to flood is negligent if there is less probability or chance of occurrence of flood. The probability of occurrence (i.e., the extreme events and associated probability) of flood is also termed as "flood hazard" by many researchers [7,13,47,50,56]. In the present study however, the available results on pe performed by the UTRCA are used. Already available 100-year and 250-year flood lines are used for risk-vulnerability analysis.

Table 1. List of the municipalities and FSAs (PSEPC* 2005).

Municipality	FSA	Area (in km^2)	Municipality	FSA	Area (in km^2)
London	N5V	46.7	London	N6L	38.0
	N5W	15.7		N6M	28.7
	N5X	27.0		N6N	70.0
	N5Y	10.0		N6P	103.0
	N5Z	11.0	Mitchell	N0K	1510.0
	N6A	7.1	Wood-Stock	N4S	326.0
	N6B	3.2		N4T	4.1
	N6C	13.5		N4V	2.1
	N6E	20.6	St. Marys	N4X	327.0
	N6G	25.8	Stratford	N4Z	0.5
	N6H	41.5		N5A	233.0
	N6J	12.0	Ingersoll	N5C	149.0
	N6K	28.4			

*Public Safety and Emergency Preparedness Canada.

Table 2. Information on data layers used in GIS.

Indicator	Source*	Layer	Type
Wetlands	OFD	wetlands_unit	Polygon
Roads	OFD	road	Line
Railways	OFD	transport_line	Line
Unpaved roads	OFD	road	Line
Intersections	RL	ren	Point
Critical facilities	OFD	building_symbol	Point
Bridges	RL	rll	Line
Land use	RL	ONland_use	Polygon
Soil permeability	SGSO	sgu_poly	Polygon
100-year flood line	UTRCA	--	Line
250-year flood line	UTRCA	--	Line
Grid	BM	obmindex	Polygon
FSAs	RL	zip	Polygon

*OFD—Ontario Fundamental Dataset (OGDE-Alymer_Guelph.mdb) from the Serge A. Sauer Map Library at UWO; RL—CanMaps Route Logistics, 2006 dataset from the UWO IDLS; SGSO—Surficial Geology of Southern Ontario, 2003 cd-rom dataset from Serge A. Sauer Map Library aUWO; UTRCA—the layers obtained from the Upper Thames River Conservation Authority in 2007; BM—Ontario Base Map Sheet, 2001 from the UWO IDLS.

The probability or likelihood of occurrence of flooding is described as the chance that a location will be flooded in any one year. For example, 1.3% chance of flooding each year implies 1 in 75 chance of flooding at that location in any year. Exceedance probability (p_e) of a flood is represented as [31,57]:

$$P[X \geq x] = p_e = 1 - F(x) \tag{2}$$

where F(x) denotes the value of Cumulative Distribution Function (CDF) of river flow x. The return period (T_x) of flood flow x is the reciprocal of exceedance probability, which is mathematically represented as [57]:

$$T_x = 1 / P[X \geq x] = 1/[1 - F(x)] \tag{3}$$

A flood line of a particular return period is the extreme boundary of the region exposed to a flood of the same return period. It represents the spatial extent of threat from the flood of a particular return period. The flood lines for a particular return period are evaluated by using physical, hydraulic and hydrologic characteristics of a particular location in the watershed. The present study utilizes 250- year flood line data for all FSAs being considered and 100-year flood line data for FSAs within the City of London, as per the availability. A flood hazard map with 100 and 250-years flood lines is used as one of risk components depicting spatial extent of flooding with exceedance probability of 0.01 and 0.004, respectively. The following procedural steps are followed in GIS for incorporating the information on probability of occurrence in this study: 1) the FSA, 100 and 250-years flood line shapefile layers are imported into ArcMap [58]; 2) the FSA layer and 100-yr flood line layers are turned "on" so that they are displayed in the Data Viewing window; 3) twenty five FSAs of interest in this study are highlighted using the selection tool and then converted into their own layer (FSA layer of interest); 4) these map features are then observed in the "layout" view where it is possible to insert map elements such as north arrows, legends and scale bars using the Insert Map Elements feature; 5) the map is then exported to ".jpeg" format. The same procedure is repeated for the 250-year flood line layer.

VULNERABILITY TO FLOOD

Vulnerability to flood is defined as measure of a region's susceptibility to flood damage [51-53]. It also includes population susceptibility to physical, mental or emotional damage due to flooding. Vulnerability could be influenced by individual emotions, seriousness of the current situation, and previous experiences with natural disasters. Traditionally, vulnerability has considered only biophysical factors. More recently, social factors have also been incorporated into assessment of vulnerability to disasters [31].

In this study, vulnerability to flood has been defined as a combination of four distinctive types of vulnerabilities: physical, economic, infrastructure and social [15]. The physical vulnerability generally incorporates only those indicators susceptible to biological sensitivity. Wetlands are for example, considered regions of physical vulnerability in this study. Wetlands are among the most productive ecosystems on earth. The richness of these transitional ecosystems relates mostly to the diversity of ecological niches created by the variability of seasonal and interannual cycles. Modifications in the hydrologic regime that disturb these cycles have been found to be the main stress factor threatening shoreline wetlands in all the world's major rivers [59]. The regulation of water levels has also caused the shrinkage of wetlands, and an incidental reduction in the diversity of plant communities and the number of plant species [60]. These regions have high biodiversity and sensitive life, which would experience higher damages, longer, slower recovery times due to flooding. Economic vulnerability includes flood damage indicators which can be expressed in monetary terms. Infrastructure vulnerability includes civil structure such as road networks, railways, and road bridges. Infrastructure components are important to movement of population, communications, and safety. Their inundation impedes traffic and hinders communications, increasing stress in the exposed population. Inundation may also block important emergency routes and cause physical damage to roads. Social vulnerability focuses on the reaction, response, and resistance of a population to a disastrous event. Vulnerable population may require special attention in an evacuation situation for example. The indicators are chosen based on a review of existing literature assessing vulnerability to current hazards [25-27, 31,32,51].

The vulnerability index (V_{Ii}) corresponding to each indicator for ith FSA is calculated using the following equation, which standardizes [61] each vulnerability index value ranging from 0.0 to 1.0 as done by Wu et al. [51]:

$$VI_i = \frac{V_i - V^{min}}{V^{max} - V^{min}} \quad (4)$$

where V^{min} and V^{max} are the minimum and maximum values of the indicator for all FSAs, respectively, and Vi is the actual value of the indicator for i^{th} FSA. All physical, economic, infrastructure (including the numbers of critical facilities and road bridges) and social vulnerability indices are directly calculated using (4). For example, vulnerability index for ith FSA of the social vulnerability indicator "Population under 20yrs of age" is calculated from the data set of 25 values (for 25 FSAs) on population under 20yrs of age using (4). The GIS is not used for determination of economic and social vulnerability indices, as they are directly calculated in the spreadsheet using Statistics Canada Census data. The economic and social vulnerability maps are created in ArcMap on the basis of calculated values of vulnerability indices. The following procedure is followed for calculating the sum of the length of roadways/railways: 1) roads/railways layer is imported into GIS; 2) the length of each road/railway is contained in the attributes table; 3) those road/railway segments from the attribute table are selected which fall within and intersect a particular FSA; 4) the "statistics" option from the "length" field in the attribute table options is selected; 5) the program calculates the sum of lengths and display them in that particular FSA; 6) the values are stored in a table and finally 4) is used for determining the vulnerability indices—"length of roads" and "length of railways". The similar procedure is followed for the vulnerability index—"unpaved roads", but the unpaved, dirt and gravel roads are selected for each FSA individually using the "Select by Location" or "Select by Attributes" tool.

This calculation of vulnerability index using 4) [61] offers an improvement over the traditional calculation [26,31,51] of vulnerability index, i.e., dividing all values by the maximum value, $VI_i = (V_i/V^{max})$, as it considers both the maximum and minimum values and ensures that the vulnerability indices are within [0, 1] interval and always non-negative. It is not necessary to use the scale [0, 1] for standardization. Montz and Evans [25], Grosshans et al. [54] and Odeh et al. [62] standardize the values of vulnerability and plotted the maps considering [0, 10], [0, 5] and [0, 100] scales, respectively. In the assessment of infrastructure vulnerability, the present study considers: a) the impact of flooding of critical facilities (schools, hospitals, and fire stations) and b) the spatial impact of flooding of main communication routes and road bridges. The developed methodologies for considering these impacts are discussed in next two subsections. The main objectives of the analysis presented in these two subsections are: 1) to model the impact of inundation of critical facilities (e.g., schools, hospitals and fire stations) of an FSA on its total infrastructure vulnerability, which is achieved by considering the "number of critical facilities (i.e., number of schools or hospitals within an FSA)" as vulnerability indicators [determined by using 4), as done for physical, economic, social and other infrastructure vulnerability indicators]; and 2) to model the impact of inundation of an FSA (which may contain one or more than one critical facilities) on its surrounding FSAs, which is modeled using a grid system. There are numerous studies that have addressed the impact of inundation of critical facilities on infrastructure vulnerability based on the number of critical facilities within an FSA [27,51,62], but none that explore the impact on surrounding FSAs, because people dwelling outside the flooded FSA also may depend on the critical facilities situated within the flooded FSA. This impact on infrastructure vulnerability of surrounding FSAs is modeled using a grid system.

Infrastructure Vulnerability Due to Inundation of Critical Facilities

Vulnerability of critical facilities is an indicator of infrastructure vulnerability [27,51,62]. Emergency shelters, nursing homes, public buildings, schools, hospitals, fire and rescue stations, police stations, water treatment or sewage processing plants, utilities, railroad stations, airports and government facilities are identified as critical facilities by Odeh et al. [62] and Flax et al. [27]. Critical facilities are those that play an integral role in public safety, health, and provision of aid [27]. As per the availability of data, the critical facilities considered in this study include schools, fire stations and hospitals, and are given special attention in vulnerability analysis in order to provide a more accurate estimate of flood risk.

Schools can be used for both education and as a place of refuge and a center for aid distribution during a flood. Fire stations provide the source of response to an emergency in the area near the station, and aid in disaster relief. Flooding of a fire station causes the population in close proximity to be more vulnerable. Hospitals represent another type of critical facilities that require special attention during flooding. Hospitals may have patients that need special attention in the case of an emergency. Procedure for assessment of infrastructure vulnerability due to inundation of critical facilities includes the use of a GIS tool. As per the availability of data, the FSAs of London are considered for the demonstration of the methodology. The impact of inundation of critical facilities of an FSA on its total infrastructure vulnerability is determined by considering the "number of critical facilities (i.e., number of schools or hospitals within an FSA)" as vulnerability indicators [determined by using 4), as done for physical, economic, social and other infrastructure vulnerability indicators]. To model the impact of inundation of an FSA (which may contain one or more than one critical facilities) on its surrounding FSAs, a 6x6 grid layer is placed over the FSAs of London, which breaks the entire city into 36 cells, as illustrated in Figure 3(a). The size of each grid cell is 25 km^2 (5 km × 5 km). The cell area for each FSA is calculated using area calculation function provided by the GIS tool. The layer containing the information on critical facilities is placed onto the grid layer and FSA layer to determine areas more susceptible to damage. The process used in assigning infrastructure vulnerability due to the inundation of critical facilities is based on the assumption that the people closest to the facility are its primary users. Thus, the spatial shape, termed as "vulnerability shape" in this study, is square as shown in Figure 3(b). There are four different color designations (red, orange, yellow and white) representing assigned Degree of Importance (DI).

The presence of just one of these is sufficient to classify the cell as important. All "important" cells are equally important. The DI values indicating vulnerability levels decrease equally in all directions with the distance from the inundated cell. Procedure implemented using GIS tool is as follows:

1) Divide the area under consideration into a grid—the grid should be regular in shape. In this analysis, a 6 ′ 6 square grid is used for demonstration purpose.

2) Use the DI to quantify the importance of a critical facility for each FSA. Red, orange, yellow and white color codes correspond to 1.0, 0.75, 0.2 and 0.0 DI values, respectively. The colors are reflecting the DI of each cell: red (high), orange (medium), yellow (low), white (no influence), which indicates the importance of the presence of critical facilities. The grid cells within an FSA that contain one or more critical facilities are identified. These grid cells are assigned red color, the highest DI of 1, assuming that the people closest to the facility are its primary users.

3) Assign a white color, indicating "zero" DI value to the remaining grid cells. The result is a square-shaped representation of vulnerability, which decreases with distance from the red (center) cell.

4) Following the previous three steps, assign DI values for all grid cells separately for each case of a grid cell with red color. For example, if 10 (ten) grid cells contain critical facilities, the grids cells would be assigned appropriate DI values 10 times. Finally, the Overall DI (ODI) for a grid cell is calculated by averaging these ten DI values. 5) The vulnerability for an FSA—area shown in bold solid line in Figure 3(c) —is calculated as:

$$Vul_{e_i} \text{ of i}^{th} \text{ FSA} = \sum_{j=1}^{k}(ODI_k \times A_k) \Big/ \sum_{j=1}^{k} A_k \quad (5)$$

where ODI_k is overall DI for kth grid cell, A_k is the area of ith FSA.

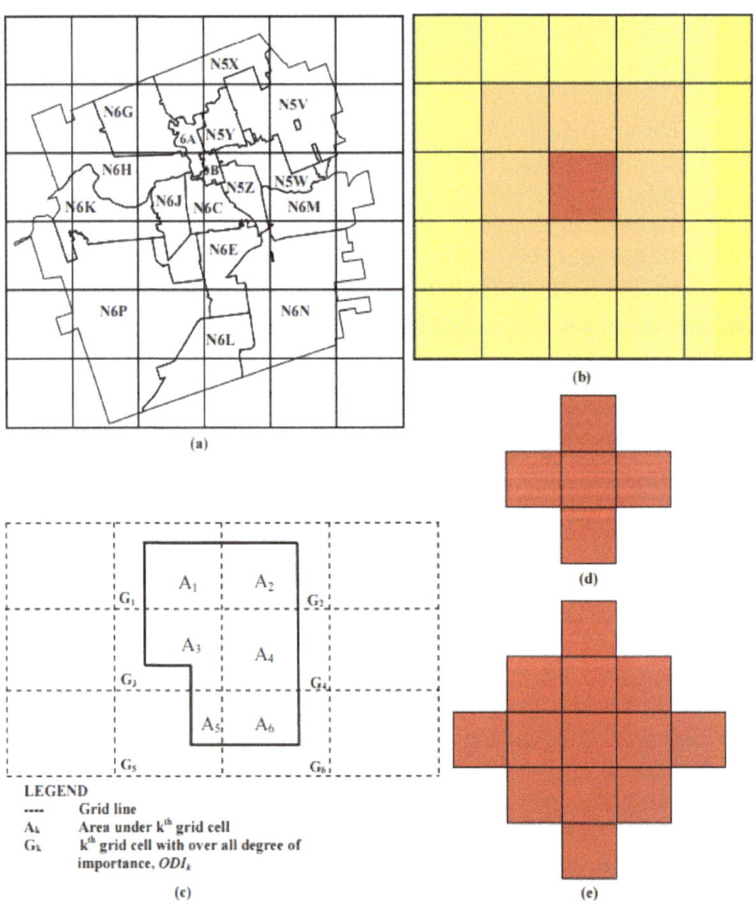

Figure 3. Determination of the vulnerability due to inundation of critical facilities and bridges. (a) 6 × 6-grid layered over the FSAs of London; (b) square vulnerability shape; (c) example FSA region divided in grid cells; (d) vulnerability shapes for cells with 1-5 bridges; (e) with 6-10 bridges (not to scale).

6) Determine the standardized vulnerability index value using:

$$Vul_{e_i}^{std} = \frac{Vul_{e_i} - Vul_e^{min}}{Vul_e^{max} - Vul_e^{min}} \quad (6)$$

where Vul_e^{max} and Vul_e^{min} are the maximum and minimum vulnerability values of critical facilities, i Vul_e is the value of vulnerability for critical facilities pertaining to the ith FSA. Thus the standardized infrastructure vulnerability values are obtained at FSA level.

Infrastructure Vulnerability Due to Inundation of Road Bridges

The infrastructure vulnerability is also affected by the inundation of road bridges. Vulnerability of an area due to the inundation of a bridge includes the interruption of traffic and formation of communication barriers between different locations in the affected region. Inundation of, or damage to a particular bridge affects not only the FSA in which it is located, but also all other FSAs that depend on the use of the bridge. In this study, only bridges over the water bodies are considered significant, because these bridges have limited alternate routes associated with them, and are necessary for safe crossing of the water body. They are frequently used as means for transporting commercial goods, a route to and from the workplace, and as emergency routes in case of a disaster.

The impact of inundation of road bridges of an FSA on its total infrastructure vulnerability is modeled by considering the "number of road bridges" as vulnerability indicator [determined by using 4), as done for physical, economic, social and other infrastructure vulnerability indicators]. To model the impact of inundation of an FSA (which may contain one or more than one road bridges) on its surrounding FSAs, the same procedure (steps 1 through 6) as described in previous subsection is followed, with the use of the new vulnerability shapes as shown in Figures 3(d) and (e). Again, 6 ′ 6 grid is used in the assessment of vulnerability as shown in Figure 3(a). However, the shape used in assessing vulnerability due to the inundation of road bridges is not a box, but rather cross-like. The shape varies with the number of bridges in any particular grid cell. Figures 3(d) and (e) illustrate the shapes of vulnerability for cells containing 1-5 and 6-10 road bridges, respectively. The number of bridges over water that is contained in each cell determines the shape that would be used in assessment of vulnerability. As the number of bridges increases, the more likely it is that inundation of that cell would affect more people. The vulnerability shape due to inundation of bridges is mainly based on a basic assumption: the need for crossing any given bridge decreases with distance from the bridge (i.e., the need for crossing the bridge is highest in areas that are closest to the bridge), because people further away from a particular bridge may have other alternatives for crossing the water body with equal convenience. The proposed method assumes that the whole cell being considered is flooded, and that bridges in that cell are unavailable for use. The cells are assigned a DI based on the vulnerability mapping in proximity to the inundated cell. The DI assignment is similar to the one used in assessing the infrastructure vulnerability due to inundation of critical facilities. However, the road bridges scenario designates a DI as either red/high (1.0) or yellow/low (0.2) for demonstration of the proposed methodology. In both analyses it was assumed that the whole grid cell is equally affected by the flooding, thus damage is assumed to be uniform across the cell area. The population density within a portion of the FSA covered by a grid cell is not known. Therefore a uniform distribution of population is assumed throughout each FSA. The following procedural steps are followed in GIS for incorporating the information on infrastructure vulnerability due to inundation of critical facilities: 1) the "buildings" layer is imported; 2) in the attribute table the type of building is specified. In the options for the attribute table "select by attribute" is used and then the category (e.g., school/hospital/fire station) is used to select only those buildings which are schools/hospitals/fire stations; 3) these buildings are saved as a separate layer for referencing in the critical facilities analysis of the study. The same procedure is followed for road bridges but the "bridge" layer is imported.

The calculation of the vulnerability indices following the procedures described in this section provides input for mapping each category of vulnerability in GIS. Table 3 shows the values of four components of vulnerability in the Upper Thames River basin. The seventh column of Table 3 indicates the standardized average vulnerability values of four vulnerability components (presented in columns 3-6) for all FSAs. In physical vulnerability, the FSA-N0K in Michell is identified as the most vulnerable due to large wetland areas in the region. The FSAs with "zero" values in the column of physical vulnerability indicate absence of wetlands. The FSA–N4S in Woodstock is the most vulnerable in economic sense due to the presence of large number of older houses in this region. The FSA–N0K in Michell is also identified as the most vulnerable in regards to infrastructure component due to its largest land area, which includes the longest road and railway networks. It is also identified that the FSA–N4Z in Stratford is the least vulnerable due to the absence of railway and minimum length of paved and unpaved roads. The column for social vulnerability shows high values for most of the FSAs within the city of London due to high population in these FSAs. The FSA–N5Y is the most vulnerable in social sense due to high values of indicators such as "differential access to resources", "social status" and "ethnicity". The present study incorporates a unique consideration of inundation of road bridges and critical facilities in assessing infrastructure vulnerability. Figure 4 displays the difference in infrastructure vulnerability due to inundation of critical facilities and road bridges. In most cases, the standardized values of infrastructure vulnerability of the FSA increase with the addition of impacts due to the inundation of road bridges and critical facilities. The GIS generated map may be produced for each component of vulnerability values as presented in Table 3 for identifying spatial variability of vulnerabilities to flood. More details of the processed numerical data and graphical results of the present vulnerability analysis to flood can be obtained from Peck et al. [63].

Calculation of the Vulnerability to Flood

Although maps of individual component of vulnerability can be useful, it is easiest to assess vulnerability throughout the watershed if the multidimensional components can be integrated into a single measure [32]. In the present study, the simplest way to combine the four components of vulnerability into a single measure would be to average the values of indices

[31,51] for each component [as given in the seventh column of Table 3]. Figure 5 shows the GIS generated map of vulnerability to flood obtained by averaging and standardizing the four components of vulnerability. The darker color indicates larger vulnerability. Map in Figure 5 provides for easy comparison of vulnerability between different FSA regions of six major damage centers, and insight into the spatial variability of vulnerability. It is identified from Table 3 and Figure 5 that FSA–N0K in Michell is the most vulnerable in the basin, as the area under this postal code is much larger than other FSAs and consequently it contains a number of wetlands and more infrastructure. It is found that the vulnerability values for FSAs in London vary between 0 and 0.562. The FSA–N5Z is identified as the most vulnerable, whereas N6M is identified as the least vulnerable within the city of London. The Table 3 and Figure 5 give a general description of region's vulnerability, and can be used for emergency flood management, disaster mitigation activities and planning future disaster protection infrastructure.

Table 3. Vulnerability analysis to flood of the Upper Thames River basin.

Damage Center	FSA	Components of Vulnerability to Flood				Overall vul*	Rank
		Phy.	*Eco.*	*Infras.*	*Soc.*		
London	N6A	0.000	0.264	0.432	0.383	0.296	15
	N6B	0.000	0.282	0.467	0.399	0.316	12
	N6C	0.014	0.613	0.477	0.794	0.498	6
	N6E	0.000	0.128	0.256	0.767	0.303	13
	N6G	0.000	0.495	0.556	0.703	0.471	9
	N6H	0.015	0.828	0.356	0.784	0.503	5
	N6J	0.000	0.876	0.447	0.732	0.296	14
	N6K	0.004	0.457	0.299	0.613	0.527	4
	N6L	0.000	0.000	0.022	0.000	0.353	11
	N6M	0.009	0.234	0.172	0.012	0.000	25
	N6N	0.044	0.004	0.055	0.020	0.110	19
	N6P	0.005	0.084	0.089	0.072	0.032	22
	N5V	0.002	0.815	0.305	0.813	0.061	20
	N5W	0.000	0.515	0.421	0.608	0.486	8
	N5X	0.034	0.288	0.293	0.390	0.405	10
	N5Y	0.001	0.707	0.461	1.000	0.268	16
	N5Z	0.005	0.539	0.502	0.736	0.562	3
Michell	N0K	1.000	0.749	1.000	0.632	1.000	1
Wood-stock	N4S	0.000	1.000	0.510	0.715	0.572	2
	N4T	0.000	0.110	0.011	0.082	0.041	21
	N4V	0.004	0.027	0.008	0.030	0.010	23
St. Marys	N4X	0.002	0.266	0.299	0.181	0.200	18
Stratford	N4Z	0.000	0.048	0.000	0.021	0.009	24
	N5A	0.134	0.673	0.384	0.690	0.494	7
Ingersoll	N5C	0.092	0.293	0.275	0.261	0.249	17

*Overall vulnerability: standardized average values

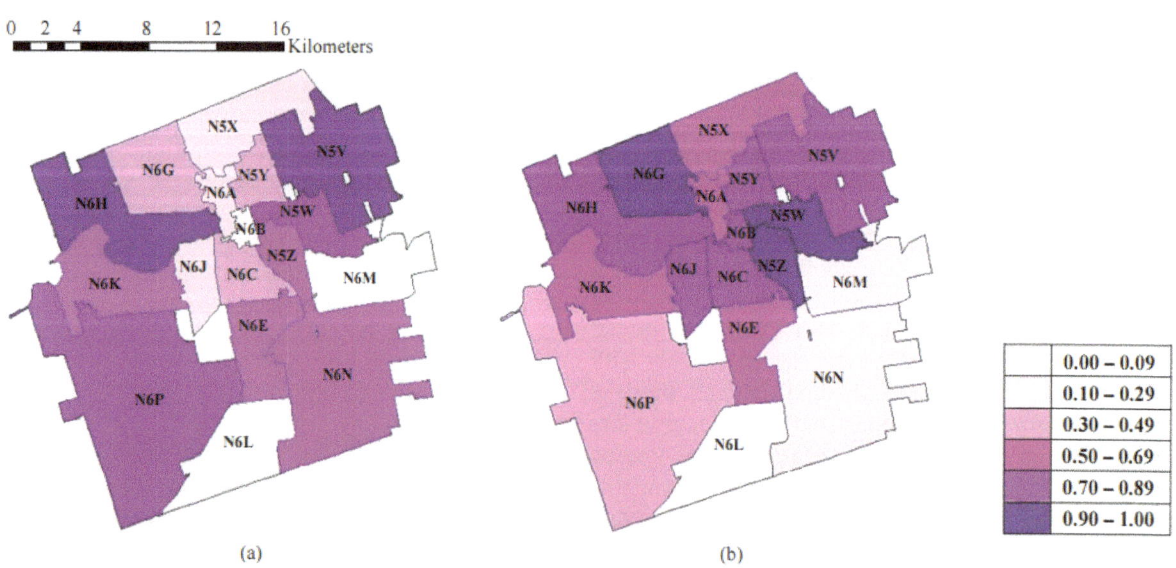

Figure 4. Infrastructure vulnerability for the FSAs of London. (a) infrastructure vulnerability not considering the impact of critical facilities and bridges; (b) infrastructure vulnerability including the critical facilities and bridges.

It should be noted for clarification that, there may be some correlation among vulnerability indicators under different components of vulnerability. In this study the chance of involving correlated indicators is very less, as the indicators are chosen based on a review of existing literature [25-27,31,32,51]. If the number of indicators is too high Principal Components Analysis (PCA) [32] can be applied to select the set of uncorrelated indicators. The basic aim of a PCA is to reduce a complex set of many correlated variables into a set of fewer, uncorrelated components.

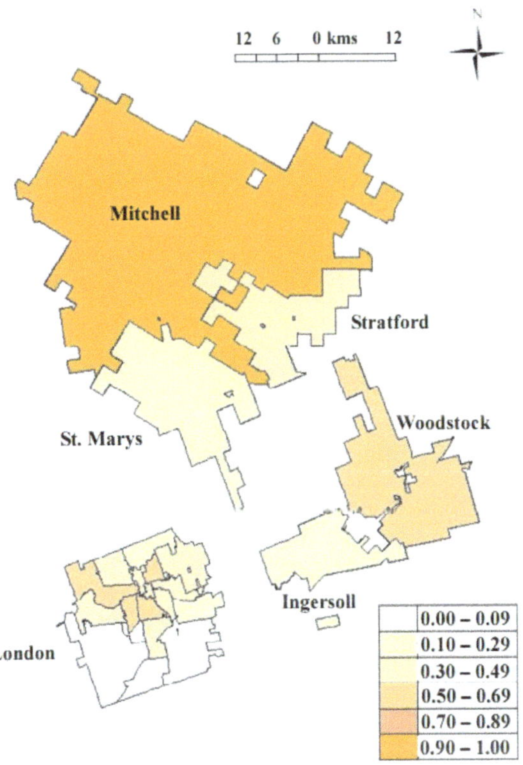

Figure 5. GIS generated map of standardized average vulnerability to flood.

EXPOSURES OF LAND USE AND SOIL PERMEABILITY TO FLOOD

The present study utilizes 250-year flood line data for all FSAs and 100-year flood line data for FSAs within the City of London, as obtained from UTRCA, Canada, for considering the values of probability of occurrence in the calculation of flood risk. The impacts of land use and soil type are not considered during generation of the flood lines and probability of occurrences. To incorporate the impact of exposures of land use and soil permeability into the analysis a separate component of the flood risk is considered as expressed in (1), following Kron [50] and Barredo et al. [13]. The indices of vulnerability to flood, as discussed in previous section, have no influence on flood flow and river channel characteristics. The exposures of land use and soil permeability are two physical watershed characteristics which affect the flood flow [64], and are considered as the important characteristics of flood risk in the Upper Thames River watershed. This study considers the impact of exposures of land use and soil permeability only for those FSAs within the municipality of London as per the availability of data. An exposure value of 1 is assigned to the regions outside of the City of London.

Land Use

The land use map used in the present study is obtained from the UWO Internet Data Library System (IDLS). Their CanMap Route Logistics 2006 dataset contains the Ontario land use GIS layer. This layer is designated as "ONland_use" and is of type "polygon" as mentioned in Table 2. The available land use data include seven different categories of use: open space, commercial, residential, parks and recreational, government and institutional, resource and industrial, and water body. Each of these land use categories has been assigned a DI value. These values, while estimated by the research team, can be changed by decision-makers with more extensive knowledge on how different land use influences flood runoff characteristics. Overdeveloped and highly commercialized areas include more pavement and impervious surfaces. They increase runoff quantity and shorten the time of concentration. On the other side, open land (including agricultural land) is exposed to direct infiltration of rainfall which decreases runoff quantity and lengthens the time of concentration. With this knowledge, the DI values are assigned to each category of land use, which are as follows: water body (0.1), parks & recreational (0.2), open area (0.3), Government and institutional (0.7), commercial (0.8), residential (0.8), resources & industrial (0.8).

Area under each land use type is expressed as a fraction of the FSAs total area. Summation of the fraction of each type multiplied by its DI provides an exposure value representative of the land use for an FSA. Therefore, mathematically the exposures of land use to flood for ith FSA is expressed as:

$$E_i^{Land} = \sum_{l=1}^{n} [DI_l \times (A_i^l / A_i)] \qquad (7)$$

where E_i^{Land} is the exposure of land use to flood, DI_l is the DI of land use type "l". "l" may be any of the land use types. Area under each land use type l is expressed as Al i for ith FSA. Total area of the ith FSA is denoted as A_i.

Soil Permeability

Soil permeability refers to the hydrological drainage characteristic of soil to allow water movement through its pores, which is inversely proportional to soil density. The more permeable the soil is, the more water can be transmitted through it. A soil with low permeability, such as clay, doesn't permit much water flow. This could cause "puddling" of water and thus higher accumulation of water on the soil surface. Regions which are composed primarily of these types of soils are prone to a higher flood risk because the water requires a longer time to drain or infiltrate into the ground [54]. Using a GIS dataset known as Surficial Geology of Southern Ontario, it was possible to spatially assess the soil permeability characteristics of the region. The data is available with different designations of permeability: low, medium-low, high or variable. A DI is assigned to each permeability category based on the ability of soil to infiltrate water, facilitate its transmission, and decrease flooding. DI values assigned to each category of soil permeability are as follows: low (0.8), low-medium (0.6), variable (0.5), high (0.3).

Area under each permeability category is expressed as a fraction of the FSAs total area. Summation of the fraction of each category multiplied by its DI provides an exposure value representative of soil permeability for an FSA. Therefore, mathematically the exposure of soil permeability to flood for i^{th} FSA is expressed as:

$$E_i^{Soil} = \sum_{p=1}^{m} [DI_p \times (A_i^P / A_i)] \quad (8)$$

where E_i^{Soil} is the exposure of soil permeability to flood, DI_p is the DI of permeability category "p". "p" may be low, medium-low, variable and high. Area under each permeability category (p) is expressed as A_i^P for ith FSA. Total area of the ith FSA is denoted as A_i. The standardization is being performed following the equation similar to (4) and (6), expressed as:

$$E_i^{std} = \frac{E_i - E^{\min}}{E^{\max} - E^{\min}} \quad (9)$$

where E^{max} and E^{min} are the maximum and minimum exposure values pertaining to land use/soil permeability, E_i is the value of exposure of land use/soil permeability pertaining to the ith FSA. The following procedural steps are followed in GIS for incorporating the information on exposures of soil permeability and land use to flood: 1) the Ontario Surficial Geology dataset for Zone 17 in Ontario is imported into ArcMap; 2) the layer features are symbolized by categorical attributes; 3) to symbolize each soil type with its own color, the Layer Properties dialogue box is opened, "Symbology" and "Categories" tabs are selected from the left menu, and "unique values" is highlighted; 4) in the Value Field, "SINGLE_PRI" is selected, which represents the single primary material of the soil composition. A color scheme is selected and then each soil is represented by its own color in the ArcMap data viewer; 5) the "Select by Location" tool of GIS is used to isolate those soils which fall within a particular FSA; 6) an area calculation is then performed on these "soils of interest" which provides the area of each soil type and the FSA boundaries that the soil area is within. An additional field is entered into the layers attribute table; the name of the field (column) and its type (floating point, integer etc.) are specified. All values are initialized to 0. The calculator option is then selected to compute the areas of the selected attributes. An advanced Visual Basics Application (VBA) area statement is used to calculate the required areas; 7) the type of each soil can be found in the layers attribute table along with the soils characteristic permeability which varied from "low" to "high"; 8) the results are summarized in a table. A similar procedure is followed for land use, but "category" is symbolized by: Commercial, Government & Institutional, Open area, Parks and Recreational, Residential, Resource and Industrial, and Water body. The impacts of exposures of land use and soil permeability to flood for FSAs of London are tabulated in Table 4. All the values are standardized with in [0, 1] to produce an indicator scores. Table 4 indicates that the exposure of land use for FSA-N6B, which is located in the central part of London, has maximum impact to flood due to presence of more commercial, residential and industrial areas; whereas the exposure of soil permeability has minimum impact to flood, as N6B has high permeable soil with high drainage capacity. The table also indicates that the exposure of land use for FSA-N6L, which is located in the southern part of London, has minimum impact to flood due to presence of more open and recreational areas; whereas the exposure of soil permeability has more impact to flood, as N6L has low permeable soil with less drainage capacity.

Table 4. Impact of exposures of land use and soil permeability to flood.

FSA	Impact of Exposures (standardized value)	
	Based on land use	*Based on soil permeability*
N6A	0.7362	0.2139
N6B	1.0000	0.0000
N6C	0.7562	0.9403
N6E	0.4311	0.9497
N6G	0.3954	0.5836
N6H	0.2481	0.5144
N6J	0.7067	0.8668
N6K	0.3339	0.4643
N6L	0.0000	1.0000
N6M	0.0386	0.7239
N6N	0.0125	0.9149
N6P	0.0122	0.7982
N5V	0.3049	0.3906
N5W	0.7427	0.1372
N5X	0.2357	0.3267
N5Y	0.8250	0.2760
N5Z	0.7938	0.4290

DEVELOPMENT OF THE INFORMATION SYSTEM

Providing a website for people to access flood risk information is an effective way of informing the public about the susceptibility to flooding that they may otherwise not be aware of. The study of Barredo et al. [13] is a contribution to the discussion about the need for communication tools between the natural hazard scientific com- munity and the political & decision making players in this field. The website can serve as an information center and may provide analysis tools for interactive processing of available flood information. It also provides the opportunity to tailor the presentation of the same information to different types of users according to their needs. According to the program evaluation glossary of USEPA [65], an information system is an organized collection, storage, and presentation system of data and other knowledge for decision making, progress reporting, and for planning and evaluation of programs. It can be either manual or computerized, or a combination of both. The information from the present risk-vulnerability analysis to flood is systematically kept in a computerized information system for more efficient use. The Adobe Dreamweaver Creative Sweet 3 software (http://www.adobe.com/ ap/products/dreamweaver) is used for creating the flood information system. The whole website is based off the Cascading Style Sheets (CSS) template provided in Adobe CS3. The developed information system is easy to navigate. The process starts by providing access to different FSAs of 6 damage centers in the Upper Thames watershed. After selecting the damage center, typing in the first three digits of an FSA will direct the user to information about that FSA. Selected three digits of the FSA activate the search engine that is created using a search engine composer. Information page is available for every FSA region. The information that is displayed for all the users; includes maps, numerical data, and an analysis tool in Microsoft excel spreadsheet format for calculation of flood risk as a function of change in land use. After typing the first three digits of an FSA in an identified cell of the spreadsheet, the user will be directed to the information on flood risk of the FSA, considering the present land use pattern and area under different categories of land use. Area under each land use type can be changed by the user to find out the flood risk under future scenario of land use pattern. It will calculate risk by using (1). The prototype information system created for this riskvulnerability analysis to flood targets 3 different user categories: 1) general public, 2) decision-makers and 3) water management professionals. The general public has access to a simple explanation of flood risk terminology, tables providing values of vulnerability to flood and a description of what they mean, 100-year and 250-year flood lines, as well as a simple analysis tool for flood risk calculation. Decision-makers are provided with a more detailed description of flood risk terminology and the implications of flooding. They have access to the same flood hazard maps as the general public. Decision-makers are provided with a more detailed and flexible analysis tool which allows the user to change the land use and compare the present level of flood risk with the one obtained under changed land use scenario. This may assist in the analyses of different land development initiatives and their consequences on flood risk. Water management professionals are presented with the most detailed descriptions and the most technical flood related information. They are provided a very detailed numerical breakdown of vulnerability and exposures of land use and soil permeability, including a list of all indicators used in the analyses. They also have access to the flood hazard maps similar to those provided to the general public and the decision-makers. The analysis tool available to professionals is the same as one provided to the decision-makers. The professionals are the only user with access to a "raw data" containing all of the numerical data used for the flood risk analyses. Screenshots of the opening page of prototype information system and analysis tool are shown in Figures 6(a) and (b). The information system is user-friendly and the details can be found in Black et al. [66].

Conclusions

The present study analyzes flood risk and vulnerability to flood in the Upper Thames River basin, Ontario, Canada. It deals with a large region as a case study with six major damage centers in the watershed for flood risk mapping considering probability of occurrence, four components of vulnerability and exposures of land use and soil permeability to flood. The impact of inundation of critical facilities and road bridges on infrastructure vulnerability is analyzed. New indices are introduced in the infrastructure vulnerability to flood, for example—length of railway, length of road, number of major intersections, number of critical facilities and road bridges. Typically, exposures of land use and soil permeability have been included as a component of risk. The minimum and maximum values of vulnerability are considered in the standardization process instead of using the conventional formula for standardization. A user-friendly information system is designed to systematically represent all flood information. The study provides an "analysis tool" for estimation of flood risk as a consequence of change in land use.

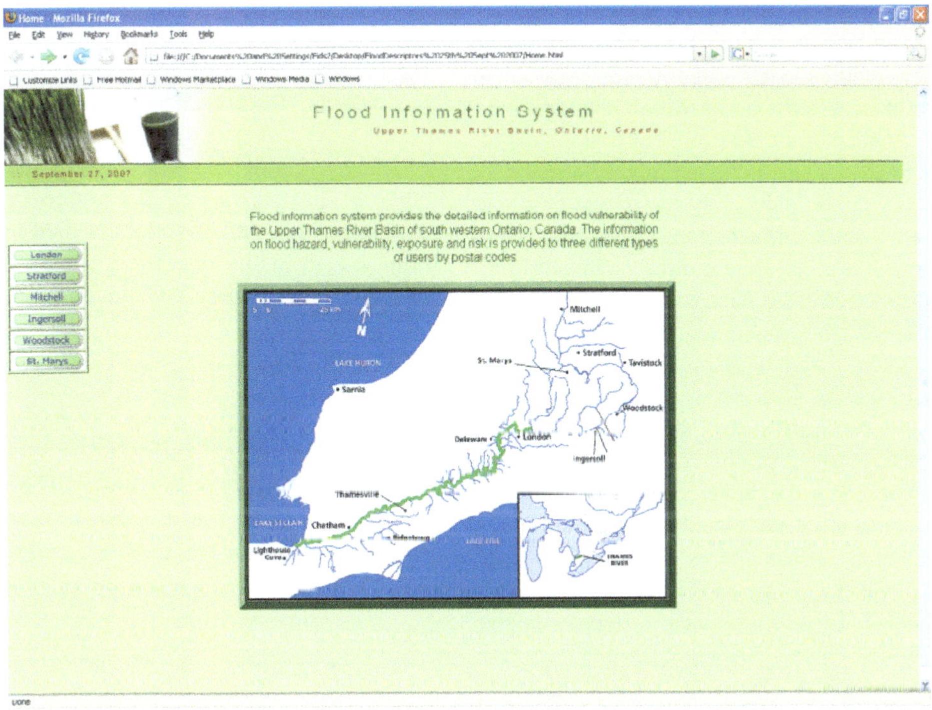

(a)

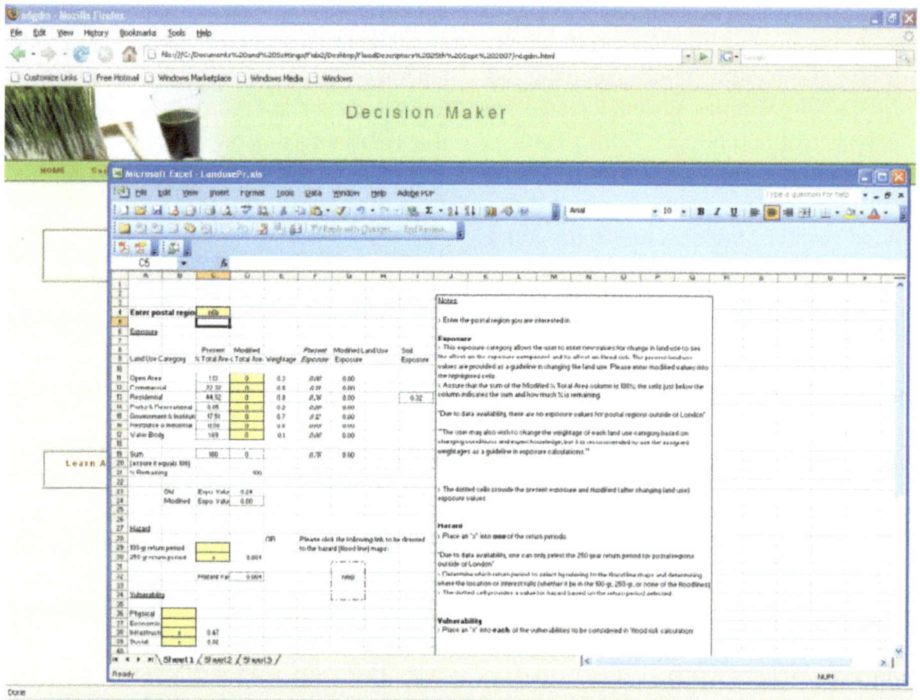

(b)

Figure 6. (a) Opening page of the information system; (b) analysis tool for decision makers.

The present study has some limitations and offers some benefits for future development. In the flood information system all the indices of infrastructure vulnerability for critical facilities are not considered due to unavailability of data. For example, emergency shelters, nursing homes, public buildings, police stations, water treatment or sewage processing plants, utilities, railroad stations, airports and government facilities; which are identified critical facilities [27,62]. The assignment of Degree of Importance (DI) for calculation of impact inundation of important service buildings, emergency service stations and road bridges across the river on infrastructure vulnerability, and in calculation for exposures of land use and soil permeability is dependent on the perspective of decision-makers or floodplain planners, which introduces some uncertainty due to vagueness or imprecision in the model. This uncertainty due to imprecision in the assignment of DI may be addressed in the flood risk calculation. In the present system only two flood lines are available, e.g., 100- and 250-years flood lines, which limit the calculation of flood risk. The impact of critical facilities and road bridges across the river on infrastructure vulnerability is calculated only for the City of London as per the availability of data. The same analysis may be performed for other damage centers in the watershed. In future studies, different shapes and sizes of "vulnerability shapes" with finer grid system and actual population density can be considered for determining the impact of inundation of critical facilities and road bridges. The impact of climate change is not considered in the current version of the system. The hazard maps or the position of flood lines will change if the climate change impacts are taken into consideration [67]. The values of flood risk for different postal codes may be easily updated to include the impact of climate change. No hydrologic calculation is performed in the present study to find out current position of flood lines. A sophisticated hydrologic modeling may be implemented for finding out the current positions of flood lines and result in more accurate calculation of flood risk. The proposed methodologies of flood risk mapping are not limited to the present case study and may be easily applied to other watersheds.

ACKNOWLEDGEMENTS

The authors sincerely thank the Upper Thames River Conservation Authority, Statistics Canada, Canadian Homebuyers Guide, Serge A. Sawyer map library & the IDLS library at The University of Western Ontario for providing data used in this study. Work presented in this paper has been conducted under the research grant by the Natural Sciences and Engineering Research Council of Canada.

REFERENCES

1. P. J. Floyd, "Reducing Flood Risks," Floods and Flood Management, 1992, pp. 419-435.
2. E. J. Plate, "Flood Risk and Flood Management," Journal of Hydrology, Vol. 267, No. 1-2, 2002, pp. 2-11.
3. A. Becker and U. Grünewald, "Disaster Management: Flood Risk in Central Europe," Science, Vol. 300, No. 5622, 2003, p. 1099.
4. D. P. Loucks, J. R. Stedinger, D. W. Davis and E. Z. Stakhiv, "Private and Public Responses to Flood Risks," International Journal of Water Resources Development, Vol. 24, No. 4, 2008, pp. 541-553.
5. M. J. Purvis, P. D. Bates and C. M. Hayes, "A Probabilistic Methodology to Estimate Future Coastal Flood Risk Due to Sea Level Rise," Coastal Engineering, Vol. 55, No. 12, 2008, pp. 1062-1073.
6. R. J. Dawson, L. Speight, J. W. Hall, S. Djordjevic, D. Savic and J. Leandro, "Attribution of Flood Risk in Urban Areas," Journal of Hydroinformatics, Vol. 10, No.

4, 2008, pp. 275-288.

7. H. Apel, G. T. Aronica, H. Kreibich and A. H. Thieken, "Flood Risk Analyses—How Detailed do We Need to be?" Natural Hazards, Vol. 49, No. 1, 2009, pp. 79-98.

8. P. Garrett, "Assessing Flood Risk," Water Bull, Vol. 355, 1989, p. 9.

9. B. Burrell and J. Keefe, "Flood Risk Mapping in New Brunswick: A Decade Review," Canadian Water Resources Journal, Vol. 14, No. 1, 1989, pp. 66-77.

10. D. G. Morris and R. W. Flavin, "Flood Risk Map for England and Wales," Report - UK Institute of Hydrology, 1996, p. 130.

11. D. D. Shrubsole, "Flood Management in Canada at the Crossroads," Global Environmental Change Part B: Environmental Hazards, Vol. 2, No. 2, 2000, pp. 63-75.

12. J. W. Hall, R. J. Dawson, P. B. Sayers, C. Rosu, J. B. Chatterton and R. Deakin, "A Methodology for Nationalscale Flood Risk Assessment," Proceedings of the Institution of Civil Engineers: Water and Maritime Engineering, Vol. 156, No. 3, 2003, pp. 235-247.

13. J. I. Barredo, A. de Roo and C. Lavalle, "Flood Risk Mapping at European Scale," Water Science and Technology, Vol. 56, No. 4, 2007, pp. 11-17.

14. S. L. Cutter, (Ed.) "American Hazardscapes: The Regionalization of Hazards and Disasters," Joseph Henry Press, Washington, D.C., 2001, p. 211.

15. S. L. Cutter, "Vulnerability to Environmental Hazards," Progress in Human Geography, Vol. 20, No. 4, 1996, pp. 529-539.

16. S. Bender, "Development and Use of Natural Hazard Vulnerability-Assessment Techniques in the Americas," Natural Hazards Review, American Society of Civil Engineers, 2002, pp. 136-138.

17. L. Roy, R. Leconte, F. Brissette and C. Marche, "The Impact of Climate Change on Seasonal Floods of a Southern Quebec River Basin," Hydrological Processes, Vol. 15, No. 3, 2001, pp. 3167-3179.

18. N. Nirupama and S. Simonovic, "Increase in Flood Risk Due to Urbanization: A Canadian Example," Natural Hazards, Vol. 40, No. 1, 2007, pp. 25-41.

19. M. Morris-Oswald and S. Simonovic, "Assessment of the Social Impacts of Flooding for Use in Flood Management in the Red River Basin," Report Submitted to the International Red River Basin Task Force, International Joint Commission, Winnipeg, Manitoba, 1997. http://www.ijc. org/php/publications/html/assess. html

20. E. Enarson, "Women, Work, and Family in the 1997 Red River Valley Flood: Ten Lessons Learned," Disaster Preparedness Resources Centre, University of British Columbia, British Columbia, Canada, 1999.

21. E. Enarson and J. Scanlon, "Gender Patterns in Flood Evacuation: A Case Study in Canada's Red River Valley," Applied Behavioural Science Review, Vol. 7, No. 2, 1999, pp. 103-124.

22. Natural Hazard Center, "Evaluation of a Literature Review of the Social Impacts of the 1997 Red River Flood," Report Submitted to the International Red River Basin Task Force, International Joint Commission. University of Colorado, Boulder, Colorado, USA, 1999, p. 11.

23. P. Blaikie, R. Cannon, I. Davis and B. Wisner, "At Risk: Natural Hazards, People's Vulnerability, and Disasters," Routledge, New York, 1994, p. 284.

24. P. M. Kelly and W. N. Adger, "Theory and Practice in Assessing Vulnerability to Climate Change and Facilitating Adaptation," Climatic Change, Vol. 47, No. 4, 2000, pp. 325-352.

25. B. Montz and T. Evans, "GIS and Social Vulnerability Analysis," In: E. Gruntfest and J. Handmer, Eds., Coping with Flash Floods, Kluwer Academic Publishers, Netherlands, 2001, pp. 37-48.

26. S. L. Cutter, J .T. Mitchell and M. S. Scott, "Revealing the Vulnerability of People and Places: A Case Study of Georgetown County, South Carolina," Annals of the Association of American Geographers, Vol. 90, No. 4, 2000, pp. 713-737.

27. L. K. Flax, R. W. Jackson and D. N. Stein, "Community Vulnerability Assessment Tool Methodology," Natural Hazards Review, American Society of Civil Engineers, Vol. 3, No. 4, 2002, pp. 163-176.

28. S. L. Cutter, B. J. Boruff and W. L. Shirley, "Social Vulnerability to Environmental Hazards," Social Science Quarterly, Vol. 84, No. 2, 2003, pp. 242-261.

29. R. Blong, "A New Damage Index," Natural Hazards, Vol. 30, No. 1, 2003, pp. 1-23.

30. N. T. Carter, "Flood Risk Management: Federal Role in Infrastructure," CRS Report for Congress, 2005, http:// fpc. state.gov/documents/organization/56095.pdf

31. J. Chakraborty, G. A. Tobin and B. E. Montz, "Population Evacuation: Assessing Spatial Variability in Geophysical Risk and Social Vulnerability to Natural Hazards," Natural Hazards Review, American Society of Civil Engineers, Vol. 6, No. 1, 2005, pp. 23-33.

32. L. Rygel, D. O'Sullivan and B. Yarnal, "A Method for Constructing a Social Vulnerability Index: An Application to Hurricane Storm Surges in a Developed Country," Mitigation and Adaptation Strategies for Global Change, Vol. 11, No. 3, 2006, pp. 741-764.

33. A. Werritty, D. Housto, T. Ball, A. Tavendale and A. Black, "Exploring the Social Impacts of Flood Risk and Flooding in Scotland," Scottish Executive Social Research, Edinburgh, 2007.

34. S. K. Sinnakaudan, A. Ab Ghani, M. S. S. Ahmad, and N. A. Zakaria, "Flood Risk Mapping For Pari River Incorporating Sediment Transport," Environmental Modelling and Software, Vol. 18, No. 2, 2003, pp. 119-130.

35. J. Bai, C. Wang, Z. Niu, S. Qi and G. Li, "Utilizing Remote Sensed TM Images and Meteorological Data to Plan Flood Risk Area," Proceedings of The International Society for Optical Engineering, Vol. 5232, 2006, pp. 370- 377.

36. F. Forte, L. Pennetta and R. O. Strobl, "Historic Records and GIS Applications for Flood Risk Analysis in the Salento Peninsula (Southern Italy)," Natural Hazards

and Earth System Science, Vol. 5, No. 6, 2005, pp. 833-844.

37. R. Abdalla, C. V. Tao, H. Wu and I. A. Maqsood, "GISSupported 3D Approach for Flood Risk Assessment of the Qu'Appelle River, Southern Saskatchewan," International Journal of Risk Assessment and Management, Vol. 6, No. 4-6, 2006, pp. 440-455.
38. D. G. Hadjimitsis, "The Use of Satellite Remote Sensing and GIS for Assisting Flood Risk Assessment—A case study of Agriokalamin Catchment Area in Paphos-Cyprus," Proceedings of The International Society for Optical Engineering, Vol. 6742, No. 67420Z, 2007.
39. K. Smith, "Environmental Hazards: Assessing Risk and Reducing Hazards," 3rd Edition, Routledge (Taylor & Francis Group), New York, 2001, p. 392.
40. J. Lowry, H. Miller and G. Hepner, "A GIS-Based Sensitivity Analysis of Community Vulnerability to Hazardous Contaminants on the Mexico/U.S. Border," Photogrammetric Engineering & Remote Sensing, Vol. 61, No. 11, 2005, pp. 1347-1359.
41. L. A. Varga, D. Radulescu and R. Drobot, "Romanian National Strategy for Flood Risk Management," IAHSAISH Publication, Vol. 323, 2008, pp. 75-86.
42. H. Chang, J. Franczyk and C. Kim, "What is Responsible for Increasing Flood Risks? The Case of Gangwon Province, Korea," Natural Hazards, Vol. 48, No. 3, 2009, pp. 339-354.
43. P. Tran, F. Marincioni, R. Shaw, M. Sarti and A. L. Van, "A Flood Risk Management in Central Viet Nam: Challenges and Potentials," Natural Hazards, Vol. 46, No. 1, 2008, pp. 119-138.
44. S. Forster, B. Kuhlmann, K. E. Lindenschmidt and A. Bronstert, "Assessing Flood Risk for a Rural Detention Area," Natural Hazards and Earth System Science, Vol. 8, No. 2, 2008, pp. 311-322.
45. M. Dilley, R. S. Chen, U. Deichmann, A. L. Lerner-Lam, M. Arnold, J. Agwe, P. Buys, O. Kjekstad, B. Lyon and G. Yetman, "Natural Disasters Hotspots: A Global Risk Analysis," Synthesis Report, The World Bank, Washington, D.C., 2005.
46. M. Arnold, R. Chen, U. Deichmann, M. Dilley, A. Lerner-Lam, R. Pullen and Z. Trohanis, "Natural Disaster Hotspots: Case Studies," Disaster Risk Management Series, No. 6, The World Bank, Washington, D.C., 2006.
47. UN, "Internationally Agreed Glossary of Basic Terms Related to Disaster Management," United Nations Department of Humanitarian Affairs, Geneva, 1992.
48. B. Merz and A. H. Thieken, "Flood Risk Analysis: Concepts and Challenges," O¨sterreichische Wasser und Abfallwirtschaft, Vol. 56, No. 3-4, 2004, pp. 27-34.
49. EU, "European Flood Directive: Richtlinie 2007/60/EG des Europäischen Parlaments und des Rates vom 23. Oktober 2007 über die Bewertung und das Management von Hochwasserrisiken," Amtsblatt der Europäischen Union, L288, pp. 27-34.
50. W. Kron, "Flood Risk = Hazard•Values•Vulnerability," Water International, Vol. 30, No. 1, 2005, pp. 58-68.
51. S. Y. Wu, B. Yarnal and A. Fisher, "Vulnerability of Coastal Communities to Sea-Level Rise: A Case Study of Cape May County, New Jersey, USA," Climate Research, Vol. 22, No. 4, 2002, pp. 255-270.
52. Y. Huang, Y. Zou, G. Huang, I. Maqsood and A. Chakma, "Flood Vulnerability to Climate Change through Hydrological Modeling: A Case Study of the Swift Current Creek Watershed in Western Canada," Water International, Vol. 30, No. 1, 2005, pp. 31-39.
53. A. Hebb and L. Mortsch, "Floods: Mapping Vulnerability in the Upper Thames Watershed under a Changing Climate," Project Report XI, University of Waterloo, 2007, pp. 1-53.
54. R. Grosshans, H. Venema and S. Barg, "Geographical Analysis of Cumulative Threats to Prairie Water Resources: Mapping Water Availability, Water Quality, and Water Use Stresses," International Institute for Sustainable Development, Winnipeg, Manitoba, 2005, p. 40.
55. M. Helsten and D. Davidge, "Flood Damage Estimation in the Upper Thames River Watershed CFCAS Project: Assessment of Water Resources Risk and Vulnerability to Changing Climatic Conditions," Project Report VII, Upper Thames River Conservation Authority, 2005, pp. 1-46.
56. H. Apel, A. H. Thieken, B. Merz and G. A. Blöschl, "Probabilistic Modelling System for Assessing Flood Risks," Natural Hazards, Vol. 38, No. 1-2, 2006, pp. 79-100.
57. W. Feller, "An Introduction to Probability Theory and its Applications," 3rd Edition, John Wiley and Sons, New York, Vol. I, 1968.
58. Environmental Systems Research Institute (ESRI), "Arc 9.2. ESRI," Redlands, California, 2006.
59. S. E. Mustow, R. F. Burgess and N. Walker, "Practical Methodology for Determining the Significance of Impacts on the Water Environment," Water and Environment Journal, Vol. 19, No. 2, 2005, pp. 100-108.
60. P. A. Keddy and A. A. Reznicek, "Great Lakes Vegetation Dynamics: The Role of Fluctuating Water Levels and Buried Seeds," Journal of Great Lakes Research, Vol. 12, No. 1, 1986, pp. 25-36.
61. S. Efromovich, "Nonparametric Curve Estimation: Methods, Theory and Applications," Springer-Verlag, New York, 1999.
62. D. Odeh, "Natural Hazards Vulnerability Assessment for Statewide Mitigation Planning in Rhode Island," Natural Hazards Review, ASCE, Vol. 3, No. 4, 2002, pp. 177-187.
63. A. Peck, S. Karmakar and S. P. Simonovic, "Physical, Economical, Infrastructural and Social Flood Risk—Vulnerability Analyses in GIS," Water Resources Research Report No. 057, Facility for Intelligent Decision Support, Department of Civil and Environmental Engineering, London, Ontario, Canada, 2007.
64. A. Sullivan, J. L. Ternan and A. G. Williams, "Land Use Change and Hydrological Response in the Camel Catchment, Cornwall," Applied Geography, Vol. 24, No. 2, 2004, pp. 119-137.

65. U. S. Environmental Protection Agency (USEPA), "Program Evaluation Glossary". http://www.epa.gov/evaluate/ glossary/i-esd.htm

66. J. Black, S. Karmakar and S. P. Simonovic, "A WebBased Flood Information System," Water Resources Research Report No. 056, Facility for Intelligent Decision Support, Department of Civil and Environmental Engineering, London, Ontario, Canada, 2007.

67. P. Prodanovic and S. P. Simonovic, "Inverse Flood Risk Modelling of the Upper Thames River Basin," Water Resources Research Report No. 052, University of Western Ontario, London, Ontario, Canada, 2006.

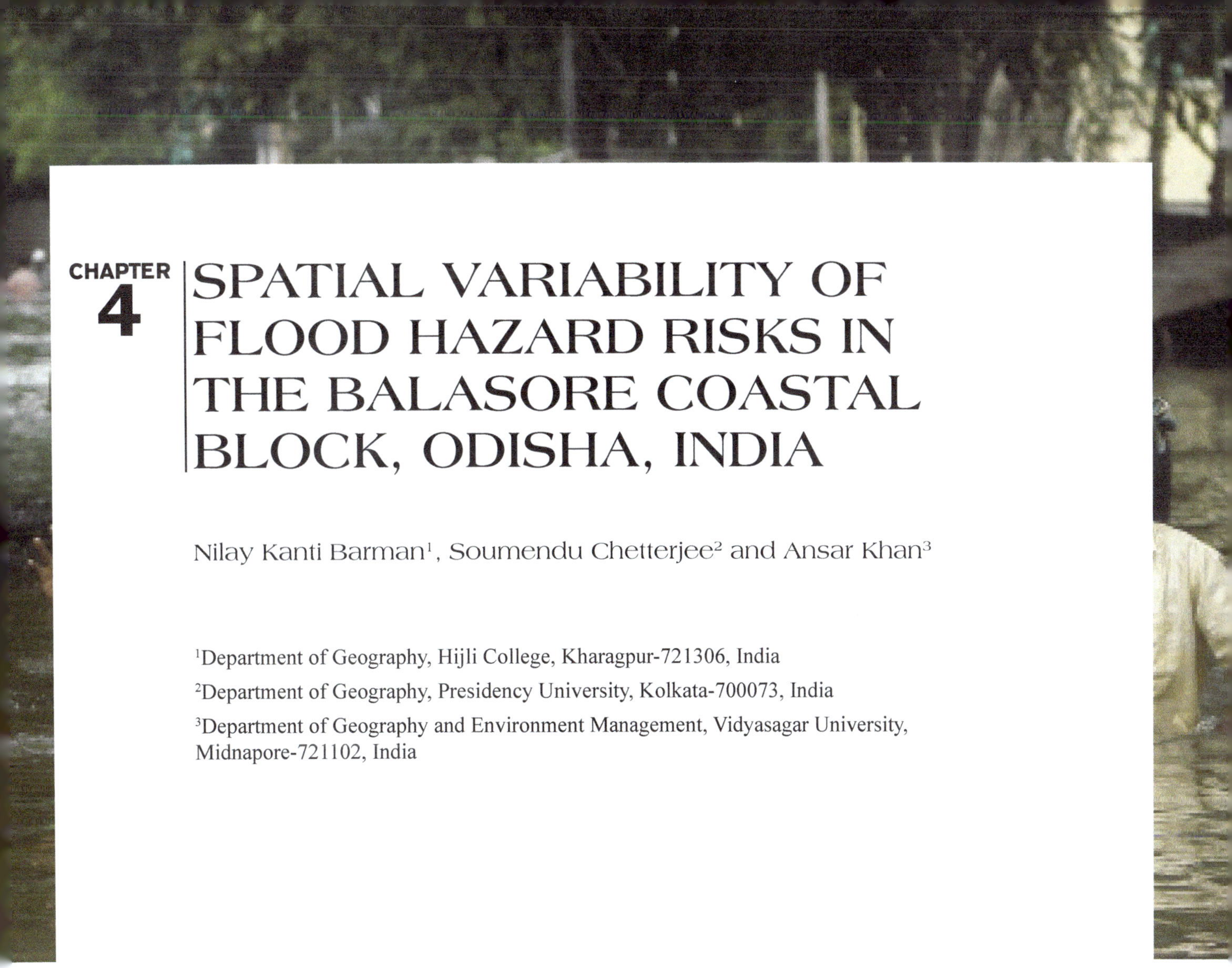

CHAPTER 4

SPATIAL VARIABILITY OF FLOOD HAZARD RISKS IN THE BALASORE COASTAL BLOCK, ODISHA, INDIA

Nilay Kanti Barman[1], Soumendu Chetterjee[2] and Ansar Khan[3]

[1]Department of Geography, Hijli College, Kharagpur-721306, India

[2]Department of Geography, Presidency University, Kolkata-700073, India

[3]Department of Geography and Environment Management, Vidyasagar University, Midnapore-721102, India

ABSTRACT

This paper pays attention on measuring and assessing the coastal flood hazard risk through quantification of flood intensity and impacts across the different Gram Panchayats (GPs) in Balasore block, Odisha, India. With respect to the June, 2008 flood event, enormity of flooding has been calculated for each GP using normalised values of measurable parameters relating flood characteristics. Thus, a Flood Magnitude Rank (FMR) has been assigned to each of the GPs according to degree of flooding severity. Similarly, Flood Impact Rank (FIR) for each GP has been derived from damage database. The product of FMR and FIR gives Flood Severity Score (FSS) of a GP which when multiplied by probability of flood event occurrence yields Flood Hazard Score (FHS) for the GP. The analysis helps dividing the study area into five flood risk zones viz. a) Very Low (FHS Below 2.0653); b) Low (FHS 2.0653 – 3.2755); c) Moderate (FHS 3.2755 – 4.4857); d) High (FHS 4.4857 – 5.6959) and e) Very high (FHS Above 5.6959), respectively. Joydebkasba, Parikhi, Bahabalpur, Sartha, Srikona falls in very high flood hazard risk class while, Patrapada, Sindhia, Rasalpur 2 tend to have very low risk from flood hazards. The rest 19 GPs along with the Balasore town of the area under study come under different risk classes in between the above two extreme classes according to their flood hazard scores.

INTRODUCTION

The coastal areas are exposed to variety of hazards as these are the zones of interaction between marine and terrestrial systems and hazardous processes that originate from both land and sea of coastal areas. Diversity makes them very sensitive to those processes and responses are highly complex also. In the context of growing importance of the coastal zones because of high productivity of the ecosystem, increasing concentration of population, industrial development, more intensive resource exploitation, expanding recreational activities etc. – concern about coastal hazards has increased and there is a need for effective coastal management to reduce these impacts of disaster events. The Swaminathan committee has recommended vulnerability as an important consideration in coastal zone management. Assessment of the physical sensitivity and exposure of coasts to hazards is an essential component for any comprehensive coastal vulnerability study. During the last few decades, a plethora of literatures on coastal

Citation: Barman NK, Chetterjee S, Khan A (2014) Spatial Variability of Flood Hazard Risks in the Balasore Coastal Block, Odisha, India. J Geogr Nat Disast 4:120. doi: 10.4172/2167-0587.1000120.

risk assessment methods have been produced consequent upon the recognition of global climate change and resultant sea level rise to put the coastal habitats and coastal communities into real threats [1,2].

IPCC's coastal zone management sub group (CZMS) developed a common methodology (CM) that involves seven steps which was employed to assess vulnerability of various coastal nations to predicted sea level rise (IPCC–CZMS, 1992). This method considered probable impacts of global sea level rise on population, economic sector, ethnicity, and social assets and on agricultural productions. But the data for one or many parameters considered in it were either inadequate or not easily available [3]. Kay and waterman [4] developed a four step methodology to overcome this limitation associated with CM. The four stages were physical and biological environment study of the area under consideration; vulnerable and cultural system; links between different parts of the area and finally, formulation of management strategy. This method was criticized by Harvey et al. [5], on the ground that physical, biological environment of the study area was poorly defined and man induced coastal hazards were not properly considered. Hervey and his colleagues developed an eight-step methodology in which the above discrepancies were removed. Most important contribution in this regard came from Gronitz et al. [6] that considered parameters like relief, rock type, landform, tectonics and shoreline shift for calculating Coastal Vulnerability Index (CVI). This method has been employed by United States Geological Survey (USGS) to map vulnerability of coastal stretches throughout USA. But no consideration of socioeconomic data has been logically criticized [7-9]. Any assessment of coastal vulnerability without reference to social aspect is not useful [3]. Accordingly, the social vulnerability index [10] and CVI were combined to develop Coastal Social Vulnerability Index (CSoVI) where poverty, population, development, ethnicity, age and urbanization were emphasized along with the physical parameters [11]. Furthermore, exposure of a place to physical hazards has been measured in terms of Place Vulnerability Index (PVI) [12].

In the coastal parts of the state of Odisha, India, flood hazards are generally caused by tropical cyclones and very rarely by tsunamis. The degree of flooding depends upon scale of the storm, height of storm surge and the tide level at the time of the event. Global sea level rise will be an increasingly important factor if predicted rise in sea level do occur. River estuaries may witness severe estuarine flooding with combined effects of a storm surge and river flood caused by rain storm inland. Coastal flooding is the most severe hazard in many coastal locations around the Bay of Bengal.

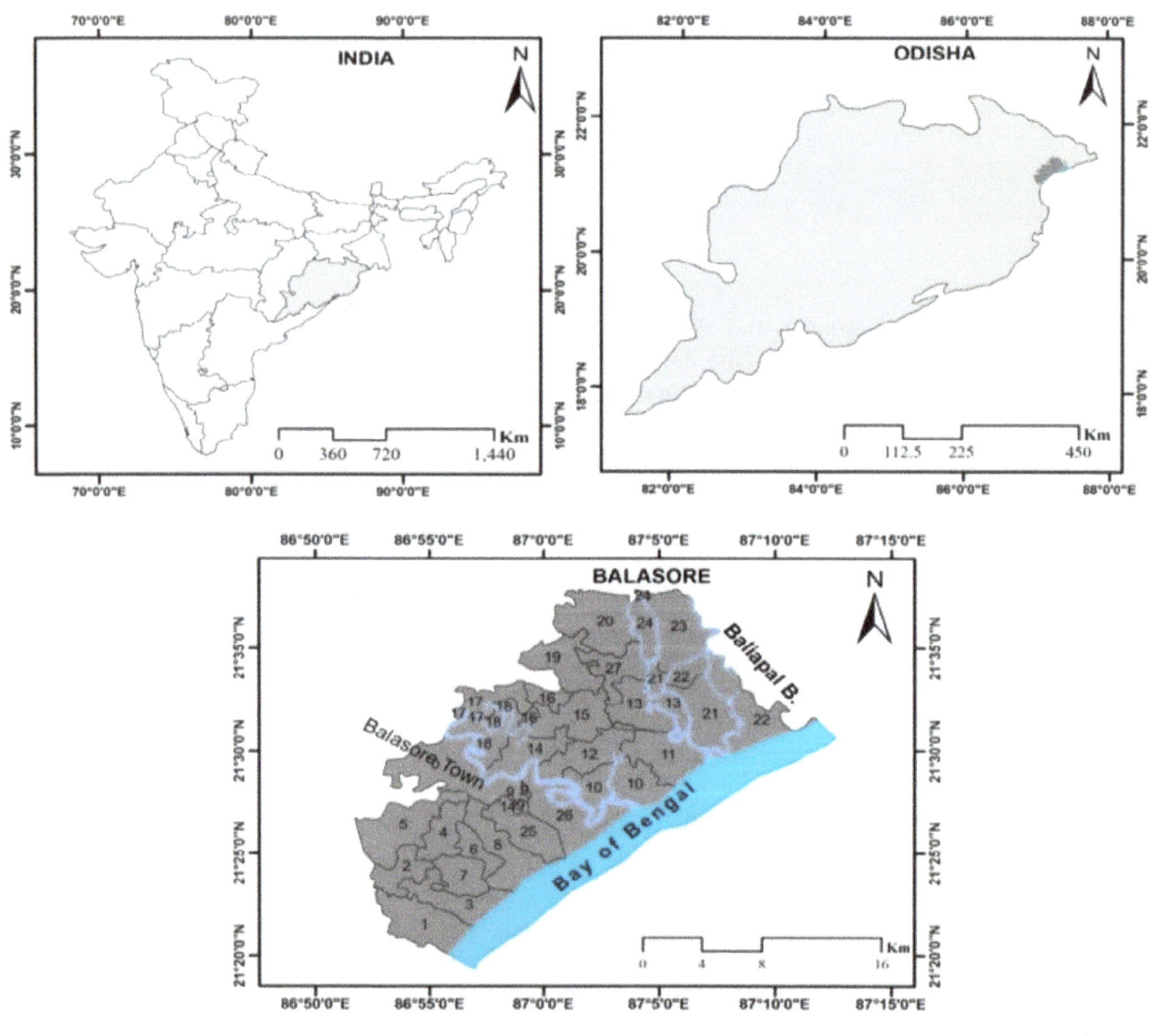

Figure 1: Location Map of the Study Area.

STUDY AREA

The present study area constitutes a part of the alluvium coast of Dugdugi and Burahbalam delta plain. It extends from Dugdugi river mouth to Rasalpur 1 along the Bay of Bengal coast of Odisha. The study area lies between 21°20'25"N to 21°39'55"N and 85°50'45"E to 87°12'37"E (Figure 1). The area is a coastal alluvial tract with an unconsolidated substrate, which is also geomorphologically dynamic, rich in habitat diversity and locationally prone to hazard like tropical cyclone induced tidal waves, storm surges and consequent coastal flooding. It is a monotonously flat alluvium surface lying between 2.5 m. to 3.5 m. above MSL. Geologically the area is characterized by ordinary alluvium deposits of the Holocene to recent origin brought down by the Dugdugi, Burahbalam and also Subarnarekha river. The area has a natural gradient to the east and south east direction which has been followed by the river Subarnarekha. The study area is mostly covered by the sandy clay and silty loam soils developed under brackish environment. The pH of the soil varies between 6.5 and 8.0 (pre monsoon) and 6.2 and 8.2 (post monsoon). This type of soil has led to a high water retaining capacity of the area. The climatic variations of the study area are more significant between monsoon and pre monsoon seasons. Temperature varies from a minimum of 9°C in winter to a maximum of 38°C in summer. Relative humidity ranges between 90% – 96% in most of the months. Low atmospheric pressure is recorded during summer and monsoon period. Wind dominantly blows from the offshore areas. There is no forest land in the study area. Some grasses e.g. Sesuvium Portolacrustum, Ipomia Bioloba and some herbs like Lantena camera, Akanthesia, Calatropis gigantia are found in the study area. Trees like Casuarina, Eucalyptus and Acacia auriculiformis are planted in this area while Coconut, Banana, Bamboo and Mango are indigenous floral species.

DATA BASE AND METHODS

Extensively low gradient coastal plain topography (1:18000) and relatively lowlying flat surface (1.7m to 2.5 m. surface elevation) of Balasore coastal area are highly susceptible to flooding activities. The range of flooding are also in the deltaic lowlands, as the high and saline water table and poor drainage circumstances of the region give buoyancy the flood water to spread over large areas. The duration of flooding is related to the spread of drainage over the tidal marshes or swamps and the duration of the flood itself. The micro tidal condition as well as the Balasore tidal swamps retards the speed of drainage during floods. Seasonality is often a vital factor in flood risk. In Odisha coastal plain 48 percent of cyclones that cause flooding on the coastal lowlands occur between October and November. However in the month of May few cyclones (7%) that cause coastal flooding occur due to the coincidence of onshore strong winds with the spring high tides. Local high seas, high rate of downstream discharge in the coastal plain, and the higher annual tidal level amplify the flooding activities in the cyclones that occur particularly in the season of retreating monsoon (October- November) in this region.

Table 1: Ranking of Flood Magnitude & Flood Impact from Standard Scores.

Standardised Value	< -3	-3 to -2	-2 to -1	-1 to 0	0 to 1	1 to 2	2 to 3	>3
Flood Magnitude Rank (FMR) Flood Impact Rank (FIR)	1	2	3	4	5	6	8	10

Assessment of the flood hazard includes characterization of the flooding in the terms of depth, duration of inundated condition, extent and velocity etc. Furthermore, the height of storm surge in low lying coastal areas is another important criterion for evaluating flood hazards. Damage to human life, properties and infrastructure caused by flood hazard is another easily measurable component of the flood hazard intensity. Field data sheet has been prepared in the proper format to generate database for flood intensity assessment in accordance to random sampling survey. Substantial number of samples (28×5) within the study area has been considered from each Gram Panchayat (GP) as the sample unit, depending on the scale of study.

The present study is based on Gram Panchayat (GP) wise secondary data collected from Gram Panchayat office, Block Development Office, District Natural Hazard Management Office of Balasore district, Odisha and primary data collected aftermath of June, 2008 flood event through intensive field survey using predesigned questionnaire at randomly sampled households to represent the concerned GP. The study has been conducted considering Gram Panchayat (GP) as the spatial unit. A Gram Panchayat (GP) is the smallest administrative unit for which disaster damage database is maintained and variations in flooding characteristics is not expected below this scale of geography when the natural setting of the terrain under study is considered. Moreover, the socio-economic and demographic features of the population exhibit an acceptable degree of homogeneity at the Gram Panchayat level.

In estimating the Flood Magnitude Rank (FMR) of a GP four parameters have been considered, namely depth of flood; flood velocity; percentage of area under inundation and distance from sea shore. On the other hand, seven indicators have been selected to enumerate Flood Impact Rank (FIR) for a GP. They are percentage of fully damaged houses; percentage of partly damaged houses; number of people died per thousand; number of cattle died per thousand of cattle population; monetary equivalent of crop damage per hector of net cropped area; monetary equivalent of fishery damage as a percentage of total value of fish production; length of road damaged as a percentage of the total length of roads in the these GPs. By use of these data, Flood Hazard Score for each GP has been calculated which gives a quantitative measure of flood risk associated with a GP. Calculation of Flood Hazard Score (FHS) involves following steps

Step I: Computing flood magnitude and flood intensity score

Flood magnitude similar to that of 2008 flood event has been characterized quantitatively using the four parameters mentioned flood velocity, percentage of area under inundation and distance from sea shore have been obtained from secondary sources supplemented by field survey. Those data for all the 32 GPs have been standardized to make them dimension and scale free and also to make them comparable. Then those standardized scores for the aforesaid parameters have been averaged to get GP wise Flood Magnitude Score (FMS)

$$FM_{GP} = \frac{\sum_{j=1}^{k} z_j}{k} \quad (1)$$

Where, j, physical parameters of hazard; k, number of physical parameters of hazard to be taken; z, normalized physical parameter for the hazard considered.

GP wise damage data can similarly be combined into a Flood Intensity Score as follows

$$FI_{GP} = \frac{\sum_{i=1}^{n} z_i}{n} \quad (2)$$

Where, i, damaged parameters of hazard; n, number of damaged parameters of hazard to be taken.

Step II: Obtaining flood severity score

For the given flood episode, GPs wise standardised values of FMGP and FIGP can be ranked on a 10 point scale (Table 1) to get the Flood Magnitude Rank and also the Flood Impact Rank (FIR)

Flood Severity Score of each and every GP has been quantified by the product of FMR and FIR.

FSS = FMR× FIR (3)

Step III: Assessing Probability of flooding

Probability (p) of the occurrence of said flood of a given magnitude is computed from recurrence interval (also called the return period). Flood recurrence interval has been computed on the basis of last 40 years worth of data. The recurrence interval (T) is defined as the average number of years between two successive floods or similar severity. Recurrence Interval (T) is given by

$$T = \frac{n+1}{m} \quad (4)$$

Where, n, No. of years in record; m, No. of occurrences of flood of similar severity

The probability (p) of occurrence of said floods of a given severity is expresses by taking the inverse of recurrence interval (T)

$$P = \frac{1}{T} \quad (5)$$

Step IV: Calculating Flood Hazard Score

Finally, the Flood Hazard Score (FHS) is calculated by multiplying the Flood Severity Score (FSS) (Table 2) and the associated probability value

FHS = FSS × P (6)

RESULTS AND DISCUSSION

Geomorphologically, the present study area belongs to the Subarnarekha delta Chenier plain along the western bank of the Burahbalam River of Odisha are arrangements of complex, diverse and fragile ecosystems which are simply unique in nature. Coastal ecosystem, such as wetlands, beaches, lagoon, estuaries, sand dunes, and mangroves, besides performing crucial coastal protections, provide rich swamping and breeding grounds for marine creatures, feeding grounds for birds, recreational grounds for tourists and available resources for the local communities. These ecosystems, are threatened by development related activities along the Balasore coast. The area is represented by regressive younger beach ridges and mudflats and floodplains appearing as depressed zones, after altered into agricultural field. The southernmost sea front part of Balasore block is composed of beach barrier complex and wash over deposits. Generally speaking, the Balasore block is dominantly a part of Burahbalam flood plain formed due to westward avulsion of the river. Interactions among marine transgression processes, huge supply of sediments and predominant wave tide dynamics have been responsible for the development of this sandy flat area surrounded by the Bhadrak district in the west, young chenier complex to the east and north and beach barrier complex and wash over deposits to the south.

Table 2: Result of Average Standard Score, Flood Magnitude Rank (FMR), Flood Impact Rank (FIR), Flood severity Score (FSS), Probability value (p) and Flood Hazard Score (FHS).

GP Code	GP Name.	Average standard score of hazard magnitude	Flood Magnitude Rank (FMR)	Average standard score of hazard damage	Flood Impact Rank (FIR)	Flood Severity Score (FSS)	Probability (p)	Flood Hazard Score (FHS)
1	Rasalpur 1	0.70206644	5	-1.08738	3	15	0.07317	1.0975
2	Sashanga	0.06822948	5	-0.374754	4	20	0.09756	1.9512
3	Joydebkasba	0.69124481	5	1.0237712	6	30	0.14634	4.3902
4	Saragaon	-0.674384	4	-0.754703	4	16	0.09756	1.5609
5	Genguti	-0.5040695	4	-0.430088	4	16	0.09756	1.5609
6	Gudu	0.3194201	5	-0.017723	4	20	0.09756	1.9512
7	Padmapuri	0.08017966	5	-0.463611	4	20	0.09756	1.9512
8	Ranasahi	0.41308439	5	-0.000134	4	20	0.09756	1.9512
9	Patrapada	-1.0838472	3	-1.151759	3	9	0.07317	0.6585
10	Parikhi	0.93988099	5	1.5748355	6	30	0.14634	4.3902
11	Bahabalpur	0.79483691	5	1.2433755	6	30	0.14634	4.3902
12	Gopinathpur	-0.2723026	4	-0.105180	4	16	0.09756	1.5609
13	Chhanua	-0.3384952	4	-0.186472	4	16	0.09756	1.5609
14	Sindhia	-0.6615708	4	-0.231161	4	16	0.09756	1.5609
15	Olanda Saragaon	-0.5988353	4	-0.095099	4	16	0.09756	1.5609
16	Odangi	-0.7014093	4	-0.893605	4	16	0.09756	1.5609
17	Nagram	-0.7476162	4	-0.875540	4	16	0.09756	1.5609
18	Baunla	-0.7351087	4	-0.823503	4	16	0.09756	1.5609
19	Haldipada	-1.056316	4	-0.939552	4	16	0.09756	1.5609
20	Kasipada	-0.9827968	4	-0.848826	4	16	0.09756	1.5609
21	Sartha	1.2745164	3	2.1505299	8	24	0.19512	4.6829
22	Kashaphala	0.8717318	5	-1.186939	3	15	0.07317	1.0975
23	Srirampur	0.6848103	5	0.5890416	5	25	0.12195	3.0487
24	Rasalpur 2	-0.7077841	4	-1.148350	3	12	0.07317	0.8780
25	Hidigaon	0.6405789	5	-1.248117	3	15	0.07317	1.0975
26	Srikona	0.9681347	5	1.5495563	6	30	0.14634	4.3902
27	Kuradiha	-0.6420300	4	-0.725454	4	16	0.09756	1.5609
28	Balasore Town	-1.4024427	3	-1.080589	4	12	0.09756	1.1707

Table 3: Gram Panchayat Wise Distribution of Flood Hazard Score with Their Assigned.

Flood Hazard Score	Assigned Attribute	G.P Code No.	Identified G.P
< 0.8551 – 2.0653	Very low	9,14,24	Patrapada, Sindhia, Rasalpur-I
2.0653 – 3.2755	Low	1,22,25, Balasore Town	Rasalpur-II, Kashaphala, Hidigaon, Balasore Town
3.2755 – 4.4857	Moderate	2, 4, 5, 6, 7, 8, 12, 13, 14, 15, 16, 17, 18, 19, 20,	Sashanga, Saragaon, Genguti, Gudu, Padmapuri, Ranasahi, Gopinathpur, Chhanua, Sindhia, Olanda Saragaon, Odangi, Nagram, Baunla, Haldipada, Kasipada
4.4857 – 5.6959	High	23	Srirampur
> 5.6959	Very high	3, 10, 11, 21, 26	Joydebkasba, Parikhi, Bahabalpur, Sartha, Srikona

Geomorphological signatures suggest that this coastal area has probably started to witness a phase of marine transgression. Frequency and intensity of the cyclones have increased to a certain extent. Cyclone induced storm surges and torrential rain in upper catchment of Burahbalam and Dugdugi river have been found to be responsible for flooding of the study area intensity and severity of which are experienced to have increased, may be due to recent climate and environment changes.

In this study it has been tried to understand, the degree of flood hazard risk across the Gram Panchayats in Balasore block, Odisha. All the Gram Panchayats (27 in number along with Balasore municipal area) of the studied C.D. block have been classed under five categories of flood hazard risk - starting from Very Low through intermediate classes to Very High (Table 3) and accordingly a map (Figure 2) has been prepared on the basis of calculated flood hazard score of each of the GPs to visualised spatial variability of risk within the block. Result shows that Joydebkasba, Parikhi, Bahabalpur, Sartha, Srikona GPs belong to very high flood risk zone which may be attributed to their vulnerable geomorphic locations. Joydebkasba, Parikhi, Bahabalpur, Srikona is located at the sea front and hence, prone to coastal flooding. Moreover, river Burahbalam and Dugdugi carries large volume of discharge loaded with huge quantity of sediments. This flow of discharge receives resistance in its drainage from the strong southwesterly monsoon wind and resultant cross-shore current, waves and high magnitude tide inflow. This causes accumulation of huge water at and near Burahbalam and Dugdugi river mouth and flooding at Joydebkasba, Parikhi, Bahabalpur, Srikona GP. On the other hand this area is only 0.5m - 1m high above the sea level which makes the area more vulnerable to flooding. The landward margin of the block under study is characterised by intricate network of tidal inlets along which sea waters can enter into the said GP in question and cause flooding even in the event of a low storm surge or a wave of moderate magnitude. The above stated GP is barely exposed to sea without any sand dune which generally acts as a natural buffer against sea. Sparse mangrove patches, which could be found in this area even few years back, have disappeared due to change in sedimentological properties of the shore deposits that constitute substrate for mangrove swamps. Land use pattern of the area has undergone such changes that have increased probability of flooding. The area gets flooded in two different ways – firstly due to spilling of river Burahbalam and Dugdugi (Sweet water flood) and coastal flooding (Saline flood) due to high magnitude wave or storm surge. These GPs are thickly populated because people here have easy access to marine resources which these coastal dwellers utilize for earning livelihood. Rasalpur-1, Sashanga, Gudu, Padmapuri, Ranasahi, Kashaphala, Rasalpur-2 and

Hidigaon. Therefore, damage volume is very high even in the event of a moderate intensity flood. Sartha and Srirampur GP located along the eastern bank of river Dugdugi where the river follows a meandering course. Due to high degree of sinuosity in the estuarine section of Dugdugi, gradient flow of river discharge gets largely obstructed particularly during monsoon. Torrential rainfall in its gigantic catchment contributes a huge volume of discharge that fails to get drained seamlessly. As such, this river often spills in rainy season to cause reverine flood in those two GPs. Moreover, during high astronomical tide phase, considerable huge volume of ocean water ingresses through this funnel shaped estuary of Dugdugi river in the form of tidal bore which restricts river discharge to be drained into sea. As a consequence, water level in river valley becomes high enough to spill over its banks causing flood. Funnelling effect of tidal water as it enters into the estuary is a major cause of flooding in Sartha and Srirampur.

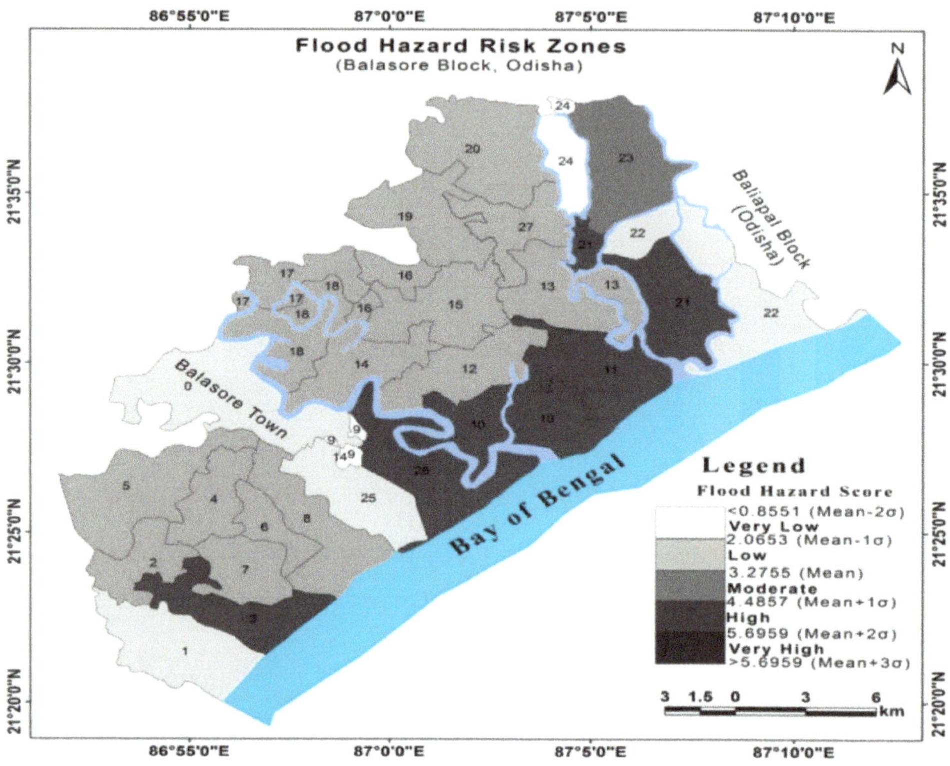

Figure 2: Flood Hazard Risks Zone.

Patrapada, Sindhia, Rasalpur 2 G.P lies in the very low flood hazard class. The 19 G.Ps along with the Balasore Township in balance belong to other intermediate flood hazard classes according to the flood hazard score, they have received. It is fascinating to note that, in spite of being located far away from the sea, Sartha and Srirampur G.Ps has very high to high risk of flood hazard. The GPs under low and very low flood hazard risk zone experience low intensity and low magnitude flood episodes spilling of river. Duration of flood water stay in those areas is also considerably short.

Land use alteration in the block has been responsible for frequent flooding in the area. Aquaculture has emerged as a profitable economic activity currently. Hence, vast stretches of laws have been given to fish farming ponds where high earthen embankments have been constructed around the ponds which restrict the flow water to spread over flood plain and causing the flood situation to become more severe. River engineering in the form of embankment construction along the both banks of river Burahbalam and Dugdugi have caused sedimentation to be restricted between the banks along leaving no scope for sediment distribution over the floodplain. This has reduced capacity of the river valley and increased a huge quantify of sediment to deposit near its mouth causing gradual narrowing of the channel thereby.

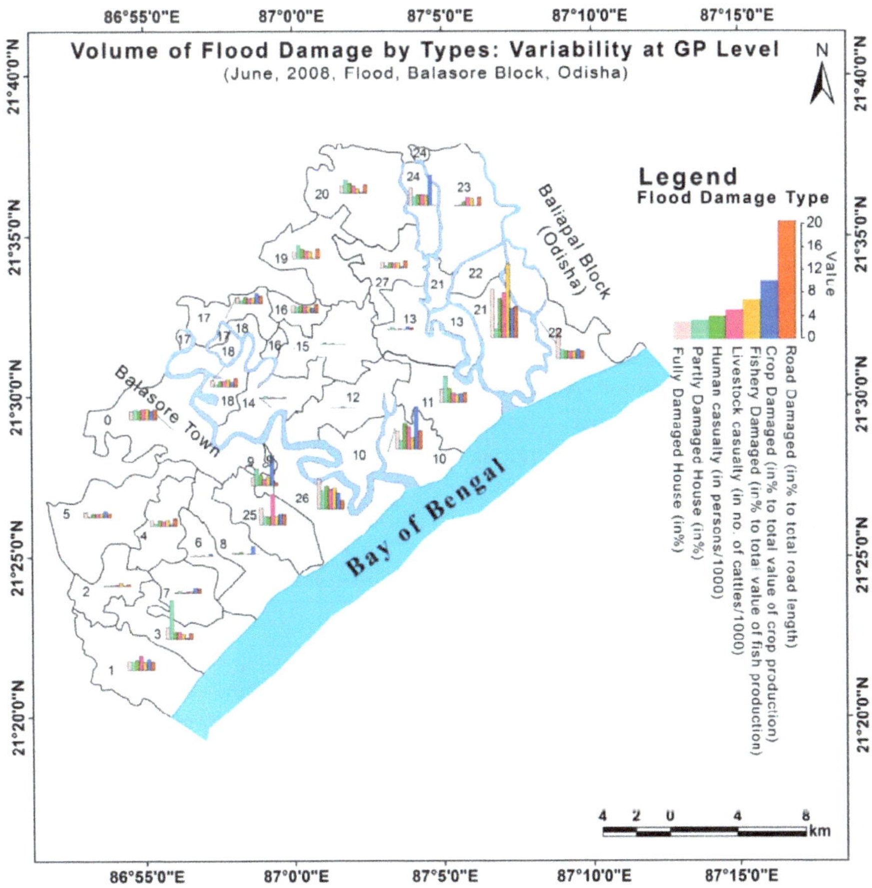

Figure 3: Flood Impact Variability across Gram Panchayats.

The flood impact assessment results have been summarized in a graph plot of averaged standard values of seven considered hazard impact parameters calculated for every GP (Figure 3). Out of the seven indices fisheries, crop and road damages intensity is very high at the coast facing GPs along with the GPs located along the river bank because there are no such settlement sites are present in the active river flood plain and also the foreshore region. These mentioned regions have mainly been used for primary activities like agriculture and fisheries. All the settlements have been developed on the top of the back barrier dune to defend the frequent natural disaster. On the other hand, the coastal GPs having interior locations are mostly suffers from crop and road damage and in terms of number of population affected because this area is densely populated.

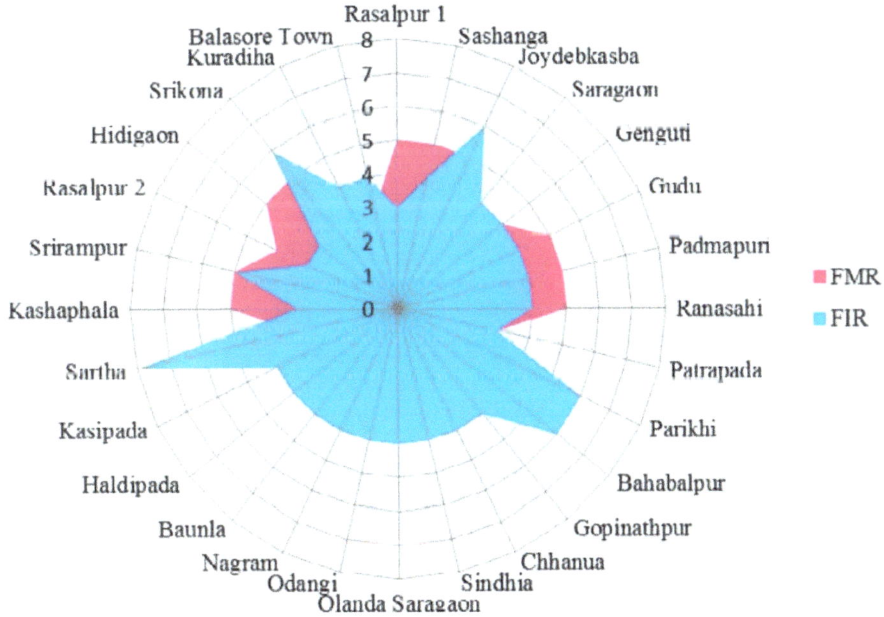

Figure 4: Radar Chart shows the spatial gradient between FMR and FIR.

Flood Magnitude Rank (FMR) and Flood Intensity Rank (FIR) received by each of the GPs have been represented by radar chart which compares between the two aspects of flood hazard for the GPs. It shows that in cases of GPs like Rasalpur-1, Sashanga, Gudu, Padmapuri, Ranasahi, Kashaphala, Rasalpur-2 and Hidigaon flood intensity or damage impacts of flooding is much less in spite of high magnitude of flood (Figure 4). It may be explained in terms of the protective measures undertaken for mitigation of flood impacts in those GPs. For example, in Hidigaon GP sea wall has been constructed for shore line protection. Besides, number of cyclone and flood shelters has been installed near Ranasahi and Kashaphala which immensely help in reduction of loss due to flood. Awareness programme and training camps are being organised to make local communities capable of combating the flood situation. In Kashaphala and Rasalpur-2 GPs reverine embankments are heightened under National Rural Employment Guarantee Act programme. Above all, the floods and cyclones are now being predicted precisely by use of modern technology and preparedness has also been strengthened than ever before. All these have contributed largely in reducing the impacts of high intensity flood in many of the GPs of Balasore block.

The Balasore block, Odisha, India consisting of 27 Gram Panchayats along with a municipal area is located at the sea front on the right bank of river Burahbalam. Due to its geomorphologically vulnerable situation, the area is prone to frequent flooding induced by hazardous processes that organised both from land and ocean. Spatial variability of flood hazard risk among the Gram Panchayats has been assessed quantitatively considering both magnitude of flood represented by it physical characteristics and intensity of flood measured from damage impacts. Combination of these two aspects of flood hazard along with its probability of occurrence yields a flood hazard score for each of the Gram Panchayats. The analysis clearly demonstrates that the GPs located along the river bank and those exposed to the sea, due to absence of any natural barrier, have high risk of flooding while GPs of interior location are in zone of lower risk. Progressively declining capacity of river Dugdugi and Burahbalam to hold huge volume of water received from high magnitude rain storm events has augmented this flooding situation. In flow of tidal water along the river channel at the time of monsoonal discharge makes this situation critical. Storm surge during cyclone episodes, high astronomical tide lead ocean water to enter along tidal inlets that floods many of the GPs under study. Moreover, embankments given around the fish ponds are also responsible for intensifying the flood situations. The study too shows that protective measures against floods have effectively reduced the damage impacts of even high magnitude floods. The GPs also vary with represent to types of damage caused by flood.

The essence of this work lies in the fact that it explores the causes, nature and types of flooding along with degree of flood hazard impacts at a reasonably lower scale of geography. The study helps to understand the varying nature of flood in different Gram Panchayats of the block. At the same time it reveals the volume of loss of lives and properties suffered in each Gram Panchayat which is not always linearly dependant on physical severity of the flood. As such the study is helpful for taking GP specific planning decisions of flood management. Moreover, this type of study can be carried out in any coastal block for coastal hazard mapping and also for assessing risk associated with other coastal hazards along with the flood.

REFERENCES

1. IPCC–CZMS (1992) A Common Methodology for Assessing Vulnerability to Sea Level Rise Second Revision. In: Global Climate Change and the Rising Challenge of the Sea. Report of the Coastal Zone Management Sub-Group, IPCC Response Strategies Working Group, Ministry of Transport, Public Works, and Water Management, The Hague.
2. The MS Swaminathan Committee Report on Coastal Management Burnt (2006) Integrated Coastal Zone Management Plan. Trivandrum, Kerala Independent Fish Workers Federation (KSMTF).
3. Klein RJT, Nicholls RJ (1999)Assessment of Coastal Vulnerability to Climate Change. Ambio 28:182–187.
4. Kay RC, Waterman P (1993) Review of the Applicability of the ‘Common Methodology for Assessment of Vulnerability to Sea Level Rise’ to the Australian Coastal Zone. In: Mclean, R.F. And Mimura, N. (Eds.), Vulnerability Assessment to Sea Level Rise and Coastal Zone Management (Proceedings of The IPCC Eastern Hemisphere Workshop).Department of Environment, Sport, and Territories, Tsukuba. 237–248.
5. Harvey N, Clouston B, Carvalho P (1999) Improving Coastal Vulnerability Assessment Methodologies for Integrated Coastal Zone Management: An Approach from South Australia Australian Geographical Studies 37: 50–69.
6. Gornitz V, Kouch S, Hartig EK (2001) Impacts of Sea Level Rise on New York City Metropolitan Area. Global and Planetary Change 32:61–88.
7. Shaw J, Taylor RB, Forbes DL, Ruz MH, Solomon S (1998) Sensitivity of the Coasts of Canada to Sea-Level Rise. Geological Survey of Canada, Bulletin 505: 1–79.
8. Abuodha PAO, Woodroffe CD (2010) Assessing Vulnerability to Sea Level Rise Using A Coastal Sensitivity Index: A Case Study From Southeast Australia. J Coast Conserv 14:189–205.
9. Kumar TS, Mahendra RS, Nayak S, Radhakrishnan K, Sahu KC (2010) Coastal Vulnerability Assessment for Orissa State, East Coast of India. Journal of Coastal Research 26:523–534.
10. Cutter SL, Boruff BJ, Shirley WL (2003) Social Vulnerability to Environmental Hazards. Social Science Quarterly 84:242-261.
11. Boruff BJ, Emrich C, Cutter SL (2005) Erosion Hazard Vulnerability of US Coastal Counties. Journal of Coastal Research 21: 932–942.
12. Cutter SL (1996) Vulnerability to Environmental Hazards. Progress in Human Geography 20:529-539.

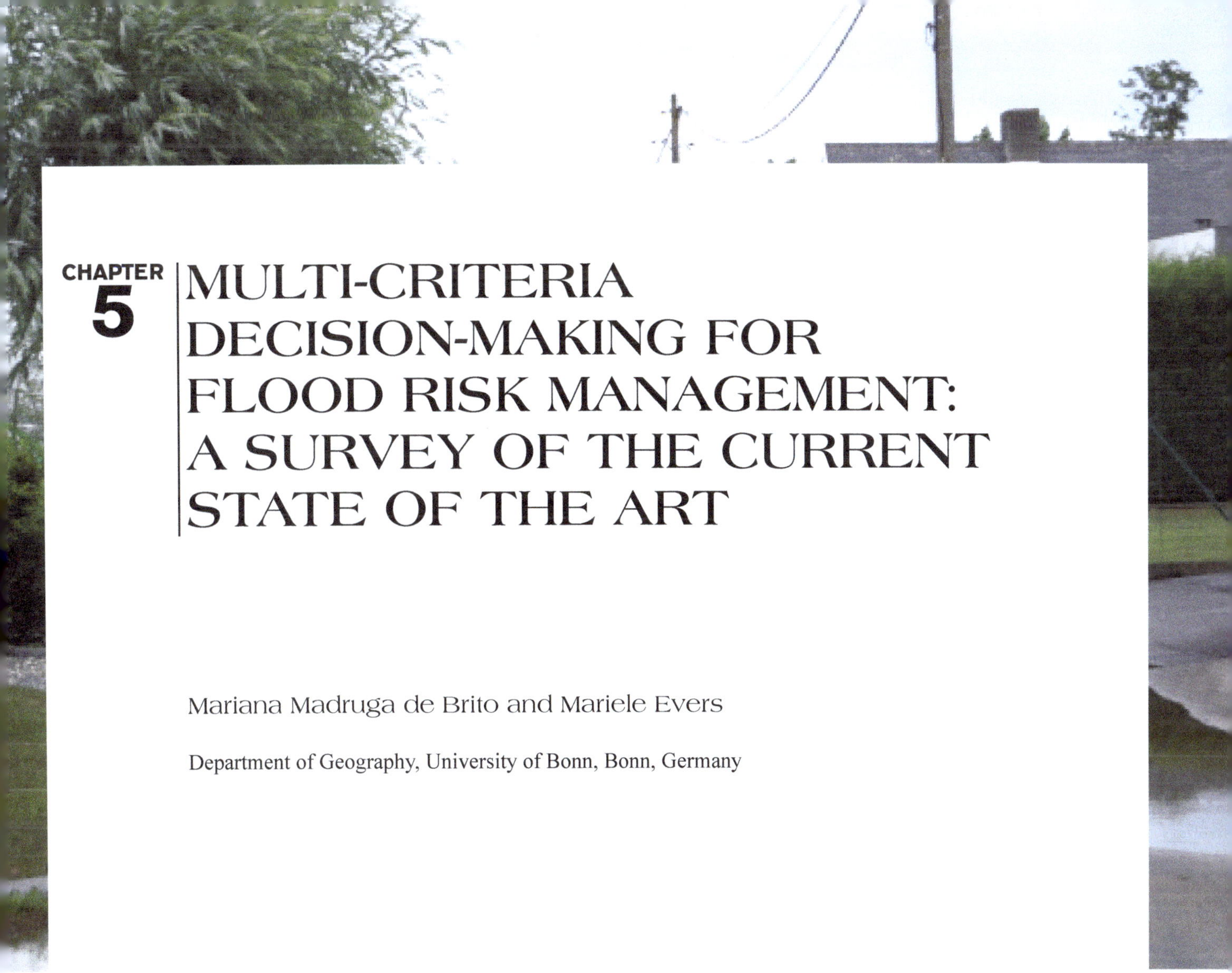

CHAPTER 5

MULTI-CRITERIA DECISION-MAKING FOR FLOOD RISK MANAGEMENT: A SURVEY OF THE CURRENT STATE OF THE ART

Mariana Madruga de Brito and Mariele Evers

Department of Geography, University of Bonn, Bonn, Germany

ABSTRACT

This paper provides a review of multi-criteria decision-making (MCDM) applications to flood risk management, seeking to highlight trends and identify research gaps. A total of 128 peer-reviewed papers published from 1995 to June 2015 were systematically analysed. Results showed that the number of flood MCDM publications has exponentially grown during this period, with over 82 % of all papers published since 2009. A wide range of applications were identified, with most papers focusing on ranking alternatives for flood mitigation, followed by risk, hazard, and vulnerability assessment. The analytical hierarchy process (AHP) was the most popular method, followed by Technique for Order Preference by Similarity to an Ideal Solution (TOPSIS), and Simple Additive Weighting (SAW). Although there is greater interest in MCDM, uncertainty analysis remains an issue and was seldom applied in flood-related studies. In addition, participation of multiple stakeholders has been generally fragmented, focusing on particular stages of the decision-making process, especially on the definition of criteria weights. Therefore, addressing the uncertainties around stakeholders' judgments and endorsing an active participation in all steps of the decision-making process should be explored in future applications. This could help to increase the quality of decisions and the implementation of chosen measures.

INTRODUCTION

Floods can be regarded as one of the most costly natural hazard both in developing and developed countries all over the world (Balica et al., 2013; Uddin et al., 2013). According to the Emergency Events Database (EM-DAT), these processes were the most frequent natural disaster worldwide between 2000 and 2014, causing at least 85 000 fatalities and affecting about 1.4 billion people. Apart from the loss of lives and physical damage, floods have resulted in approximately USD 400 billion in damage since 2000 (CRED and OFDA, 2015). In order to mitigate these impacts, a set of flood reduction measures need to be taken. The decision-making process related to flood risk management, especially in the prevention and emergency phases, tends to be rather complex and uncertain (Akter and Simonovic, 2005; Kenyon, 2007). Part of this complexity arises from the involvement of multiple stakeholders, each one with different views, background knowledge, interests, and frequently with competing objectives (Evers, 2008). In

Citation: de Brito, M. & Evers, M. (2015). Review Article: Multi-criteria decision making for flood risk management: a survey of the current state-of-the-art. Natural Hazards And Earth System Sciences Discussions,3(11), 6689-6726. http://dx.doi.org/10.5194/nhess-sd-3-6689-2015.

addition, the exact flood magnitude and damage are generally unknown and surrounded by considerable uncertainties (de Kort and Booij, 2007). As a consequence, making these decisions can rarely be solved with intuition alone. Thus, flood risk management requires the use of decision support tools, which can consider multiple stakeholders' views, objectives, trade-offs, feasible alternatives, and evaluation criteria.

Flood risk management can benefit from the use of multicriteria decision-making (MCDM) tools. MCDM is an umbrella term used to describe a set of methods for structuring and evaluating alternatives on the basis of multiple criteria and objectives (Voogd, 1983). These methods provide targeted decisions, as they can handle the inherent complexity and uncertainty of such problems as well as the knowledge arising from the participation of several actors (Yan et al., 2011; Zagonari and Rossi, 2013).

MCDM can enhance the quality of decisions, by making the process more explicit, rational, and efficient, leading to justifiable and explainable choices (Mateo, 2012a). Furthermore, MCDM promotes the role of participants in the decision process, facilitates compromise and group decisions, and provides an adequate platform for stakeholders to communicate their personal preferences (Pohekar and Ramachandran, 2004). The combination of these characteristics enables the development of real participatory processes, which are crucial for the implementation of successful and long-lasting flood management programs (Affeletranger, 2001).

Therefore, MCDM provides a powerful tool for flood management and has received a great deal of attention in solving such problems, not only from researchers but also decision makers and practitioners outside the scientific community. Since the mid-1990s, MCDM has been successfully applied to select the best strategies for flood risk mitigation, helping to optimize the allocation of available resources (e.g. Tkach and Simonovic, 1997; Ghanbarpour et al., 2013; Malekian and Azarnivand, 2015). In recent years, MCDM has also been used to access the flood risk and coping capacity (e.g. Guo et al., 2014; Roy and Blaschke, 2015; Yang et al., 2013). Several authors have reviewed MCDM techniques in various fields of study. For example, Stewart (1992) conducted a theoretical review by identifying potential advantages and pitfalls in the usage of various MCDM methods. Hajkowicz and Collins (2007) analysed over 134 papers in the field of water resource planning and management, focusing on problems such as water policy evaluation, strategic planning, and infrastructure selection. More recently, Estévez and Gelcich (2015) presented a concise literature survey, exploring the challenges behind participatory MCDM in marine conservation. However, despite practical experiences and methodological advances, there is no comprehensive literature review that explores the use of MCDM for flood risk management.

Hence, we believe that there is a need for a systematic survey to consolidate and synthesize recent research conducted in this area. Therefore, this paper aims to provide a literature review of the state of the art regarding the application of MCDM as a decision support tool for flood risk management, seeking to assess emerging trends and identify issues for future investigation. In addition, it attempts to provide a better understanding of the current status of how participatory MCDM is being conducted and the way uncertainties are considered in the decision-making process. With this review, we attempt to answer the following questions.

1. Which flood risk management problem has used MCDM approaches most?
2. Where was the research undertaken?
3. Which MCDM method was most commonly applied?
4. Were multiple stakeholders explicitly included in the decision-making process?
5. To which extent did these studies apply uncertainty and sensitivity analysis?

For reader's convenience, the remainder of this paper is structured as follows. In Sect. 2, the basic features of the MCDM methods are briefly described. Section 3 outlines the search strategy and the procedure used to classify the literature. Section 4 covers the discussion of the outcomes and provides answers to the research questions. In Sect. 5, limitations of this study and recommendations for further research are provided. Finally, Sect. 6 presents concluding remarks. We hope that this review will serve as a useful and ready source of information for scholars and practitioners working with MCDM and flood risk management.

OVERVIEW OF MULTI-CRITERIA DECISION-MAKING METHODS

MCDM is a broad term used to describe a set of methods that can be applied to support the decision-making process by taking into account multiple and often conflicting criteria through a structured framework (Cinelli et al., 2014). Since the 1960s, dozens of MCDM techniques have been developed (Mendoza and Martins, 2006). Generally, they can be classified into the following groups (Hajkowicz and Collins, 2007).

- Multi-attribute utility and value functions: the goal of these methods is to define an expression for the decision maker's preferences through the use of utility/value functions. Based on this, all criteria are transformed into a common dimensionless scale (Linkov et al., 2004). Popular methods include MAUT (Multi-attribute utility theory) and MAVT (Multi-attribute value theory), which have a compensatory nature. This implies that the poor performance of one criterion (e.g. high loss of lives) can be compensated by the better performance of another (e.g. financial cost). Although MAUT and MAVT have well-established theoretical foundations, the preference elicitation can be cognitively challenging and time-consuming (Schuwirth et al., 2012).
- Pairwise comparisons: this approach involves comparing pairs of criteria by asking how much more important one is than the other according to a predefined scale. Pairwise comparisons are particularly useful when it is not possible to define utility functions; otherwise MAUT is recommended (Ishizaka and Nemery, 2013). Common techniques include AHP (analytical hierarchy process), ANP (analytical network process) and MACBETH (measuring attractiveness by a categorical based evaluation technique). Due to its simplicity and flexibility, AHP is the most applied MCDM tool. Nevertheless, AHP has a limitation when dealing with interde pendence among the criteria as it assumes that they are independent (Li et al., 2011). In

addition, only a limited number of alternatives can be considered at the same time.

- Outranking approaches: unlike MAUT, MAVT, and AHP, outranking methods are based on the principle that one alternative may have a degree of dominance over another (Kangas et al., 2001), rather than assuming that a single optimal solution exists. Common methods include ELECTRE (Elimination et choix traduisant la realité), PROMETHEE (preference ranking organization method for enrichment of evaluations) and ORESTE (Organization, rangement et synthese de donnes relationnelles). An advantage of outranking approaches is that they avoid compensation between criteria and any normalization process, which alters the original data (Ishizaka and Nemery, 2013). Therefore, they are appropriate when criteria metrics are not easily aggregated, measurement scales vary over wide ranges, and units are incommensurate or incomparable (Linkov et al., 2004).
- Distance to ideal point methods: the alternatives are evaluated and ordered based on their distance from the ideal point, which represents a hypothetical alternative that best suits the decision makers' goals. Hence, the alternative that is closest to the ideal point is the best solution (Malczewski, 1999). Well-known methods include TOPSIS (Technique for Order Preference by Similarity to an Ideal Solution), CP (Compromise programming) and VIKOR (VlseKriterijumska Optimizacija I Kompromisno Resenje). The main characteristic and advantage of this family of approaches is the ability to consider a non-limited number of alternatives and criteria.
- Other methods: there are a large number of miscellaneous techniques that cannot be placed under any of the described categories. These include, for example, tailored methods which usually extend or adapt a fundamental method to a particular application, as well as fuzzy and hybrid approaches.

Despite the large number of MCDM methods, none is perfect and applicable to all decision problems. Therefore, the selection of an appropriate tool will depend on the problem type and decision makers' objectives. Guidelines such as the one proposed by Guitouni and Martel (1998) can be followed to choose from available MCDM techniques. Table 1 provides an outline of the fundamental properties of the MCDM methods that have been cited throughout the paper. A comprehensive and detailed description of the theoretical foundations of these techniques alongside with their main strengths and weaknesses can be found in Triantaphyllou (2000), Tzeng and Huang (2011), and Ishizaka and Nemery (2013).

FRAMEWORK FOR SYSTEMATIC LITERATURE REVIEW

Search Strategy

A comprehensive literature review was undertaken, aiming to identify peer-reviewed papers that apply MCDM to floodrelated problems. With this scope in mind, the systematic quantitative approach outlined in Pickering and Byrne (2014) was used since this method is explicit, reproducible, and has fewer biases when compared to traditional narrative reviews. To ensure that potentially relevant papers were not missed, six databases were systematically searched, including Scopus, ProQuest, Science Direct, SpringerLink, Emerald Insight, and Web of Science. Publications such as doctoral dissertations, book chapters, reports, and conference proceedings were not considered. Furthermore, only papers written in English were included. To find eligible papers in the mentioned databases, Boolean functions were applied to combine the following keywords:

Keywords (Multi-criteria OR MCDM OR multi criteria decision-making OR MCDA OR MCA OR AHP OR analytic hierarchy process OR ANP OR analytic network process OR MAUT OR multi-attribute utility theory OR MAVT OR multiattribute value theory OR ELECTRE OR TOPSIS OR MACBETH OR PROMETHEE OR NAIADE OR VIKOR OR weighted sum method OR simple additive weighting OR DSRA OR ORESTE OR DEMATEL OR goal programming) AND (flood OR floods).

Distinct combinations of these terms were used, taking into consideration the syntax requirements of each search engine. When possible, only the abstract, title, and keywords were searched. This narrowed the search space substantially and enabled to exclude papers that mention the keywords only in the references or literature review sections. These queries elicited over 1350 references published between September 1989 and June 2015. In order to have a 2 decades review, which is considered to be long enough to arrive at consistent conclusions (Jato-Espino et al., 2014), 1995 was chosen as a starting date for this survey. At first, the title, abstract, and keywords were screened manually to exclude irrelevant references. After this preselection, the full text of 207 selected papers was revised in detail. Of this total, 74 papers were found to be beyond the scope of the inquiry and five were not available through the library system. In the end, 128 papers met all inclusion criteria and were included in the analysis.

The review covers articles published in 72 different journals, in several areas of knowledge, suggesting that a diversity of publishers share an interest in flood risk management. Journals with the most papers were Natural Hazards, followed by Natural Hazards and Earth System Sciences, Water Resources Management, and Stochastic Environmental Research and Risk Assessment, with 16, 11, 10, and 6 articles, respectively. The remaining journals account mainly for one or two papers each.

Table 1. Description of different MCDM methods cited in the reviewed papers.

Abbreviation	Method	Description	Reference
AHP	Analytic hierarchy process	Structured technique for analysing MCDM problems according to a pairwise comparison scale, where the criteria are compared to each other	Vaidya and Kumar (2006)
ANP	Analytic network process	Generalization of the AHP method which enables the existence of interdependences among criteria	Saaty (2004)
CP	Compromise programming	Method based on the use of different distance measures to select the most suitable solution	Ballestero and Bernabeu (2015)
ELECTRE	Elimination et choix traduisant la realité	Group of techniques addressed to outrank a set of alternatives by determining their concordance and discordance indexes	Figueira et al. (2013)
MAUT	Multi-attribute utility theory	Method in which decisions are made by comparing the utility values of a series of attributes in terms of risk and uncertainty	Wallenius et al. (2008)
MAVT	Multi-attribute value theory	Simplification of MAUT that does not seek to model the decision makers' attitude to risk	Belton (1999)
PROMETHEE	Preference ranking organization method for enrichment of evaluations	Family of outranking methods based on positive and negative preference flows for each alternative that is used to rank them according to defined weights	Behzadian et al. (2010)
TOPSIS	Technique for order preference by similarity to an ideal solution	Technique based on the concept that the best alternative is the one which is closest to its ideal solution and farthest from the negative ideal solution	Behzadian et al. (2012)
VIKOR	Vlsekriterijumska optimizacija i kompromisno resenje	Method that uses aggregating functions and focuses on determining compromising solutions for a prioritization problem with conflicting criteria	Mateo (2012b)
SAW*	Simple Additive Weighting	Tool that aims to determine a weighted score for the alternatives by adding each attribute multiplied by their weights	Abdullah and Adawiyah (2014)

* Other terms such as weighted linear combination (WLC), weighted summation, weighted linear average, and weighted overlay are also used to describe SAW.

Classification Scheme

Following the selection, all included papers were classified according to some key domains: publication year; area of application; country of application; MCDM method; whether or not it was carried out in a participatory process; participatory techniques applied; and if uncertainty and sensitivity analysis were performed. With regard to the MCDM method, only techniques that were used thrice or more have their own category, whilst the rest were grouped in "others". In terms of research area, the papers were classified based on the overall emphasis of the application discussed. A total of eight types of MCDM applications were identified as follows.

- Ranking of alternatives for flood mitigation: this comprises the selection of the best combination of structural and/or non-structural mitigation solutions from a set of potential alternatives to reduce flood impacts and magnitude.
- Reservoir flood control: this consists in selecting operational options among a range of alternatives to ensure safe operation of reservoirs during high inflow events, aiming to reduce the floods intensity to acceptable levels.
- Susceptibility assessment: this expresses the likelihood that a flood will occur in an area on the basis of local terrain conditions (e.g. slope, elevation, lithology). It does not consider the flood temporal probability or return period (i.e. when or how frequently floods may occur) (Santangelo et al., 2011).
- Hazard assessment: this comprehends a qualitative or quantitative assessment of the spatial and temporal probability of the occurrence of potentially damaging floods of a certain magnitude in a given area within a specific period of time (Dang et al., 2011).
- Coping capacity assessment: this comprises the evaluation of the ability of people, organizations, and systems, using available skills and resources, to face and manage adverse conditions and emergencies resulting from floods (UNISDR, 2009).
- Vulnerability assessment: this refers to articles that assess the propensity of exposed elements such as

human beings, their livelihoods, and assets to suffer adverse effects when impacted by floods (UNISDR, 2009).

- Risk assessment: this consists in analysing potential flood hazards combined with existing conditions of vulnerability that together could potentially harm exposed people, property, services, livelihoods, and the environment (UNISDR, 2009).
- Emergency management papers: these are concerned with the organization and management of resources and responsibilities for addressing all aspects of emergencies, in particular, preparedness and response steps (UNISDR, 2009).

RESULTS AND DISCUSSION

This section presents a systematic analysis of 128 peerreviewed papers published between 1995 and June 2015. To help readers extract quick and meaningful information, the results are summarized in various charts and tables. A complete list of the reviewed papers, including their classification scheme, is provided in the Supplement.

Trends by Year of Publication

In an attempt to model the evolution of MCDM in time, the data gathered were organized by year of publication. As can be seen from Fig. 1, there has been a continuing growth in the number of flood MCDM studies from 1995 to June 2015. In fact, over 82 % of the compiled papers have been published since 2009. Until 2004, the number of publications was equal to or less than one per year. Surprisingly, from 2010 to 2013, the use of MCDM dramatically increased, from 5 to 22 papers. Accordingly, it can be estimated that in the coming years, these numbers will keep growing. This indicates that MCDM has a good vitality and acceptance for flood risk management.

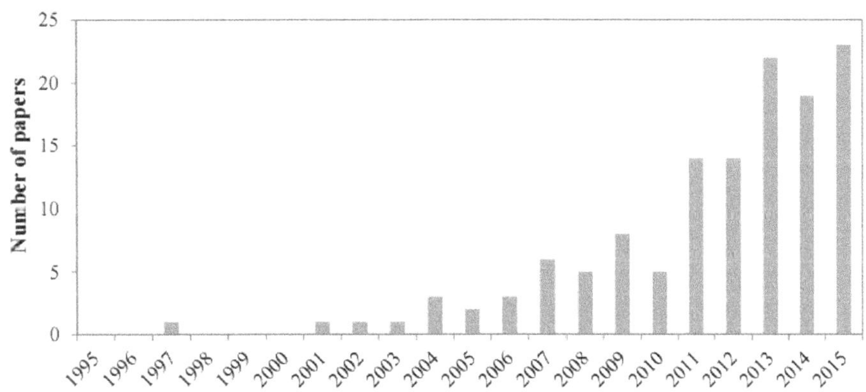

Figure 1. Number of MCDM flood papers published over the period 1995–June 2015.

A reason for the increasing number of publications could be a reflection of a growing awareness of natural disaster prevention and reduction policies. Secondly, the availability of easy-to-use and inexpensive MCDM software packages may also be an influencing factor. Alternatively, this increase may just match a general rise in published papers related to floods as a whole. To correctly measure the interest in MCDM for flood risk management, an increase of MCDM papers in relative terms needs to be calculated. Thus, a normalization was made according to the number of flood publications in the Web of Science and Science Direct databases, found through searches using only "flood" as keyword. Figure 2 shows that the increase of flood MCDM publications is significantly greater than the increase of flood publications, especially after 2011. This confirms the hypothesis that the use of MCDM to solve flood-related problems has been growing considerably since 1995.

Trends by Area of Application

During the last 2 decades, ranking alternatives for flood mitigation was the most widespread flood management topic, with more than 22 % of all applications (Table 2). These studies focus mainly on selecting traditional engineering measures to reduce flood risk (e.g. Azibi and Vanderpooten, 2003; Tkach and Simonovic, 1997). Nevertheless, in recent years, they emphasize not only the so-called structural measures, which are still relevant, but also incorporate a wide range of non-structural options such as the development of evacuation plans, enforcement of building codes, and insurance schemes (e.g. Zagonari and Rossi, 2013; Chitsaz and Banihabib, 2015).

Table 2. Distribution of applications by flood risk management topic.

Area of application	N	%
Ranking of alternatives for flood mitigation	41	22.78
Risk assessment	38	21.11
Vulnerability assessment	27	15.00
Hazard assessment	27	15.00
Susceptibility assessment	21	11.67
Coping capacity	11	6.11
Reservoir flood control	8	4.44
Emergency management	7	3.89
Total	180	100

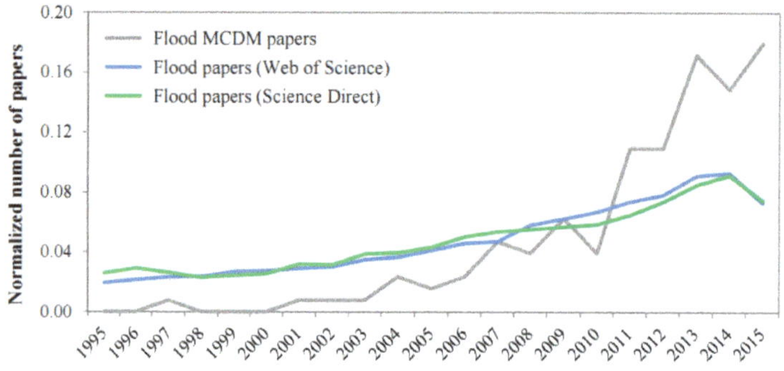

Figure 2. Normalized number of MCDM and flood papers published between 1995–June 2015, based on data from the Web of Science and Science Direct.

The second most common theme was risk assessment (21.11 %), followed by vulnerability and hazard analysis, both with 15.00 % of all applications. In this regard, it is worth noting that several papers evaluate the vulnerability, hazard, and risk simultaneously (e.g. Lee and Chung, 2007; Zou et al., 2013; Wu et al., 2015). Few papers used MCDM as a decision support tool in reservoir flood control and emergency management problems. This is probably because managing emergencies, both in rivers and reservoirs, is a complex task, requiring effective coordination and communication among teams involved as well as reliable information regarding the current situation of emergency (Shan et al., 2012).

In order to have a complete overview of works published through time, Fig. 3 presents a temporal breakdown of the different flood topics. As can be seen, flood risk management has recently shifted its main focus from ranking alternatives for flood mitigation towards a risk-based perspective, which includes the assessment of risk and its components. This finding is in agreement with a worldwide trend, where disaster prevention is emphasized over assistance or relief, and evaluating the risk becomes a key element (World Bank, 2006). Another interesting result is that coping capacity studies are quite new in comparison to other topics, with the first paper published in 2009. In addition, the graph reveals that since 2010, the trend in the other flood problems has remained fairly stable. This diversity of applications shows MCDM flexibility to support decision-making in all stages of the flood management cycle.

Table 3. Distribution of applications by country of application.

Country	*N*	%	Country	*N*	%
China	26	19.40	the Netherlands	2	1.49
Germany	13	9.70	Finland	2	1.49
South Korea	10	7.46	Italy	2	1.49
Iran	7	5.22	Kenya	1	0.75
Greece	6	4.48	Kuwait	1	0.75
India	6	4.48	Vietnam	1	0.75
Canada	6	4.48	Taiwan	1	0.75
Malaysia	5	3.73	Bhutan	1	0.75
Bangladesh	5	3.73	Switzerland	1	0.75
USA	5	3.73	South Africa	1	0.75
UK	5	3.73	Poland	1	0.75
France	4	2.99	Spain	1	0.75
Slovakia	3	2.24	Portugal	1	0.75
Egypt	2	1.49	Serbia	1	0.75
Turkey	2	1.49	Nigeria	1	0.75
Japan	2	1.49	Chile	1	0.75
Australia	2	1.49	Argentina	1	0.75
Croatia	2	1.49	Romania	1	0.75
Austria	2	1.49	Total	134	100.00

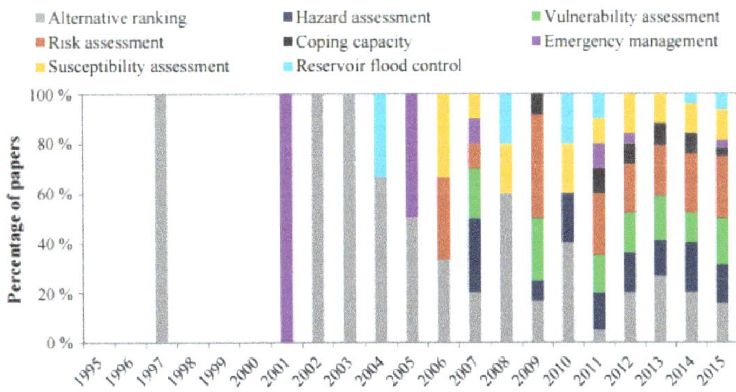

Figure 3. Distribution of MCDM papers by application area between 1995 and June 2015.

Trends by Country of Application

A total of 37 countries on all populated continents have contributed to this survey (Table 3), showing that the spread of MCDM is truly global. China accounts for 19.40 % of all applications, which is not too surprising. Indeed, similar results were obtained by other MCDM review papers (e.g. Jato-Espino et al., 2014). In contrast to previous surveys (e.g. Govindan and Jepsen, 2016), Germany and South Korea were found to be prolific users of MCDM tools.

Surprisingly, South American countries such as Brazil, Colombia, and Venezuela, which are severely affected by floods (CRED and OFDA, 2015), were not represented in the literature. The limited use of MCDM in these countries could be explained by restrictions, such as lack of expertise, resources, and technology. On the other hand, it could be that the existing studies are published in non-English journals (e.g. Drozino et al., 2015; Magalhães et al., 2011). Unlike other MCDM review papers (e.g. Behzadian et al., 2010; Mosadeghi et al., 2013), MCDM tools were rarely applied to solve flood-related problems in Australia. The reason could be that potentially relevant studies are published in conference papers, government reports, non-indexed journals, or in other grey literature. Half of the MCDM studies were conducted in Asia, followed by Europe (35.07 %), North America (8.21 %), Africa (3.73 %), and finally by Australia and South America, each with 1.49 % of all applications. Therefore, it is clear that when we analyse the findings of the present study, we are providing a predominantly Asiatic and European view of flood risk management.

Furthermore, only three papers report cross-country investigations (e.g. Ceccato et al., 2011; Evers et al., 2012; Almoradie et al., 2015). For example, Ceccato et al. (2011) analysed five case studies in Austria, Germany, India, Bhutan, and China. The authors found out that although the studied watersheds were characterized by distinct ecological, social, and economic dimensions, the criteria selected by the stakeholders were rather similar. In this regard, multiple-case studies allow findings to be compared, parallels to be drawn, and differences across diverse cultural, environmental, and governmental contexts to be examined.

Trends by MCDM method

Results showed that AHP and its family of methods were by far the most used MCDM approach (Table 4). One reason for this might be that its structure is straightforward, flexible, and easily understandable (Cinelli et al., 2014). Thanks to these characteristics, it can be adapted to different problems without requiring previous knowledge from the analyst. Moreover, several software packages incorporate AHP (e.g. DECERNS, ExpertChoice, MakeItRational, DECERNS, EasyMind and Super decisions), including GIS (Geographic Information System) software (e.g. ArcGIS, Idrisi, and ILWIS). The second most employed method was TOPSIS, closely followed by SAW. These results, with a few differences and similarities, were confirmed by other MCDM review papers such as Jato-Espino et al. (2014) and Broekhuizen et al. (2015) that ranked AHP as the first and TOPSIS as the second method with more applications.

Note that the sum of the applications (165 items) in Table 4 does not match the number of papers (128 items) since some articles used several MCDM techniques to analyse differences in scoring and ranking. For example, Chitsaz and Banihabib (2015) compared seven MCDM tools and concluded that ELECTRE III stood superior to select flood management options. On the other hand, Chung and Lee (2009) employed five methods and found out that there is no clear methodological advantage to any of the considered techniques. Apart from comparative studies, several researchers have combined two MCDM approaches to complement each other (e.g. Margeta and Knezic, 2002; Lee and Chung, 2007; Zhou et al., 2014). For instance, Zhou et al. (2014) applied AHP to assign relative weights to each criterion and TOPSIS to rank the risk. Overall, 106 out of 128 papers (82.81 %) used one MCDM method while 12.50 % used two, 3.13 % used three, and 1.56 % applied more than three.

Table 4. Distribution of applications by MCDM method.

MCDM method	N	%
AHP, fuzzy AHP, trapezoidal fuzzy AHP and ANP	70	42.42
TOPSIS, fuzzy TOPSIS and modified TOPSIS	22	13.33
SAW	21	12.73
Others (MACBETH, NAIADE, goal programming, etc.)	20	12.12
CP, spatial CP and fuzzy CP	10	6.06
ELECTRE I, II, III and TRI	7	4.24
MAUT and MAVT	7	4.24
PROMETHEE I and II	5	3.03
VIKOR and fuzzy VIKOR	3	1.82
Total	165	100

The survey also showed that MCDM techniques are not

used only in a stand-alone mode, but are commonly extended and combined with soft computing technologies, including fuzzy set theory (e.g. Chen and Hou, 2004; Guo et al., 2014), artificial neural network (e.g. Radmehr and Araghinejad, 2014; Liu et al., 2014), and tools such as SWOT (Strengths, Weaknesses, Opportunities, and Threats) analysis (e.g. Vafaei and Harati, 2010; Miyamoto et al., 2014). Furthermore, there are also numerous hybrid methods, developed to address gaps in classical techniques (e.g. Yang et al., 2013; Shams et al., 2014). This suggests that MCDM is versatile, enabling researchers to combine it effectively with different tools according to the requirements of the decision to be taken. Overall, AHP is the most prominent MCDM method in all application areas, except for reservoir flood control (Table 5). The primary reason for the popularity of AHP for mapping the risk and its components is that the implementation of this technique within the GIS environment is straightforward, enabling the users to quickly derive the weights associated with criteria map layers (Malczewski, 2007). For reservoir flood control, miscellaneous methods such as fuzzy hybrid approaches were the preferred techniques. This is probably because reservoir operations involve a large number of uncertain factors that can be properly addressed by fuzzy set theory. Additionally, TOPSIS is highly popular for ranking alternatives for flood mitigation, which emphasizes the effectiveness of this technique to deal simultaneously with con- flicting objectives.

Although the most widespread MCDM methods were used at least once, no study has used DEMATEL (Decision Making Trial and Evaluation Laboratory) or ORESTE (Organization, Rangement Et Synthese De Donnes Relationnelles). A likely explanation is that these methods are cognitively demanding when compared to classical approaches, especially when numerous criteria are involved (Dou et al., 2014; Moffett and Sarkar, 2006). For instance, DEMATEL needs to be coupled with other MCDM tools, such as ANP to generate criteria weights, which makes its application dif- ficult. In addition, there is a limited amount of software available, and most of it costs (e.g. Decision Era). However, DEMATEL was specifically developed to address limitations of traditional techniques regarding interdependence between criteria. Likewise, ORESTE is suitable for problems with limited information and with incommensurable criteria (Moffett and Sarkar, 2006).

Table 5. Distribution of applications by MCDM method and area of application.

Area of application/ number of applications	AHP	TOPSIS	SAW	Others	CP	ELECTRE	MAUT	PROMETHEE	VIKOR
Ranking of alternatives for flood mitigation	14	10	9	8	9	5	2	3	1
Risk assessment	27	10	5	6	1	1	3	1	2
Vulnerability assessment	21	3	5	4	1	1	2	0	0
Hazard assessment	25	3	2	5	1	1	0	0	0
Susceptibility assessment	18	0	4	0	0	0	0	0	0
Coping capacity	8	4	2	0	0	0	1	0	0
Emergency management	5	0	1	0	0	0	0	1	0
Reservoir flood control	1	1	0	5	0	1	0	0	0
Total*	119	31	28	28	12	9	8	5	3

* It is important to highlight that some papers analysed two or more flood problems simultaneously by using the same MCDM method. Thus, the number of applications in Table 5 is higher than in Table 4.

Trends Regarding Stakeholders' Involvement

Flood risk management decisions may be designed without the direct participation of multiple stakeholders. However, they cannot be implemented without them (Affeletranger, 2001). Therefore, flood management decisionmaking should be ideally carried out in a participatory process, where the knowledge and preferences of interested actors are integrated into the process from the beginning. According to Evers et al. (2014), this creates trust among decision makers and stakeholders, which often lead to a successful implementation of the chosen measures. The survey revealed that 65 (50.78 %) studies have explicitly acknowledged the involvement of multiple actors in the decision-making process. Policy makers and experts were the stakeholders that participated most. This was expected since they are often responsible for the selection and implementation of chosen measures and have a broad knowledge of the problem of interest. Additionally, some papers mentioned the involvement of local community members (e.g. Kandilioti and Makropoulos, 2012; Sahin et al., 2013; Roy and Blaschke, 2015). According to Affeletranger (2001), the consideration of community members' opinion may improve their resilience as well as their response capacity when confronting natural disasters.

Nevertheless, participation was generally fragmented and restricted to consultation at specific stages, such as the selection of evaluation criteria (e.g. Haque et al., 2012) and the definition of criteria weights (e.g. Kienberger et al., 2009; Sahin et al., 2013). This segmentation may be related to methodological and time constraints since participatory decision-making is time-consuming and costly, particularly when the decisions are made in a group where proper facilitation is required. Crucial aspects of the decision-making process like the definition of objectives, identification of the alternatives, and estimation of its consequences were usually constrained to analysts and experts, which inhibit the achievement of genuine participation. Only in exceptional cases was the input from the stakeholders a critical element in the entire process (e.g. Ceccato et al., 2011; Evers et al., 2012). For example, Ceccato et al. (2011)

developed a methodological proposal aimed at strengthening the communication and collaboration within the scientific community and local actors for flood management decision-making. The authors applied the NetSyMoD (Network Analysis – Creative System Modelling) framework (Giupponi et al., 2008), where the identification of relevant stakeholders, definition of the problem, establishment of objectives and criteria, and the selection of alternatives are conducted in a participatory process.

Another interesting result is that only four studies seek to obtain consensus (e.g. Haque et al., 2012; Lee et al., 2013, 2014, 2015), in which participants make decisions by agreement rather than by majority vote or averaging approaches. Nevertheless, enhancing mutual understanding for consensus building is essential for a long-lasting and successful flood management program, especially for selecting alternatives for flood mitigation and emergency management. It allows decision makers to derive meaningful solutions that fulfil their own needs while at the same time satisfying the requirements of other actors, legitimating the participation as a learning process to solve complex problems. A total of 43 out of 65 studies provided unambiguous descriptions of the participatory decision-making techniques applied. Figure 4 shows that questionnaires (e.g. Giupponi et al., 2013; Taib et al., 2016) and face-to-face interviews (e.g. Deshmukh et al., 2011; Jun et al., 2011) were the most applied tools. These methods allow for opinions to be conveyed without influence from dominant participants and are simple and fast to realize. On the other hand, the participants are not able to share and hear different perspectives through open dialogue, which is essential for achieving common agreement.

In this sense, Mendoza and Martins (2006) argue that group elicitation methods involving open discussion offer several advantages, including the consistency in the information obtained, and a better definition of the preferences. On the other hand, the results can be influenced by dominant stakeholders and noise in the responses (Hsu and Sandford, 2007). Group elicitation methods such as workshops (e.g. Kenyon, 2007; Porthin et al., 2013), group meetings (e.g. Azibi and Vanderpooten, 2003; Marttunen et al., 2013), and focus group discussions (e.g. Rahman and Saha, 2007; Haque et al., 2012) were less applied in the reviewed papers. Recently, researchers have used the Delphi technique to overcome shortcomings of conventional group elicitation methods regarding dominant individuals and time constraints (e.g. Chung et al., 2014; Lee et al., 2014). This method provides anonymity to respondents, a structured feedback process, and is suited for consensus building (Hsu and Sandford, 2007). Additionally, it is advantageous when the stakeholders live some distance apart, and it is prohibitive to bring them together for a workshop or group meeting (Lee et al., 2013). It is interesting to highlight that two studies reported the use of collaborative web-based platforms in which stakeholders select and rank alternatives interactively (e.g. Evers et al., 2012; Almoradie et al., 2015). These platforms have the potential to overcome hindrances in participatory MCDM such as the limitation of financial resources and stakeholders' spatial distribution, providing full transparency of information and results. By taking this approach, the confidence in the decision-making process is increased as well as the level of acceptance of negotiated measures, which are crucial conditions for successful participatory flood risk management.

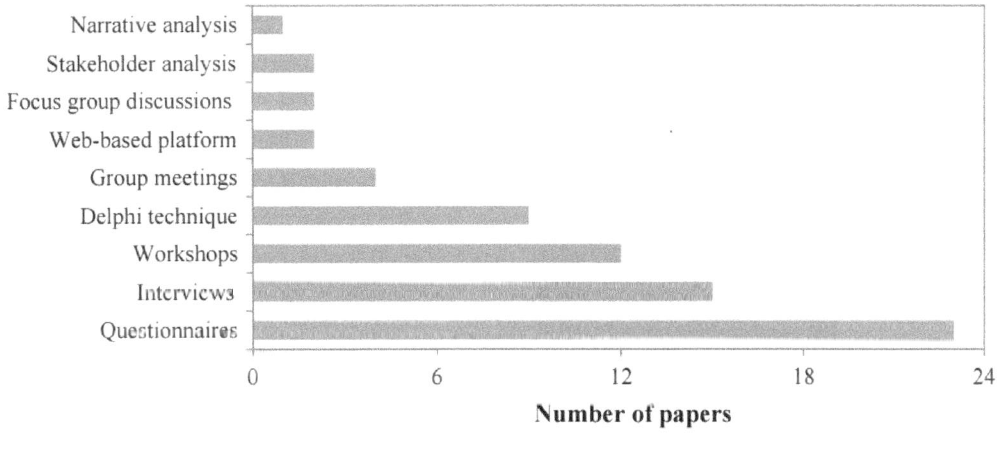

Figure 4. Methods used to incorporate multiple stakeholders' views in the decision-making process.

Trends Regarding Sensitivity and Uncertainty Analysis

Flood decision-making is subjected to multiple sources of uncertainty, including the assessment of criteria weights, the parameters' uncertainties, and structural uncertainty (Broekhuizen et al., 2015). In addition, there are uncertainties associated with the inherent randomness of flood events (Merz et al., 2008), which, in principle, cannot be reduced. Thus, in order to improve the quality of decisions and verify the robustness of the model outputs, flood risk management should be based on a comprehensive assessment of the sensitivity combined with a thorough investigation of the uncertainties involved. In this review, 93 (72.65 %) papers do not report any kind of sensitivity analysis, thereby ignoring the impact of changes in input weights on model results. The remaining articles (35 or 27.34 %) applied mainly one-way sensitivity analysis, where one criteria weight or performance score is modified at a time and the variation of the alternatives' ranking is observed. If the induced variation does not change the rank order of alternatives, the decision is considered robust. This technique is intuitively appealing and requires little time, making it a practical way to assess the sensitivity. Even though this method is sufficient for most flood applications, the range over which weights are varied is normally arbitrarily defined, and the commutative impact of uncertainty is not considered. Hence, these drawbacks may lead to a biased view of the influence of uncertainty on the final decision (Broekhuizen et al., 2015).

Two papers performed global sensitivity analysis (GSA) by applying the FAST (Fourier Amplitude Sensitivity Test) procedure, where two or more evaluation criteria are varied at the same time (e.g. Fernández and Lutz, 2010; Chen et al., 2015). Although GSA allows for the full uncertainty range of the criteria to be explored and analysed, it can become an extremely time-consuming task, as a large number of criteria are included in the analysis. Additionally, four papers elaborated best- and worst-case scenarios to incorporate decision makers' attitude to risk (e.g. Kandilioti and Makropoulos, 2012; Penning-Rowsell et al., 2013; Ghanbarpour et al., 2013; Alipour, 2015). Finally, two studies used a probabilis tic approach (e.g. Yazdandoost and Bozorgy, 2008; Fernández and Lutz, 2010), which is the most rigorous form of sensitivity analysis. This approach requires the estimation of a maximum percentage that the actual criteria weight may differ from the estimated value.

Several authors have listed the uncertainty as a major drawback (e.g. Bana e Costa et al., 2004; Edjossan-Sossou et al., 2014; Godfrey et al., 2015; Almoradie et al., 2015). However, only eight papers (6.25 %) perform uncertainty analysis in an attempt to describe the entire set of possible outcomes, together with their associated probabilities of occurrence. In situations where uncertainty is mainly due to randomness, the methods used were probability-based. This is the case of Qi et al. (2013) and Li (2013) who used Monte Carlo simulation to convert uncertainties in input criteria into probability distributions. Another approach applied was the Taylor's series error propagation method (e.g. Fernández and Lutz, 2010), which analyses how the uncertainty in input data propagates through the model and affects its outputs. In addition, three papers assessed the uncertainty qualitatively, by describing its main sources (e.g. Cozannet et al., 2013) or by analysing the degree of confidence related to stakeholders' opinion (e.g. Ceccato et al., 2011; Penning-Rowsell et al., 2013). Apart from uncertainty and sensitivity analysis, fuzzy set theory is widely combined with AHP, TOPSIS, and CP to handle uncertainty and incomplete information about the decision situation. For instance, Lee et al. (2013) integrated TOPSIS and fuzzy set theory to fuzzify the weighting values and all criteria maps. In the same sense, the approach proposed by Yang et al. (2012) combines AHP and triangular fuzzy number to assess the flood risk and its components. Fuzzy set theory is widespread in MCDM due to its intuitiveness and computational requirements. Nevertheless, some studies have shown that fuzzy AHP do not provide better results than regular AHP since the judgments in AHP are already fuzzy (Saaty, 2006). Therefore, the additional complexity of utilizing fuzzy numbers may be unnecessary in some cases.

Finally, it is relevant to note that some MCDM methods explicitly account for uncertain input criteria scores. For instance, ELECTRE and PROMETHEE adopt the pseudocriterion model that introduces indifference and preference thresholds. Likewise, MAUT considers imprecise data input with probabilistic approaches (Cinelli et al., 2014). Also, AHP allows the generation of an inconsistency index, which can be considered as an indirect measure of the uncertainty in the criteria weighting step.

RESEARCH LIMITATIONS AND RECOMMENDATIONS FOR FUTURE RESEARCH

Limitations

There are some caveats that should be considered when interpreting the results obtained in this review. One of the main shortcomings is that the papers' quality was not evaluated since they were all published in peer-reviewed journals. Thus, some applications were superficial, while others were detailed, including intensive stakeholder participation, validation of results, and probabilistic-based uncertainty and sensitivity analysis. Some studies were carried out with real data, involving real decision makers and stakeholders, while others discussed hypothetical applications or were secondary studies that re-examined empirical work. A future review paper can address this limitation by applying heuristic checklists (e.g. Beecham et al., 2008) to assess the overall quality of the study. In addition, defining the flood application area turned out to be a subjective process, especially when it came to distinguishing between susceptibility, hazard, and risk assessment. There is a misunderstanding about these terms in the literature, which are used in slightly different ways by researchers with different backgrounds. Thus, in some cases, it was diffi- cult to define a clear line for when it was susceptibility, hazard, or risk. Whenever possible, the term used by the authors was respected.

The exclusion of non-English literature can also be understood as a limitation (Behzadian et al., 2010). The results of our preliminary searches showed that several MCDM French school authors have published in French language journals. Furthermore, there are a significant number of research papers published in German, Chinese, and Korean. Thus, it should be emphasized that, when feasible, searches using multiple languages are advantageous (Pickering and Byrne, 2014). Nevertheless, despite these potential limitations, this paper is the first to present a literature review of the state of the art of the use of MCDM for flood-related problems. The sample of papers analysed provides sufficient information to stimulate discussion and research that addresses challenges in this area of knowledge.

Recommendations for Future Research

This review enabled us to identify gaps in the knowledge of MCDM for flood risk management regarding several aspects. First, classical MCDM methods such as MAUT, MAVT, PROMETHEE, and DEMATEL were overlooked. Almost half of reviewed applications used AHP to elicit criteria weights, which is a relatively easy and flexible method, requiring fewer skills than other tools. In this sense, exploring the implications of methodological differences in existing MCDM methods for flood risk management is an interesting research challenge. Similarly, future research can focus on understanding advantages and limitations of each method for handling different sources of uncertainty. Secondly, there were surprisingly few studies that effectively considered stakeholders' participation throughout the entire decision-making process. Therefore, greater rigour in endorsing an active participation in all stages of the decisionmaking process should be undertaken, in order to increase the feasibility and subsequent implementation of chosen measures. Future research could be directed towards developing web platforms to elicit stakeholders' preferences, aiming to reach consensus in a simpler and easily accessible way. In addition, this course of action can be combined with other participatory techniques such as cognitive mapping, Delphi technique, and voting theory. Conversely, it should be noted that intensive participation is time-consuming. Thus, in reallife applications, trade-offs have to be made between the available resources and the expected outcomes of the MCDM process.

The third challenge, and perhaps the most relevant research gap, refers to fully considering the uncertainties around decision makers' judgments. Although uncertainty in MCDM is not a new problem and significant improvements have been made over the last decades, it remains a major open issue. Previous studies suggest that properly addressing the uncertainties can substantially improve MCDM applications, assisting stakeholders to make better decisions. Potential exists to apply Bayesian framework methods (e.g. Bayesian networks and Dempster–Shafers' theory), possibility theory, and evidence theory. Additionally, innovative approaches may be required to account for special characteristics in developing countries, where limitations in resources and technology exist. However, regardless of the uncertainty method applied, considering all sources of uncertainty in the decision-making process might not be a feasible task (Mowrer, 2000). Nevertheless, it is essential to identify as many sources of uncertainty as possible, and attempt to reduce or handle them. Lastly, a significant gain can be made if flood MCDM applications are able to consider climate and socioeconomical changes, which have potential to aggravate existing risks. This has been tackled in a recent study by Giupponi et al. (2013) that assessed the flood vulnerability within the broad context of climate change adaptation.

CONCLUSIONS

This study has presented a systematic review of 128 papers that apply MCDM to flood-related problems, aiming to provide an overall picture of what has motivated researchers and practitioners in 37 different countries over the past 2 decades. Our findings suggest an increasing interest in flood MCDM since 2009, as compared to the previous 14 years. A wide range of applications were identified, with most papers focusing on ranking alternatives for flood mitigation, followed by risk, hazard, and vulnerability assessment. This highlights the utility of MCDM as a decision support tool in all stages of the flood management process.

Nearly 85 % of the applications were conducted in Asian and European countries, mainly in China, Germany, and South Korea. Hence, potential exists to develop crosscountry investigations, especially in South America and Australia. Overall, AHP was the most widespread method, indicating that other methods may be overlooked. The review also shows that fuzzy and hybrid approaches (e.g. triangular fuzzy AHP, hybrid fuzzy AHP-TOPSIS, AHP-SWOT, modified TOPSIS) are being increasingly applied to overcome limitations of classical methods. About half of the studies have acknowledged the involvement of multiple stakeholders. However, participation was fragmented and focused on particular stages of the decisionmaking process. Most of the reviewed studies rely on the use of questionnaires and interviews to capture stakeholders' perspectives, with few applications seeking to obtain consensus. In addition, shortcomings remain in handling the uncertainty. Thus, greater rigour in considering the uncertainties around stakeholders' preferences and endorsing an active participation are important research gaps. Additionally, sensitivity analysis should be conducted as a primary method to check the stability of the results and identify the most critical criteria. This could help to increase the quality of decisions as well as the transparency and credibility of the MCDM outcomes. It is clear from the literature that the challenge for further research is to foster the development of true collaborative MCDM applications that take the uncertainty around stakeholders' judgments into account. We believe that this paper can provide valuable information for guiding future research and that it can serve as a ready reference for researchers and practitioners working with flood risk management and MCDM.

Acknowledgements

This work was supported by the Brazilian Coordination for the Improvement of Higher Education Personnel (CAPES) through the grant 13669-13-3. The authors would like to thank the editor Paolo Tarolli and two anonymous referees for providing helpful comments and feedback. We are also thankful for the suggestions made by Adrian Almoradie.

REFERENCES

1. Abdullah, L. and Adawiyah, C. W. R.: Simple additive weighting methods of multi criteria decision making and applications: a decade review, Int. J. Inf. Process. Manage., 5, 39–49, 2014.
2. Affeletranger, B.: Public participation in the design of local strategies for flood mitigation and control, International Hydrological Programme, Technical Documents in Hydrology no. 48, UNESCO, Paris, 2001.
3. Akter, T. and Simonovic, S. P.: Aggregation of fuzzy views of a large number of stakeholders for multi-objective flood management decision-making, J. Environ. Manage., 77, 133–143, doi:10.1016/j.jenvman.2005.02.015, 2005.
4. Alipour, M. H.: Risk-informed decision making framework for opperating a multi-purpose hydropower reservoir during flooding and high inflow events, case study: Cheakamus River System, Water Resour. Manage., 29, 801–815, doi:10.1007/s11269-014- 0844-3, 2015.
5. Azibi, R. and Vanderpooten, D.: Aggregation of dispersed consequences for constructing criteria: the evaluation of flood risk reduction strategies, Eur. J. Oper. Res., 144, 397–411, doi:10.1016/S0377-2217(01)00400-3, 2003.
6. Balica, S. F., Popescu, I., Beevers, L., and Wright, N. G.: Parametric and physically based modelling techniques for flood risk and vulnerability assessment: A comparison, Environ. Model. Softw., 41, 84–92, doi:10.1016/j.envsoft.2012.11.002, 2013.
7. Ballestero, E. and Bernabeu, A. G.: Compromise programing and utility functions, in: Socially responsible investment – a multicriteria decision making

approach, edited by: Ballestero, E., Pérez-Gladish, B., and Bernabeu, A. G., Springer International Publishing, Cham., 155–175, 2015.

8. Beecham, S., Baddoo, N., Hall, T., Robinson, H., and Sharp, H.: Motivation in software engineering: a systematic literature review, Inf. Softw. Technol., 50, 860–878, doi:10.1016/j.infsof.2007.09.004, 2008.

9. Behzadian, M., Kazemzadeh, R. B., Albadvi, A., and Aghdasi, M.: PROMETHEE: a comprehensive literature review on methodologies and applications, Eur. J. Oper. Res., 200, 198–215, doi:10.1016/j.ejor.2009.01.021, 2010.

10. Belton, V.: Multi-criteria problem structuring and analysis in a value theory framework, in: Multicriteria Decision Making Advances in MCDM Models, Algorithms, Theory and Applications, edited by: Gal, T., Stewart, T. J., and Hanne, T., Springer US, New York, 335–366, 1999.

11. Broekhuizen, H., Groothuis-Oudshoorn, C. G. M., van Til, J. A., Hummel, J. M., and IJzerman, M. J.: A review and classification of approaches for dealing with uncertainty in multi-criteria decision analysis for healthcare decisions, Pharmacoeconomics, 33, 445–455, doi:10.1007/s40273-014-0251-x, 2015.

12. Ceccato, L., Giannini, V., and Giupponi, C.: Participatory assessment of adaptation strategies to flood risk in the Upper Brahmaputra and Danube river basins, Environ. Sci. Policy, 14, 1163– 1174, doi:10.1016/j.envsci.2011.05.016, 2011.

13. Chen, H., Ito, Y., Sawamukai, M., and Tokunaga, T.: Flood hazard assessment in the Kujukuri Plain of Chiba Prefecture, Japan, based on GIS and multicriteria decision analysis, Nat. Hazards, 78, 105–120, doi:10.1007/s11069-015-1699-5, 2015.

14. Chitsaz, N. and Banihabib, M. E.: Comparison of different multi criteria decision-making models in prioritizing flood management alternatives, Water Resour. Manage., 29, 2503–2525, doi:10.1007/s11269-015-0954-6, 2015.

15. Chung, E., Won, K., Kim, Y., and Lee, H.: Water resource vulnerability characteristics by district's population size in a changing climate using subjective and objective weights, Sustainability, 6, 6141–6157, doi:10.3390/su6096141, 2014.

16. Cozannet, G. L., Garcin, M., Bulteau, T., Mirgon, C., Yates, M. L., Méndez, M., Baills, A., Idier, D., and Oliveros, C.: An AHPderived method for mapping the physical vulnerability of coastal areas at regional scales, Nat. Hazards Earth Syst. Sci., 13, 1209– 1227, doi:10.5194/nhess-13-1209-2013, 2013.

17. Dang, N. M., Babel, M. S., and Luong, H. T.: Evaluation of food risk parameters in the Day river flood diversion area, Red River Delta, Vietnam, Nat. Hazards, 56, 169–194, doi:10.1007/s11069- 010-9558-x, 2011.

18. Deshmukh, A., Oh, E. H., and Hastak, M.: Impact of flood damaged critical infrastructure on communities and industries, Built Environ. Proj. Asset Manage., 1, 156–175, doi:10.1108/20441241111180415, 2011.

19. Dou, Y., Sarkis, J., and Bai, C.: Government Green Procurement: A Fuzzy-DEMATEL Analysis of Barriers, in: Supply Chain Management Under Fuzziness: Recent Developments and Techniques, edited by: Kahraman, C. and Öztay¸si, B., SpringerVerlag, Berlin, Heidelberg, 567–589, 2014.

20. Drozino, B., Oliveira, F. B. R., and Samed, M. M. A.: AHP method applied to the location problem of an assistance central for emergency situations in the state of Paraná, Rev. Tecnol., 24, 131–141, doi:10.4025/revtecnol.v24i1.27410, 2015.

21. Edjossan-Sossou, A. M., Deck, O., Al Heib, M., and Verdel, T.: A decision-support methodology for assessing the sustainability of natural risk management strategies in urban areas, Nat. Hazards Earth Syst. Sci., 14, 3207–3230, doi:10.5194/nhess-14- 3207-2014, 2014.

22. Estévez, R. A. and Gelcich, S.: Participative multi-criteria decision analysis in marine management and conservation: Research progress and the challenge of integrating value judgments and uncertainty, Mar. Policy, 61, 1–7, doi:10.1016/j.marpol.2015.06.022, 2015.

23. Evers, M., Jonoski, A., Maksimovic,ˇ C., Lange, L., Ochoa Ro- ˇ driguez, S., Teklesadik, A., Cortes Arevalo, J., Almoradie, A., Eduardo Simões, N., Wang, L., and Makropoulos, C.: Collaborative modelling for active involvement of stakeholders in urban flood risk management, Nat. Hazards Earth Syst. Sci., 12, 2821–2842, doi:10.5194/nhess-12-2821-2012, 2012

24. Fernández, D. S. and Lutz, M. A.: Urban flood hazard zoning in Tucumán Province, Argentina, using GIS and multicriteria decision analysis, Eng. Geol., 111, 90–98, doi:10.1016/j.enggeo.2009.12.006, 2010

25. Ghanbarpour, M. R., Salimi, S., and Hipel, K. W.: A comparative evaluation of flood mitigation alternatives using GIS-based river hydraulics modelling and multicriteria decision analysis, J. Flood Risk Manage., 6, 319–331, doi:10.1111/jfr3.12017, 2013.

26. Giupponi, C., Giove, S., and Giannini, V.: A dynamic assessment tool for exploring and communicating vulnerability to floods and climate change, Environ. Model. Softw., 44, 136–147, doi:10.1016/j.envsoft.2012.05.004, 2013.

27. Godfrey, A., Ciurean, R. L., van Westen, C. J., Kingma, N. C., and Glade, T.: Assessing vulnerability of buildings to hydrometeorological hazards using an expert based approach – An application in Nehoiu Valley, Romania, Int. J. Disaster Risk Reduct., 13, 229–241, doi:10.1016/j.ijdrr.2015.06.001, 2015.

28. Guitouni, A. and Martel, J.-M.: Tentative guidelines to help choosing an appropriate MCDA method, Eur. J. Oper. Res., 109, 501– 521, doi:10.1016/S0377-2217(98)00073-3, 1998.

29. Guo, E., Zhang, J., Ren, X., Zhang, Q., and Sun, Z.: Integrated risk assessment of flood disaster based on improved set pair analysis and the variable fuzzy set theory in central Liaoning Province, China, Nat. Hazards, 74, 947–965, doi:10.1007/s11069-014- 1238-9, 2014.

30. Hajkowicz, S. and Collins, K.: A review of multiple criteria analysis for water resource planning and management, Water Resour. Manage., 21, 1553–1566, doi:10.1007/s11269-006-9112-5, 2007.

31. Haque, A. N., Grafakos, S., and Huijsman, M.: Participatory integrated assessment of flood protection measures for climate adaptation in Dhaka, Environ. Urban., 24, 197–213, doi:10.1177/0956247811433538, 2012.

32. Hsu, C. and Sandford, B.: The delphi technique: making sense of consensus, Pract. Assess. Res. Eval., 12, 1–8, doi:10.1016/S0169-2070(99)00018-7, 2007.

33. Jato-Espino, D., Castillo-Lopez, E., Rodriguez-Hernandez, J., and Canteras-Jordana, J. C.: A review of application of multi-criteria decision making methods in construction, Autom. Constr., 45, 151–162, doi:10.1016/j.autcon.2014.05.013, 2014.

34. Jun, K. S., Chung, E. S., Sung, J. Y., and Lee, K. S.: Development of spatial water resources vulnerability index considering climate change impacts, Sci. Total Environ., 409, 5228–5242, doi:10.1016/j.scitotenv.2011.08.027, 2011.

35. Kangas, J., Kangas, A., Leskinen, P., and Pykäläinen, J.: MCDM methods in strategic planning of forestry on state-owned lands in Finland: applications and experiences, J. Multi-Crit. Decis. Anal., 10, 257–271, doi:10.1002/mcda.306, 2001.

36. Kienberger, S., Lang, S., and Zeil, P.: Spatial vulnerability units – expert-based spatial modelling of socio-economic vulnerability in the Salzach catchment, Austria, Nat. Hazards Earth Syst. Sci., 9, 767–778, doi:10.5194/nhess-9-767-2009, 2009.

37. Lee, G., Jun, K. S., and Chung, E. S.: Integrated multi-criteria flood vulnerability approach using fuzzy TOPSIS and Delphi technique, Nat. Hazards Earth Syst. Sci., 13, 1293–1312, doi:10.5194/nhess-13-1293-2013, 2013.

38. Lee, G., Jun, K. S., and Chung, E.: Robust spatial flood vulnerability assessment for Han River using fuzzy TOPSIS with α-cut level set, Expert Syst. Appl., 41, 644–654, doi:10.1016/j.eswa.2013.07.089, 2014.

39. Li, F., Li, Z.-K., and Yang, C.-B.: Risk Assessment of Levee Engineering Based on Triangular Fuzzy Number and Analytic Network Process and Its Application, in Modeling Risk Management in Sustainable Construction, edited by: Wu, D. D., SpringerVerlag, Heidelberg, Berlin, 415–426, 2011.

40. Li, Q.: Fuzzy approach to analysis of flood risk based on variable fuzzy sets and improved information diffusion methods, Nat. Hazards Earth Syst. Sci., 13, 239–249, doi:10.5194/nhess-13- 239-2013, 2013.

41. Magalhães, I. A. L., Thiago, C. R. L., Agrizzi, D. V., and Santos, A. R.: Geotechnology for use mapping of flood risk areas in Guaçuí, ES: a comparative analysis between two methods, Cad. Geociências, 8, 63–70 available at: http://www.portalseer.ufba.br/index. php/cadgeoc/article/view/5288 (last access: 24 April 2016), 2011.

42. Malekian, A. and Azarnivand, A.: Application of Integrated Shannon's Entropy and VIKOR Techniques in Prioritization of Flood Risk in the Shemshak Watershed, Iran, Water Resour. Manage., 30, 409–425, doi:10.1007/s11269-015-1169-6, 2015.

43. Marttunen, M., Mustajoki, J., Dufva, M., and Karjalainen, T. P.: How to design and realize participation of stakeholders in MCDA processes? A framework for selecting an appropriate approach, EURO J. Decis. Process., 3, 187–214, doi:10.1007/s40070-013- 0016-3, 2013.

44. Merz, B., Kreibich, H., and Apel, H.: Flood risk analysis: uncertainties and validation, Osterr. Wasser Abfallwirt., 60, 89–94, doi:10.1007/s00506-008-0001-4, 2008

45. Miyamoto, M., Osti, R., and Okazumi, T.: Development of an integrated decision-making method for effective flood early warning system, J. Disaster Res., 9, 55–68, 2014.

46. Moffett, A. and Sarkar, S.: Incorporating multiple criteria into the design of conservation area networks: A minireview with recommendations, Divers. Distrib., 12, 125–137, doi:10.1111/j.1366- 9516.2005.00202.x, 2006.

47. Mosadeghi, R., Warnken, J., Tomlinson, R., and Mirfenderesk, H.: Uncertainty analysis in the application of multi-criteria decision-making methods in Australian strategic environmental decisions, J. Environ. Plan. Manage., 56, 1097–1124, doi:10.1080/09640568.2012.717886, 2013.

48. Pickering, C. M. and Byrne, J.: The benefits of publishing systematic quantitative literature reviews for PhD candidates and other early career researchers, High. Educ. Res. Dev., 33, 534–548, doi:10.1080/07294360.2013.841651, 2014.

49. Pohekar, S. D. and Ramachandran, M.: Application of multicriteria decision making to sustainable energy planning – A review, Renew. Sustain. Energy Rev., 8, 365–381, doi:10.1016/j.rser.2003.12.007, 2004.

50. Qi, H., Qi, P., and Altinakar, M. S.: GIS-based spatial Monte Carlo Analysis for integrated flood management with two dimensional flood simulation, Water Resour. Manage., 27, 3631–3645, doi:10.1007/s11269-013-0370-8, 2013.

51. Radmehr, A. and Araghinejad, S.: Developing strategies for urban flood management of Tehran city using SMCDM and ANN, J. Comput. Civ. Eng., 28, 05014006, doi:10.1061/(ASCE)CP.1943- 5487.0000360, 2014.

52. Rahman, M. R. and Saha, S. K.: Flood hazard zonation – A GIS aided multi criteria evaluation (MCE) approach with remotely sensed data, Int. J. Geoinf., 3, 25–37, 2007.

53. Sahin, O., Mohamed, S., Warnken, J., and Rahman, A.: Assessment of sea-level rise adaptation options: Multiple-criteria decisionmaking approach involving stakeholders, Struct. Surv., 31, 283– 300, doi:10.1108/SS-01-2013-0006, 2013.

54. Santangelo, N., Santo, A., Di Crescenzo, G., Foscari, G., Liuzza, V., Sciarrotta, S., and Scorpio, V.: Flood susceptibility assessment in a highly urbanized alluvial fan: The case study of Sala Consilina (southern Italy), Nat. Hazards Earth Syst. Sci., 11, 2765–2780, doi:10.5194/nhess-11-2765-2011, 2011.

55. Schuwirth, N., Reichert, P., and Lienert, J.: Methodological aspects of multi-criteria decision analysis for policy support: A case study on pharmaceutical removal from hospital wastewater, Eur. J. Oper. Res., 220, 472–483, doi:10.1016/j.ejor.2012.01.055, 2012.

56. Shan, S., Wang, L., Li, L., and Chen, Y.: An emergency response decision support system framework for application in e-government, Inf. Technol. Manage., 13, 411–427, doi:10.1007/s10799-012-0130-0, 2012.

57. Taib, C. M. I. C., Yusoff, B., Abdullah, M. L., and Wahab, A. F.: Conflicting bifuzzy multi-attribute group decision making model with application to flood control Project, Gr. Decis. Negot., 25, 157–180, doi:10.1007/s10726-015-9437-7, 2016.

58. Uddin, K., Gurung, D. R., Giriraj, A., and Shrestha, B.: Application of remote sensing and GIS for flood hazard management: a case study from Sindh Province, Pakistan, Am. J. Geogr. Inf. Syst., 2, 1–5, doi:10.5923/j.ajgis.20130201.01, 2013.

59. Vafaei, F. and Harati, A. N.: Strategic management in decision support system for coastal flood management, Int. J. Environ. Res., 4, 169–176, doi:10.1016/j.jconhyd.2010.08.009, 2010.

60. Wallenius, J., Dyer, J. S., Fishburn, P. C., Steuer, R. E., Zionts, S., and Deb, K.: Multiple Criteria Decision Making, Multiattribute Utility Theory: Recent Accomplishments and What Lies Ahead, Manage. Sci., 54, 1336–1349, doi:10.1287/mnsc.1070.0838, 2008.

61. Wu, Y., Zhong, P., Zhang, Y., Xu, B., Ma, B., and Yan, K.: Integrated flood risk assessment and zonation method: a case study in Huaihe River basin, China, Nat. Hazards, 78, 635–651, doi:10.1007/s11069-015-1737-3, 2015.

62. Yan, H. Bin, Huynh, V. N., and Nakamori, Y.: A probabilistic model for linguistic multi-expert decision making involving semantic overlapping, Expert Syst. Appl., 38, 8901–8912, doi:10.1016/j.eswa.2011.01.105, 2011.

63. Yang, X., Zhou, J., Ding, J., Zou, Q., and Zhang, Y.: A fuzzy AHPTFN based evaluation model of flood risk analysis, J. Comput. Inf. Syst., 8, 9281–9289, 2012.

64. Zhou, Z., Wang, X., Sun, R., Ao, X., Sun, X., and Song, M.: Study of the comprehensive risk analysis of dam-break flooding based on the numerical simulation of flood routing. Part II: model application and results, Nat. Hazards, 73, 675–700, doi:10.1007/s11069-013-1029-8, 2014.

65. Zou, Q., Zhou, J., Zhou, C., Song, L., and Guo, J.: Comprehensive flood risk assessment based on set pair analysis-variable fuzzy sets model and fuzzy AHP, Stoch. Environ. Res. Risk A., 27, 525–546, doi:10.1007/s00477-012-0598-5, 2013.

CHAPTER 6

REVIEW ARTICLE: A REVIEW AND CRITICAL ANALYSIS OF THE EFFORTS TOWARDS URBAN FLOOD RISK MANAGEMENT IN THE LAGOS REGION OF NIGERIA

U. C. Nkwunonwo[1,2], M. Whitworth[2], and B. Baily[3]

[1]Department of Geoinformatics and Surveying, University of Nigeria, Enugu Campus, Nigeria

[2]School of Earth and Environmental Sciences, University of Portsmouth, Portsmouth, UK

[3]Department of Geography, University of Portsmouth, Portsmouth, UK

ABSTRACT

Urban flooding has been and will continue to be a significant problem for many cities across the developed and developing world. Crucial to the amelioration of the effects of these floods is the need to formulate a sound flood management policy, which is driven by knowledge of the frequency and magnitude of impacts of these floods. Within the area of flood research, attempts are being made to gain a better understanding of the causes, impacts, and pattern of urban flooding. According to the United Nations office for disaster reduction (UNISDR), flood risk is conceptualized on the basis of three integral components which are frequently adopted during flood damage estimation. These components are: probability of flood hazard, the level of exposure, and vulnerabilities of elements at risk. Reducing the severity of each of these components is the objective of flood risk management under the UNISDR guideline and idea of "living with floods". On the basis of this framework, the present research reviews flood risk within the Lagos area of Nigeria over the period 1968–2012. During this period, floods have caused harm to millions of people physically, emotionally, and economically. Arguably over this period the efforts of stakeholders to address the challenges appear to have been limited by, amongst other things, a lack of reliable data, a lack of awareness amongst the population affected, and a lack of knowledge of flood risk mitigation. It is the aim of this research to assess the current understanding of flood risk and management in Lagos and to offer recommendations towards future guidance.

INTRODUCTION

Flood events and impacts in recent times have arguably been unprecedented and affected the lives of hundreds of millions of people across the world. These impacts have been shared by both developed and developing countries (DCs) with rapid urban expansion taking place in many flood-prone areas. Concerns for flooding and the associated human impacts are clearly of global significance, especially when allied with the fears of climatic change and associated changes in rainfall events and sea level rise (Kundzewicz et al. 2014). The rapidly growing urban environments in many areas correspond with a lack of urban planning strategies, the deterioration and lack of capacity of urban drainage infrastructure and an increased rate of development on floodplains (Gill 2004; CII, 2001). Additionally, the increasing densities of populations

Citation: Nkwunonwo, U., Whitworth, M., & Baily, B. (2016). Review article: A review and critical analysis of the efforts towards urban flood risk management in the Lagos region of Nigeria. Nat. Hazards Earth Syst. Sci., 16(2), 349-369. http://dx.doi.org/10.5194/nhess-16-349-2016.

(particularly in the urban areas of most DCs such as Nigeria), alongside the poor level of awareness and the limited efforts of many stakeholders towards flood risk reduction are critical issues undermining possible efforts towards addressing the hazard (Action Aid, 2006; McMichael et al., 2006; Raaijmakers et al., 2008). The present research attempts to clarify these issues using a synthesis and analysis of available historical flood data in the Lagos area of Nigeria, spanning the period 1968 to 2012. This research also reviews the current status and awareness of pluvial flood risk within Lagos and makes recommendations to help improve the management of these events.

DESCRIPTION OF THE LAGOS METROPOLIS OF NIGERIA

The Lagos metropolis is a densely populated low-lying coastal area in the southwestern part of Nigeria, West Africa. The city is located within geographical coordinates of 3.1 to 3.4° E longitude and 6.5 to 6.8° N latitude and covers a land area of approximately 1100 km^2 (or 425 sq. miles). It is bordered in the south by the Atlantic Ocean (see Fig. 1). With a dense network of roads and buildings, and several inland waterways including the Lagos Lagoon which empties into the Atlantic, the conurbation serves as a major hub for transportation, tourism, and economic activities in Nigeria. With a population of over 20 million people (LSG, 2012), the Lagos metropolis is the biggest city in Nigeria, (although the smallest land area), the second largest city in Africa, and the seventh largest city in the world. The population growth rate in the Lagos metropolis is estimated at 3.2 % (World Bank, 2013). Against this background, the United Nations predicts that Nigeria will be one of the eight countries expected to account collectively for half of the total population increase in the world from 2005 to 2050, and will by 2100, have a population of between 505 million and 1.03 billion (United Nations, 2004). High population density is a major impasse in the Lagos region, subjecting the area to lack of space for the myriad of human activities, which often manifests itself in muddled human settlements, overcrowding, slum envelopments, pollution, illegal structures, and other social and environmental disorder. These factors, either singularly or in combination, have the potential to increase exposure to hazard, especially the risks associated with pluvial urban flooding events.

FLOODING IN LAGOS – FREQUENCY, CAUSES, AND IMPACTS

Flooding and flood risk management are issues of grave significance in Lagos (Aderogba, 2012a, b). It is clear from previous studies (for example, Ajibade et al. 2013, 2014; Adelekan, 2013), that flooding in the area can be devastating, affecting hundreds of thousands of people and causing considerable economic damage. A typical example is the July 2011 flooding event, which affected approximately 5 thousand people and resulted in about 25 deaths. The direct economic losses resulting from the event totalled about 50 billion Nigerian naira (i.e. USD 250 million). Public utilities including road networks, bridges, and schools were destroyed. In addition, houses collapsed, private homes were submerged, and several cars were swept away by flood water (IFRC, 2011; Oladunjoye, 2011).

Frequency of Occurrence

From a historical perspective, Lagos has always been susceptible to various types of flooding which are well documented from the 1960s onwards (Odunuga, 2008; Oyebande, 1974; Etuonovbe, 2011). However, in recent years pluvial flooding events (rainfall-related), have arguably been more widespread (Olajuyigbe et al., 2012). With the exception of 1973, the drought year, flooding in Lagos has occurred annually (usually between July and October rainy season) with an increasing intensity and an increased severity of impacts from 1960 onwards (Oyebande, 1974). According to FME (2012), Lagos is one of the few locations in Nigeria with more frequent flooding events (see Fig. 2). A number of floods have occurred in the Lagos area, although keeping track of events in the Nigerian context is challenging due partly to lack of relevant data collection capacities. As a result, data relating to hydrodynamics and historical flooding events are often lacking (Ajibade et al., 2013). According to previous studies, more severe flooding has been recorded in select areas of Lagos including Lagos Island, Apapa, Ikeja, Mushin, Surulere, and parts of Ikorodu (Oyebande, 1974; Odunuga, 2008). There is no clear explanation for this peculiar circumstance, but it seems to follow a pattern of spatial and temporal distribution of rainfall within and around Lagos. Table 1 shows a summary of major flooding events and associated consequences in the Lagos metropolis of Nigeria from 1968 to 2012. These data which appear to represent generalized flooding situations were obtained from a wide range of sources including EM-DAT and the Nigerian FME (Federal Ministry of Environment). It is argued that the conclusions that can be drawn about flooding in Lagos from these data sets relate to events of higher magnitudes and return periods (Guha-Sapir et al., 2013). Only journalistic and non-quantitative evidence is available for lesser impacts and more frequent flooding events (see for example, IFRC, 2012). The problem with these forms of evidence is that they often do not have ethical and empirical groundings. For most of the events considered, data relating to flood duration and impacts in terms of number of people displaced, mortality, and economic losses were not available. On the basis of this inconsistency, the effectiveness of flood management policies is being queried (Adelekan, 2015). In many cases, different types of flood damage were aggregated. Indeed, this situation adversely affects accurate flood damage estimation since a critical understanding of Lagos flood risk in the context of flood damage typology is difficult.

To qualitatively identify risk levels in this area, an approach used in the 2005 World Bank Hotspot project was adopted (Dilley et al., 2005). In this approach, records of flood events and affected areas were coupled with the population density of the local enumeration areas in Lagos and mapped in a GIS. The result of this analysis as shown in Fig. 3, indicates that Ajeromi-Ifelodun and Mushin areas are at a higher risk of flooding than other areas. This approach, regardless of its simplicity, offers a potentially valuable insight into flooding patterns in Lagos. However, much uncertainty lies within the results and this suggests the need for more detailed research that will investigate the flood risk levels at Lagos using a more detailed quantitative or qualitative data set.

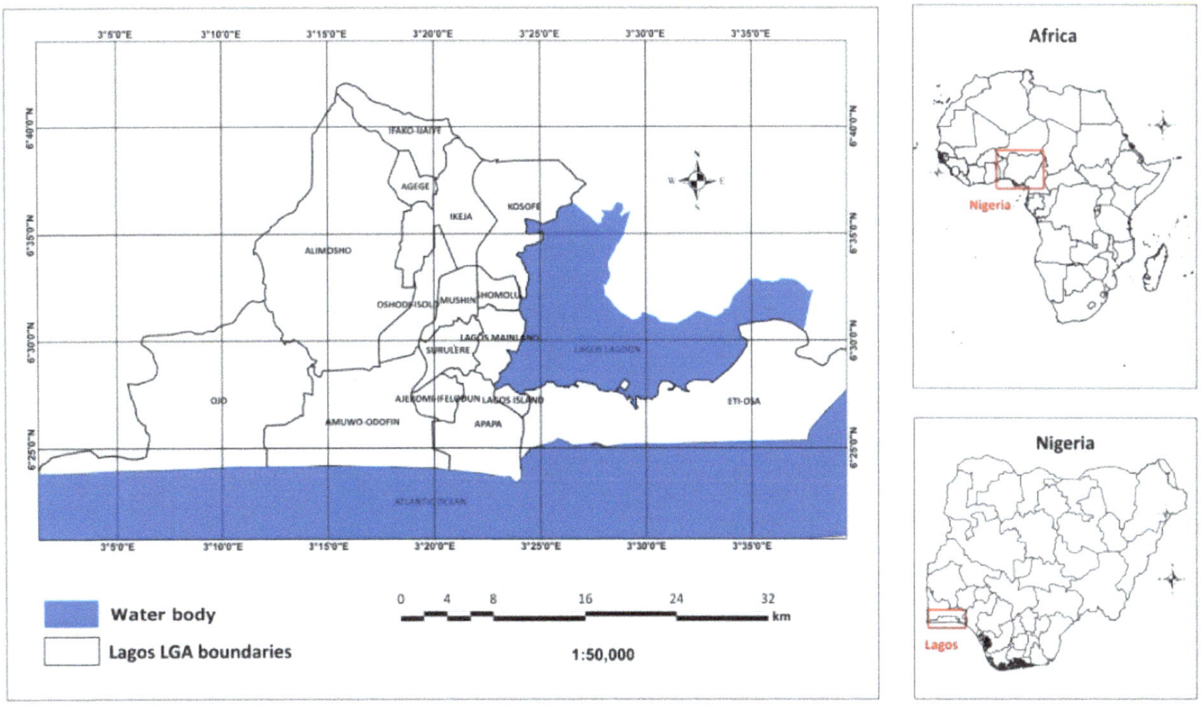

Figure 1. The Lagos metropolis of Nigeria. Inset shows the location of Lagos State in Nigeria.

Causes of Flooding

Over the past 2 decades, the causes of flooding in Lagos have received significant attention in the literature (Ayoade and Akintola, 1980; Action Aid, 2006; Adeloye and Rustum, 2011; Adelekan, 2013; Aderogba, 2012a; Oshodi, 2013; Ajibade et al., 2013, 2014; Soneye, 2014). Debates arising from the literature indicate that Lagos floods are mainly the consequences of climate-change-induced short-durationhigh-intensity or long-duration-low-intensity rainfall (Houston et al., 2011). This is unsurprising considering that climate change has arguably influenced regional precipitation patterns in recent history. Odjugo (2006) concluded that there are now more high-intensity short-duration rainfall events and more low-intensity long-duration rainfall events than there were 3 decades ago. Other factors have also been investigated with reference to the causes of these floods in Lagos. These include the topography of the area, land use (LU) and land cover (LC) modifications, and influence of canals, lagoons, and beaches (Aderogba, 2012a; Aderogba et al., 2012; Odunuga, 2008). Other factors considered are urbanization and population growth, poor urban planning, and poor environmental management and the indiscriminate disposal of solid waste (Lamond et al., 2012; Adeloye and Rustum, 2011). It is also suggested that tidal and co-tidal influences and frequent incursion from the Atlantic into the lowlands during heavy storms also play important roles (Ojinnaka, 2013).

These factors (schematized in Fig. 4), seem to influence the occurrence of the hazard and the exposure of elements at risk. However, in relation to the vulnerabilities of social systems to flooding in the area, the development of slum settlements and poor perception of flooding among local communities, urban residents and the general public are critical factors (Agbola and Agunbiade, 2007; Nkwunonwo, 2013; Ayoade and Akintola, 1980; Odunuga et al., 2012; Oloke et al., 2013; BNRCC, 2008).

Impacts of Flooding

To date, flooding in Lagos is characterized by severe consequences, which raise concerns about a lack of early warning and evacuation systems. Impacts from flooding (as illustrated in Fig. 5) are compounded by population density and the upward trends of urban growth in the area (Aderogba, 2012b). Also, flood water depth, inundation extent and duration as well as water flow velocity play contributory roles. The general impacts (such as displacement from homes, mortality, physical injuries, disruption of economic activities, destruction of urban infrastructure, and submergence of buildings) that relate to social systems directly have been extensively considered in the literature (Ugwu and Ugwu, 2013; Adigun et al., 2013; Ajibola et al., 2012; Aderogba, 2012b; Olajuyigbe et al., 2012). However, there are reports that Lagos flooding causes severe additional impacts including the loss of social values, spread of vector-borne diseases, as well as air and water pollution (Adelekan, 2010; Olajuyigbe et al., 2012; Bashir et al., 2012).

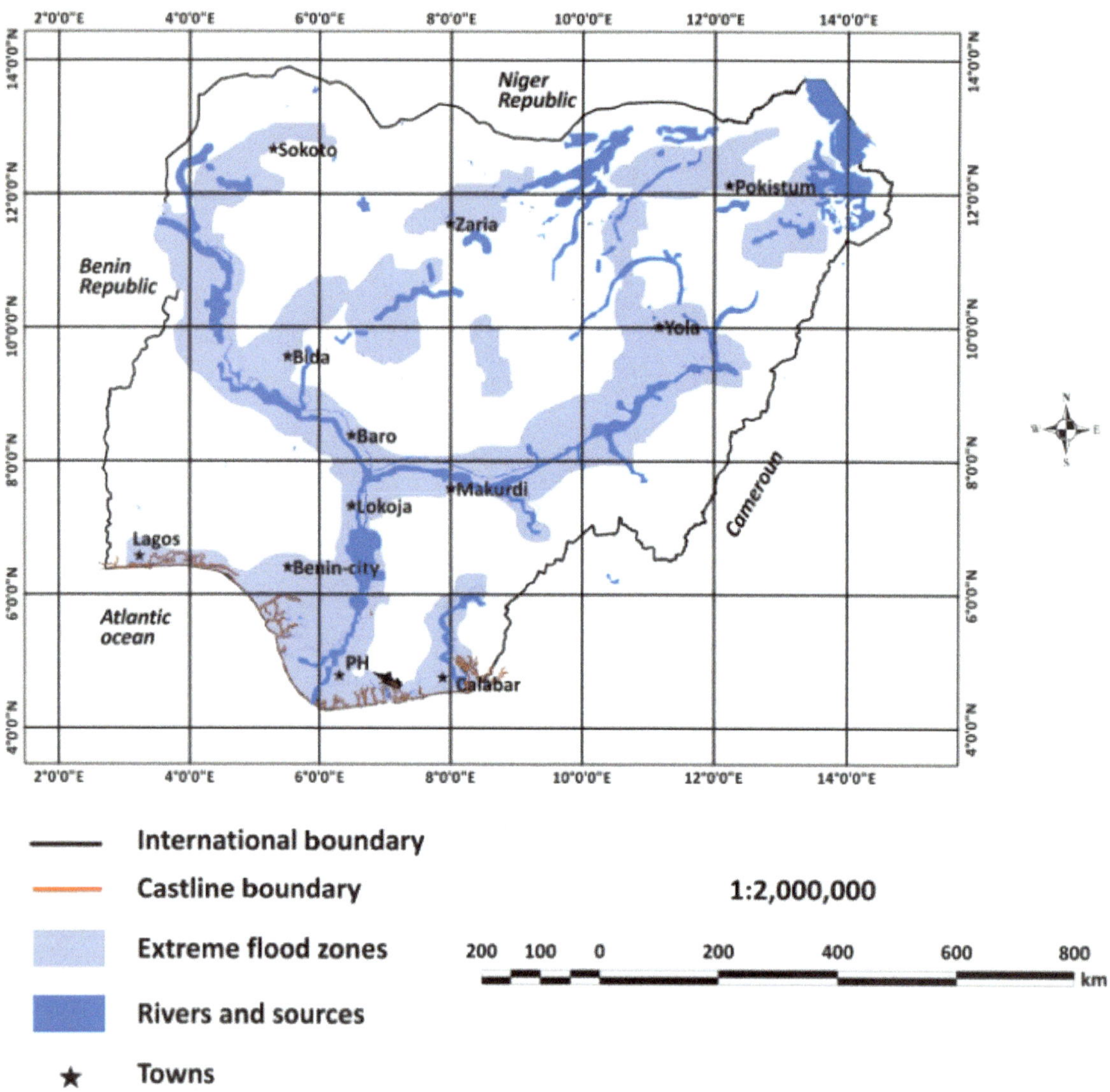

Figure 2. Spatial distribution of areas affected by extreme floods in Nigeria between 2000 and 2012. Adapted from the Federal Ministry of Environment (2012).

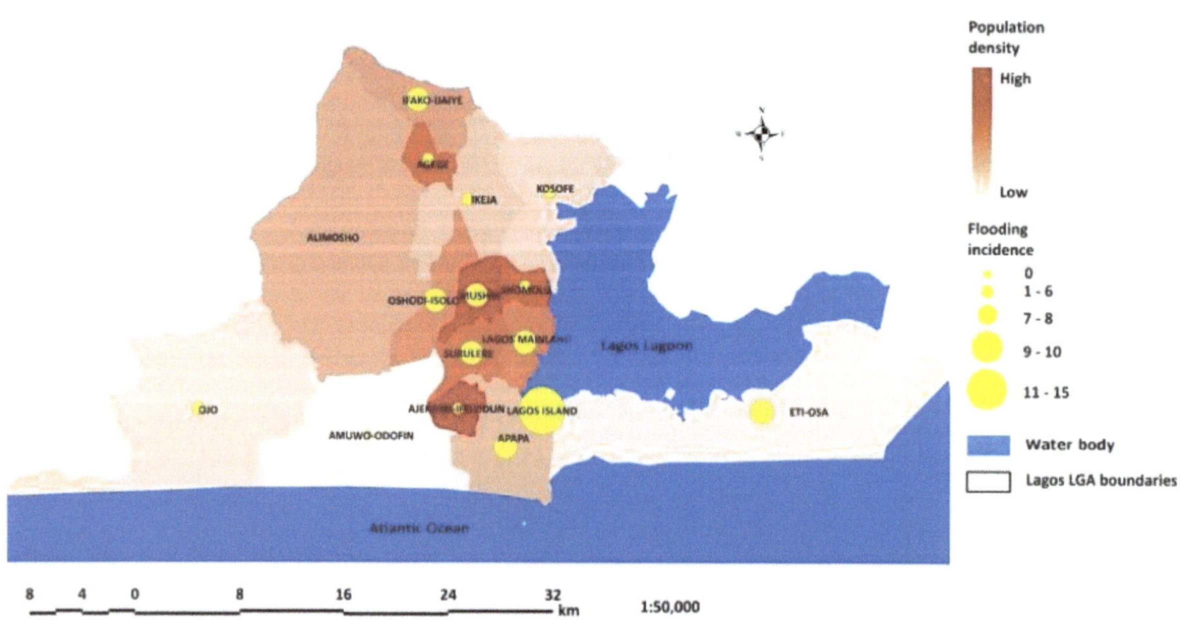

Figure 3. Flood risk levels in the Lagos area qualitatively determined by coupling population density with list of flooding events and locations.

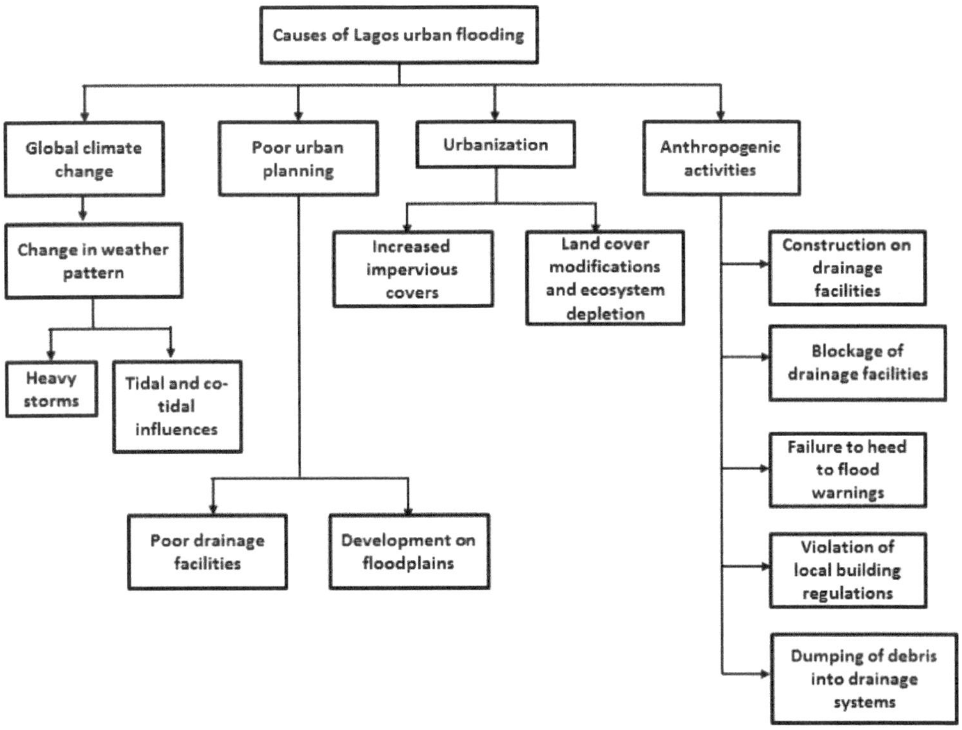

Figure 4. Main causes of urban flooding in the Lagos area of Nigeria showing global climate change, poor urban planning, urbanization, and anthropogenic activities.

Olajuyigbe et al. (2012) report that flood hazard increases city-wide poverty as a result of the farmlands which are destroyed and essential services which are often interrupted. Adelekan (2010) investigated these impacts using four poor urban communities in Lagos as case studies and identified three significant scales: individual, household, and community. At the individual scale, the reluctance of friends and family to visit one another while in flooded houses affects social relationships. This has broad adverse implications on community lifestyle and further compounds depression among flood victims in Lagos. Food insecurity is equally an important issue at this scale as food items stored in individual homes are often lost during flooding. In addition, there can be numerous health impacts including chronic skin infections from exposure to contaminated environmental systems and increased effects on those with an already poor health history.

Household and community scales of impacts are mainly indicated by the secondary effects of flooding in Lagos. Household impacts include deterioration of building quality, intrusion of contaminated water into apartments, lack of good drinking water, and loss or damage to household properties including sanitation facilities. The community impacts include an unclean environment, disruption of movement, and damage to public utilities. Urgent needs arise where community schools were flooded and schooling for children has been interrupted. This is an important issue within the context of human development. In many other DCs where it is also applicable, community leaders and the local authorities have often instigated measures to ensure that children's schooling is not interrupted despite the magnitude of flooding. In Bangladesh for example, a strategy known as "floating schools" in which classrooms are constructed on boats is being put in place during flooding (Huq and Aslam, 2003). This enables provision of uninterrupted education for children who have been impacted and whose education has been disrupted by flood catastrophes.

The impacts of flooding in Lagos also trigger concerns for environmental management, urban sustainability, and the development, governance, and the vulnerability of urban residents and local communities. Other factors of concern are humanitarian needs and services, especially primary health delivery (Soneye, 2014; Ajibade, 2013; Lamond et al., 2012). Needless to say, concerns for solid waste management are crucial as long as the indiscriminate dumping of wastes into drainage systems remains prevalent within Lagos. One example of this is the water sold in polythene sachets which is the major source of drinking water for residents. It is perceived that many residents dispose of the containers which end up in drainage facilities. Being a non-degradable waste, it accumulates over time and blocks these drainage facilities. Unfortunately, little has been done to address such an important issue in flood control management.

Figure 5. Some flooding scenes examples in the Lagos metropolis of Nigeria: (a) living room submerged by flood water, (b) residential building submerged, (c) local community affected by flood waters, and (d) expressway overwhelmed by flood water. Source: Online images of flooding in Lagos, Nigeria.

PRESENT RESEARCH AND GAPS IN KNOWLEDGE

Despite the frequency of occurrence, causes, and severity of impacts of Lagos urban flooding alongside the understanding of a potentially increased risk in the future a better understanding of flooding in Lagos is lacking. Whilst most of the data relating to these impacts are at local levels and limited to small geographical areas, wide-ranging crossdisciplinary intervention measures are constrained. Lack of data and poor data scale present an important limitation to comprehensive urban flood impact assessment and management. Flood risk management efforts in Lagos considered in the light of global current flood risk management practices (see Sect. 5) reveals significant gaps. A number of nonstructural approaches which promote flood risk assessment and management are frequently ignored. Efforts to manage and assess flood risk in Lagos are aimed at both prevention and control of flooding. Moreover, such efforts have been unprecedented in Nigeria and clearly demonstrate a practical commitment to fighting the flood hazard (Njoku and Udeagha, 2013; Obeta, 2014). However, results from various studies which highlighted the increasing vulnerabilities to flooding indicate that such efforts have so far been of little assistance to the victims of flooding.

The problems of a lack of data and poor data scale in Lagos are exacerbated by the general lack of funds and inadequate access to improved technology alongside a lack of political will (Nkwunonwo et al., 2014; Adeloye and Rustum, 2011). Unfortunately, the level of existing knowledge regarding the state of affairs is unsatisfactory and fails to assist in providing a potential solution to ways of reducing the hazard or its impacts on human populations. The more critical and disturbing scenario is that the Lagos area is a fast-growing city within which a great deal of the population currently lives within areas prone to flooding. Locally, the majority of these areas are slums which provide provisional dwelling places to poor urban residents. Adelekan (2010) argued that these residents often do not have the capacity to cope with flood hazard or the ability to quickly recover from losses. Moreover, the vast spatial distribution of such flood-prone areas in Lagos increases the exposure of the human population to flooding. A clearer picture taken from a global and regional perspective shows that Lagos is among the top 20 cities with increasing numbers of the present and future population exposed to flooding (see Table 1). Within such an environment, destruction of biodiversity and depletion of ecosystem functions to earn a livelihood are common practices. Given the issues of coping capacity and resilience which remain unresolved for the burgeoning vulnerable population, sustainable development under such circumstances is difficult and almost impossible.

Table 1. A summary of major flooding events and associated threats in the Lagos metropolis of Nigeria from 1968 to 2012. Source: EM-DAT (2014), FME (2012) and published works

S/No.	DATE	LGA(S) AFFECTED	DURATION (DAYS)	CAUSE (S)	NO OF PEOPLE DISPLACED	MORTALITY	ECONOMIC LOSS (N)	AFFECTED HOUSES/ OTHERS
1	Oct 2012	Lagos city*	Many days, unspecified	Heavy rain	Thousands	>50	Millions, unspecified	Many*, including interruption of traffic and other activities
2	Jul 2011	Lagos island, Mainland, Mushin	2 days	Heavy rain	10 000	100	Millions, unspecified	Many*
3	Oct 2010	Lagos island, Apapa, Kosofe,	Many days, unspecified	Heavy rain	Thousands	20	Millions, unspecified	Many* including interruption of traffic and other activities
4	Jul 2009	Lagos city*	Many days	Heavy rain	Many	Nil	Millions, unspecified	Many*
5	Oct 2008	Lagos city*	N/A	Heavy rain	Not specified	No data	Millions, unspecified	Many* including interruption of traffic and other activities
6	Aug 2007	Ikorodu, Kosofe and Abeokuta	15	Heavy rain	5000	17	Millions, unspecified	5000
7	Jul 2005	Lagos city	5	Heavy storm	3000	25	Millions	N/A
8	Jun 2004	Lagos city	2	Heavy rain	1000	Nil	Millions	Drainages
9	Jul 2002	Lagos city	3	Heavy rain	200	2	Millions	Many*
10	Jun, Jul, Sep 2000	Victoria Island and Ikoyi	2	Brief Torrential Rain	500	Nil	Millions, unspecified	Tens of thousands
11	May, Jun, Jul 1999	Mushin and Idiaraba	N/A				70 000 000	
12	Jul 1990	Lagos city	2	Heavy rain	3000	5	Thousands	Many*, not specified
13	Jul 1990	Lagos city	2	Heavy rain	500	Nil	N/A	Hundreds of inhabitants
14	Jun 1974	Idiaraba, Ikorodu, Surulere, and Yaba	Many days, unspecified	Heavy rain	Thousands	Nil	N/A	
15	Jun 1972	Lagos Island	N/A	Heavy rainfall	Not specified	Nil	N/A	Traffic was disrupted, few houses
16	Jul 1971	Lagos Island	5	Heavy rainfall	Not specified	Nil	N/A	Traffic was disrupted, few houses
17	Jul 1970	Lagos Island	N/A	Winds, accompanied by short-duration, high-intensity rain	Nil	Nil	5000	Few
18	Jun 1969	Surulere and Yaba	10	Short-duration, high-intensity rain	Nil	Nil	N/A	Many*, not specified
19	Jun 1968	Lagos Island and Ijora	N/A	Heavy storm	Nil	Nil	6000	Traffic was disrupted, Few houses

* Grouped instead of treating as separate variables due to lack of data. N/A: Not available. Nil: No information.

The present research attempts to address the challenges of flooding within and around Lagos through a review of literature and flood information covering the hazard in the area and how it has so far been managed. Available historical flood data in the area, spanning the period 1968 to 2012, were synthesized and analysed. These data were obtained from various online databases (for example: NEST, 1991; Guha-Sapir et al., 2013) and published data on historical flooding in Lagos, Nigeria. Sourcing data from a wide range of databases is an attempt to clarify the issue of lack of flood data which creates significant gaps in knowledge relating to flood management efforts in Lagos. The research discusses the challenges faced by Lagos in managing and reducing flooding impacts. It contextualizes the current situation and puts forward relevant recommendations for more effective remedies to alleviate the threats of flooding in the area. Ultimately, it is argued that the lack of more robust techniques such as flood modelling and assessment of vulnerability accounts for a stark limitation in the efforts towards addressing the challenges of flooding in the area. In particular, the importance of flood modelling in flood risk reduction and the need for it to be included in the present and future efforts at reducing the impacts of flooding are emphasized. This research in general and the recommendations in particular are driven by three key aims: firstly, to understand the unique situation which exists in Lagos in relation to flooding; secondly, to align the focus of flood risk reduction in the Lagos area with that of more developed countries such as the US, the Netherlands, and the United Kingdom; and finally, to suggest measures for possible improvements.

FLOOD RISK MANAGEMENT: THE GENERAL CONCEPT

In recent times, a shift from flood hazard control to flood risk reduction has been found necessary (Samuels et al., 2006). This is due to the inexorable nature of the hazard and the general perception that climate change, demographic, and economic activities would influence regional and global occurrences and consequences of flooding in the future (Kundzewicz et al., 2010, 2014). The importance of this new framework lies in the way in which risk is contextualized, which enables strategic and adaptable policies to be easily formulated.

There are various interpretations of the concept of risk in the literature (see for example Renn, 2008) and this complicates a cross-disciplinary approach to risk reduction. An agreed perception of risk is still lacking despite decades of research. In an attempt to address this conceptual conflict, the United Nations in 1992 proposed a seminal definition: "risk is the expected loss (of lives, persons injured, property damaged, and economic activity disrupted) due to a particular hazard for a given area and reference period". In addition, the United Nations International Strategy for Disaster Reduction defined risk as the combination of probability of a hazard and its negative consequences (UNISDR, 2004). Within these frameworks, hazard-specific and place-based concepts of risk are highlighted (see for example Penning-Rowsell and Chatterton, 1997).

Table 2. Top 20 countries ranked in terms of population exposed to coastal flooding in the 2070s, including both climate change and socioeconomic change and showing present-day exposure. (Source: Nicholls et al., 2008; OECD, Paris.) (*Highlight of Nigeria is by the authors.)

Rank	Country	Urban agglomeration	Exposed population (current)	Exposed population (future)
1	India	Calcutta	1 929 000	14 014 000
2	India	Mumbai	2 787 000	11 418 000
3	Bangladesh	Dhaka	844 000	11 135 000
4	China	Guangzhou	2 718 000	10 333 000
5	Vietnam	Ho Chi Minh City	1 931 000	9 216 000
6	China	Shanghai	2 353 000	5 451 000
7	Thailand	Bangkok	907 000	5 138 000
8	Myanmar	Rangoon	510 000	4 965 000
9	USA	Miami	2 003 000	4 795 000
10	Vietnam	Hai Phòng	794 000	4 711 000
11	Egypt	Alexandria	1 330 000	4 375 000
12	China	Tianjin	956 000	3 790 000
13	Bangladesh	Khulna	441 000	3 641 000
14	China	Ningbo	299 000	3 305 000
15*	Nigeria	Lagos	357 000	3 229 000*
16	Cote d'ivoire	Abidjan	519 000	3 110 000
17	USA	New York	1 540 000	2 931 000
18	Bangladesh	Chittagong	255 000	2 866 000
19	Japan	Tokyo	1 110 000	2 521 000
20	Indonesia	Jakarta	513 000	2 248 000

Regardless of conceptual disparity, on the basis of hazard, risk is too often hypothesized as the product of hazard and its consequences (Samuels, 2000; Samuels et al., 2006; De Wrachien et al., 2011; Solín and Skubincan, 2013). This concept of risk generally embodies three components which are considered fundamental to various methods used today in flood risk assessment and management and they include: "probability of hazard occurrence", "exposure", and "vulnerabilities" of elements at risk. Global current flood risk management practices or "best practices" in flood risk management can be identified in a wide range of sources (see for example Ashley et al., 2007; Fratini et al., 2012; Sayers et al., 2013) which aim to reduce the likelihood and/

or the consequences of floods. The idea of "living with floods rather than fighting them" is the underlying framework. This philosophy tends towards a policy whereby societies adapt to floods by being prepared and having the right attitude towards damage reduction (van Ogtrop et al., 2005). It stems from three key considerations: (1) the limitations or failures of traditional flood control structural measures; (2) the need for a people-friendly means of tackling flooding; and (3) the goal to lessen all impacts of extreme floods while at the same time exploiting all benefits of ordinary floods (UNISDR, 2004; Di Baldassarre and Uhlenbrook, 2012). These considerations are often perceived in the context of a holistic management cycle which covers prevention, protection, preparedness, emergency response, recovery, and lessons learned (Samuels, 2000). Within this framework, a sound understanding of flooding, accurate and actionable assessment of flood hazard and flood risk, knowledge-based decision making, and strong political leadership are indispensable (UNISDR, 2004). As a result, integrated approaches which combine structural and non-structural measures are extensively applied (EC, 2004; Pitt, 2008; Kazmierczak and Carter, 2010; Sayers et al., 2013).

Structural measures which aim at reducing the occurrence of flood hazard are technically based and involve channelization and the use of natural and man-made barriers to contain waters in rivers and seas. Non-structural measures have the aim of reducing vulnerabilities and developing the capacities of ecological systems to cope with floods through multidisciplinary approaches (Few, 2003; Schanze, 2006; Miceli et al., 2008; Sayers et al., 2013). Non-structural measures include flood risk mapping, land use zoning and planning, flood vulnerability assessment, flood proofing, flood modelling, institutionalization of policies, flood awareness cam paign, flood insurance, flood forecasting, relocation of properties, resettlement of human population, and green infrastructure planning (Merz et al., 2007; Jha et al., 2012; Smith, 2013). Fundamental to these measures is the assessment of flood risk which is a procedure to estimate the damage or negative impacts resulting from flooding. Flood negative impacts are often classified in terms of tangible and intangible damage. Tangible damage such as damage to properties and infrastructure are those that can be associated with monetary values. Intangible damage is difficult to express in monetary terms and may include depression, anxiety, and loss of life. Categorization of flooding impacts as direct and indirect damage is also applicable. Direct damage includes those which result from direct contact with flood water. When direct flood damage is tangible, a broader classification, direct tangible flood damage, applies. Indirect tangible flood damage is a secondary consequence of direct flood impacts and may include economic disruption, individual misfortune, and an increase in water-borne diseases (Samuels et al., 2006; Hammond et al., 2015).

The assessment of flood damage on the basis of damage typology is characterized by significant limitations. By means of heterogeneous methodologies, assessment of direct tangible flood damage has received significant attention in the literature (Hammond et al., 2015; Samuels et al., 2006). Damage functions, especially stage–damage curves which have been applied in many case studies around the world, remain potential tools for assessing direct and indirect tangible flood damage (Smith, 1994; Merz et al., 2004). A major limitation of stage–damage curves arises from the neglect of water flow velocity which also influences flood damage. To date, the difficulty in assessing intangible flood damage remains unresolved. Methods are being proposed in the current literature, for example the concept of "anxiety–productivity–income" which assumes that anxiety is a function of flood water depth and duration (Lekuthai and Vongvisessomjai, 2001; Price and Vojinovic, 2008). The "Risk-to-life" model, which uses flood characteristics and an estimate of the number of people exposed to flooding to assess the possible mortality, was proposed by Jonkman et al. (2008). A comprehensive review of other current methods to assess intangible flood losses can be found in Hammond et al. (2015).

Despite the progress made so far, uncertainties and the lack of time series stage data and good-quality damage data are major factors constraining the ubiquitous assessment of flood risk. In line with predictions of worsened future risk scenarios, these constraints underscore the importance of developing capacities for estimating flood hazard (in terms of water depth, velocity, etc.), exposure, and vulnerabilities of elements at risk as discrete components. The review presented in this research considers various aspects of Lagos flood risk assessment and management in light of the general context of flood risk assessment. With regard to flood risk assessment, procedures to estimate flood hazard are reviewed. Fundamental to the research is knowledge about the extent to which current and future ecosystem exposure to flood hazard have been influenced. Given reported trends in urbanization and population growth (summarized in Sect. 8.1), it can be assumed that exposure of systems to flooding in Lagos is at a critical level.

THE MANAGEMENT AND REDUCTION OF FLOOD RISK IN LAGOS

General Measures

General measures to tackle the challenges of flooding in Lagos have been discussed by Oshodi (2013). Recent practices have included:

- The expansion of drainage infrastructure within the city heartland.
- The annual debris removal from principal drainage facilities within the city heartland.
- Providing advice to the inhabitants of flood plains and wetlands to relocate.
- The demolition of homes in flood-prone areas.
- Proposed resettlement scheme for the residents of Ogun River catchment areas.

Oshodi (2013) claimed that these practices are being carried out, although it could be argued that they are politically influenced. One example of this is the expansion and upgrading of the primary drainage facilities mainly being carried out in Bariga, Surulere, and Gbagada. The annual clearing of primary and secondary channels by Lagos State Government through the Ministry of the Environment is carried out principally in the metropolitan areas. Evacuation and resettlement is carried out for residents who live in flood-prone areas. A proposed resettlement scheme for residents of Ajegunle community near

Ikorodu was undertaken between October 2011 and January 2012 (Oshodi, 2013). The move was necessitated based on the belief that the current location of the community in the Lagos urban Master Plan was originally zoned as wetland for agricultural use. The area was a major catchment of Ogun River. It could be argued that the lack of clear implementation policy for the plan with enormous housing shortfall in the city to cope with rapid population growth and urbanization has led to the conversion of the area into residential use. Significant environmental impacts are often associated with the use of land for purposes it was not originally allotted for. Flooding can be the case when such inappropriate use affects ecological equilibrium. However, a sound investigation is needed to validate this claim, especially within the context of Lagos.

Major failures of these general measures arise from the issues of continuity and the scope of application. These measures are often limited to the core urban areas of the city, excluding the majority of the outskirts (Oshodi, 2013). Many areas within Lagos did not benefit from the expansion of drainage facilities; moreover, the project was not completed because of a change in administration. These projects often suffer when new governments abandon uncompleted projects of their predecessors. This shows the potential for urban mismanagement and the degree of limitation in urban development given that there are many abandoned projects while new ones are being considered. There are occasional inconsistencies in the annual cleaning of the channels. Sometimes this operation is delayed, which allows debris to accumulate in the channels. Coupled with careless attention to the channels after cleaning, accumulation of debris has led to early collapse of the channels. This causes potholes on motorways and retention surfaces for water during flooding.

The problems of where to relocate to and the sparcity of support for relocation are overwhelming. Given the vast distribution of flood-prone areas and the limited financial resources to facilitate resettlement, the choice of which slum locations to resettle is complicated. Communities which cannot be resettled often face the risk of having individual homes demolished. Examples of this include the Agege and Ijeshatedo demolitions in August 2011 and that of Ijora-Badia in 2010, 2012, and 2013. The demolition of homes is understandably a distressing measure to those affected. It could be argued that a particularly controversial aspect is the failure and neglect of the state government to provide any form of alternative housing arrangements to those whose houses have been demolished. There have been a few exceptional cases, for example the Ijora-Badia (World Bank assisted) drainage channels project, which necessitated the demolition and burning of homes. Following a community-led protest the affected families were awarded relocation assistance costs. On the basis of these potential government inadequacies, Ahonshi (2002), Kamunyori (2007), and Basinsksi (2009), cited in Lawanson (2015), argued that the achievement of government's urban sustainability goals (which include general flood management measures) in Lagos are often without regard to the needs of the poor residents of the area. At the same time, demolition and eviction frequently leads to forced social disconnection amongst families, and compensation and legal process on behalf of those affected can take a significant amount of time to complete. This is in part an aspect of social vulnerability which needs to be investigated within the context of Lagos urban flooding. Under the circumstances of eviction and resettlement uncertainties, most people occupying the flood-prone areas seem to ignore government's flood risk mitigation proposals and flood warnings. This influences large-scale impacts of urban flooding in Lagos. However, despite numerous tensions, this issue has only received limited attention in the literature and in governance.

Institutional Efforts

One critical aspect of flood response is the institutional efforts which have been undertaken by local authorities and stakeholders. Odunuga (2008) recognized several flood preventive and curative initiatives ranging from community selfassistance actions to World Bank assisted programmes. Recently, key initiatives, which include the Drain Dock and The Emergency Flood Abatement Gang (EFAG), were launched by the government of Lagos state to improve current efforts towards addressing the challenges of flooding. The ministries of Environment, Works, and Health as well as the Lagos Metropolitan Development and Governance Project (LMDGP) have a number of initiatives aimed at controlling flood hazard in the area, and these include shoreline protection, low carbon emissions, the school advocacy programme, and the climate change club. Lagos is also the first region in Nigeria to carry out a detailed topographic mapping of the area with LiDAR (Light Detection and Ranging) data acquisition and GIS-based analysis aimed at addressing the challenges of flooding. In addition to these efforts, the Nigerian government and international community have been active with measures to address the challenges of flooding at various locations within the country including the Lagos area (Olorunfemi, 2011; NIHSA, 2013). As well as engineering works such as dams, bridges, and sustainable urban drainage systems there has also been financial assistance to victims of flooding and this appears to be a common practice. These are undertaken by the National Emergency Management Agency (NEMA), the Nigeria Hydrological Services Agency (NIHSA), the Nigerian Meteorological Agency (NIMET), and the National Environmental Standards and Regulations Enforcement Agency (NESREA) which in 2009 superseded the Federal Environmental Protection Agency (FEPA). It is not intended to discuss the structure, specific roles, and the unique position of these agencies with regard to flood management in Lagos. These aspects have been comprehensively discussed by Obeta (2014). However, it is important to mention that the activities of these institutions with regards to disaster management are generally coordinated by NEMA.

Although detailed data are not available, a key point will be highlighted in terms of the historical perspective of disaster management institutions, to give clarity to a synthetic description of the temporal evolution of flood awareness in Lagos. The institutional framework in Nigeria can be traced back more than four decades in history. The federal government of Nigeria has since the first, second, and third National Development Plans of 1962–1968, 1970–1974, and 1975–1980 initiated plans for the management of all disasters including flooding. This was organized through the federal and state ministries of works. On the basis of flooding and associated hazards, the primary aim of these initiatives was to create awareness among the citizenry and to develop sound response strategies. This development has evolved to the present time into what is now known as institutional approaches to managing disaster. Lately, the institutional framework has incorporated operations such as flood warning through the NISHA, improving general flood aware ness through the National Orientation Agency (NOA), and integration of local state and federal disaster emergency management agencies.

Despite the recent initiatives, these developments have been criticised as weak, while the roles of the institutions are not clearly defined (Adeaga et al., 2005; Oshodi, 2013; Soneye, 2014; Nkwunonwo et al., 2014; Adelekan, 2015). Critically, current measures undertaken by these agencies appear aimed to control flood rather than mitigate its impacts on ecological systems. Efforts are being made to facilitate evacuation and provide flood victims with urgent humanitarian needs. Environmental sustainability and policy, social responses, physical intervention, and environmental management are also critical issues requiring attention (Aderogba et al., 2012; Olajuyigbe et al., 2012; Aderogba, 2012b; Adeaga, 2008, Ilesanmi, 2010). However, despite the increasing number of people affected, the effectiveness of these efforts is challenged. Whilst it is unreasonable to claim that the weakness of these flood mitigation measures probably leads to more frequent flooding in the area, in can be argued that such measures have improved the experience of the general population with regard to flooding.

Recent Research Efforts

From the literature, researchers have suggested several options in relation to possible flood hazard mitigation and adaption responses. Adedeji et al. (2012) highlighted the importance of building the capacity for flood preparedness through spatial planning and land management. Ogunsote et al. (2011) suggested combating environmental degradation through sustainable landscaping. The need for sustainable management of solid waste which was emphasized in section 3 of this paper was recommended by Folorunsho and Awosika (2001). Komolafe et al. (2014) argued for the adoption of proactive measures to risk management and adaptation whilst constant geophysical and hydrological evaluation of rising groundwater levels was emphasized by Oyedele et al. (2009). Adelekan (2013) reiterated the UNISDR recommendation which calls for the participation of the private sector in risk management through investment decision making in building and construction. Other factors besides flood prevention are also important to reduce the potential impacts of flood events. The humanitarian relief supply chain for victims of flooding in the Lagos area was investigated by Soneye (2014). This study identified the need for more empirical investigation into such crucial components of flood risk management in Lagos. In relation to the planning framework, sustainable housing development and functionality of planning laws and regulations as well as the role of governance in flood management in the Lagos area and indeed in Nigeria have been examined by a number of authors including Aluko (2011) and Oshodi (2013).

As well as the general discussions in the literature, there is other research which focuses on integral components of flood risk such as probability of flooding, and exposure and vulnerability to flooding. But these are insufficient, and leave significant gaps in knowledge. Insufficient knowledge of the vulnerabilities to flooding of local communities, urban residents, and the general public constrains effective flood risk management in Lagos. Urban flooding in particular has arguably not received the attention it deserves. A general critique, which should provide a nuanced understanding of the strengths and limitations of present efforts to address the threats of flooding in the Lagos area, is lacking. The lack of flood data and other ancillary data, as discussed in Sects. 3 and 5, is a major setback towards containing these threats and has not been fully addressed. Importantly, the focus of these studies which solely rested on general knowledge of the causes, impacts, and remedies of flooding suggests that the global view of the situation has been imperfect. More scientific approaches such as the flood modelling and flood vulnerability assessment which drives recent approaches to flood risk management in more developed countries are generally lacking. Little action has been undertaken to raise public awareness of flood risk or to address gendered vulnerability, as highlighted by Odunuga et al. (2012), Ajibade et al. (2013), and Adelekan (2010). As an unprecedented measure, the Lagos state government has made significant efforts at providing high-resolution air-borne LiDAR data and topographic maps which promote research towards flood risk in the area. However, since many of these data sets are produced and sold commercially, the limited access of researchers to them arguably undermines their usefulness. A possible solution to such a limited access to LiDAR data is to apply global data sets such as ASTERGDEM (Advanced Spaceborne Thermal Emission and Reflection Radiometer Global Digital Elevation Model) and STRM (Shuttle Radar Topographic Mission). However, the resolution of spatial data is a crucial factor in flood modelling and flood risk assessment. Due to the geomorphology of urban features which significantly influence hydrodynamics, global data sets often do not provide realistic results when used to model flood hazard in urban areas (van de Sande et al., 2012). Flood modelling research is still looking into possible ways of simulating accurate flood variables on the basis of low-scale global data sets.

FLOOD RISK ASSESSMENT IN LAGOS

There has been a great deal of research about the assessment of flood risk in Lagos in general. However, it can be argued that none of these studies explicitly assessed the level of exposure to flooding of social and environmental systems. This aspect of flood risk assessment can be derived from studies that consider LU and LC change analyses, and population expansion and urbanization (see Table 3). From the literature, it is clear that vulnerability is being assessed under varying contexts and objectives. No study known to the authors has considered flood hazard estimation, although Odjugo (2006) and Adelekan (2010) investigated rainfall patterns over spatial and temporal scales. Odjugo (2006) investigated the changing rainfall pattern in Lagos over 3 decades. Adelekan (2010) on the basis of qualitative survey and secondary data analyses attributed increasing flood risks in Lagos Island mainly to changes in the frequency and intensity of rainstorms between 1971 and 2005. Other studies that estimate flood hazard were embodied within the general framework of climate change research (see for example Aderogba, 2012a). The lack of an explicit study on flood hazard estimation makes it difficult to appreciate how the probability of flood hazard occurrence is being determined in Lagos. Moreover, a clear understanding of the evolutionary trend of the hazard is denied. A consensus view is that the hazard has been increasing over the last 5 decades (Odunuga, 2008; Adelekan, 2015). It is expected that future research should be directed towards this crucial aspect of flood risk assessment.

Exposure

LU and LC analyses are procedures to classify land use/cover and to identify rates and patterns of change. However, the outcome of these analyses to Lagos has continuously suggested expansion in the urban areas. LU and LC analyses provides a basis for researchers and policy formulators to analyse the current and future urban exposure to environmental hazards especially flooding. Flood risk is heavily influenced by human and economic developments which evolve within spatial and temporal scales (Samuels et al., 2006). These developments are known to influence the hydrological cycle particularly through the development of more impervious surfaces which reduce soil infiltration capacity and lead to increased surface water runoffs. The spatial and temporal dimensions of flood risk can in part be directly related to LU and LC modifications. Barredo and Engelen (2010) and Gupta and Nair (2010) have shown that modelling LU and LC scenarios is integral to flood risk management. Identification of the population and urban growth pattern is also fundamental (Barredo and Demicheli, 2003). Many studies covering this crucial aspect of flood risk assessment within the context of the Lagos area exist in the literature (Table 3). Some of the major conclusions emerging from these studies show that significant sprawl and development in residential and industrial or commercial land classes are progressively taking place. There has been also been increasing pressure on arable lands (Akpomrere and Nyorere, 2012) and depletion of wetlands, mangroves, and swamps are significant issues (Obiefuna et al., 2013). The associated migration of people into Lagos city from other rural communities and states in Nigeria escalates residential needs and challenges. Barredo and Demicheli (2003) predicts that up to 27 million people will inhabit Lagos by 2020. Based on these studies, it can be shown that the level of human and economic resources exposed to urban flooding in Lagos is undoubtedly high. Figure 6 shows the 2012 LU and LC of the Lagos metropolis of Nigeria. Despite this, Lagos is relatively low on a global scale when compared to Calcutta, Dhaka, and Miami (Nicholls et al., 2007). However, the need to identify and address the vulnerabilities of exposed systems should be prioritized.

Vulnerability

There are numerous studies that have considered the vulnerabilities of social and environmental systems to flooding within the Lagos context (for example, Action Aid, 2006; Douglas et al., 2008; Adelekan, 2010; Ajibade et al., 2013, 2014; Nkwunonwo et al., 2015; Nsorfon, 2015; Olokesusi et al., 2015). Action Aid (2006) investigated vulnerability but tied Lagos with five other African cities. The study was carried out on the basis of key management criteria including local people's perceptions of the causes of flooding, adaptation, and the community's social coping capacity. Some of the limitations in the study were addressed by Douglas et al. (2008), who considered the vulnerability on the basis of climate change and adaptation strategies for the urban poor in Africa. Adelekan (2010) investigated the vulnerability of coastal communities in Lagos and responses to changing climatic conditions. The patterns of flood vulnerability and resilience amongst women were investigated by Ajibade et al. (2013). On the basis of political ecology, Ajibade et al. (2014) argued that the two crucial factors responsible for vulnerabilities of social systems in Lagos are limited access to housing and weak housing rights. The sources of social vulnerability to floods in informal settlements of Lagos was investigated in Nsorfon (2015) while Nkwunonwo et al. (2015) highlighted the relevance of assessing such vulnerability for the whole of Lagos. Olokesusi et al. (2015) investigated the influence of awareness of and responses to flood warnings on physical vulnerabilities of the affected communities in Lagos.

Table 3. Studies on exposure to urban flooding in the Lagos area of Nigeria.

S/No.	Author(s)	Study	Context	Major findings
1	Nwafor (1986)	Physical environment, decision-making and land use development in metropolitan Lagos	Urban growth	Rapid changes in land use classes were detected. Major upward trend in urban growth since 1960. Urban renewal and highway development radiating from the inner city to the hinterland were the major catalysts.
2	Abiodun (1997)	The challenges of growth and development in metropolitan Lagos.	Urban growth	Extensive and persistent urban growth. Key growth factors are the city's economic vitality and pivotal position in Nigeria's economy. Major growth challenges are survival of urban residents and city's sustainable development.
3	Barredo and Demicheli (2003)	Urban sustainability in developing countries' megacities: modelling and predicting future urban growth in Lagos.	Urban growth	By 2020, Lagos will experience astronomical spatial growth as a direct consequence of population expansion. Up to 27 million people will inhabit Lagos by 2020.
4	Adepoju et al. (2006)	Land use/land cover change detection in metropolitan Lagos (Nigeria):	LU and LC change detection	Lagos urban growth is phenomenal. Between 1984 and 2002, about 35 % increase in urban areas was recorded
5	Sunday and Ajewole (2006)	Spatial determinants of urban land use change in Lagos, Nigeria.	LU and LC change detection	Changing pattern of LU and LC in Lagos is characterized by significant socio-economic and environmental implications. Flooding and other implications are expected to worsen in the future given this present trend.
6	Braimoh and Onishi (2007)	Implications of the changing pattern of LC of the Lagos coastal area of Nigeria.	LU and LC change detection	Remote sensing was used to investigate LU changes, while binary logistic regression was used to model the probability of observing urban development as a function of spatially explicit independent variables.
7	Odunuga (2008)	Urban LU change and the flooding patterns in Ashimowu Watershed, Lagos, Nigeria	LU scenario and flooding (Alimosho)	A progressive increase in built-up area at the rate of 28 ha yr^{-1} between 1965 and 2003.
8	Olaleye et al. (2009)	LU change detection and analysis using remotely sensed data in Lekki Peninsula area of Lagos, Nigeria.	LU and LC change detection	Between 1964 and 2003 built-up areas grew from 40.93 to 1271.19 ha. Evolution in new classes of LU was observed, which include industrial, commercial and recreational LU class
9	Nwokoro and Dekolo (2012)	LU change and environmental sustainability: the case of Lagos metropolis.	LU change and environmental sustainability	Between 1990 and 2006, built-up areas increased by approximately 17 %. There was an obvious loss of forest resources and agricultural land to urban development.
10	Akpomrere and Nyorere (2012)	LU patterns and economic development of Ikeja in Lagos state, Nigeria: the GIS approach.	LU patterns and economic development	From 1962 to 1994, the built-up area in Lagos rose from 6.55 to 63.90 and from 63.90 to 67.99 % between 1994 and 2004. Decreases in vegetation cover and undeveloped areas were recorded.
11	Adebayo (2012)	Impact of urban LU changes on property values in metropolitan Lagos.	Urban LU changes and property	Significant changes in LU pattern from residential to commercial. Significant implication in property values in the area. These created social and environmental problems such as traffic congestion and noise pollution. The need for adequate LU planning was identified.
12	Nkwunonwo (2013)	LU/LC mapping of the Lagos metropolis of Nigeria using 2012 SLC off Landsat ETM+ Satellite Images	LU and LC mapping	By 2012, four major LU/LC themes were identified in Lagos: water body, vegetation, residential, and industrial areas. Urban areas, namely residential and industrial areas, account for more than 54 % of the whole Lagos metropolis.
13	Obiefuna et al. (2013)	Spatial changes in the wetlands of Lagos/Lekki Lagoons of Lagos, Nigeria.	Wetland changes	Mangrove wetlands decreased from 88.51 to 19.95 km^2 between 1984 and 2006 at $-3.12\ km^2\ yr^{-1}$. Swamps decreased from 344.75 to 165.37 km^2 between 1984 and 2006 at $-8.15\ km^2\ yr^{-1}$. Built-up areas increased from 48.97 to 282.78 km^2 at 10.61 $km^2\ yr^{-1}$. Water body decreased from 685.58 to 654.98 km^2 at $-0.16\ km^2\ yr^{-1}$. Bare land increased from 24.32 to 72.73 km^2 at 2.2 $km^2\ yr^{-1}$. Vegetation decreased marginally from 1369.15 to 1361.08 km^2 at $-0.37\ km^2\ yr^{-1}$. Most of the increase in built-up area occurred in the Eti-Osa Local Government Area (LGA) and then in the Kosofe LGA. The key implication is of flood risk on affected areas.

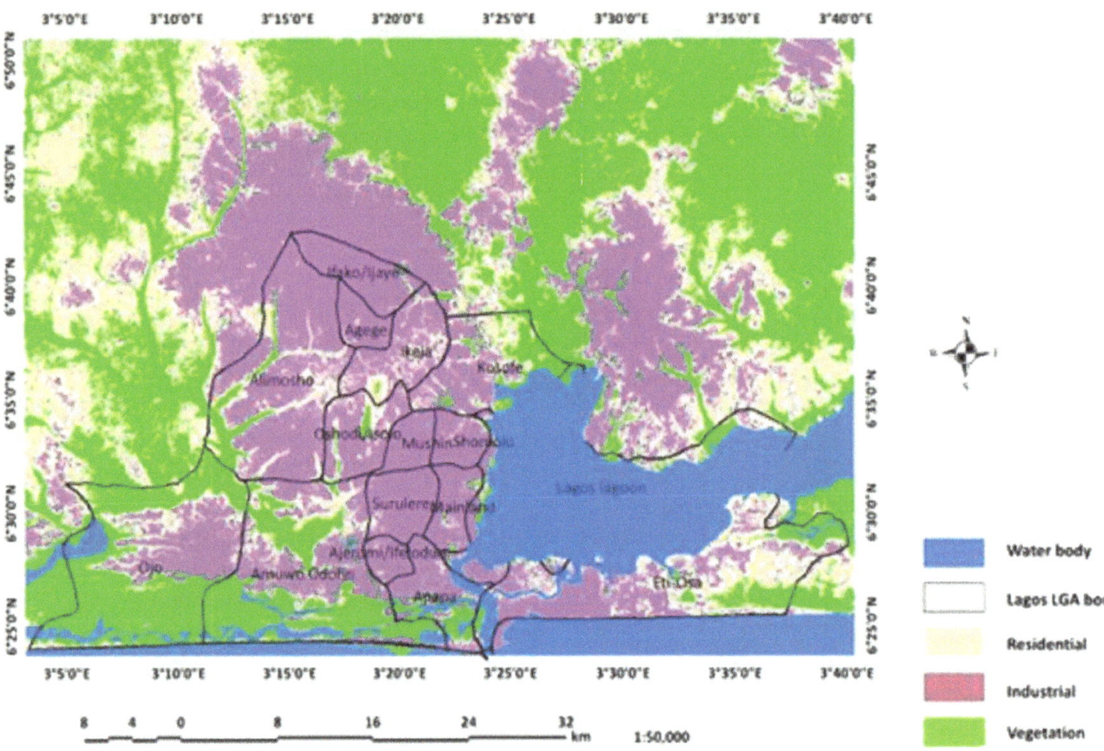

Figure 6. 2012 Land use (LU) and Land Cover (LC) map of the Lagos metropolis of Nigeria. Source: Nkwunonwo (2013).

Although these studies are major contributions to knowledge which provide some evidence to suggest that vulnerability is an important issue of flood risk in Lagos, signifi- cant gaps still exist in the literature with regard to accurate assessment of vulnerability in the area. Most of these studies were based on a limited random sample of data which is insufficient to make accurate generalization. Critical issues of vulnerability analyses such as choice and measurement of vulnerability indices were not addressed. This contradicts a widely accepted philosophy that accurate assessment of vulnerability is related to these critical issues (Adger, 2006). Classical analyses of vulnerability should be able to consider every available factor that undermines the chances of resistance available to social and environmental systems.

THE CASE FOR FLOOD MODELLING IN LAGOS

Based on the issues identified by this research with regard to Lagos, three significant issues emerge which are: (1) lack of data relating to flooding and its consequence, (2) lack of accurate flood risk assessment, and (3) the "unidentifiable method" of flood hazard estimation. To address these issues, flood modelling which is widely being applied in hazard research (see for example Samuels et al., 2006; Merz et al., 2010) to estimate flood hazard should be considered. Attention should be given to the specific objectives that can be achieved in Lagos. It is also important to relate these objectives to specific factors which account for the reasons flood modelling has been largely ignored in the area.

Whilst it is widely acknowledged that stage–damage functions (presented in Sect. 5) are viable tools for flood risk assessment (Samuels et al., 2006; Merz et al., 2010; Hammond et al., 2015), the benefits which Lagos flood risk assessment can derive from such a tool should be investigated. This depends on the availability of various data relating to water depth and extent of flooding events. Even the use of probability concepts to estimate flood hazard requires data relating to a large sample of historical flooding events. As these data are not readily available for Lagos, a flood modelling approach can be used to reconstruct particular historical events (such as 1 in 50, 1 in 100, 1 in 200, 1 in 500, and 1 in 1000 flood return periods) in terms of inundation depth, extent, and water flow velocity, and on the basis of simulated data predict future occurrences. This will largely improve warning systems currently in place. However, when flooding occurs, and a predicted return period is the case, field survey should be carried out to determine properties and assets that are exposed. If the present market values of the exposed elements are known, damage functions on the basis of ex post flood hazard estimate can be used to assess the level of risk following the flooding event.

Theoretically, flood modelling is a procedure to investigate how rainfall or any other hydrological input into a drainage area is transformed to a streamflow (Chow et al., 1988). Since many of the floods in Lagos are pluvial based, accurate flood modelling results will assist in delineating areas likely to get increased runoff volumes and flooding. Flood modelling will play an important role in producing Lagos flood hazard and flood risk maps which do not currently exist. Needless to say, a comprehensive flood risk and flood hazard map for Lagos will increase public awareness of flood risks and inform stakeholder decision about flood mitigation options and policies in terms of structural and/or non-structural measures. From the point of view of cost benefit analysis (CBA), Lagos area will derive maximum benefits in the long run if flood modelling is the basis of any flood management operation.

Flood modelling is used to promote flood risk reduction in the US, the Netherlands, and the United Kingdom, but in

Nigeria it is arguably too often ignored. This has continued to raise the question of how actionable flood risk assessment can be achieved. A number of structural measures are being put in place. However, an issue that has been less considered in the Nigerian context is the reliability of structural measures in the absence of accurate flood data and scientific means of acquiring such data. Flood modelling has been largely ignored in Lagos for reasons such as data requirement and availability, lack of specialist technology, funding, and the applied skills required. Academic research seems to be the most likely option in terms of the responsibility to develop flood models. The present research has been unable to refer to specific contributions of academic research towards flood modelling. However, the lack of bespoke or generalized flood models and poor application in flood risk management in the area seem to suggest that flood modelling procedures considered by academic curricula in Lagos are mainly theoretical and on the periphery. As NIHSA (2013) pointed out, such an intensive research should be funded by the government and other interested bodies. To date, the researchers are not aware of any such development. Moreover, the uncertainties associated with flood modelling can discourage investment of resources into it. To the best of the authors' knowledge, no public agency in Lagos undertakes the procedure as a specific role.

It can be argued that relatively few studies have highlighted the relevance of flood modelling and its implications with the paucity of topographic data (Nkwunonwo et al., 2014; van de Sande et al., 2012), although Adeaga (2008) implemented a flood hazard mapping and risk management in the northeastern part of Lagos. Although flood modelling was mentioned in these studies, solutions to the problems raised remain largely unanswered. Critically, current measures undertaken by flood management agencies appear to control flood rather than mitigate its impacts on human populations and urban infrastructure.

FLOODING IN LAGOS – THE WAY FORWARD

The review of flood management covered in this research prompts several recommendations for reducing flood risk in Lagos. These recommendations are based on three key considerations: first, is the understanding and demonstration of the roles more scientific approaches (such as flood modelling,) can play in flood risk reduction within the context of Lagos; second, is the need to align the focus of flood risk reduction in the Lagos area to the objectives of similar measures in more developed countries such as the US, the UK, and the Netherlands. This should be linked to improving collaboration between Lagos and indeed Nigeria and the developed countries in terms of promoting a more effective flood risk management philosophy. In hazard research, the need for collaboration to strengthen learning and information exchange is increasingly being identified (Thomalla et al., 2006). Moreover, similar to the European Union flood risk directive, regional collaborations are being advocated for countries in Africa (Holloway, 2003). Thirdly, there is a need to promote awareness of flooding among local communities, urban residents and the general public and to delineate more suitable locations for relocation of human populations during flooding events. Further specific recommendations include:

- The government of Lagos state should as a matter of urgency prioritize legislation and provision of resources towards flood hazard and flood risk mapping for the whole of Lagos state. This is the basis of flood risk mitigation within the European Union framework, which requires all constituting states to prepare flood hazard and flood risk maps to promote the concept of living with floods (EC, 2004).
- Flood risk reduction under the "living with floods" idea is multi-disciplinary, indicating that various industries can assist in reducing the impacts of flooding. This is the case in the UK in particular (EA, 2010). In view of the widening of public awareness of flooding, there is a need for improved collaboration between the Lagos state government and federal ministries, departments, and agencies such as NEMA, NESREA, and NIHSA.
- Flood alert and flood early warning systems should be improved within the Lagos area. Globally, there is a growing concern regarding the dissemination of and response to flood warnings (Moore et al., 2005; Parker et al., 2009). On the basis of this, researchers in the UK and other European countries are anxious that flood warning systems are failing to yield the expected scale of damage savings (Penning-Rowsell et al., 2000; Parker et al., 2009). A review of 20 years of the progress of hazard warning systems in the US was presented by Sorensen (2000) although this research shows that no warning system is perfect for any hazard.
- A series of issues need to be addressed with regard to the design of flood warning systems, and improved forecasting and response systems which correspond to the needs of the public and are often influenced by social, economic, and environmental variability. These issues should underlie improvements in any flood warning systems in Lagos. The cultural, social, and religious values of the people (which are important to many residents of Lagos) may also need to be taken into consideration when disseminating flood warnings. The reluctance of many people living in flood-prone areas to heed flood warnings should also be thoroughly investigated.
- Flood insurance is a non-structural approach which many property owners have benefitted from in developed countries following flood disasters. It is argued that the benefits of flood insurance far outweigh the limitations which exist (Treby et al., 2006; Crichton, 2008). Flood insurance is still being applied in places like the UK, the Netherlands, and Indonesia, although the coverage is at a relatively low scale (Jha et al., 2012). The need to enable flood victims to recover more quickly underscores flood insurance (Lamond and PenningRowsell, 2014). Purchase of insurance is highly dependent on a number of factors, including its availability and cost, the level of the provision of disaster relief, general risk awareness, and attitudes to collective and individual risk (Lamond et al., 2009). For the Lagos area, poor awareness of flood insurance is a major issue. Urban residents need to know that a means of assisting them recover from flood losses is possible. They must be given the opportunity to try flood insurance based on individual volition. To support the roles of flood insurance in Lagos, it is recommended

that the role of FEMA in this regard should be extended to the state, whilst encouraging insurance companies to commence an awareness campaign for property owners to take positive steps towards purchasing insurance cover.

- The enforcement of environmental standards and laws is often a key factor towards containing adverse effects of climate change including flooding (UN/ISDR, 2004). Indiscriminate waste disposal, construction along the flood plain, and the blockage of drainage facilities, among other anthropogenic activities influence flooding in Lagos and are often illegal. It can be argued that NESREA should embark on enforcement measures to address these issues.
- The reaction to the 1953 floods in the Netherlands has arguably made the Dutch an exemplar in terms of flood management (Vis et al., 2003). Invariably, the success of flood risk reduction in the Netherlands is built on a strong commitment to avoid a repeat occurrence. The population in general and the Dutch government are committed to the collective efforts which underlie success towards addressing the challenges of flooding. Collective flood management costs each Dutch adult about USD 110 annually (Kazmierczak and Carter, 2010). For this reason, we recommend that urban residents and the general public in Lagos need to engage more fully in flood management and control. This would include raising awareness of the risks and educating the wider public about potential solutions. This would include encouraging adherence to environmental laws and a commitment to engage with flood alerts and early warning systems. To support this, more education is required to make the public more aware about flooding and its consequences. Qualitative research such as CBA, involving the local community should also be encouraged.
- Globally, it appears that research is proportional to success towards addressing the threats of flooding. From scoping investigations it is clear that literature relating to flooding in Lagos appears embryonic compared to literature in the US, the Netherlands, China, and the UK for example. This is a strong indicator that more research is required for Lagos. Equally, the universities and research agencies should be encouraged to include flood awareness and management in their curricula and programmes. More research should be directed towards developing bespoke hydrologic and hydraulic flood models for simulating flood hazard and other hydrological parameters in the area.
- Globally, accurate data play a key role in flood risk mitigation and the government of Lagos state has made an unprecedented attempt to improve its data by the acquisition of LiDAR data sets, although access to these data sets has been limited due to funds. Given the importance of these data sets and the need to optimize their usefulness for the Lagos area, the state government should consider subsidizing these data sets for universities and research institutions. Additionally, we also recommend further improvement on data acquisition such as SAR (Synthetic Aperture Radar) for flood modelling within the Lagos area. It is argued that the key to wider applied research requires easy access to data and a collaborative approach between researchers and government agencies.

CONCLUSION

Flooding is a global phenomenon, but the impacts in many urban areas in the developing countries can be overwhelming. It is easier to understand the threats of flooding in Lagos by the attention generated both in the media and in academic literature relating to social and environmental science. Flood waters have impacted upon the local population, destroyed critical infrastructure, and disrupted economic activities. However, based on "current practice" in flood management and flood risk reduction in the context of "living with floods", the actions of the state government and other stakeholders towards addressing the challenge of flooding in the Lagos area have arguably been limited. Unfortunately, relevant data on flood events are not readily available and the means of building a community resilient to flood threats have continued to elude present efforts.

This research is an attempt towards addressing the challenges of flooding in and around Lagos. Looking to the future, this research argues that flood modelling and assessment of vulnerability are necessary requisites for more effective results towards addressing the challenges of flooding in the Lagos area and indeed in Nigeria. The wider availability of high-quality data sets (including surface modelling and detailed demographic data) is a crucial step towards addressing this problem. It is recognized that a major limitation of this study and many other studies is the level of information available with regard to flooding in the Lagos area. The majority of the research offers a generalized overview of flooding rather than scaling the hazard to the local level. If flood data had been available for LGAs, it would have offered a better understanding of the spatial distribution of flooding over the epoch considered. In addition, there are places within various LGAs which may never have flooded over the period considered. It is important to investigate these locations with a view to finding out how they defend themselves from flooding. It was not intended within this study to address flood modelling and vulnerability assessment. However, the authors recommend future research that focuses on developing bespoke flood models for simulating flood hazard in the Lagos area of Nigeria.

Flooding is and will continue to be a hazard for the people of Lagos. The key to ameliorating the effects of flooding on the population is a more integrated, informed flood management philosophy which is linked to improved data collection and provision. This alongside greater public engagement and education as well as improved government coordination will greatly reduce the potential impacts of flood events on the population of Lagos.

ACKNOWLEDGEMENTS

This publication derives from a PhD research which is being funded by the Tertiary Institutions Education Trust Fund (TETFund) programme for staff of university of Nigeria. The Surveyors Council of Nigeria (SURCON) is acknowledged for providing some emergency grants. Data relating to flooding were sourced mainly from the Center for Research in Epidemiology of Disasters (CRED), Nigeria's Ministry of the Environment and the Nigeria Environmental Study Action/Team (NEST). Previous works in this area are equally acknowledged and so are the anonymous reviewers of this paper.

REFERENCES

1. Abiodun, J. O.: The challenges of growth and development in metropolitan Lagos. The urban challenge in Africa: growth and management of its large cities, 192–222, 1997.
2. Action Aid: Climate Change, Urban Flooding and the Rights of the Urban Poor in Africa: Key Findings From Six African Cities, Action Aid International, London, 2006.
3. Adeaga, O.: Flood Hazard Mapping and Risk Management in Parts of Lagos N. E., available at: http://www.gsdi.org/gsdiconf/ gsdi10/papers/TS13.3paper.pdf (last access: 10 March 2015), 2008.
4. Adeaga, O., Savic, D. A., Bertoni, J. C., Mariño, M. A., and Savenije, H. H. G.: A sustainable flood management plan for the Lagos environs, in: Sustainable Water Management Solutions for Large Cities: The Proceedings of the International Symposium on Sustainable Water Management for Large Cities (S2): Held During the Seventh Scientific Assembly of the International Association of Hydrological Sciences (IAHS) at Foz Do Iguaçu, Brazil, 3–9 April 2005, No. 293, International Assn of Hydrological Sciences, 226–229, 2005.
5. Adedeji, O. H., Odufuwa, B. O., and Adebayo, O. H.: Building capabilities for flood disaster and hazard preparedness and risk reduction in Nigeria: need for spatial planning and land management, Journal of Sustainable Development in Africa, 14, 45–58, 2012.
6. Adelekan, I.: Private Sector Investment Decision in Building and Construction: Increasing, Managing and Transferring Risks: Case Study of Lagos, Nigeria, The United Nations Office for Disaster Risk Reduction, Global Assessment Report on Disaster Risk Reduction, 1–9, 2013.
7. Aderogba, K.: Qualitative studies of recent flood and sustainable growth and development of cities and towns in Nigeria, International Journal of Academic Research in Economics and Management Science, 1, 1–25, 2012a.
8. Adigun, F., Abolade, O., and Yusuf, A. A.: Incidence of flood and its impacts: empirical evidence from Ajeromi-Ifeledun, Lagos State, Nigeria, International Journal of Innovative Research and Studies, 2, 239–254, 2013.
9. Agbola, T. and Agunbiade, E.: Urbanization, slum development and security of tenure: the challenges of meeting millennium development goal (MDG 7) in metropolitan Lagos, Nigeria, in: PRIPODE Workshop, Nairobi, Kenya, 11–13 June 2007.
10. Ajibade, I., Armah, F. A., Kuuire, V. Z., Luginaah, I., McBean, G., and Tenkorang, E. Y.: Assessing the bio-psychosocial correlates of flood impacts in coastal areas of Lagos, Nigeria, J. Environ. Plann. Man., 58, 445–463, doi:10.1080/09640568.2013.861811, 2014.
11. Bashir, O., Oludare, H., Johnson, O., and Aloysius, B.: Floods of fury in Nigerian cities, Journal of Sustainable Development, 5, 69–79, 2012.
12. Barredo, J. I. and Demicheli, L.: Urban sustainability in developing countries' megacities: modelling and predicting future urban growth in Lagos, Cities, 20, 297–310, 2003.
13. Braimoh, A. K. and Onishi, T.: Spatial determinants of urban land use change in Lagos, Nigeria, Land Use Policy, 24, 502–515, 2007.
14. BNRCC (Building Nigeria's Response to Climate Change): Vulnerability, Impacts and Adaptation, Climate Change in Nigeria, available at: www.infoo@climatechange.org (last access: 10 March 2015), 2008.
15. CII (Climate Change and Insurance): Chartered Institute for Insurance, Research report of the Chartered Institute for Insurance, London, 2001.
16. Crichton, D.: Role of insurance in reducing flood risk, Geneva Pap. R. I.-Iss. P., 33, 117–132, 2008.
17. De Wrachien, D., Mambretti, S., and Schultz, B.: Flood management and risk assessment in flood-prone areas: Measures and solutions, Irrig. Drain., 60, 229–240, 2011.
18. Di Baldassarre, G. and Uhlenbrook, S.: Is the current flood of data enough? A treatise on research needs for the improvement of flood modelling, Hydrol. Process., 26, 153–158, 2012.
19. Dilley, M., Chen, R. S., Deichmann, U., Lerner-Lam, A., Arnold, M., Agwe, J., Buys, P., Kjekstad, O., Lyon, B., and Yetman, G.: Natural Disaster Hotspots, A Global Risk Analysis, Washington, D.C., The World Bank, 2005.
20. Douglas, I., Alam, K., Maghenda, M., Mcdonnell, Y., McLean, L., and Campbell, J.: Unjust waters: climate change, flooding and the urban poor in Africa, Environ. Urban., 20, 187–205, 2008.
21. EA (Environment Agency): Working With Natural Processes to Manage Flood and Coastal Erosion Risk, Environment Agency, London, 2010.
22. EC (European Commission): Flood Risk Management – Flood Prevention, Protection and Mitigation, Communication From the Commission to the Council, the European Parliament, the European Economic and Social Committee and the Committee of the Regions, Brussels, 2004.
23. EM-DAT: The International Disaster Database – Centre for Research on the Epidemiology of Disasters – CRED: Flooding Data for Nigeria, available at: www.emdat.be/ (last access: 10 March 2015), 2014.
24. Etuonovbe, A. K.: The devastating effect of flooding in Nigeria, in: FIG Working Week, May 2011, available at: http://www.fig. net/pub/fig2011/papers/ts06j/ts06j_etuonovbe_5002.pdf (last access: March 2015), 2011
25. FME (Federal Ministry of Environment): Bulletin on Ecological Disasters, FME, Abuja, Nigeria, 25, 2012.
26. Folorunsho, R. and Awosika, L.: Flood Mitigation in Lagos, Nigeria through Wise Management of Solid Waste: a case of Ikoyi and Victoria Islands, Nigerian, UNESCO-CSI workshop, 19–23 November 2001,

Maputo, 2001.

27. Gill, S.: Literature Review: Impacts of Climate Change on Urban Environments – Draft 30 Copy, with contributions from: Pauleit, S., Ennos, R., Lindley, S., Handley, J., Gwilliam, J., and Ueberjahn-Tritta, A., The Impact of Flooding on Urban and Rural Communities, Centre for Urban and Regional Ecology, University of Manchester, UK, 2004
28. Guha-Sapir, D., Hoyois, P., and Below, R.: Annual Disaster Statistical Review 2012: The Numbers and Trends, CRED, Brussels, 2013.
29. Holloway, A.: Disaster risk reduction in Southern Africa: hot rhetoric-cold reality, African Security Studies, 12, 29–38, 2003.
30. Huq, S. and Alam, M.: Flood management and vulnerability of Dhaka City. Building Safer Cities: The Future of Disaster Risk, Washington, D.C., 121–135, 2003.
31. Ilesanmi, A. O.: Urban sustainability in the context of Lagos megacity, Journal of Geography and Regional Planning, 3, 240–252, 2010.
32. Isong, M., Eyo, E., Eyoh, A., Nwanekezie, O., Olayinka, D. N., Udoudo, D. O., and Ofem, B.: GIS cellular automata using artificial neural network for land use change simulation of Lagos, Nigeria, Journal of Geography and Geology, 4, p. 94, 2012.
33. Jha, A. K., Bloch, R., and Lamond, J.: Cities and flooding: a guide to integrated urban flood risk management for the 21st century, World Bank Publications, Washington, D.C., 2012.
34. Jonkman, S. N., Vrijling, J. K., and Vrouwenvelder, A.: Methods for the estimation of loss of life due to floods: a literature review and a proposal for a new method, Nat. Hazards, 46, 353–389, 2008.
35. Kazmierczak, A. and Carter, J.: Adaptation to Climate Change Using Green and Blue Infrastructure, a Database of Case Studies, GRaBS project, University of Manchester, Manchester, UK, 2010.
36. Komolafe, A. A., Adegboyega, S. A. A., Anifowose, A. Y., Akinluyi, F. O., and Awoniran, D. R.: Air pollution and climate change in Lagos, Nigeria: needs for proactive approaches to risk management and adaptation, American Journal of Environmental Sciences, 10, 412–423, 2014.
37. Kundzewicz, Z. W., Kanae, S., Seneviratne, S. I., Handmer, J., Nicholls, N., Peduzzi, P., Mechler, R., Bouwer, L.M., Arnell, N., Mach, K., Muir-Wood, R., Brakenridge, G. R., Kron, W., Benito, G., Honda, Y., Takahashi, K., and Sherstyukov, B.: Flood risk and climate change: global and regional perspectives, Hydrolog. Sci. J., 59, 1–28, 2014.
38. Lagos State Government (LSG): Abstract of Local Government Statistics, Lagos: Lagos Bureau of Statistics, Ministry of Economic Planning and Budget Secretariat, Alausa, Ikeja, 2012.
39. Lamond, J., Bhattacharya, N., and Bloch, R.: The role of solid waste management as a response to urban flood risk in developing countries, a case study analysis, in: Flood Recovery Innovation and Response, edited by: Proverbs, D., Mambretti, S., Brebbia, C., and de Wrachien, D., WIT Press, Southampton, 193–205, 2012.
40. Lamond, J. E., Proverbs, D. G., and Hammond, F. N.: Accessibility of flood risk insurance in the UK: confusion, competition and complacency, J. Risk Res., 12, 825–841, 2009.
41. McMichael, A. J., Woodru, R. E., and Hales, S.: Climate change and human health: present and future risks, Lancet, 367, 859– 869, 2006.
42. Merz, B., Kreibich, H., Thieken, A., and Schmidtke, R.: Estimation uncertainty of direct monetary flood damage to buildings. Natural Hazards and Earth System Sciences, 4,153–163, 2004.
43. Merz, B., Kreibich, H., Schwarze, R., and Thieken, A.: Review article "Assessment of economic flood damage", Nat. Hazards Earth Syst. Sci., 10, 1697–1724, doi:10.5194/nhess-10-1697- 2010, 2010.
44. Miceli, R., Sotgiu, I., and Settanni, M.: Disaster preparedness and perception of flood risk: A study in an alpine valley in Italy, J. Environ. Psychol., 28, 164–173, 2008.
45. NEST (Nigeria Environmental Study/Action Team): Nigeria's Threatened Environment: a National Profile, Ibadan, Nigeria: NEST, IRC Main, Abuja, Nigeria, 824 NG91, 9582, 1991.
46. Nicholls, R., Hanson, J. S., Herweijer, C., Patmore, N., Hallegatte, S., Corfee-Morlot, J., Château, J., and Muir-Wood, R.: Ranking of Port Cities With High Exposure and Vulnerabilities to Climate Extremes, OECD Environment Working Papers, No. 1, OECD Publishers, University of Southampton, UK, 2008.
47. Njoku, J. D. and Udeagha, M.: Assessing the flooding potentials of Oguta Lake Watershed using remote sensing technology, Paper presented at 5th Annual National Conference, Abuja, organized by the Nigerian Association of Hydrological Sciences (NAHS) at University of Nigeria, Nsukka, 21–30, 2013.
48. Nkwunonwo, U. C.: Land use/land cover mapping of the Lagos metropolis of Nigeria using 2012 SLC-off Landsat ETM+ Satellite Images, International Journal of Scientific and Engineering Research, 4, 1217–1223, 2013.
49. Nkwunonwo, U. C., Whitworth, M., and Baily, B.: Relevance of social vulnerability assessment to flood risk reduction in the Lagos metropolis of Nigeria, British Journal of Applied Science & Technology, 8, 366–382, 2015.
50. Obeta, C. M.: Institutional approach to flood disaster management in Nigeria: need for a preparedness plan, British Journal of Applied Science & Technology, 4, 4575–4590, 2014.
51. Obiefuna, J. N., Nwilo, P. C., Atagbaza, A. O., and Okolie, C. J.: Spatial changes in the wetlands of Lagos/ Lekki Lagoons of Lagos, Nigeria, Journal of Sustainable Development, 6, p. 123, 2013.
52. Odjugo, P. A. O.: An analysis of rainfall pattern in Nigeria, Global Journal of Environmental Sciences, 4, 139–145, 2006.

53. Odunuga, S., Oyebande, L., and Omojola, A. S.: Social-economic indicators and public perception on urban flooding in Lagos, Hydrology for Disaster Management: Special Publication of the Nigerian Association of Hydrological Sciences, NAHS, Abuja, 82–96, 2012.

54. Parker, D. J., Priest, S. J., and Tapsell, S. M. Understanding and enhancing the public's behavioural response to flood warning information, Meteorol. Appl., 16, 103–114, 2009.

55. Penning-Rowsell, E. C. and Chatterton, J. B.: The Benefits of Flood Alleviation: A Manual of Assessment Techniques, Gower Publishing Company Limited, Aldershot, UK, 1977.

56. Penning-Rowsell, E. C., Tunstall, S. M., Tapsell, S. M., and Parker, D. J.: The benefits of flood warnings: real but elusive, and politically significant, Water Environ. J., 14, 7–14, 2000.

57. Raaijmakers, R., Krywkow, J., and van der Veen, A.: Flood risk perceptions and spatial multicriteria analysis: an exploratory research for hazard mitigation, Nat. Hazards, 46, 307–322, 2008.

58. Samuels, P., Klijn, F., and Dijkman, J.: An analysis of the current practice of policies on river flood risk management in different countries, Irrig. Drain., 55, S141–S150, 2006.

59. Samuels, P. G.: An overview of flood estimation and flood prevention. In invited paper presented at the first international symposium on flood defense, Kassel Reports of Hydraulic Engineering, 9, G1–G11, 2000.

60. Sayers, P., Li, Y., Galloway, G., Penning-Rowsell, E., Shen, F., Wen, K., Chen, Y., and Le Quesne, T.: Flood Risk Management: a Strategic Approach, UNESCO, Paris, 2013.

61. Soneye, A.: An overview of humanitarian relief supply chains for victims of perennial flood disasters in Lagos, Nigeria (2010–2012), Journal of Humanitarian Logistics and Supply Chain Management, 4, 179–197, 2014.

62. Sunday, O. A. and Ajewole, A. I.: Implications of the changing pattern of land cover of the Lagos Coastal Area of Nigeria, American-Eurasian Journal of Scientific Research, 1, 31–37, 2006

63. Thomalla, F., Downing, T., Spanger-Siegfried, E., Han, G., and Rockström, J.: Reducing hazard vulnerability: towards a common approach between disaster risk reduction and climate adaptation, Disasters, 30, 39–48, 2006

64. Treby, E. J., Clark, M. J., and Priest, S. J.: Confronting flood risk: implications for insurance and risk transfer. J. Environ. Manage., 81, 351–359, 2006.

65. Ugwu, L. I. and Ugwu, D. I.: Gender, floods and mental health: the way forward, International Journal of Asian Social Science, 3, 1030–1042, 2013.

66. UN (United Nations): World Population to 2300, United Nations Department of Economic and Social Affairs, Population Division, New York, 2004.

67. UN/ISDR: United Nations International Strategy for Disaster Reduction: Living with Risks: a global Review of Disaster Reduction Initiatives, 2004 Version Volume 1, available at: http:// www.unisdr.org/files/657_lwr1.pdf (last access: 10 March 2015), 2004.

68. van Ogtrop, F. F., Hoekstra, A. Y., and van der Meulen, F.: Flood management in the lower Incomati River Basin, Mozambique: two alternatives, J. Am Water Resour. As., 41, 607–619, 2005

69. Vis, M., Klijn, F., De Bruijn, K. M., and Van Buuren, M.: Resilience strategies for flood risk management in the Netherlands, International Journal of River Basin Management, 1, 33–39, 2003.

70. World Bank: World Bank Population Growth Rate 2009–2013, available at: http://data.worldbank.org/indicator/SP.POP.GROW (last access: 10 March 2015), 2013.

CHAPTER 7

CREATIVITY AND THINKING SKILLS INTEGRATED INTO A SCIENCE ENRICHMENT UNIT ON FLOODING

Audrey C. Rule[1], Jean Suchsland Schneider[1], Denise A. Tallakson[1], Diane Highnam[2]

[1] Department of Curriculum and Instruction, University of Northern Iowa, Cedar Falls, USA

[2] Area Education Agency 267, Cedar Falls, USA

ABSTARCT

Floods that used to happen every hundred years are now occurring more frequently. Human influences on the damage inflicted by flooding need to be well-understood by future voters and property-owners. Therefore, the timely topic of flooding was used as the focus of a special multi-grade enrichment short course taught by two university education professors for 26 preK-8th grade high-achieving and creative students. During the course, students listened to guest speakers (city council member, meteorologist, and environmentalist), watched two flood-related videos, read books on floods, viewed electronic presentations related to dams and recent floods, discussed causes, effects, and mitigations of flooding, and devised creative games from recycled materials to teach peers about flood concepts. The de Bono CoRT Breadth thinking skill system was used to organize many of the course activities. The flood lesson activities were relevant to these students who had experienced a flood of the city's river the previous year and challenged students more than their typical classroom activities, an important finding considering that many gifted students drop out of school because of irrelevant and non-demanding class work. The course broadened students' knowledge of floods and assisted them in thinking beyond the immediate situation.

INTRODUCTION

Elementary and middle school students who are high-achieving in science and who exhibit creativity are often not challenged or given the opportunity to fully use their abilities in the regular classroom. Many gifted students drop out because school is boring, repetitious, and lacks relevance to real life (Hansen & Toso, 2007). Students who consider dropping out want more exciting, challenging tasks and coursework (Snyder, 2003). Unfortunately, many classroom teachers lack sufficient background knowledge to design stimulating, advanced science projects for these students; some avoid science altogether. A national survey (Fulp, 2002) revealed that multi-subject teachers (mostly elementary teachers) are ten times more likely as single-subject teachers to be unprepared to teach science. Another study (Dorph et al., 2007) found that eighty percent of K-5th grade multiple-subject teachers who were responsible for teaching science in their classrooms reported they spent an hour or less per week on science, with sixteen percent of teachers spending no time at all on science. Special programs for students that dig deeply into relevant science topics and highlight careers fill an important need, as motivated high-performers or exceptionally creative students are the ones most likely to become future scientists,

Citation: Rule, A., Schneider, J., Tallakson, D. & Highnam, D. (2012). Creativity and Thinking Skills Integrated into a Science Enrichment Unit on Flooding. Creative Education, 3, 1371-1379. doi: 10.4236/ce.2012.38200.

engineers, and inventors. The attractiveness of science careers and self-confidence in science are strong factors in the choice of science or engineering coursework and careers (Robertson, 2000; Woolnough et al., 1997).

Zhao (2009) asks what knowledge is of most worth in the current global and digital economy that has changed the old economic rules by enlarging and connecting markets with workforces across the world, by meeting a larger variety of needs, and by reaching customers with unique wants. Preparing students for the global and digital economy requires that future workers acquire 21st century skills of core subjects such as science; but also learning and innovation skills; information, media, and technology skills; and life and career skills (Partnership for 21st Century Skills, 2011). Besides a need for Science, Technology, Engineering, and Mathematics (STEM) skills for economic growth from innovative processes and products, the current global climate change phenomenon requires more scientific knowledge to mitigate and solve the effects of severe weather, flooding, tornadoes, hurricanes, blizzards, drought, and wild fires on energy supply, transportation and infrastructure, residential and commercial buildings, agriculture, forestry, waste management, and industry, among other issues (Barker et al., 2007).

To meet the need for a science enrichment short course focused on 21st century skills at a local elementary school, two education professors designed, in collaboration with school faculty, a short course for exceptional preschool through middle school students in which the students delved deeply into the science content of flooding. Students practiced learning and innovation skills as they devised creative games to teach their classmates about the information they had learned. They utilized information, media, and technology skills by accessing online information and developing game boards and components with drawing software. Three guest-speakers, who were working in the flood-related areas of meteorology, environmental science, and city planning, provided students with a glimpse of possible career paths related to this real-world topic. This article reports the results of this successful endeavor to provide ideas and guidance to others designing a similar program or teaching about the topic of flooding.

National Standards

Technology and invention are important components of the elementary and middle school science curricula. Science and Technology Content Standard E of the National Science Education Standards (National Research Council, 1996) states that, as a result of activities in grades K-4, all students should develop abilities of technological design. Similarly, Science and Technology Content Standard E for grades 5 - 8 states that all students should develop an understanding of abilities of technological design: "In the middle school years, students' work with scientific investigations can be complemented by activities in which the purpose is to meet human need, solve a human problem, or develop a product rather than to explore ideas about the natural world" (p. 161). Invention supports scientific inquiry, allowing students to make connections to the real world and to other subject areas. During this flood unit, all students worked either individually or in teams to invent a game using many recycled items that presented flood concepts to players.

The authors of the Benchmarks for Science Literacy (American Association for the Advancement of Science, 2008, 1993) in Chapter 3, The Nature of Technology, discuss how elementary and middle school students need to know about the connections between science and technology, the nature of engineering and design, and societal issues related to technology. Besides students considering the suitability of recycled items for game parts, students also considered the use of human technology in flood prevention and mitigation throughout the short course. Additionally, the Benchmarks, in Chapter 4, The Physical Setting, state that elementary students should know how water shapes the earth's surface through erosion and deposition. The topic of flooding explores these ideas. Invention lessons address the often-neglected domains in science education of "Imaging and Creating" (Domain III) and "Using and Applying" (Domain V) as defined by Yager (2000). Invention allows students to combine objects in new ways, to produce alternate or unusual uses for objects, and to design and test devices and machines. Through invention, students apply learned science concepts and problem-solving skills to everyday technological problems and household devices. These operations occurred as students used cast-offs in their game-making. Students involved in invention also work in Domain VI, "Viewing Science and its History as Human Enterprises." In this science domain, students considered the motives of scientists, engineers, and technologists, along with investigating the history of technology and its effects on our society. These concepts were addressed as students learned of beaver dams, historic human dams, and modern hydroelectric dams during the project.

Objective of the Study and Rationale

The purpose of this project was fourfold: 1) to present highachieving elementary and middle school students with a challenging special enrichment short course; 2) to involve students with the real world problem of flooding directly connected to their community; 3) to provide students with the opportunity to study a science topic in-depth with expert instruction that provided models of possible careers; and 4) to ask students to create an authentic product—a flood game—that would allow them to practice creative thinking and invention while applying recently-learned information about floods.

The invention of a flood game allows students to apply their new knowledge while practicing important creativity skills. Globalization since World War II has challenged America's role as economic and strategic world leader. Much of our workforce now competes for employment with lower-wage earners from other countries. Workers in other nations conduct innovative scientific and engineering work, and are easily employed through the Internet (National Research Council, 2005). Inventors and innovators will help Americans to compete with other nations and to craft solutions for our serious environmental problems.

Sustainable development is the practice of protecting the environment while improving living standards for all, and invention and innovation is the key to its success. Invention and innovation for sustainable development isn't just developing new technology, but includes new processes and new ways of solving old problems—creative thinking is the rubric...

Despite the fact that people everywhere have an innate ability to be creative, rich countries are not doing enough to stimulate and harness invention and creative thinking… due to a combination of factors … [including] education systems that don't inspire or value creativity… (Lemelson-MIT Program, 2003: p. 4).

Educational Theory

The Schoolwide Enrichment Model (Renzulli, 1976; Renzulli & Reis, 1997) has three main instructional components: Type I General Exploratory Activities, Type II Group Training Activities, and Type III Individual and Small Group Investigations of Real Problems. These three types of enrichment activities were implemented in the flood unit discussed here. The Type I General Exploratory Activities included the three speakers' presentations, the flood videos, a water table (with containers, sponges, blocks, and toys) for exploring how flooding works, many books on floods, and electronic slide presentations of various aspects of flooding. The Type II Group Training Activities were the de Bono CoRT thinking skill exercises. Finally, the Type III investigations centered on producing a game from found or recycled materials to teach peers flood concepts (games were later housed in the school library for checkout) and to provide models for teachers through conference presentations and publication.

Students learning to develop their thinking skills and creativity need a structure that scaffolds their learning and allows them to practice skills. Therefore, the popular CoRT (Cognitive Research Trust) Breadth thinking skill system (de Bono, 2000) of Edward de Bono was chosen for this project. The CoRT system consists of 6 sets of thinking skills with the Breadth set being the most basic. These Breadth thinking skills have been used in numerous schools around the globe for several decades, are well-known in the field of gifted education, and have been employed by businesses and governments in solving problems (Barak & Doppelt, 1999; Gardyasz, 2007; Melchior, Kaufold, & Edwards, 1988; Rule & Barrera, 2006, 2008). The CoRT Breadth skills form the foundation of de Bono's lateral thinking, a creative thinking approach that helps thinkers generate new ideas in a way that may not be obtainable by traditional stepby-step logic (Carter, 2007). The set contains ten thinking skills generally referred to by initials or one-word titles. Table 1 shows the skills and how each was applied during the unit on floods.

Table 1. Edward de Bono's CoRT thinking skills applied to the flood unit.

Thinking Skill	Brief Explanation	Application to Flood Unit
CAF	Consider all factors	Make a list of all the factors involved in flooding to consider.
APC	Generate alternatives, possibilities, choices	Determine alternatives, possibilities and choices of games on which to model flood game final products.
C and S	Determine consequence and sequel of an action	Determine consequences and sequels of flooding at different time intervals: immediate, short term, and long-term.
Planning	Constructing a formal plan	Plan a flood game, including the object of the game, directions, flood facts, game board, and playing pieces.
PMI	Rate aspects of ideas as plus, minus, interesting	Consider the pluses, minuses and interesting aspects (neither positive nor negative ideas that arise as consequences or causes) of specific game ideas.
Decision	Make a decision based on criteria	Decide the basic components of the game that make it engaging and that teach flood concepts.
Rules	State rules	Write rules for the game.
FIP	Rate ideas for first important priority	List new learning and determine the most important ideas learned. Tell why they are most important.
AGO	Determine aims, goals, objectives	Check if any additional components could be added to bolster the aims, goals, or objectives of the game.
OPV	Consider other people's views	Obtain other people's views on the game and use this feedback to improve it.

Real World Connection

The topic of flooding is a timely one, as America is experiencing a steady increase in the number of record-breaking natural disasters connected to global warming with resources being overwhelmed by "hundred-year" events occurring much more frequently (Huber & Gulledge, 2011; Huq, Kovats, Reid, & Satterthwaite, 2007) Young citizens need to know how to vote to design communities and how to build homes in a sustainable way, harmonious with flood plain environments. A unit that helps students to understand the causes, effects, and mitigation of floods addresses important issues. In particular, the students in this investigation lived through a record-breaking flood of the river in their city that occurred less than a year before the students participated in this project. The students were highly motivated to learn how floods work, as many had experienced inundated homes or sandbagging efforts, firsthand. Therefore, this project had a very real-world, authentic focus for the students involved. The products—the games—generated by students were also effective because, through these games, students were able to review the information they learned and transfer it to peers who had not participated in the flood unit. The audience for the games extended beyond the school as the authors made two scholarly presentations of the project, highlighting the games students made to science educators and gifted educators.

Flood-Related Instruction Literature Review

In an editorial in the Journal of Geoscience Education, the editors (Libarkin, Elkins, McNeal, & St. John, 2010) stressed the importance of education for persons living in geologically hazardous areas so that they understand the causes, effects, and ways to mitigate disaster, making informed decisions about where to live (i.e., not on a floodplain), how to build, and how politics enter into the situation. In further exploration of the politics of disaster, Freudenburg, Gramling, Laska, and Erickson (2008) focused on the ways humans alter nature through levee- and canal-building, causing later human suffering during floods. They noted that, although many of these construction projects were promoted as economic development, they are more accurately characterized as "removal of money from the many for the benefit of the few" (p. 1016). Therefore, students who are to become voting citizens need to understand the political and economic factors that create disasters. Students need to better understand how global events affect local happenings, such as how the El Niño/Southern Oscillation phenomena is associated with flooding, droughts, and increased hurricane activity. Mjelde, Litzenberg, Hoyle, Holochwost, and Funkhouse (2007) presented a learning module combining science and mathematics that focused on students' understanding the probabilities of above or below normal precipitation, dependent upon Pacific Ocean anomaly patterns. Authors of another article (Mahaya, Tippins, Mueller, & Thomson, 2009) examined global water quality, highlighting the devastating effects of flooding on a pure drinking water supply. These authors presented classroom-tested activities including a case study of a Kenya water supply, along with water sanitation and bacteria testing methods.

Several authors have presented ideas for helping students who have experienced floods understand their experiences. Shreve, Danbom and Hanhan (2002) examined the artwork and writings of children who were victims of a Red River flood. They recommended that teachers ask children about their thoughts and experiences, explain how other people (rescue workers, firefighters, volunteers) help during the disaster, and initiate lessons that show how floods actually work so as to aid understanding and alleviate anxieties. Similarly, Zevenbergen, Sigler, Duerre, and Howse (2000) examined teachers' spontaneous changes to the next school year's curriculum after a flooding disaster occurred at the end of the previous academic year, finding that most teachers chose flood-related examples to illustrate information they taught, along with more discussion, drawing, and writing activities related to floods. Because many of the students in the enrichment class had been flood victims themselves, we added in opportunities for emotional expression and community-building.

THE FLOOD UNIT

A description of the setting and participants, the course content, the special speakers, the materials, and the pretest/posttest follows.

Setting and Participants

Twenty-six high-achieving and creative students participated in the special science enrichment class on flooding. The class was divided into two multi-age sections of equal numbers of students from grades prekindergarten through third grade (9 females, 4 males; 4 third graders, 4 second graders, 3 first graders, 1 kindergartener, and 1 preschooler) and fourth through eighth grade (6 females, 7 males; 1 eighth grader, 4 sixth graders, 5 fifth graders, and 3 fourth graders), each group meeting separately for an hour and a half each day for eight days near the end of the school year. Students were chosen through teacher recommendations, standardized test scores, and scores on a short creativity test. The authors obtained approval from their university's human subjects committee to collect data on the efficacy of the project. All students and parents gave permission for participation, data collection, and for photographs to be used in any subsequent publications.

Course Content

The course consisted of hour-and-a-half classes, eight in all, for each of the two age-groups, which occurred over a two week period. Course content was delivered via guest speakers, videos, reading of books, and electronic slide presentations. Exercises and discussions using the CoRT thinking skills helped students mentally process and better understand the information they received, as shown in Table 1. Each day, students were asked to contribute their ideas on several large posters as a way of recording ideas that could be discussed or used in the games being developed. Topics of these fact sheets were the causes, effects, vocabulary, exciting facts, and mitigations of flooding. Following each guest speaker or presentation, the instructors prompted students to add additional ideas to the posters.

The three guest speakers were a meteorologist, an environmentalist, and a city council member. A former local television meteorologist was the first speaker who explained how rainfall is measured, how river levels are monitored, and how these levels, gathered throughout an area, inform flood predictions. He brought a rain gauge and other instruments for the students to explore, demonstrating how they work. He also talked about flood safety, telling how one of the most common causes of death during flooding is drowning in a car that has been swept away by flood waters. Many people are unaware of the danger of driving into or across flooded streets; only two feet of water can float a car and allow it to be swept away. Rolling down windows and escaping immediately may save the occupants' lives. The second speaker was the director of the Center for Energy and Environmental Education at the local university. He provided an effective demonstration of soil permeability and water storage by use of a clever analogy. As a demonstration, he poured a bucket of water onto an empty tabletop allowing students to verbally react as the impermeable surface caused the water to cascade onto the floor as runoff. Then, covering the table with a folded beach towel simulating prairie grasses, he again poured a bucket of water onto the table. This time, however, the "grasses" absorbed the water, showing how deeprooted vegetation can store water and prevent flooding. He also explained how rivers naturally overflow their banks inundating their floodplains regularly; structures such as artificial channels and levees confine the river to a narrow path but produce overflow elsewhere because the volume of the river stays the same. Therefore, allowing the river to expand onto its floodplain and halting construction in this area would prevent much damage to human structures.

The third guest speaker was a city council member who showed a documentary video of the city's flood and discussed the current city issues of buying property that had been underwater during the recent flood. Students noted the familiar landmarks and remembered the awesome impact of the flood. The other video shown to students was a NOVA presentation called The Mystery of the Megaflood (NOVA, 2005a). An odd landscape located about 200 miles east of Seattle includes vast, deep gorges, huge boulders dropped in a seemingly random pattern, rippled hills, and gigantic, waterless waterfalls (NOVA, 2005b). An insightful geologist, J. Harlan Bretz hypothesized that a huge flood had created the unique landscape, caused by a glacial dam giving way to a huge glacial lake. Later, other geologists agreed with Bretz, after a similar, smaller glacial dam burst in 1996 in Iceland. The flood-related books were written at several reading levels and contained exciting stories with interesting visuals (Kusky, 2008; Sipiera & Sipiera, 1998; Thompson, 2000; Woods & Woods, 2007). Students read them in their free time, especially to find additional flood facts for their games. Several electronic slide presentations illustrated and explained current floods across the nation (images and information were drawn from Internet news sites); how flood control and hydroelectric dams work, including the differences between historic and modern dams; beaver dams and their role in natural flood control; and another flood of a nearby community that involved an upstream town on the same river.

Pretest-Posttest

At the start of the unit, students were asked to respond in writing (younger students could dictate their ideas) to several open-ended criterion-referenced questions with these same questions being asked at the end of the unit as a way of comparing pretest and posttest knowledge of floods. These included the following questions:

1) What is a flood?
2) Tell as many different effects of flooding as you know.
3) How can people prevent or lessen (mitigate) the negative effects of floods?

A survey of "Attitudes about School and Learning" (Rogers, 2002) was administered to students at the start and end of the unit to gauge the attitudinal impact of the unit and to determine whether students perceived work during this short course differently than their regular schoolwork (This survey was read to younger students). Twenty statements related to perceptions about school in general were provided. Students responded on a four point scale: 1—always agree; 2—usually agree; 3—sometimes agree; and 4—disagree. A few example statements are listed: "3. School is exciting, every day is great," "7. My work at school makes me feel proud," and "20. The harder the work, the more interesting it is" (Rogers, 2002: pp. 454-455). The wording of some of the attitude statements was changed slightly on the posttest to reflect work during the flood unit rather than toward school in general. Cronbach's alpha (Cronbach, 1951; Cronbach & Shavelson, 2004), a commonly used measure of internal consistency of test scores, was .83, indicating reasonably strong reliability.

RESULTS

Results of the effectiveness of the unit were measured through the students' products of creative flood games to teach others what they had learned and a comparison of the pretests and posttests. Examples of the games are described in the next section. Results of the study are illustrated in Table 2 through 5.

Creative Flood Games

Students used creativity skills to put recycled and found materials to new uses or to use parodies of existing games in their flood games. Flood game products were of several types: fine motor skill activities, a word find, board games, a flood simulation game, a memory game, and matching puzzle games. The fine motor skill games were made by younger students and involved using tongs to move animals or people from flooded areas to safety or to place miniature sandbags along a river to build a flood levee. A fine motor skills game was made by a kindergarten student who was very concerned about the animals that would become engulfed in flood waters. She found clipart images of her favorite animals, mirrored them on the computer, and cut them out. She glued a pompom in between the two sides to make them three-dimensional, so they could stand and be easily manipulated using tongs. She also designed sorting boards for the animals, as every individual animal was a different clip-art image. Figure 1 shows her in action with the game. The second fine motor skills game was created by a first grade student and is shown in Figure 2. It had a village fashioned from cut pieces of craft foam, trees represented by green flat glass marbles, and hills made of trimmed foam packing material, and the river represented by shredded blue paper. Her game involved mostly sandbagging because her family had sandbagged their home before the flood; she wanted to honor their efforts by making a game of it.

The student wrote the directions for her game called "The Big Flood:"

There can be 2 to 3 players. Use tongs to pick the sand bags up. You have to be careful. Use the board carefully so you do not break the board. Block the water from the houses with sandbags. DO NOT USE WATER! When all your sand bags are placed, you win.

In the fine motor skills game shown in Figure 3, a player draws a card from a stack of 20 scenarios and reads the information, such as, "Floodwater fills home—people need rescue," "Live power lines down—rescue people," or "Man drives car through flooded streets; car floats away—rescue people," and then acts out the drama by picking up plastic people with tongs and placing them on the rooftop for helicopter or boat rescue. Blue shredded paper was used to represent the flood waters, while brown rocks represented boulders and debris carried by the flood waters. Students made the homes out of used paper cartons with clip art house fronts glued to them. For all these fine motor skill games, the tongs were made with two complete Popsicle sticks with several partial sticks glued between them at one end, allowing lever action when pinched.

Table 2. Pretest and posttest responses of 26 students to, "What is a flood?"

Component of Flood Definition	Frequency	
	Pretest	Posttest
Consists of water	24	25
Excess or over-abundant amount of water	12	10
Overflowing usual confines	10	11
Rising or surging	10	7
Causing damage to human structures	6	4
Caused by rain	6	4
Puts land usually dry under water	5	4
Caused by dam break	2	2
Dangerous/can kill	1	5
Contains polluted water	0	1
Carries debris or ice	0	1
Causes erosion	0	1
Total	76	75

Figure 1. Fine motor skills game of rescuing and sorting animals made by first grade student.

A word find, made by a kindergarten student, focused on new flood-related vocabulary, such as mudflow, jumble ice, and debris. Words were printed backwards and on diagonals with other non-related words (e.g., happy, joy, sweet) used to fill space and make it more entertaining. Clues for the words were complex and attempted to direct the player to a certain place of the board.

A variety of board games was created. One activity made by second and third graders combined fractions with house-rebuilding in a cooperative format of moving around a board, landing on spaces, and giving fractional house pieces to flood victims attempting to rebuild their homes (See Figure 4).

Figure 2. The big flood fine motor skills sandbagging game.

Figure 3. Fine motor skills game flood rescue game made by second grade students.

Others focused on answering questions about flood fact information to move around a board or to rescue people from flooded homes. For example, one pair of sixth grade boys made a parody of a Monopoly® game that focused on flood issues. There were interesting properties and spaces such as "levee", "river", "river gauge", "drainage pump", and "foot-deep water". "Rescue" and "Flood" cards told flooding scenarios and consequences for the player.

Another unique game involved players matching questions to answers, resulting in raising or lowering the river level. This game consisted of a three-dimensional topographic contour river scene. The landscape of this game, shown in Figure 5, was made with layers of thick cardboard cut to show the contours of hills surrounding a river valley. The parts that had been cut away from each cardboard layer fit exactly into the river valley and were colored blue to represent different stages of the river that could be placed onto the landscape. These pieces were added or subtracted in response to players' work of matching cards showing terms and definitions or explanations, thereby raising or lowering the water level of the river. The game was made by three fifth graders collaborating with a sixth grader who assumed a leadership position. Students disassembled some pompoms and colored the fibers with a marker to make vegetation. They made houses from bits of sponge foam with paper roofs.

Figure 4. Game combining fractions in rebuilding flood-damaged homes.

Figure 5. Topographic contour map game of rising and falling river levels

Two other games required players to place puzzle pieces (identical sectors of a circle) into a tray (actually, a decorated label-covered compact disk (CD) cut into 8 pieces labeled A-H placed into a CD case with areas marked 1 - 8 with a permanent marker) as players matched vocabulary to definitions or identified the order of events in flooding. The reverse sides of these pieces, when the tray was flipped over, revealed a pattern that matched the pictorial key if the player had answered all of the questions correctly. Two sixth grade students working on this game are shown in Figure 6; a page of questions and pictorial key that fit neatly into the CD case is shown in Figure 7; the glued-on labels for the sectors of the cut-apart CD are shown in Figure 8.

Figure 6. Two middle school students working on a fact puzzle game made from a compact disk and its case.

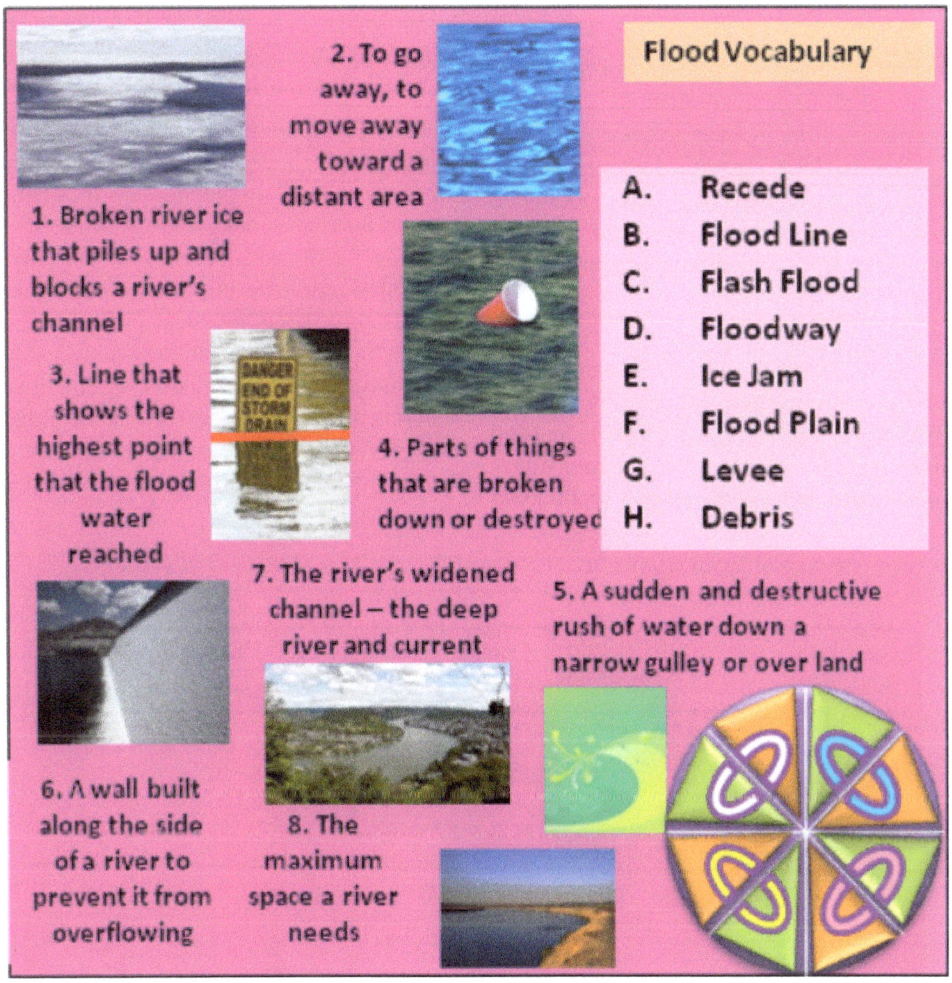

Figure 7. Matching questions for modified compact disk game.

Figure 8. Fronts and backs of the eight game pieces made from a cut-apart compact disk with glued-on labels.

Posttest Outcomes

Table 2 shows that most students, even the younger group of kindergarten and first graders, had a good concept of what a flood was at the beginning of the program, with ideas shifting slightly with the experience of the flood instruction unit. This is understandable, as many students had witnessed the flooding of the city's river the previous summer. New ideas, addressed during instruction, that were mentioned were the ideas of polluted water, debris piles, erosion, and the many dangers of flood waters.

Table 3 shows that on the pretest, students focused on the immediate effects of flood damage, inundation, danger, and loss of property, while on the posttest, students recognized the after-flooding problems of debris piles and jumble ice, and looked ahead to clean-up and rebuilding. This demonstrates a broadening of student time perception related to flooding effects from the present to the near and more distant future. Some secondary effects of flooding such as more flooding, emotional toll, and building of dams are additional indications of broadening the ability to recognize longer-term effects.

Sandbagging was the most common response of flood mitigation methods on both the pretest and posttest. Students knew about sandbagging because they had seen it on television news reports and many had participated or had relatives who did. On the pretest, several students, as shown in Table 4, suggested currently impractical ideas such as trying to stop rainstorms or keep flood waters from coming into the area. On the posttest, students exhibited more knowledge of the effects dams have in regulating flood water and the importance of not building on the floodplain. New knowledge of the role of deep-rooted vegetation and safety rules such as evacuating when required and never driving into flooded streets, appeared on the posttest.

Table 3. Pretest and posttest responses of 26 students to, "Tell as many different effects of flooding as you know".

Flood Effect	Frequency	
	Pretest	Posttest
Building damage and destruction	11	11
Water invades homes/things underwater	10	2
Drowning/injury/death of people	9	7
Dirt/ mud/mold everywhere	8	8
Lost possessions/cars	8	5
Danger and evacuation	5	3
Trees/telephone/electric poles fall over	5	1
Homeless people	3	4
Water carries dirt/soil/sewage/disease	3	4
Power outage/surges damage items	3	2
Erosion of topsoil/mudslides	2	4
Economic cost/business loss/crop loss	2	4
Rebuild roads	2	0
Animals die	2	3
Debris washed into piles	1	10
No safe drinking water	1	1
Items moved by flood waters	1	0
Drains overflow	1	0
Jumble ice left behind	0	6
Knock houses off foundations	0	3
Rebuilding/other people help	0	3
Clean up	0	3
Snakes in water	0	2
Scabland landscape	0	1
More flooding	0	1
Emotional toll-sadness	0	1
Dams regulate water	0	1
Total	77	90

Table 4. Responses of 26 students to, "How can people prevent or lessen the negative effects of (mitigate) floods?"

Flood Mitigation Action	Frequency	
	Pretest	Posttest
Use sandbags to make levees or walls to protect property	22	21
Build higher than flood plain or away from shore	5	13
Stop rain storms and water from coming	5	0
Have stronger walls	3	0
Have a drainage system for water	3	0
Go to high ground	3	0
Go upstairs/Move items upstairs	3	0
Lock doors and seal cracks	2	0
Purify water	1	0
Sing rain song to lift spirits	1	0
Stop urbanization	1	0
Use amphibious vehicles	1	0
Use dams to regulate water	0	11
Plant prairie grasses creating spongy soil to absorb runoff	0	6
Be prepared	0	4
Evacuate when told to	0	3
Educate people about flood s	0	2
Don't waste water	0	2
Don't litter or dump debris in water	0	2
Don't change a river's shape	0	2
Don't drive cars into flood waters	0	2
Work as a team	0	1
Plant more trees and forests	0	1
Have plumber fix drains	0	1
Total	50	71

Table 5 shows the mean scores of the responses to the attitude questions that exhibited significant change from the beginning to the end of the program. These questions show that students felt more challenged during this special program than during their regular classroom activities where many were used to taking the spotlight as the most knowledgeable students in the class (see questions numbered 6 and 8 in Table 4). The fact that students were less satisfied with grades and checked their work more carefully before turning it in indicates the more rigorous nature of the flood unit. The overall means of attitude statements on the pretest and posttest were almost identical, corresponding to answering "usually agree" to most statements. This indicates an overall satisfaction and feeling of motivation toward both regular schoolwork and the special flood unit.

Table 5. Student responses to attitude statements that showed significant pretest to posttest changes and overall mean scores on all questions.

Statements	Mean Rating[1]		T-Test Result
	Pretest	Posttest	*p* value
6. Teachers think I am one of the best students (pretest). The professors think I am one of the best students (posttest).	1.92	2.35	.02
8. Teachers count on me for correct answers (pretest). The professors count on me for correct answers (posttest).	1.96	2.42	.04
10. I am satisfied with the grades I get (pretest). I am satisfied with the assessment of the work I did on this project (posttest).	1.44	1.88	.02
17. I check my work carefully before handing it in (retest). I checked my work carefully before handing it in on this project (posttest).	2.04	1.63	.01
Overall mean of all 20 questions	1.83	1.85	.84[2]

Note: [1]indicates always agree, 4 indicates disagree. [2]this *p* value does not indicate a statistically significant difference.

CONCLUSION

This project satisfied its four planned objectives. First, the project presented high-achieving elementary and middle school students with a challenging special enrichment short course. High-achieving elementary and middle school students were challenged by the content and skills of the unit as evidenced by their responses to the attitude survey. They felt less confident and checked their work more thoroughly during the enrichment project.

Second, the project involved students with the real world problem of flooding directly connected to their community. The enrichment topic was particularly relevant to the students who had endured the flooding the previous year.

Third, students had the opportunity to study a science topic in-depth with expert instruction that provided models of possible careers. Students enjoyed listening to and asking questions of the guest speakers about their jobs as a meteorologist, an environmentalist (also a college professor), and a city council member. Additional information was provided via videos, electronic presentations, and books. Finally, the project asked students to create an authentic product—a flood game—that would allow them to practice creative thinking and invention while applying recently-learned information about floods. Students were thoroughly engrossed in developing their flood games and finding interesting facts for game questions. They worked individually or in small groups to create an authentic product—a flood game—using recycled or found materials such as pebbles, sponges, cardboard, plastic lids, fabric scraps, string, clothespins, and egg or milk cartons. This allowed them to assign new uses to common materials, thereby practicing creativity and their newly-acquired knowledge of floods. A focus on the current, local science issue of floods with an authentic product motivated the students. Incorporation of thinking skills boosted effective processing of information, while attention to the science domain of creativity resulted in exciting game products. This enrichment unit on flooding exemplifies a successful, challenging, integrated, authentic science project.

REFERENCES

1. American Association for the Advancement of Science (2008, 1993). Benchmarks for Science Literacy. Washington: Author.
2. Barack, M., & Doppelt, Y. (1999). Integrating the Cognitive Research Trust (CoRT) programme for creative thinking into a project-based technology curriculum. Research in Science and Technological Education, 17, 139-151. doi:10.1080/0263514990170202
3. Barker T., Bashmakov, I., Bernstein, L., Bogner, J. E., Bosch, P. R., Dave, R., Davidson, O. R., Fisher, B. S., Gupta, S., Halsn?s, K., Heij, G. J., Kahn Ribeiro, S., Kobayashi, S., Levine, M. D., Martino, D. L., Masera, O., Metz, B., Meyer, L. A., Nabuurs, G.-J., Najam, A., Nakicenovic, N., Rogner, H.-H., Roy, J., Sathaye, J., Schock, R., Shukla, P., Sims, R. E. H., Smith, P., Tirpak, D. A., & Urge-Vorsatz, D., & Zhou, D. (2007). Technical summary. In B. Metz, O. R. Davidson, P. R. Bosch, R. Dave, & L. A.
4. Carter, D. (2007). Thinking ahead: Edward de Bono, the father of lateral thinking. Training Journal, 19-21.
5. Cronbach, L. J. (1951). Coefficient alpha and the internal structure of tests. Psychometrika, 16, 297-334. doi:10.1007/BF02310555
6. Cronbach, L. J., & Shavelson, R. J. (2004). My current thoughts on coefficient alpha and successor procedures. Educational and Psychological Measurement, 37, 827-838.
7. de Bono, E. (2000). Edward de Bono's CoRT thinking lessons. Oxford: Cavendish Information Products, Ltd.
8. Dorph, R., Goldstein, D., Lee, S., Lepori, K., Schneider, S., & Venkatesan, S. (2007). The status of science education in the Bay Area (Research brief). Berkeley, CA: Lawrence Hall of Science, University of California.
9. Freudenburg, W. R., Gramling, R., Laska, S., & Erikson, K. T. (2008). Organizing hazards, engineering disasters? Improving the recognition of political-economic factors in the creation of disasters. Social Forces, 87, 1015-1038. doi:10.1353/sof.0.0126
10. Fulp, S. L. (2002). Status of elementary school science teaching. Chapel Hill: Horizon Research Inc.
11. Gardyasz, J. (2007). Hats off to creativity (De Bono thinking systems) (Innova training and consulting) (Cover story). Business Record (Des Moines), 25, 13-14.
12. Hansen, J. B., & Toso, S. J. (2007). Gifted dropouts: Personality, family, social, and school factors. Gifted Child Today, 30, 30-41.
13. Huber, D. G. (2011). Extreme weather and climate change: Understanding the link and managing the risk. Arlington, VA: Center for Climate and Energy Solutions.
14. Huq, S., Kovats, S., Reid, H., & Satterthwaite, D. (2007). Editorial: Reducing risks to cities from disasters and climate change. Environment and Urbanization, 19, 3-15. doi:10.1177/0956247807078058
15. Kusky, T. (2008). Floods: Hazards of surface and groundwater systems. New York: Facts on File/Infobase Publishing.
16. Lemelson-MIT Program (2003). Invention and innovation for sustainable development: Report of a workshop sponsored by the Lemelson- MIT Program and LEAD International, London. URL (last checked 19 December 2012). http://web.mit.edu/invent/n-pressreleases/downloads/sustainable.pdf
17. Libarkin, J. C., Elkins, J. T., McNeal, K., & St. John, K. (2010). Editorial: What role do geoscientists play in society? Journal of Geoscience Education, 58, p. 1. doi:10.5408/1.3544290
18. Mahaya, E., Tippins, D. J., Mueller, M. P., & Thomson, N. (2009). Infectious disinfection: Exploring global water quality. Science Activities, 46, 25-31. doi:10.3200/SATS.46.2.25-32
19. Melchior, T. M., Kaufold, R. E., & Edwards, E. (1988). How schools teach thinking: Using CoRT thinking in schools. Educational Leadership, 45, 32-33.
20. Mjelde, J. W., Litzenberg, K. K., Hoyle, J. E., Holochwost, S. R., & Funkhouser, S. (2007). Fires,

floods, and hurricanes: Is ENSO to blame? Science Scope, 30, 38-42.

21. National Research Council (1996). National science education standards: Observe, interact, change, learn. Washington: National Academy Press.

22. National Research Council (2005). Rising above the gathering storm: Energizing and employing America for a brighter future. Washington: National Academies Press.

23. NOVA (2005a). Mystery of the megaflood: What unleashed a catastrophic flood that scarred thousands of square miles in the American northwest? Boston, MA: Mentorn, A Television Corporation Company and WGBH Educational Foundation.

24. NOVA (2005b). Mystery of the megaflood: TV Program Description. URL (last checked 19 December 2012). http://www.pbs.org/wgbh/nova/megaflood/about.html

25. Partnership for 21st Century Skills (2011). Framework for 21st Century Learning. URL (last checked 19 December 2012). http://www.p21.org/overview/skills-framework

26. Renzulli, J. S. (1977). The enrichment triad model: A guide for developing defensible programs for the gifted and talented. Gifted Child Quarterly, 21, 227-233.

27. Renzulli, J. S., & Reis, S. M. (1997). The schoolwide enrichment model: A how-to guide for educational excellence. Mansfield Center, CT: Creative Learning Press.

28. Robertson, I. J. (2000). Influences on choice of course made by university Year 1 bioscience students—A case study. International Journal of Science Education, 22, 1201-1218. doi:10.1080/09500690050166751

29. Rogers, K. B. (2002). Re-forming gifted education: Matching the program to the child. Scottsdale, AZ: Great Potential Press.

30. Rule, A. C., & Barrera III, M. T. (2006). CoRT thinking skills guide PBL science. Academic Exchange Quarterly, 10, 145-149.

31. Rule, A. C., & Barrera III, M. T. (2008). Three authentic curriculumintegration approaches to bird adaptations that incorporate technology and thinking skills. ERIC database, ED 501247.

32. Shreve, R., Danbom, K., & Hanhan, S. (2002). "When the flood km we had to lv": Children's understandings of disaster. Language Arts, 80, 100-108.

33. Sipiera, P. P., & Sipiera, D. M. (1998). Floods: A true book. New York: Children's Press.

34. Snyder, S. (2003). Boredom cited as a reason for thought of dropping out. The Philadelphia Inquirer, B1.

35. Thompson, L. (2000). Natural disasters: Floods. New York: High Interest Books/Children's Press/Grolier.

36. Woods, M., & Woods, M. B. (2007). Floods: Disasters up close. Minneapolis, MN: Lerner Publications Company.

37. Woolnough, B. E., Guo, Y., Leite, M. S., Jose de Almeida, M., Ryu, T., Wang, Z., & Young, D. (1997). Factors affecting student choice of career in science and engineering: Parallel studies in Australia, Canada, China, England, Japan, and Portugal. Research in Science and Technology Education, 15, 105-121. doi:10.1080/0263514970150108

38. Yager, R. E. (2000). A vision for what science education should be like for the first 25 years of the new millennium. School Science and Mathematics, 100, 327-341. doi:10.1111/j.1949-8594.2000.tb17327.x

39. Zevenbergen, A. A., Sigler, E. A., Duerre, L. J., & Howse, E. (2000). The impact of natural disasters on classroom curricula. Journal of Educational Thought, 34, 285-303.

40. Zhao, Y. (2009). Catching up or leading the way: American education in the age of globalization. Alexandria, VA: Association for Supervision and Curriculum Development (ASCD).

CHAPTER 8

ENVIRONMENTAL RISK IMPOSED BY DIVERTED FLOOD WATERS ON WATER AND SOILS IN EMERGENCY RETENTION BASINS

Karl Erich Lindenschmidt[1], Robert Harrison[1], Martina Baborowski[2]

[1] Manitoba Water Stewardship, Surface Water Management, Winnipeg, Manitoba, Canada

[2] Department of River Ecology, UFZ-Centre for Environmental Research, Magdeburg, Germany.

ABSTRACT

Emergency retention basins (ERB) are diked enclosures alongside rivers into which water from the main river channel is diverted during extreme floods. If the basins are operated during extreme flooding, two negative environmental impacts may occur: 1) contamination of the soils due to their transport by suspended sediments to the basin and 2) depletion of dissolved oxygen in the basin water. A computer-based methodology is presented which was used to assess the environmental risk exhibited by the operation of an ERB system proposed for the Elbe River in Germany. The August 2002 extreme flood event was used as a test case. For such a flood, the results showed that there is a 77% risk that dissolved oxygen levels fall below 2 mg/L in the water and a 48% chance of exceeding the inspection value of 500 mg zinc/kg in the soil.

INTRODUCTION

The European Union (EU) has passed a Floods Directive to provide a legislative foundation for flood protection and mitigation in Europe. Since most large river basins in Europe extend over several countries, the directive will help to orient and broaden flood management efforts beyond the municipal and state authorities to the national and international levels and establish an improved and more cost-effective flood protection scheme (see also [1]). Interestingly, the directive includes many references to the environment and the EU Water Framework Directive, e.g.:

- Article 4 (Preliminary flood risk assessment): ...[must consider] potential adverse consequences for human health, the environment
- Article 9 (Flood hazard maps and flood risk maps): ... shall be coordinated with and may be integrated into the reviews provided for 2000/60/EC [WFD]
- Article 7 (Flood risk management plans): ... shall take into account costs and benefits, flood extent and conveyance routes, environmental objectives of 2000/60/EC [WFD], soil and water management

Hence, it will be mandatory to consider what impact new flood defence measures will have on the environment. Also, these measures are not to conflict with the goals of the EU WFD of achieving a good ecological status of water bodies. After the extreme flood event along the Elbe River in August 2002, efforts were made to revamp flood management schemes in the river basin. Hence, the construction of an emergency

Citation: K. Lindenschmidt, R. Harrison and M. Baborowski, "Environmental Risk Imposed by Diverted Flood Waters on Water and Soils in Emergency Retention Basins," Journal of Environmental Protection, Vol. 1 No. 2, 2010, pp. 129-135. doi: 10.4236/jep.2010.12017.

retention basin (ERB) system has been proposed for the middle reaches of the Elbe River to reduce flood risk in the area [2]. Emergency retention basins are diked enclosures alongside rivers used to retain flood water by diverting and storing a portion of the river discharge during floods. The diversion from the main river channel to and from the ERBs is controlled in order to attain maximum capping of the peak discharge volume. The floodwater diversion reduces water levels downstream in the river and inundated areas and alleviates stress and damages in regions prone to high flood risk.

Studies on environmental risk assessment on retention basins by flood events are sparse in the literature. Therefore, the aim of the paper was to provide a possible computer modelling methodology on how such assessments can be tackled in fulfilling the proposed EU directives. Further, a preliminary assessment of the environment risk of heavy metal soil contamination and dissolved oxygen depletion in ERBs should be given. To solve these goals, excerpts have been drawn from [3-6].

In detail, the following steps were in the focus of the research:

- develop and test a quasi-2D (two-dimension) model in which the equations of motion and continuity are calculated in 1D (one-dimension) but the discretisation scheme allows for a 2D representation of flow and substance distribution; this minimises computing and pre-processing expenditure without compensating spatial differentiation of flood behaviour,
- test how effective the proposed ERB system is able to cap the peak discharge of the hydrograph from the August 2002 flood,
- investigate the environmental risk of heavy-metal contamination of ERB soils after flood water retention, and
- assess the environmental risk of dissolved oxygen depletion in the water retained in the basins.

STUDY AREA

Two ERBs were investigated in this studie, P1 and P3 (see Figure 1). They are situated between Torgau and Wittenberg. Morphological characteristics of the ERBs are given in Table 1. Several gates are to control the flow of water through the ERBs: an inlet gate at the southernmost tip of P1, an outlet gate at the westernmost point of P3 by Pretzsch and a connecting gate between P1 and P3.

Table 1. Morphological characteristics of emergency retention basins P1 and P3

Polder	Surface area (km^2)	Volume (10^6 m^3)	Head (m.a.s.l.)	Depth (m)
P1	24.5	85	77.5	3.3
P3	8.2	20	75.3	2.3

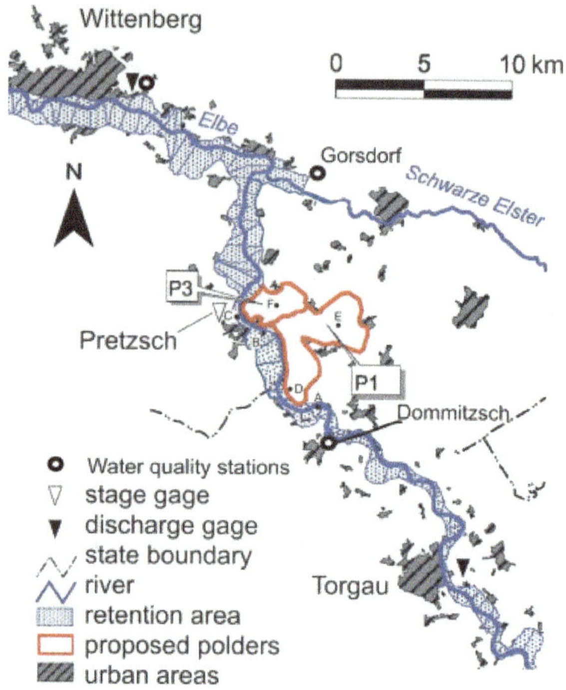

Figure 1. Investigated emergency retention basins (ERB) proposed for the Elbe River reach between Torgau and Wittenberg, modified from [2]

MODELING TOOL AND SETUP

The hydrodynamic module DYNHYD (based on St. Venant equations), the sediment and contaminant transport module TOXI and the water-quality module EUTRO from the WASP5 (Water quality Analysis Simulation Program v.5) modelling package, developed by the U.S. Environmental Protection Agency [7] was used for the floodwater simulations. Figure 2 (left panel) shows the discretisation of the river-ERB system domain in junctions and channels. The junctions ensure volume continuity, hence all junction water volumes represent the total water volume in the system. The channels are the basis for computing the flow in 1D between the junctions. Many channels may branch from a junction allowing the discretisation to take a 2D characterisation. The segment discretisation for the contaminant transport and water-quality model, shown in Figure 2 (right panel), corresponds 1-to-1 the junction discretisation.

In TOXI, processes that describe the transport and fate of heavy-metal contaminants include:

- longitudinal dispersion in the water column,
- vertical diffusion of dissolved substances between the water column and the bottom sediments,
- sedimentation of suspended solids,
- re-suspension of solids from the bottom sediments to the water column, and
- sorption of dissolved substances to suspended and deposited sediments.

EUTRO was used to simulate water quality pertaining to the oxygen balance in a river and ERBs using the cycles of dissolved oxygen (DO) decomposition and nutrient-limited phytoplankton growth. Oxygen is also consumed in the sediments, which is described in the model by the sediment oxygen demand. An important source of oxygen into the water body is re-aeration via the water surface from the atmosphere. An important source and sink of DO are phytoplankton photosynthesis and respiration, respectively. Phytoplankton growth is also light limited and its loss rate is governed by respiration, death, settling and zooplankton grazing.

The DYNHYD, EUTRO and TOXI modules were structured in a simulation platform HLA (High Level Architecture) to allow Monte-Carlo analyses (MOCA) to be carried out [8]. The MOCA allowed uncertainty bounds to be computed in order to assess the degree of risk in exceeding environmental thresholds. For the environmental risk appraisal the following were investigated: 1) the amount of zinc, with high contamination potential along the Elbe [9], sorbed and deposited from the diverted flood waters and 2) the minimum concentration of DO attained in the basin waters.

RESULTS

For a comprehensive description of the development, calibration and validation of the models with corresponding sensitivity analyses of the parameters and boundary conditions, the reader is referred to [6] for the hydrodynamic modelling, [4] for contaminant transport modelling and [5] for the eutrophication modelling. Only the key results to support the environmental assessments will be presented here.

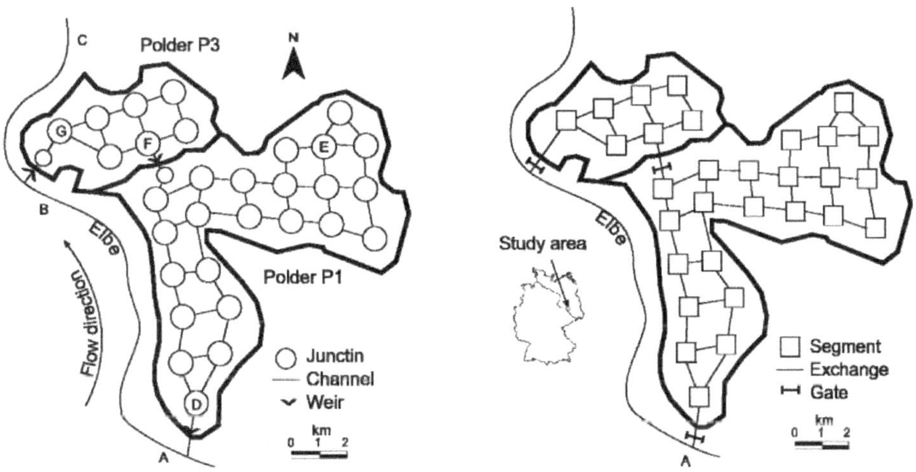

Figure 2. Discretisation for the hydrodynamic model with junctions and channels (left panel) and the water-quality model with segments (right panel), modified from [6] and [5]

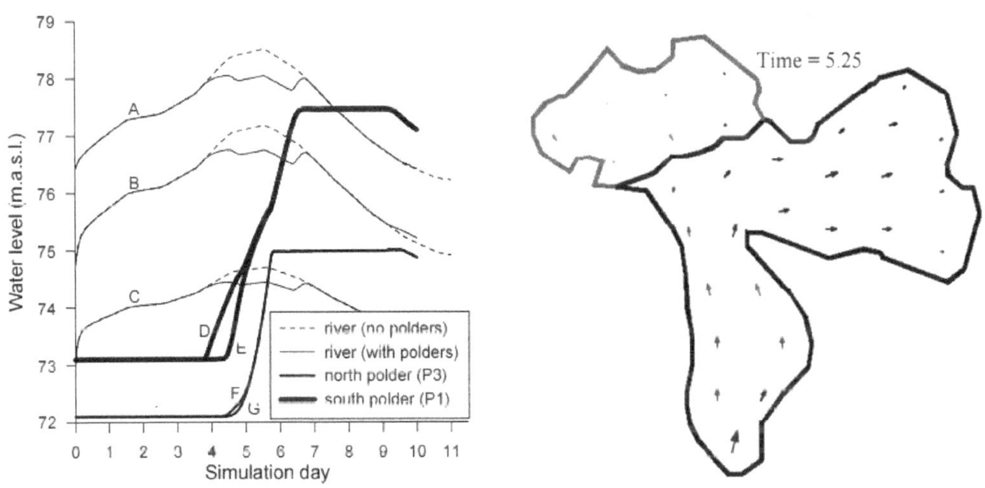

Figure 3. Left panel: water levels in the river and ERBs during operation. Right panel: flow velocity field in the ERBs during filling (longest vector length corresponds to 1.2 m/s), adapted from [6]

Hydrodynamics

Figure 3 (left panel) shows the water level hydrographs of the August 2002 flood simulated along the river at Positions A, B and C (see Figure 2 for spatial reference). The dashed line of the hydrograph crests represent water level heights that occurred without the ERB system, the solid line represents the lowering of the water levels that can be obtained when the ERBs are included in the modelling. The capping lies in the range of 35 to 40 cm and progressively recedes as the flood wave moves downstream (compare Position C). This is due to flood wave attenuation and widening of the floodplain downstream from Pretzsch towards Wittenberg (see Figure 1). Figure 3 (right panel) shows the 2D spatial differentiation of the flow field. Details on the development, calibration and validation of the model and the optimisation of the control of ERB filling are given in [4] and [6].

Soil Contamination

Figure 4 (left panel) shows the results of zinc concentrations in the river and ERBs. The initial concentrations in the ERB segments on the onset of water inflow correspond to the concentrations in the water in the immediate upstream-lying segments. Once the ERBs are filled and the flood gates are closed (approximately after Day 6 for P3 and after Day 7 for P1), there is a substantial drop in substance concentrations in the ERB water column due to sedimentation of suspended sediment. Up to 90% of the sediment settles out of the water column to the bottom sediments. Figure 4 (right panel) shows the spatial distribution of particulate zinc concentrations in the bottom sediments representing zinc deposition during the retention stage of flood water diversion. Particulate zinc is deposited quite quickly when water enters P1. Much of the particulate matter has also been trapped in P1 before the water enters P3. During flood water retention, zinc concentrations are evenly distributed throughout the ERB bottom sediments.

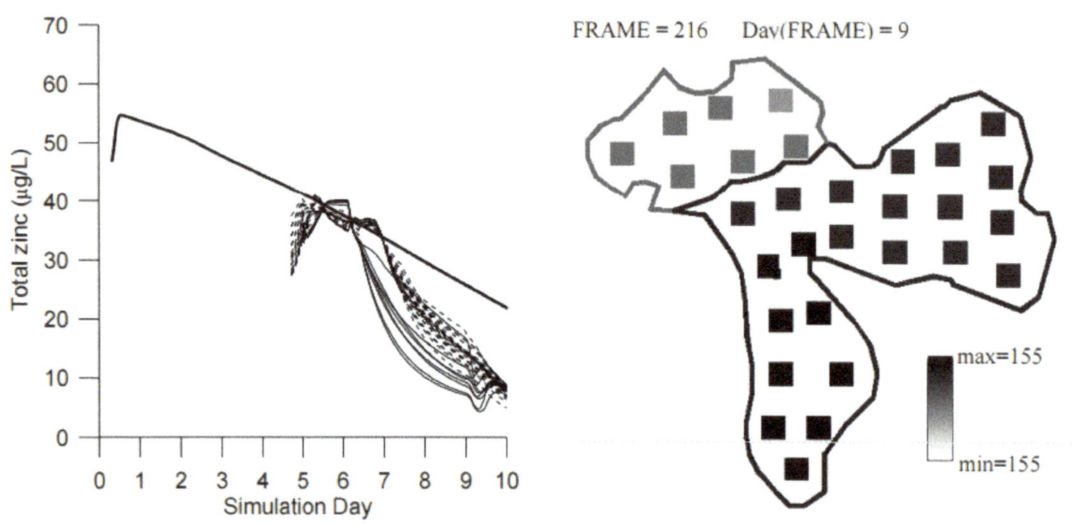

Figure 4. Total zinc (left panel) in the river at Position A (see Figure 2) and in ERBs P1 and P3. Zinc concentration (mg/kg) in bottom sediments during flood water retention [4]

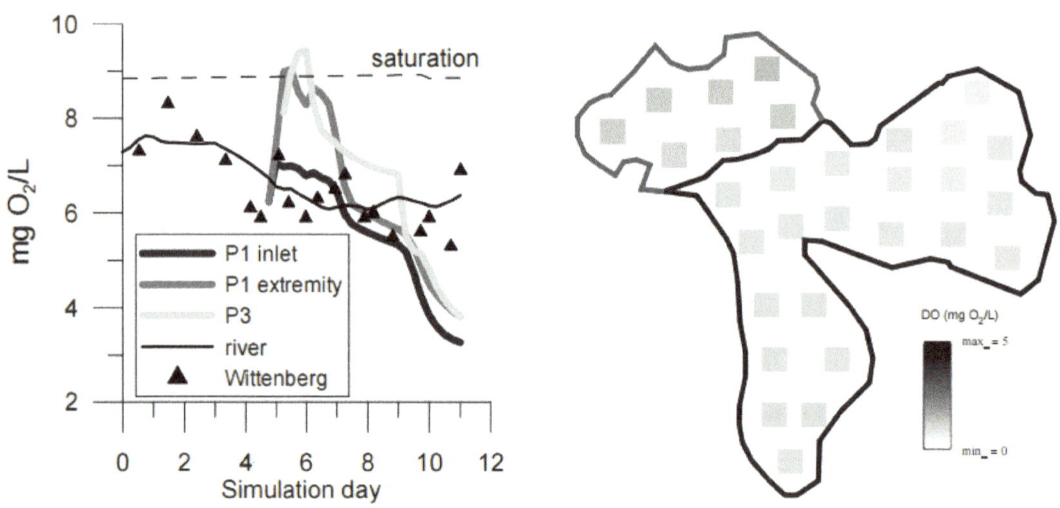

Figure 5. Left panel: DO concentrations in the river and ERBs during floodwater diversion. Right panel: lower 10% quartile of minimum DO at each segment throughout the ERB system [5]

Dissolved Oxygen Depletion

Figure 5 (left panel) shows the DO simulations in the river at Wittenberg and in the ERBs. The DO levels in the river remain substantially below saturation levels. Upon entering the ERBs, diverted flood water receives more oxygen from the hydraulic re-aeration when passing through the inlet gates. The initial flow of shallow waters during basin filling (Day 4 to 7) also contributes to the input of oxygen into the water. There is a progressive decline in DO concentrations during flood water retention (Days 8 and 9) which progresses into the emptying phase (after Day 9).

The lower decile of the minimum DO concentrations is shown spatially in Figure 5 (right panel). The DO levels are higher in P3 than in P1. The extremity (eastern bay) of P1 is generally less oxygenated than the remaining water in the ERB. It should be emphasised that the August 2002 simulation represents a best-case scenario. Additional factors that may impact DO negatively but were not included in the model are 1) higher sediment oxygen demand if surfaces are arable lands and not grasslands and 2) anoxia in the water-soil interface causing re-dissolution of inorganic nutrients from the ERB soils.

ENVIRONMENTAL RISK ASSESSMENT

Soil Contamination

In order to make an environmental risk assessment of contamination in the ERBs, the deposition of pollutants onto the ERB surfaces must be placed within the legal context of the German sewage sludge ordinance. An example is given here using zinc as the heavy-metal contaminant. The ordinance provides threshold values of zinc content allowed in the soils and sediment ("sludge") deposited onto the land surfaces within the basins (summarised in Table 2).

In accordance to the German Sewage Sludge Ordinance, the background zinc concentrations in the soils, where the ERBs are to be situated, must be determined. The geogenic value of 150 mg zinc/kg sediment [10] for the Elbe catchment is used here for illustration. This value does not exceed the maximum allowable zinc content in the soils of 200 mg/kg. Hence, sediment may be deposited from the river water diverted into the ERBs during extreme flooding. The content of zinc within the deposited sediment may not exceed 2500 mg/kg, if the land surfaces are to continue to be used for agricultural purposes. Table 3 summarises the results from the MOCA showing that the probability of reaching this value is minute (< 1%). A different approach based on standards provided by the German Soil Protection and Contaminated Sites Ordinance is given in [5].

Table 2. Threshold values of zinc content in soils and deposited sediment ("sludge") as given by the German ordinances for sewage sludge

	mg zinc / kg sediment	Comment
Sewage sludge ordinance		
max. content in soil allowed	200	soils with high clay content
max. content in sediment allowed	2500	soils with high clay content
Proposed amendments to the sewage sludge ordinance		
max. content in soil allowed	150	for loamy soils
max. content in sediment allowed	1500	

Table 3. Summary of the MOCA giving the probability of certain values of zinc in deposited sediments exceeding 50%, 16% and 1% in ERBs P1 and P3

Polder	Units	Exceedence probability		
		50% (μ)	16% $\approx(\mu+\sigma)$	1% $\approx(\mu+3\sigma)$
P1	mg/kg	460	642	1000
P3	mg/kg	510	626	858

μ - mean; σ - standard deviation

An important question that arises is how many floodings of the ERB can occur before the concentrations in the soil increase from the geogenic level of 150 mg/kg to the threshold value of 200 mg/kg. Figure 4 (right panel) shows a spatial distribution of zinc bound to deposited sediment, which corresponds to an 8% exceedence probability extracted from the MOCA. The zinc content in the soil increases by 5 mg/kg. Hence, approximately ten ERB floodings may cause zinc accumulation to reach the threshold value of 200 mg/kg. Although the environmental risk for zinc contamination is low, it should be noted that an amendment to the sewage sludge ordinance is currently being discussed in which the threshold values for allowed zinc content in soils and deposited sediment ("sludge") are lower (see Table 2). Careful measurements of the background zinc levels in the soils of the ERB sites should be carried out to determine how close these values are to the maximum allowable content in soils.

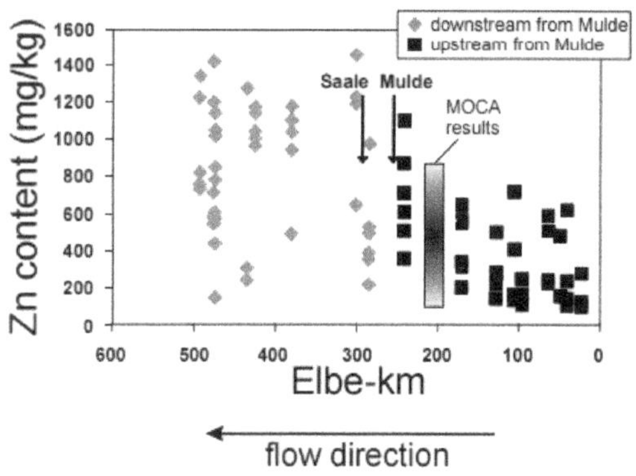

Figure 6. Longitudinal profile of zinc loading on soils along the Elbe River measured in 2003 after the extreme flood in the previous year, modified from [11], p. 242

The environmental risk analyses presented here shows that there is minimal risk of plant contamination by zinc deposited onto the ERBs during floods. However, the risk may be substantially increased if the ERBs are located further downstream. Figure 6 shows the zinc content in deposited sediment along the portion of the Elbe River in Germany. Imbedded in the graph are the results obtained from the MOCA, the range of which complies with the range of all measurements taken upstream from the Mulde River inflow. For the river stretch downstream from the Mulde and Saale tributaries, the range and the maximum of zinc content in deposited sediments is twice as high as for the upstream reach.

Dissolved Oxygen Depletion

From the MOCA runs, probability distributions of the lowest DO levels attained in the ERBs during the simulation time frame were derived. The cumulative frequency distributions for three locations in the ERB system are shown in Figure 7. The threshold value of the minimum concentration of DO in inland waters used for a risk assessment is 3 mg O_2/L—the minimum oxygen content with which most fish in the Elbe River can endure for a short term [12]. As soon as the oxygen content sinks below approximately 3 mg/L, the fish begin to search for water pockets with better oxygen supply or begin to snap for air at the water surface [13].

For a summer time extreme flood such as the August 2002 event along the Elbe, the probability of the DO values dropping below a threshold value of 3 mg O2/L is quite high, 77% for the inlet region of P1, 65% for the far eastern point of the P1 extremity and 54% for P3 (see Figure 7). In general, the DO concentrations are higher more frequently in P3 than in P1 due to the shorter residence times of the water in this ERB and allowing less time for de-oxygenation of the water. The generally higher DO content in P3 is also due to its shallower depth, making it a more favourable aquatic environment for phytoplankton growth. The low DO values are also a result of low re-aeration (wind calm period) and high nitrification.

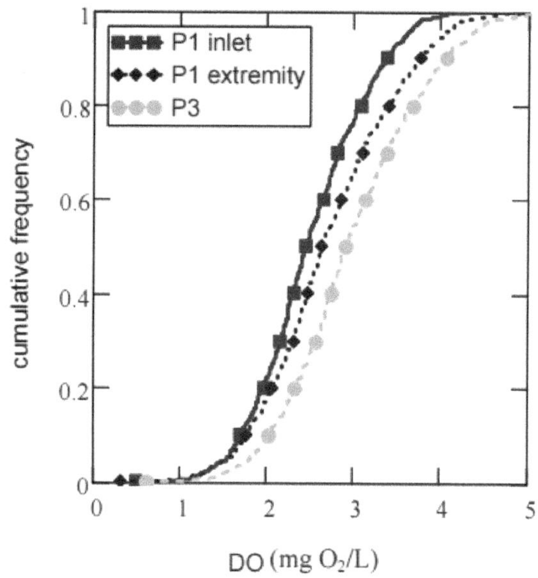

Figure 7. Cumulative frequency distributions of the lowest dissolved oxygen concentrations attained in the ERBs

In contrast to experiences gained from the Havel River during the August 2002 flood, the low oxygenated water emptying into the River Elbe does not appear to have negative implications to the oxygen levels in the river water. Although the water of the Elbe River was under the DO saturation level, DO was still above the threshold value of 3 mg/l at Wittenberg (see Figure 8). However, if these ERBs were situated further downstream where the DO concentrations of the river water fall below the threshold value, the emptied basin water could reduce the DO concentrations further and further stress the aquatic ecosystem.

CONCLUSIONS

The quasi-2D approach was successful in simulating the spatial distribution of flow and substance deposition in ERBs and determining the effectiveness of discharge capping. The environmental risk to heavy-metal contamination is minimal but may increase due to stricter threshold values. Risk may also be higher for ERBs constructed further downstream past the Saale and Mulde river confluences where exposure to contamination is higher. There is a high potential for DO levels in the water retained in the basins to drop to very critical levels (< 3 mg O_2/L), even after 4 days of flood water retention. From an ecological perspective, the basins should also be emptied as soon as possible after the flood in order to reduce the time of de-oxygenation of the water.

REFERENCES

1. K.-E. Lindenschmidt, F. Hattermann, V. Mohaupt, B. Merz, Z. W. Kundzewicz and A. Bronstert, "Large-Scale Hydrological Modelling and the Water Framework Directive and Floods Directive of the European Union," Advances in Geosciences, Vol. 11, 2007, pp. 1-6.

2. IWK, "Untersuchung von Hochwasserretentions maßnahmen entlang der Elbe im Bereich der Landkreis Wittenberg und Anhalt-Zerbst" (Kurzfassung), Institute für Wasserbau und Kulturtechnik, Universität Karlsruhe, 2004.

3. K.-E. Lindenschmidt, "Quasi-2D Approach in Modelling the Transport of Contaminated Sediments in Floodplains during River Flooding-Model Coupling and Uncertainty Analysis," Environmental Engineering Science, Vol. 25, No. 3, 2008, pp. 333-352.

4. K.-E. Lindenschmidt, S. Huang and M. Baborowski, "A Quasi-2D Flood Modelling Approach to Simulate Substance Transport in ERB Systems for Environmental Flood Risk Assessment," Science of the Total Environment, Vol. 397, No. 1-3, 2008, pp. 86-102.

5. K.-E., Lindenschmidt, I. Pech and M. Baborowski, "Environmental Risk of Dissolved Oxygen Depletion of Diverted Flood Waters in River ERB Systems—A Quasi- 2D Flood Modelling Approach," Science of the Total Environment, Vol. 407, No. 5, 2009, pp. 1598-1612.

6. S. Huang, J. Rauberg, H. Apel, M. Disse and K.-E. Lindenschmidt, "The Effectiveness of ERB Systems on Peak Discharge Capping of Floods along the Middle Reaches of the Elbe River in Germany," Hydrology and Earth System Sciences, Vol. 11, 2007, pp. 1391-1401.

7. R. B. Ambrose, T. A. Wool and J. L. Martin, "The Water Quality Simulation Program," Water Quality Analysis Simulation Program v.5, U.S. Environmental Protection Agency, Athens, GA, 1993.

8. K.-E. Lindenschmidt, J. Rauberg and F. Hesser, "Extending Uncertainty Analysis of a Hydrodynamic—Water Quality Modeling System Using High Level Architecture," Water Quality Research Journal of Canada, Vol. 40, No. 1, 2005, pp. 59-70.

9. M. Baborowski, W. von Tümpling and K. Friese, "Behaviour of Suspended Particulate Matter and Selected Trace Metals during the 2002 Summer Flood in the River Elbe at Magdeburg Monitoring Station," Hydrology and Earth System Sciences, Vol. 8, No. 2, 2004, pp. 135-150.

10. F. Krüger, A. Prange, E. Jantzen, K. Trejtnar and G. Miehlich, "Geogene Hintergrundwerte," Wasserwirtschaft Wassertechnik, Vol. 7, 1998, pp.16-19.

11. W. Geller, K. Ockenfeld, M. Böhme and A. Knöchel (Eds.) "Schadstoffbelastung nach dem Elbe-Hochwasser 2002," 2004, p. 460.

12. ARGE-Elbe, "Gewässergütebericht der Elbe 2004," Arbeitsgemeinschaft für die Reinhaltung der Elbe, 2005, p. 6.

13. M. Böhme, F. Krüger, K. Ockenfeld and W. Geller, "Schadstoffbelastung nach dem Elbe-Hochwasser 2002," 2005, p. 69.

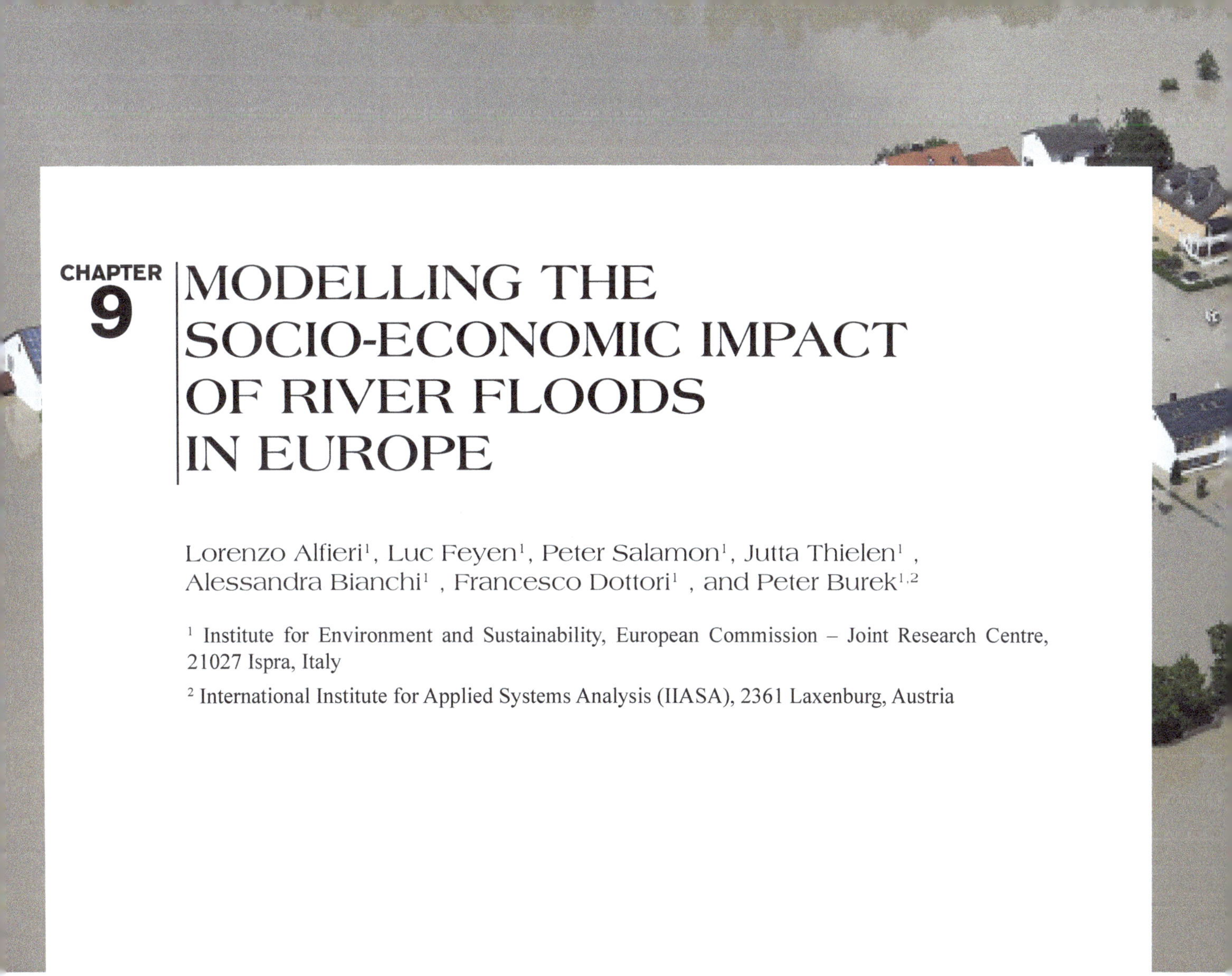

CHAPTER 9

MODELLING THE SOCIO-ECONOMIC IMPACT OF RIVER FLOODS IN EUROPE

Lorenzo Alfieri[1], Luc Feyen[1], Peter Salamon[1], Jutta Thielen[1] , Alessandra Bianchi[1] , Francesco Dottori[1] , and Peter Burek[1,2]

[1] Institute for Environment and Sustainability, European Commission – Joint Research Centre, 21027 Ispra, Italy

[2] International Institute for Applied Systems Analysis (IIASA), 2361 Laxenburg, Austria

ABSTRACT

River floods generate a large share of the socioeconomic impact of weather-driven hazards worldwide. Accurate assessment of their impact is a key priority for governments, international organization, reinsurance companies and emergency responders. Yet, available databases of flood losses over large domains are often affected by gaps and inconsistencies in reported figures. In this work, a framework to reconstruct the economic damage and population affected by river floods at continental scale is applied. Pan-European river flow simulations are coupled with a high-resolution impact assessment framework based on 2-D inundation modelling. Two complementary methods are compared in their ability to estimate the climatological average flood impact and the impact of each flood event in Europe between 1990 and 2013. The event-based method reveals key features, such as the ability to include changes in time of all three components of risk, namely hazard, exposure and vulnerability. Furthermore, it skilfully reproduces the socio-economic impact of major flood events in the past two decades, including the severe flooding hitting central Europe in June 2013. On the other hand, the integral method is capable of reproducing the average flood losses which occurred in Europe between 1998 and 2009. Strengths and limitations of the proposed model are discussed to stress the large potential for filling in the gaps of current datasets of flood impact.

INTRODUCTION

The devastation caused by severe river floods in different areas of the world is brought to the people's attention by the media on a daily basis. In its physical abstraction, "flooding" is simply a temporary covering by water of land not normally covered by water (European Commission, 2007). Hence, the interest in floods grows only when such land is occupied by dwellings and businesses, which are disrupted by the flood inundation. Yet, the history of river flooding and its socio-economic impact has shaped the development of civilizations for millennia (Chen et al., 2012; Di Baldassarre et al., 2013). The latest relevant directives on flood prevention and preparedness define the flood risk as a product of flood hazard, exposure (of population and assets) and their vulnerability (European Commission, 2007; Kron, 2005; UNISDR, 2009). Each of those three components can change in space and in time, following population growth along rivers, migration fluxes, river flood protection and changes in climatic extremes, among others. In addition, effective adaptation measures to reduce the

Citation: Alfieri, L., Feyen, L., Salamon, P., Burek, P., & Thielen, J. Modelling the socio-economic impact of river floods in Europe. Natural Hazards And Earth System Sciences Discussions, 1-14. http://dx.doi.org/10.5194/nhess-2015-347.

flood impact can be achieved by reducing one or more of those three risk components (Alfieri et al., 2016; IPCC, 2012).

Flood risk is characterized by strong spatial variability, due to its dependence on land use, local infrastructure and the elevation and distance in relation to the surrounding river network. The use of high-resolution datasets and methods are of utmost importance to achieve a meaningful mapping of the flood risk. Physically consistent delineation of the flood extent and depth for a range of event magnitudes is derived through hydraulic floodplain models, commonly used in regional flood risk assessments (Broekx et al., 2011; Falter et al., 2015; Foudi et al., 2015; Koivumäki et al., 2010; te Linde et al., 2011). The computing resources needed by these models grow fast with the simulation area, so that applications to large river basins are rare (Falter et al., 2016; Schumann et al., 2013). Because of such a constraint, pan-European and global flood hazard mapping have been traditionally performed through simplified approaches based on topographic indices (e.g. Lugeri et al., 2010), on planar approximation of water levels (Barredo et al., 2007; Rojas et al., 2013) or at coarser resolution, often coupled with methods for downscaling to finer grid resolution (Hirabayashi et al., 2013; Pappenberger et al., 2012; Winsemius et al., 2013). Furthermore, large-scale flood risk assessments produced so far were focused on evaluating the average flood impact over long periods, assuming a stationary climate, while no research has yet considered the actual statistical frequency of occurrence of past extreme events at the pan-European scale. Recent advances in large-scale flood hazard mapping led to European and global inundation maps derived through high-resolution 2-D hydraulic modelling (Alfieri et al., 2014a; Sampson et al., 2015), opening new opportunities in global and continental flood impact assessment.

In this work, we assess the impact of river floods in Europe by coupling continental-scale hydrological simulations between 1990 and 2013 with high-resolution mapping of the potential impact of floods for different return levels. Resulting flood risk is expressed in terms of direct damage and population affected and aggregated over different spatial and temporal scales. We show two complementary approaches for flood risk assessment, namely an integral method and an event-based method, using the same underlying data and models. In addition, the event-based method is used to assess the impact of the severe flood hitting a vast portion of central Europe in June 2013, and results are evaluated against reported figures from reinsurance companies and post-event reports. This research shows the first large-scale application of the impact model based on a high-resolution observational meteorological dataset. A modified version of this risk assessment framework was recently used by Alfieri at al. (2015) to project the future flood risk in Europe under climate change. Flood impact results are compared with available validation data and are then discussed by stressing the strengths and limitations of both methods.

DATA AND METHODS

The proposed approach follows a modelling framework composed of five steps (see Fig. 1), described as the following: (1) continuous hydrological simulation, (2) extreme value analysis, (3) flood inundation modelling, (4) impact modelling and (5) flood risk assessment.

Continuous Hydrological Simulation

Hydrological simulations were performed with the Lis- flood distributed model (Burek et al., 2013; van der Knijff et al., 2010). Processes simulated by Lisflood include snowmelt, soil freezing, surface run-off, infiltration, preferential flow, and redistribution of soil moisture within the soil pro- file, drainage of water to the groundwater system, groundwater storage and base flow. Run-off produced for every grid cell is routed through the river network using a kinematic wave approach. For this work we used the operational dataset of the European Flood Awareness System (EFAS, see Alfieri et al., 2014b; Thielen et al., 2009), in the form of daily streamflow in the European river network from 1 January 1990 to 31 December 2013. Streamflow maps at 5 km grid resolution covering an area of 6.5 million km2 are produced by forcing Lisflood with the EFAS-Meteo dataset (Ntegeka et al., 2013), which includes daily precipitation, temperature and evapo-transpiration maps derived through spatial interpolation of point observations. The current Lis- flood version is calibrated at 693 stations across Europe for up to 8 years of daily observed discharge. The calibration work was performed using the R package "hydroPSO", which implements a state-of-the-art version of the Standard Particle Swarm Optimization 2011 (Zambrano-Bigiarini and Rojas, 2013). Some performance of the calibrated stations are shown in the Supplement (Fig. S1 and Table S1). In addition, extensive work is continuously carried out on the model to improve the representation of hydrological processes, the parameter calibration, the simulation of key features along the river network and the improvement of the underlying spatial datasets. Recent improvements include the parameterization of 182 lakes and 34 large reservoirs and the simulation of water withdrawals through monthly maps of water use from the SCENES (Water Scenarios for Europe and Neighbouring States) project (Kamari et al., 2008).

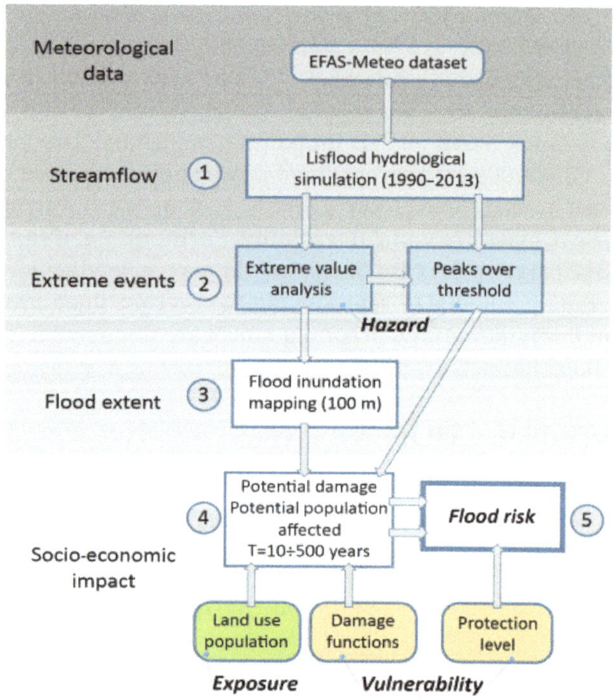

Figure 1. Schematic view of the proposed modelling framework for flood risk assessment. The three components of the risk formula, namely hazard, exposure and vulnerability, are highlighted with colours.

Extreme Value Analysis

Simulated streamflow maps from 1990 to 2013 were analysed statistically to estimate analytical curves relating extreme flow peaks to their probability of occurrence and consequently to their return period. For each grid point of the European river network, the set of maximum annual discharge peaks was fitted with a Gumbel extreme value distribution, using the L-moment approach (see Hosking, 1990). L-moment estimators are nearly unbiased for a wide range of sample sizes and distributions (Vogel and Fennessey, 1993) and are particularly useful for relatively short samples as in this study. In addition, we calculated the mean of the annual maximum discharge over different durations, from 1 up to 30 days, based on the same sample of streamflow data. The resulting information provides key descriptors of the peak flow and the related hydrograph volume, and is thus used to produce coherent flood hydrographs for any point of the river network using the procedure described by Maione et al. (2003).

Flood Inundation Modelling

Flood inundation maps for the entire European domain were produced at 100 m resolution using the Lisflood-FP floodplain model (Bates et al., 2010; Neal et al., 2012) forced by the flood hydrographs with specific return period described in the previous section. The full procedure for deriving panEuropean flood hazard maps is described in detail by Al- fieri et al. (2014a), together with some performance scores of the 1-in-a-100-year simulated map vs. official regional maps available for Germany and the UK. The procedure was semi-automated to speed up the computing time and was then repeated for six different return periods commonly considered in similar flood hazard mapping (e.g. Sampson et al., 2015; Winsemius et al., 2016): T = {10, 20, 50, 100, 200, 500} years. Each flood hazard map defines the maximum flood depth and extent caused by the corresponding flood return period and is produced by merging the results of over 37 000 hydraulic simulations along the European river network. Visual examples of the maps produced can be seen in Fig. S2.

Impact Modelling

The impact model used in this study is focused on estimating the population affected and the direct economic damage due to river floods. The potential population affected (PPA) by floods of a specific return period is estimated by overlaying the corresponding flood hazard map with the 100 m resolution map of European population density by Batista e Silva et al. (2013).

The potential damage (PD) of floods is estimated through functions relating the flood depth to the corresponding direct damage. For this task we used the country specific depthdamage functions defined by Huizinga (2007) for different land uses, while the spatial variability in exposure is determined according to the refined version of the Corine Land Use provided by Batista e Silva et al. (2013). Depth-damage functions per country and land use class comprise two damage indicators (Huizinga, 2007): an absolute damage value in EUR m^{-2} , which is attributed to all flood depths equal or larger than 6 m; and a damage factor relative to the maximum damage (i.e. between 0 and 1), which is defined by piece-wise linear functions. Those two indicators are derived through analysis of written documentation and data on the internet from 31 countries in Europe. To account for the large regional differences in exposed assets for a given land use class that exist in some countries, country specific depthdamage functions were further rescaled by the Gross Domestic Product (GDP) per capita of NUTS 2 (Nomenclature of territorial units for statistics) administrative level (see http://ec.europa.eu/eurostat/web/nuts/overview). Impact data on damage and population affected by flood peaks with selected return period are first assessed at 100 m resolution and then aggregated to 5 km resolution maps through the corresponding Areas of Influence (AoI) as defined by Alfieri et al. (2015). AoI are polygons at 100 m resolution which define a univocal link between the high- and low-resolution maps by assigning the flooded area to the 5 km grid point of the causative inflow hydrograph.

Flood Risk Assessment

Flood risk is the combination of the impact of events and their frequency of occurrence. Here, it is assessed through two different approaches: (1) an integral method and (2) an event-based method. The integral method estimates the average annual impact of floods by computing a piece-wise integral of the damageprobability curve for a selected range of return periods, as done by Feyen et al. (2012), Rojas et al. (2013) and Winsemius et al. (2013) in previous European and global risk assessments. For both damage and population affected, the integral sum is truncated at the return period of the protection level of the corresponding location, assuming that no impact occurs for events of lower magnitude. In this step, we used the European flood protection map derived by Jongman et al. (2014).

The event-based method estimates the damage of each simulated flood, rather than considering the theoretical probability of occurrence. It is based on a selection of all discharge peaks over threshold (POT) exceeding the flood protection level (by Jongman et al., 2014) at any location. For each discharge peak, first the return period is calculated through the corresponding analytical extreme value distribution estimated in Sect. 2.2. Then, it is assigned a value of direct damage and population affected by interpolating linearly between the impacts of the two closest return periods among those available. As in the previous method, discharge peaks exceeding the 500-year return period are assigned impact values corresponding to the 500-year event. Annual damage and population affected are estimated by summing the impact values of all events simulated within the year.

RESULTS

The two approaches described in Sect. 2.5 were run over the time span 1990–2013 on a large portion of Europe, including 26 countries of the European Union (all except Malta and Cyprus), Norway and the Republic of Macedonia. Impact values are based on population density estimates of 2006 and GDP Purchasing Power Standards (PPS) of 2007. Average annual estimates of damage (AD) and population affected (APA) are shown in Fig. 2, spatially aggregated for the 28 considered countries. Values plotted in Fig. 2 are expressed as ratios of the respective country GDP and country population, while absolute values are shown as labels alongside each colour bar. Blue bars are based on statistical frequencies of occurrence, hence are representative of the country's average vulnerability to extreme events. The largest relative damage is found in Hungary, at 2.6‰ of the country GDP, while about 1.9 ‰ of the Dutch population is estimated to be affected annually by river floods. On the other hand, values in green give an indication of the impact of extreme events occurring in the simulation period 1990–2013. The largest relative damage was found in the Baltic states, while the ratio of population affected was the largest in Croatia. When absolute impact values are considered, the largest annual damage is found in Italy (EUR 929 and 645 million per year for the integral- and event-based approaches, respectively) while population affected is the highest in Germany (40 000 pp yr^{-1} in both methods).

Additional information on the attribution of the flood risk in different areas of Europe can be obtained by looking at the different components contributing to the risk formula, namely hazard, exposure and vulnerability. Those are shown in Fig. 3, in the form of the maximum simulated flood return period (T) within 1990–2013 (Fig. 3a); potential population affected by a flood with a 100-year return period (Fig. 3b); potential damage of a flood with a 100-year return period (Fig. 3c) and return period of flood protection levels (Fig. 3d). Figure 3 denotes a rather complex distribution of the three risk components, though with a general trend of high exposure and low vulnerability in central Europe and England, which is opposed to a higher vulnerability and lower exposure in the outer regions. The hazard component in Fig. 3a gives information on the most extreme events simulated in the reference period 1990–2013. It is the main reason for the differences between event-based and integral method in Fig. 2, as the latter considers theoretical probabilities of occurrence of extreme events in place of their actual statistical distribution. Indeed, one can note how the most severe events are found in the Baltic states, eastern Europe and the Iberian Peninsula, thus justifying the corresponding larger impact estimates of the event-based method in Fig. 2.

Estimates of flood impact at the European Union level are compared with figures by the European Environment Agency (EEA, 2010), who reported an average damage around EUR 5 billion per year and population affected of 250 000 pp yr^{-1} over the time window 1998–2009. Aggregated figures for the same years derived through the event-based method herein proposed lead to EUR 5.4 billion per year and 220 000 pp yr^{-1}, while long-term averages of the integral method indicate EUR 5.9 billion per year and 239 000 pp yr^{-1}.

Integral Method

The integral method is based on statistical frequencies of the occurrence of extreme events derived by analytical distributions. Hence, estimates of the average impact made with such an approach are also robust when applied to relatively small spatial aggregations. Estimates of annual damage and population affected at the NUTS 2 aggregation level are shown in Fig. 4. Average estimated impact among all the considered NUTS 2 regions amounts to EUR 21 million per year of damage and 900 pp yr^{-1} of population affected. Despite the high protection standards assumed for the whole of the Netherlands (1 in 1000 years, Jongman et al., 2014), the highest estimated impacts among NUTS 2 regions are found in South Holland (EUR 255 million per year of damage and 19 000 pp yr^{-1}) as a consequence of the considerable exposure of people and assets in case of extreme events exceeding the flood protection level. Other regions at high risk are found in the north of Italy, France, Croatia, Austria and Hungary. It is worth noting that the presented approach is focused on rivers with upstream areas larger than 500 km^2 . Hence, the flood risk is likely to be underestimated in regions where the hydrography is dominated by smaller streams (e.g. coastal regions of Greece, southern Italy, Croatia, Norway, the UK, Denmark, as well as some mountainous regions in the Alps) where local storms and flash floods are major components of the overall impact of floods. Similarly, the impact of coastal floods is not modelled in the results shown.

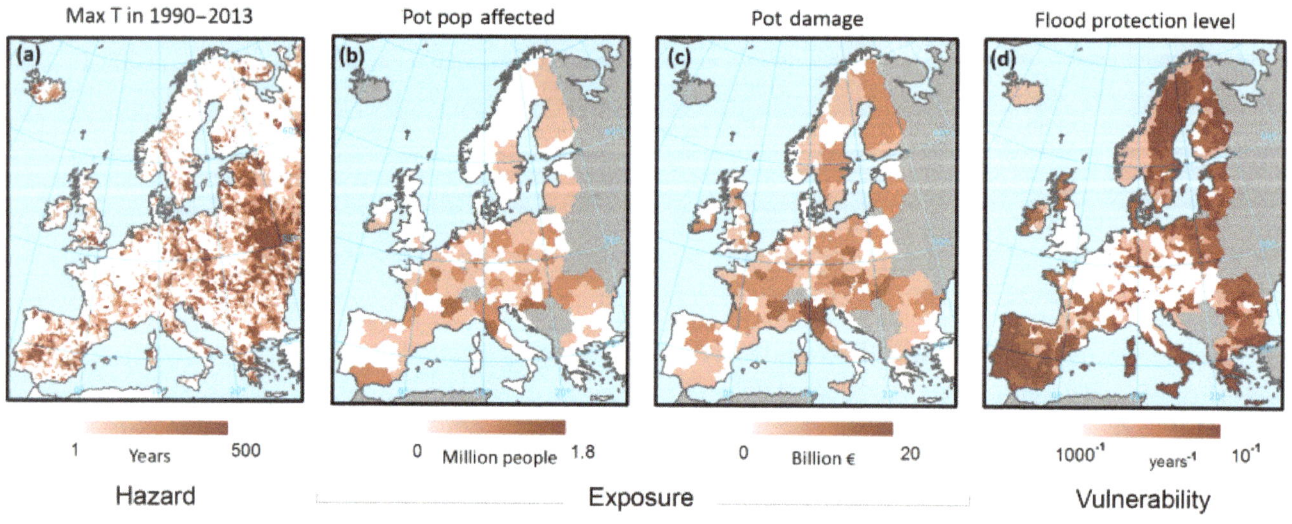

Figure 3. Modelled factors contributing to the overall flood risk: maximum simulated flood return period (T) within 1990–2013 (a); potential population affected by a flood with a 100-year return period (b); potential damage of a flood with a 100-year return period (c); return period of flood protection levels (d).

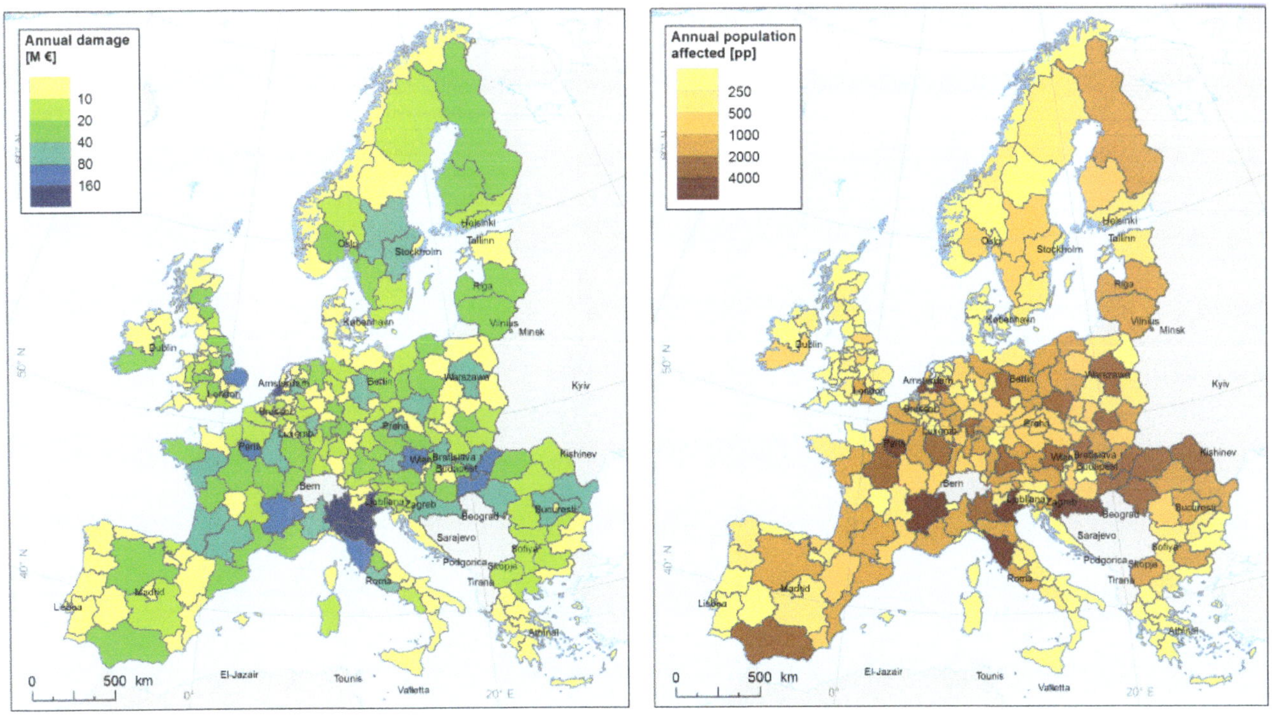

Figure 4. Mean annual damage and population affected at NUTS 2 level estimated through the integral method. Areas outside the simulation domain are masked in grey.

Event-based Method

The peculiarity of the event-based method is its applicability to any desired time window, from single events to annual aggregations. On the other hand, observed impact data of past floods are scarcely available, and in addition are usually affected by considerable uncertainty levels (Merz et al., 2010; Penning-Rowsell, 2015). A report by Fenn et al. (2014), prepared for the European Commission Directorate-General for the Environment (DG Env), includes an assessment of financial, economic and social impacts of river floods in the countries of the European Union between 2002 and 2013, based on post-event reports and estimates from insurance companies. Fenn et al. (2014) addressed the scarcity of flood impact data by extrapolating the cost of major floods in the European countries on the basis of the available data, so that the overall estimated flood impact is given by the sum of extrapolated and quantified data. Figure 5 compares annual flood damage aggregated over the European Union of the event-based method from 1990 to 2013 and data by DG Env for the available years. Data from the two datasets are in good qualitative agreement, with a Pearson correlation coefficient of R = 0.69. The largest discrepancies are found in those years when the damage was higher (i.e. in 2002, 2010, 2013), though one should consider the substantial proportion of extrapolated damage in the validation dataset by DG Env (orange bars in Fig. 5) which poses the issue of its accuracy. The event-based approach proves its ability to spot the years when the most severe flood events occurred, including the "millennium flood" of the Oder (and Vistula) river which hit Poland, Czech Republic and Germany in 1997 causing more than 100 fatalities and material damages estimated at USD 5.7 billion (EM-DAT, 2015).

Case study – central European floods in 2013

The catastrophic floods hitting central Europe in June 2013 were selected as a case study to test the performance of the event-based method for rapid risk mapping. This was a severe, large-scale event which affected several countries and led to the loss of lives as well as considerable damage in the Danube and Elbe river basins. The event was associated with a quasi-stationary upper-level low located north-east of the Alps and characterized by a significant contribution of orographic lifting (Pappenberger et al., 2013). Also, in the weeks leading up to the event, rainfall accumulations were significantly above normal in large parts of central Europe, exacerbating the run-off process. The return period of the discharge peaks was estimated to equal or exceed 100 years in various rivers including the Isar, Inn, Salzach, Danube, Elbe, Mulde, Saale, Rhine and Neckar (Zurich, 2014). Figure 6 shows maps of damage and population affected in central Europe, based on the simulated discharge maps from 25 May to 10 June 2013. Impact data in the figure are aggregated over NUTS 2 regions, while grey circles indicate hotspots of simulated damage larger than EUR 100 million and population affected in excess of 5000. Aggregated estimates of direct damage in Germany, Austria and Czech Republic amount to EUR 10.9 billion and 360 000 people affected by the flood event. These estimates are in agreement with reported figures ranging between 11.4 and EUR 16 billion (Aon Ben- field, 2014; Munich Re, 2014; Swiss Re, 2013), especially if one considers that the higher estimates from insurance companies account for total combined economic losses, rather than just direct damage. As an indication, indirect costs are commonly estimated within 5 and 40 % of direct damage (Jongman et al., 2012).

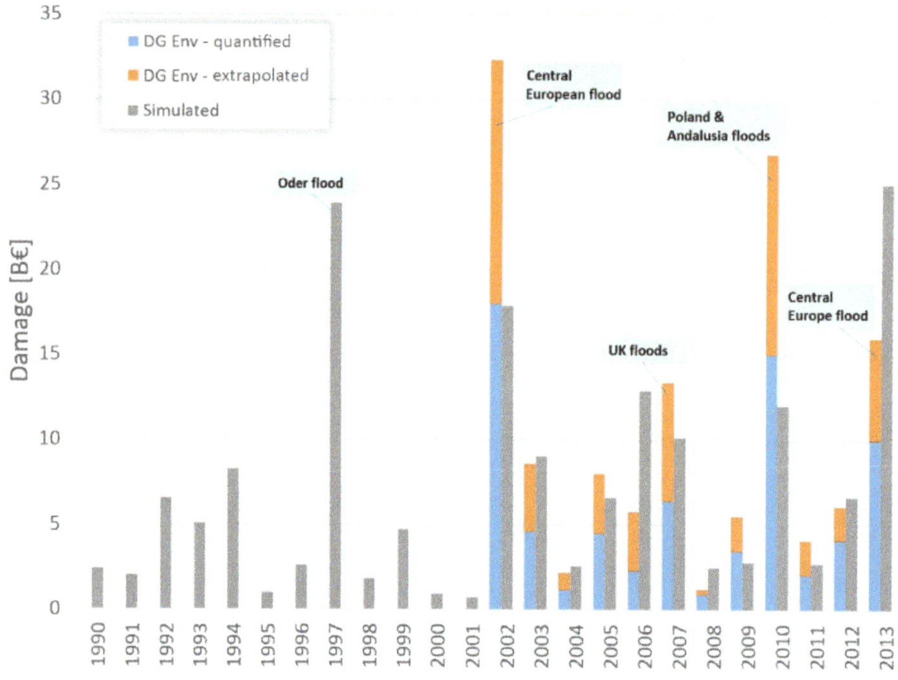

Figure 5. Annual estimates of flood damage in the EU over 1990– 2013 (PPS of 2013) using the event-based method (in grey) and comparison with data by DG Env (Fenn et al., 2014).

The spatial distribution of estimated damage and population affected shown in Fig. 6 correctly locates hotspots of observed flood damage during the event. Among the most affected areas, extensive damage from flooding was reported along the Elbe river and its tributaries Mulde and Saale in central Germany and western Czech Republic (Schröter et al., 2015); along the Danube, particularly near the German– Austrian border and its tributaries Inn, Isar, Aist, Kamp (Blöschl et al., 2013) and in the Neckar river basin in the region of Stuttgart in south-western Germany, often as a consequence of major dyke breaches (Schröter et al., 2015; Zurich, 2014). In comparison to reported figures, major differences with the proposed approach are the underestimation of the flood impact for the city of Prague (no simulated impact) and neighbouring areas along the Vltava and Elbe rivers, and the overestimation of the flood damage in the Austrian regions of Salzburg and Upper Austria (EUR 1 billion reported vs. EUR 5 billion simulated). In the first case, reasons are found for the underestimation of the simulated discharge along the Vltava and the upper Elbe rivers (see Supplement material) and of the consequent inundated area. In the case of Austria, the actual impact was lower than our estimates as a result of the significant investments to improve flood protection following the disastrous floods of 2002 (Blöschl et al., 2013; Fenn et al., 2014), which is not reflected in the flood protection map used in our simulations.

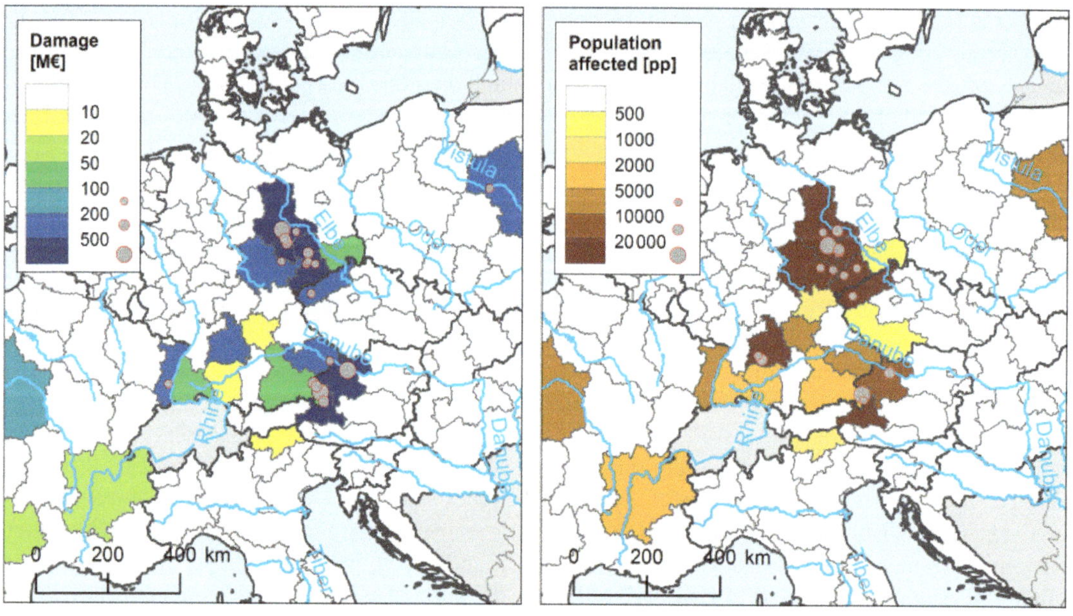

Figure 6. Estimates of damage and population affected (event-based method) in central Europe from 25 May to 10 June 2013. Grey circles indicate hotspots of simulated damage larger than EUR 100 million and population affected in excess of 5000. Areas outside the simulation domain are masked in grey

DISCUSSION

This work compares two methods of flood risk assessment based on a modelling framework and aims to stress their complementarity of use. While the integral approach has been extensively used and discussed in past research works, the event-based method has only found a few applications so far, due to the larger data requirements on the past climatic forcing in the form of seamless space–time hydrometeorological datasets. Hence, a dedicated discussion section is presented in the following.

Event-based Method

When the event-based method is applied to sufficiently long time windows adequately representing the climatic variability, it can be used to estimate average flood losses, which can be compared to estimates of the integral method, as shown in Fig. 2. Yet, the event-based approach is not suitable for estimating average flood losses in case of a small sample of extreme events, as it would give the estimate insufficient robustness. This normally occurs when the aggregation area is not large enough or when the average protection level is particularly high, making flood inundation an extremely rare event. Is these cases, one can see substantial differences in results obtained with the two proposed methods (see Fig. 2), as for Luxembourg and the Netherlands, where no event above the protection level was simulated in the years 1990–2013. It is worth noting that, being based on actual event occurrences, the event-based method is more suitable to account for non-stationary climates such as under changes in the frequency and magnitude of extreme floods (see Alfieri et al., 2015). Similarly, it can account for temporal changes of the flood exposure (e.g. through land use changes and population growth) and vulnerability (e.g. through increase in the flood protection levels or reduction of the damage potential). The proposed approach assigns each peak over threshold with specific return period a precomputed hydrograph with the same flood frequency and daily discharge values consistent with the flow duration curve of the simulated time series. In comparison to fixed hydrograph shapes, this approach is more appropriate to reproduce coherent hydrograph shapes and their statistics, occurring in large model domains featured by a variety of climate regimes. It assigns a univocal link between peak flow and corresponding hydrograph, hence neglecting the impact variability of hydrographs with same peak discharge and different flood volume (see Falter et al., 2015). Yet, it has the advantage of drastically reducing the computing resources needed, making high-resolution impact modelling possible on large-scale applications.

The Influence of Flood Protections

As previously noted by Ward et al. (2013), impact estimates of assessment frameworks based on flood threshold exceedance are very sensitive to the assumed flood protection standards. This issue may be exacerbated in the eventbased method, where the magnitude of events is estimated from the simulated peak discharge, and their frequency might deviate from the theoretical one, assumed by their statistical extreme value distribution. In the limit abstraction of peak flow magnitude close to the flood protection level, one would obtain either no impact or very high impact following minor deviations changing the relative rank of these two values. Flood risk assessments are affected by various sources of uncertainty (Apel et al., 2008; Koivumäki et al., 2010), hence a probabilistic approach is likely to bring benefits to the methodology by accounting for the uncertainty range around the central impact estimate. In this work, a quantitative characterization of uncertainty was deliberately not included, so to stress the comparison between the two approaches shown. In observed events, the impact of flood inundation is often caused by the collapse of flood protection measures at specific sites, which concentrates the damage in a certain area and reduces the likelihood of more flooding downstream due to the floodplain storage and consequent reduction of the peak discharge (e.g. Zurich, 2014). Research work by Hall et al. (2005) tackled this process by including estimates of the probability of defense failure for each river section, under a range of load conditions. This approach might be difficult to apply at continental scale, due to the difficulty in gathering detailed information for such an extensive river network. However, simplified approaches could prove beneficial in improving the impact estimates.

A further comment must be addressed to the so-called "adaptation effect" (see Di Baldassarre et al., 2015), which is related to a reduction of vulnerability following disastrous flood events, due to the increased awareness and the implementation of structural and non-structural flood protection measures. The benefits of adaptation to floods have been demonstrated in a number of research works and can be measured in a clear reduction of the impact on economy and society in regions hit by series of floods within a time frame of few years (Blöschl et al., 2013; Bubeck et al., 2012; e.g. Jongman et al., 2015; Wind et al., 1999). If the adaptation effect is not considered in the impact assessment, flood risk is likely to be overestimated in those areas hit by a series of floods within a relatively short time range, as seen in Sect. "Case study – central European floods in 2013". Yet, this process can be accounted for in the proposed event-based method, provided that reliable vulnerability information is soon made available to improve future flood risk assessment.

Sources of Uncertainty

In addition to the uncertainty related to the flood protections, the impact model discussed here is affected by a number of uncertainty sources which affect each stage of the modelling chain. Although no quantitative uncertainty analysis was performed in this work, a non-exhaustive list of potential uncertainty sources is presented in the following.

- Meteorological data: it is mostly associated to the transformation of point measurements to a seamless space– time dataset. Largest uncertainty is located in those areas with lower station density (e.g. in the Balkan countries, Ireland, Iberian peninsula), particularly in the early years of the dataset when the number of operating stations was smaller (Ntegeka et al., 2013).
- Streamflow data: it is related to the ability of the Lis- flood model to accurately estimate the discharge time series in each point of the river network. Larger uncertainty is found in river basins where no calibration of the model parameters was possible (e.g. Portugal, southern Italy, Greece, Baltic states) and in several

Spanish rivers, where the influence of regulation from reservoirs heavily affects the streamflow regime and the calibration performance (see Supplement).

- Extreme value analysis: it is linked to the choice of the extreme value distribution, which for simplicity is assumed to be of the same type (i.e. 2-parameter Gumbel distribution) for the entire European river network. In addition, larger uncertainty affects estimates of large return periods (i.e. up to 500 years) from samples of only 24 annual maxima.
- Flood depth and extent: most uncertainty in this stage is attributable to the quality of the digital elevation model, which in this case is derived from SRTM (Jarvis et al., 2008), with spatial resolution of 90 m and vertical noise of ~ 8.7 m over Europe (Rodriguez et al., 2006).
- Impact assessment: it is mainly related to the extrapolation of depth-damage functions for specific areas and countries to regional functions which depend on the land use class. Another source of uncertainty is related to the assumption of a constant population density layer over the simulation period, though it is expected to have a minor effect on the results due to the relatively low population growth (~ 5 %, see http://ec.europa.eu/eurostat/statistics-explained/index. php/Population_and_population_change_statistics) in the considered simulation period.

CONCLUSIONS

This work presents a comparison of two methods for reconstructing the socio-economic impact of river floods in Europe in the present climate. The two methods are based on the same data framework, though they provide complementary information of key importance for effective flood risk assessment at European level. The integral method is a tool used to estimate average flood losses at region (NUTS 2) and country level under a stationary climate. On the other hand, the event-based method is able to reproduce the impact of simulated events in time. Hence it is more suitable for damage assessments over specific time windows and as operational tool for real time flood risk mapping.

Results of this research show that the proposed model is a useful and relatively accurate tool for estimating the socioeconomic impact of river floods in Europe, both with regard to long-term averages, over annual aggregations and for specific events. Despite the numerous sources of uncertainty potentially affecting the modelling chain, model results compare quantitatively well with flood impact data reported by EEA and DG Env at European scale. Also, one should note that the hydrological model Lisflood is the only component calibrated against the observed data. We propose this model as a valid alternative to filling in gaps of the database of socio-economic impact of floods, particularly in consideration of the following key points:

- The model simulates the impact of all floods, independently from their magnitude.
- The methodology is capable of retrieving comprehensive and capillary coverage of all the flood losses, which is often infeasible with alternative damage assessment methods.
- Impact estimates of the proposed model can be produced as soon as the meteorological input data is available, hence potentially during the flood event itself, with a considerable time gain in comparison to post-event assessment.

As a final remark, we want to stress the large potential in coupling the event-based method with numerical weather predictions as input data, which opens the door to flood impact forecasting applications at European scale. Its implementation is currently being tested within EFAS, with the aim of providing probabilistic forecast of areas at risk of flooding in the coming 10–15 days, with direct estimation of possible consequences on people affected and economic damage. This is of crucial importance in the early assessment of the magnitude of imminent events to support the planning of emergency actions and ultimately speed up the recovery phase.

ACKNOWLEDGEMENTS

The data used in this work were provided by the European Flood Awareness System (EFAS) and by publications listed in the references. Lorenzo Alfieri and Luc Feyen received financial support from the European Union Seventh Framework Programme FP7/2007-2013 under grant agreement no. 603864 (HELIX: “High-End cLimate Impacts and eXtremes”; http://www.helixclimate.eu).

REFERENCES

1. Alfieri, L., Salamon, P., Bianchi, A., Neal, J., Bates, P., and Feyen, L.: Advances in pan-European flood hazard mapping, Hydrol. Process., 28, 4067–4077, doi:10.1002/hyp.9947, 2014a.
2. Alfieri, L., Pappenberger, F., Wetterhall, F., Haiden, T., Richardson, D., and Salamon, P.: Evaluation of ensemble stream- flow predictions in Europe, J. Hydrol., 517, 913–922, doi:10.1016/j.jhydrol.2014.06.035, 2014b.
3. Alfieri, L., Feyen, L., Dottori, F., and Bianchi, A.: Ensemble flood risk assessment in Europe under high end climate scenarios, Global Environ. Change, 35, 199–212, doi:10.1016/j.gloenvcha.2015.09.004, 2015.
4. Alfieri, L., Feyen, L., and Baldassarre, G. D.: Increasing flood risk under climate change: a pan-European assessment of the bene- fits of four adaptation strategies, Climatic Change, 136, 507–521, doi:10.1007/s10584-016-1641-1, 2016.
5. Aon Benfield: Annual Global Climate and Catastrophe Report, Impact Forecasting – 2013, Chicago, available at: http://thoughtleadership.aonbenfield.com/Documents/20140113_ab_if_annual_climate_catastrophe_report.pdf (last access: 14 June 2016), 2014.
6. Barredo, J. I., de Roo, A., and Lavalle, C.: Flood risk mapping at European scale, Water Sci. Technol., 56, 11–17, doi:10.2166/wst.2007.531, 2007.
7. Bates, P. D., Horritt, M. S., and Fewtrell, T. J.: A simple inertial formulation of the shallow water equations for efficient twodimensional flood inundation modelling, J. Hydrol., 387, 33–45, 2010.
8. Batista e Silva, F., Lavalle, C., and Koomen, E.: A procedure to obtain a refined European land use/cover

map, J. Land Use Sci., 8, 255–283, doi:10.1080/174742 3X.2012.667450, 2012.

9. Batista e Silva, F., Gallego, J., and Lavalle, C.: A highresolution population grid map for Europe, J. Maps, 9, 16–28, doi:10.1080/17445647.2013.764830, 2013.

10. Broekx, S., Smets, S., Liekens, I., Bulckaen, D., and de Nocker, L.: Designing a long-term flood risk management plan for the Scheldt estuary using a risk-based approach, Nat. Hazards, 57, 245–266, doi:10.1007/s11069-010-9610-x, 2011.

11. Bubeck, P., Botzen, W. J. W., Kreibich, H., and Aerts, J. C. J. H.: Long-term development and effectiveness of private flood mitigation measures: an analysis for the German part of the river Rhine, Nat. Hazards Earth Syst. Sci., 12, 3507–3518, doi:10.5194/nhess-12-3507-2012, 2012.

12. Burek, P., Knijff van der, J., and Roo de, A.: LISFLOOD, distributed water balance and flood simulation model revised user manual 2013, Publications Office, Luxembourg, available at: http: //dx.publications.europa.eu/10.2788/24719 (last access: 12 December 2014), 2013.

13. Chen, Y., Syvitski, J. P. M., Gao, S., Overeem, I., and Kettner, A. J.: Socio-economic impacts on flooding: A 4000-year history of the yellow river, China, Ambio, 41, 682–698, doi:10.1007/s13280- 012-0290-5, 2012.

14. Di Baldassarre, G., Viglione, A., Carr, G., Kuil, L., Salinas, J. L., and Blöschl, G.: Socio-hydrology: Conceptualising human- flood interactions, Hydrol. Earth Syst. Sci., 17, 3295–3303, doi:10.5194/hess-17-3295-2013, 2013.

15. EEA: Mapping the impacts of natural hazards and technological accidents in Europe – An overview of the last decade, European Environment Agency Technical report 13/2010, Copenhagen, Denmark, 144 pp., 2010.

16. European Commission: Directive 2007/60/EC of the European Parliament and of the Council of 23 October 2007 on the assessment and management of flood risks, available at: http://eur-lex.europa.eu/LexUriServ/LexUriServ.do?uri= CELEX:32007L0060:EN:NOT (last access: 14 June 2016), 2007.

17. Falter, D., Dung, N. V., Vorogushyn, S., Schröter, K., Hundecha, Y., Kreibich, H., Apel, H., Theisselmann, F., and Merz, B.: Continuous, large-scale simulation model for flood risk assessments: proof-of-concept, J. Flood Risk Manage., 9, 3–21, doi:10.1111/jfr3.12105, 2016.

18. Falter, D., Schröter, K., Dung, N. V., Vorogushyn, S., Kreibich, H., Hundecha, Y., Apel, H. and Merz, B.: Spatially coherent flood risk assessment based on long-term continuous simulation with a coupled model chain, J. Hydrol., 524, 182–193, doi:10.1016/j.jhydrol.2015.02.021, 2015.

19. Fenn, T., Fleet, D., Hartman, M., Garrett, L., Daly, E., Elding, C., and Udo, J.: Study on economic and social benefits of environmental protection and resource ef- ficiency related to the European semester final report, available at: http://ec.europa.eu/environment/integration/green_ semester/pdf/RPA_Final_Report-main_report.pdf (last access: 30 April 2015), 2014.

20. Hall, J. W., Sayers, P. B. and Dawson, R. J.: National-scale assessment of current and future flood risk in England and Wales, Nat. Hazards, 36, 147–164, 2005.

21. Hirabayashi, Y., Mahendran, R., Koirala, S., Konoshima, L., Yamazaki, D., Watanabe, S., Kim, H., and Kanae, S.: Global flood risk under climate change, Nat. Clim. Change, 3, 816–821, doi:10.1038/nclimate1911, 2013.

22. IPCC: Managing the risks of extreme events and disasters to advance climate change adaptation, A special report of Working Groups I and II of the intergovernmental panel on climate change, edited by: Field, C. B., Barros, V., Stocker, T. F., Qin, D., Dokken, D. J., Ebi, K. L., Mastrandrea, M. D., Mach, K. J., Plattner, G.-K., Allen, S. K., Tignor, M., and Midgley, P. M., Cambridge University Press, Cambridge, UK, and New York, NY, USA, 582 pp., 2012.

23. Jarvis, A., Reuter, H. I., Nelson, A., and Guevara, E.: Hole- filled SRTM for the globe Version 4, available at CGIARCSI SRTM 90 m Database: http://srtm.csi.cgiar.org (last access: 14 June 2016), 2008.

24. Jongman, B., Kreibich, H., Apel, H., Barredo, J. I., Bates, P. D., Feyen, L., Gericke, A., Neal, J., Aerts, J. C. J. H., and Ward, P. J.: Comparative flood damage model assessment: towards a European approach, Nat. Hazards Earth Syst. Sci., 12, 3733–3752, doi:10.5194/nhess-12-3733-2012, 2012.

25. Jongman, B., Hochrainer-Stigler, S., Feyen, L., Aerts, J. C. J. H., Mechler, R., Botzen, W. J. W., Bouwer, L. M., Pflug, G., Rojas, R., and Ward, P. J.: Increasing stress on disaster-risk fi- nance due to large floods, Nat. Clim. Change, 4, 264–268, doi:10.1038/nclimate2124, 2014.

26. Kamari, J., Alcamo, J., Barlund, I., Duel, H., Farquharson, F. A. K., Florke, M., Fry, M., Houghton-Carr, H. A., Kabat, P., Kaljonen, M., Kok, K., Meijer, K. S., Rekolainen, S., Sendzimir, J., Varjopuro, R., and Villars, N.: Envisioning the future of water in Europe – the SCENES project, E-WAter, European Water Association (EWA), 1–28, 2008.

27. Koivumäki, L., Alho, P., Lotsari, E., Käyhkö, J., Saari, A., and Hyyppä, H.: Uncertainties in flood risk mapping: a case study on estimating building damages for a river flood in Finland, J. Flood Risk Manage., 3, 166–183, doi:10.1111/j.1753- 318X.2010.01064.x, 2010.

28. Lugeri, N., Kundzewicz, Z. W., Genovese, E., Hochrainer, S., and Radziejewski, M.: River flood risk and adaptation in Europe – assessment of the present status, Mitig. Adapt. Strat. Glob. Change, 15, 621–639, doi:10.1007/s11027-009-9211-8, 2010.

29. Merz, B., Kreibich, H., Schwarze, R., and Thieken, A.: Review article "Assessment of economic flood damage", Nat. Hazards Earth Syst. Sci., 10, 1697–1724, doi:10.5194/nhess-10-1697- 2010, 2010.

30. Munich Re: Natural catastrophes 2013: Analyses, assessments, positions, Munich, available at: http://www.munichre.com/site/ corporate/get/documents_E435413367/mr/assetpool.shared/ Documents/0_Corporate_Website/_Publications/302-08121_en. pdf (last access: 14 June 2016), 2014.

31. Ntegeka, V., Salamon, P., Gomes, G., Sint, H., Lorini, V., Thielen, J. and Zambrano-Bigiarini, M.: EFAS-Meteo: A European daily high-resolution gridded meteorological data set for 1990-2011, available at: http://publicationsjrceceuropa. ourtownypd.com/ repository/handle/111111111/30589 (last access: 3 June 2014), 2013.

32. Pappenberger, F., Dutra, E., Wetterhall, F., and Cloke, H. L.: Deriving global flood hazard maps of fluvial floods through a physical model cascade, Hydrol. Earth Syst. Sci., 16, 4143–4156, doi:10.5194/hess-16-4143-2012, 2012.

33. Pappenberger, F., Wetterhall, F., Albergel, C., Alfieri, L., Balsamo, G., Bogner, K., Haiden, T., Hewson, T., Magnusson, L., de Rosnay, P., Muñoz Sabater, J., and Tsonevsky, I.: Floods in Central Europe in June 2013, ECMWF Newslett., 136, 9–11, 2013.

34. Rojas, R., Feyen, L., and Watkiss, P.: Climate change and river floods in the European Union: Socio-economic consequences and the costs and benefits of adaptation, Global Environ. Change, 23, 1737–1751, doi:10.1016/j. gloenvcha.2013.08.006, 2013.

35. Sampson, C., Smith, A., Bates, P., Neal, J., Alfieri, L., and Freer, J.: A high-resolution global flood hazard model, Water Resour. Res., 51, 7358–7381, doi:10.1002/2015WR016954, 2015.

36. Schröter, K., Kunz, M., Elmer, F., Mühr, B., and Merz, B.: What made the June 2013 flood in Germany an exceptional event? A hydro-meteorological evaluation, Hydrol. Earth Syst. Sci., 19, 309–327, doi:10.5194/hess-19-309-2015, 2015.

37. Schumann, G. J.-P., Neal, J. C., Voisin, N., Andreadis, K. M., Pappenberger, F., Phanthuwongpakdee, N., Hall, A. C., and Bates, P. D.: A first large-scale flood inundation forecasting model, Water Resour. Res., 49, 6248–6257, doi:10.1002/wrcr.20521, 2013.

38. te Linde, A. H., Bubeck, P., Dekkers, J. E. C., de Moel, H., and Aerts, J. C. J. H.: Future flood risk estimates along the river Rhine, Nat. Hazards Earth Syst. Sci., 11, 459–473, doi:10.5194/nhess-11-459-2011, 2011.

39. Thielen, J., Bartholmes, J., Ramos, M.-H., and de Roo, A.: The European Flood Alert System – Part 1: Concept and development, Hydrol. Earth Syst. Sci., 13, 125–140, doi:10.5194/hess-13-125- 2009, 2009.

40. UNISDR: UNISDR Terminology on disaster risk reduction, United Nations International Strategy for Disaster Reduction (UNISDR), Geneva, Switzerland, 35 pp., 2009

41. van der Knijff, J. M., Younis, J., and de Roo, A. P. J.: LISFLOOD: A GIS-based distributed model for river basin scale water balance and flood simulation, Int. J. Geogr. Inf. Sci., 24, 189–212, 2010.

42. Ward, P. J., Jongman, B., Weiland, F. S., Bouwman, A., van Beek, R., Bierkens, M. F. P., Ligtvoet, W., and Winsemius, H. C.: Assessing flood risk at the global scale: model setup, results, and sensitivity, Environ. Res. Lett., 8, 044019, doi:10.1088/1748- 9326/8/4/044019, 2013.

43. Wind, H. G., Nierop, T. M., De Blois, C. J., and De Kok, J. L.: Analysis of flood damages from the 1993 and 1995 Meuse floods, Water Resour. Res., 35, 3459–3465, doi:10.1029/1999WR900192, 1999.

44. Winsemius, H. C., Aerts, J. C. J. H., van Beek, L. P. H., Bierkens, M. F. P., Bouwman, A., Jongman, B., Kwadijk, J. C. J., Ligtvoet, W., Lucas, P. L., van Vuuren, D. P., and Ward, P. J.: Global drivers of future river flood risk, Nat. Clim. Change, 6, 381–385, doi:10.1038/nclimate2893, 2016.

45. Zambrano-Bigiarini, M. and Rojas, R.: A model-independent Particle Swarm Optimisation software for model calibration, Environ. Model. Softw., 43, 5–25, doi:10.1016/j.envsoft.2013.01.004, 2013.

46. Zurich: Risk Nexus, Central European floods 2013: a retrospective, available at: http://www.zurich. com/_/ media/dbe/corporate/docs/whitepapers/ risk-nexus-may-2014-central-european-floods-2013.pdf (last access: 14 June 2016), 2014.

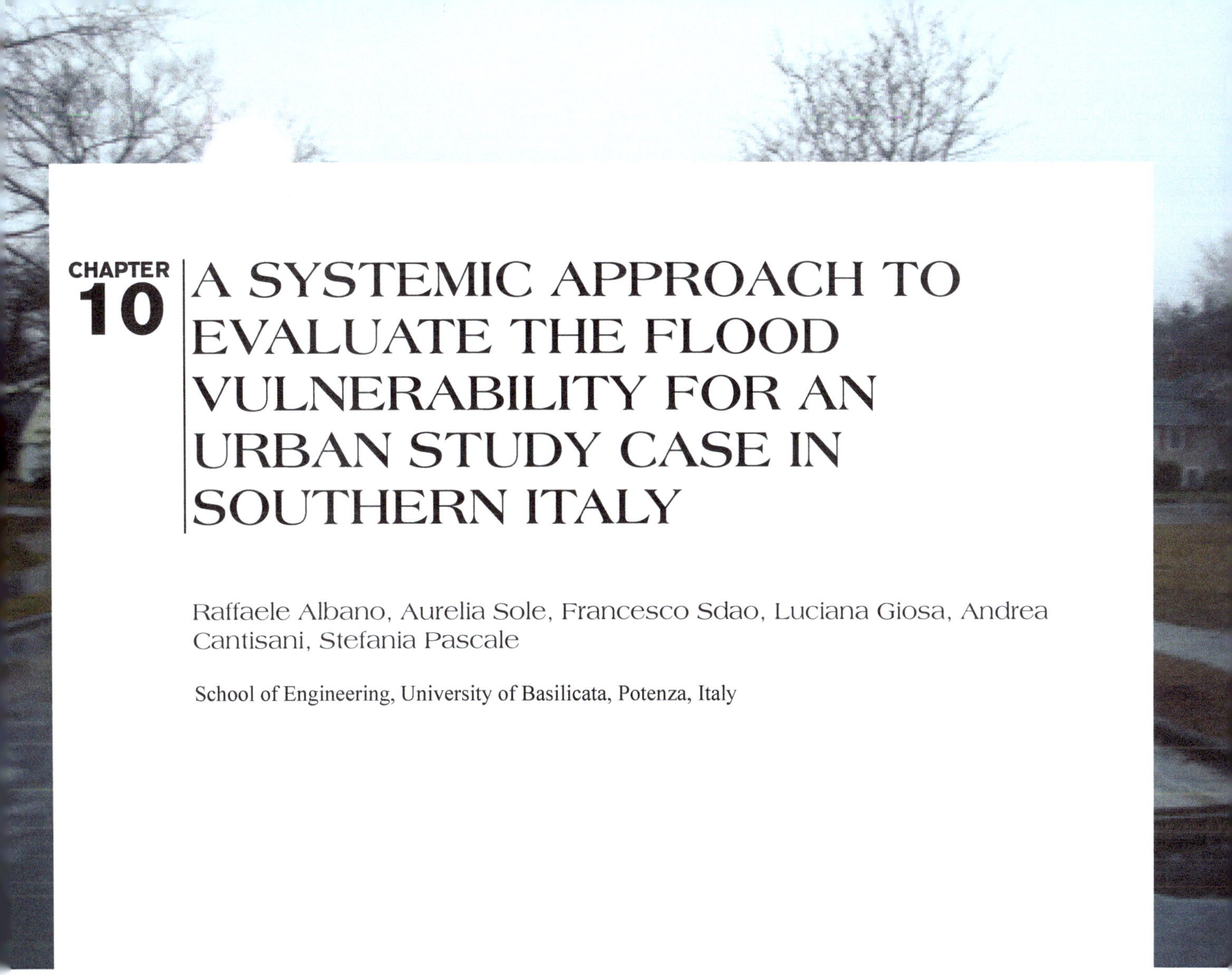

CHAPTER 10

A SYSTEMIC APPROACH TO EVALUATE THE FLOOD VULNERABILITY FOR AN URBAN STUDY CASE IN SOUTHERN ITALY

Raffaele Albano, Aurelia Sole, Francesco Sdao, Luciana Giosa, Andrea Cantisani, Stefania Pascale

School of Engineering, University of Basilicata, Potenza, Italy

ABSTRACT

Currently, the urban flooding is one of the most concerning problems in hydraulic protection, both for the enormous number of people and the different elements (buildings, roads, vehicles, and so on) potentially exposed to risk, as well as the complexity of the territory at issue. At the practical level, vulnerability indicators are often predictably too narrow in their coverage of aspects of vulnerability. An important need remains to produce more conceptually informed vulnerability indicators or parameters and more satisfactory operational tools to assess weaknesses and resilience in coping with natural risks. In this paper, we present an innovative methodology that adopts a systemic approach to evaluate the vulnerability due to a flood scenario. The operative efficiency of the proposed GIS tool is validated in pilot application site, i.e. an urban area in Puglia Region, Southern Italy, on the basis of, studies surveys and damages carried out from a recent flood event occurred in the area. The model evaluates the direct structural damages and explores the potential operating conditions of the road network in case of the flood event. The resulting vulnerability assessment tool can guide evaluators towards a comprehensive understanding of strengths and fragilities of a territory and community where a flood occurs embedding and integrating as much as possible the multifaceted and articulated nature of an urban system.

INTRODUCTION

The urban flood has more serious consequences than flooding in un-urbanized areas crossed by a river [1]. This is due to a high concentration of people and objects exposed to damage or to the loss of functionality. In fact, the urban areas are a complex system of houses, emergency centers, schools, local health centers and factories connected by roads.

Citation: Albano, R. , Sole, A. , Sdao, F. , Giosa, L. , Cantisani, A. and Pascale, S. (2014) A Systemic Approach to Evaluate the Flood Vulnerability for an Urban Study Case in Southern Italy. Journal of Water Resource and Protection, 6, 351-362. doi: 10.4236/jwarp.2014.64037.

Floods can cause both direct and indirect suspension and loss of functionality with damage to the elements in a certain area or the loss of lives. In effect, one-third of the annual natural disasters and economic losses, and more than half of all the victims, are flood-related [2] . Between 1975-2002, floods due to drainage problems, flash and river floods accounted for 9% of all deaths from natural disasters, claiming about 175,000 fatalities worldwide [3] , and affecting more than 2.2 billion people [4] . From 2000 to 2006, water-related disasters killed more than 290,000 people, affecting more than 1.5 billion, and inflicting more than US $422 billion in damage [5] .

Vulnerability analysis is a helpful tool for the evaluation and the prevention of risks which can affect people, environment and human development. Modern society considers essential the vulnerability analysis to increase the safety of the people and infrastructure. The purpose of this paper was to build a tool to analyze and evaluate the vulnerability of infrastructures and other elements of territorial systems during the emergency phase of potential flood events in a populated areas with complex demographics, infrastructure and economic activity. The territorial vulnerability during a natural catastrophe depends not only on the features of its elements but also on the relations between their parts [6] [7] [8] [9] ; in most cases such relations show critical points in a territory.

When the authorities have to operate in case of emergencies, they should consider the strategic importance of infrastructure in the territorial system, its evolutionary tendencies and its critical elements to increasing the efficacy and efficiency of the decisional activities. The model proposed is able to support the civil protection decision and management of an emergency due to natural catastrophes such as floods: the authorities can recognize the priorities of interventions, i.e. the most vulnerable elements, and can give the community a more efficient support to mitigation strategies and to delineate the emergency plans.

The validation of the method adopted was realized in the Municipality of "Ginosa Marina", located near the mouth of the "Bradano River" and bathed by the "Ionian Sea".

"Ginosa Marina" and the other neighboring municipalities are often afflicted by flood events. The last one, in March 2011, has caused a lot of damage to all the activities in the area and a lot of private houses have been made uninhabitable.

The proposed model is also being adopted in a project that aims to define the state of hydro-geological and structural instability and the relative degree of exposure and vulnerability of the "Provincial" road networks of "Basilicata Region", South of Italy [10] . After this introduction ("Heading 1"), "Heading 2" describes the methodology and calculation, "Heading 3", the validation and results on a real flood event, and "Heading 4", the overall discussion and conclusions.

METHOD

The innovative scientific strength of the research is the implementation of a framework for dependency analysis described by [7] to evaluate the systematic vulnerability, here, suitably improved and modified to put forward a model to evaluate the vulnerability of a system when its buildings and road network is affected by a possible flood event. The system functions are highly dependent on networked systems and the operability of these systems can be vulnerable to disasters [11] . The analysis and evaluation of the connections and relation-ships in the complex systems of a city are essential after the flood event in the emergency phases. Concurrently with the occurrence of physical and functional damage, which leads to the discontinuation of most of the urban functions, the operability of the strategic emergency structures, their accessibility and connection with the territorial contest become a top priority.

This method integrates traditional data modeling approach with innovative spatial approach. The modeling steps include: 1) describe the topological characteristics of territorial system; 2) assess scenarios in terms of spatial flood hazard; 3) analyze the physical vulnerability of buildings and the vulnerability of road network directly affected by the scenarios; 4) analysis of interdependencies and evaluation of systemic vulnerability. The modeling results from this paper may support the decision making process in both urban planning and emergency response.

The Topological Characterization of the Territorial System

The first phase deals with the topological characterization of the territorial system and consists in the collecting, organization and management of data concerning its components on one hand and their connections and interactions on the other.

All available data should be obtained (topographic, population, building typology, and so on), supplemented by more detailed data from building occupancy type (residential, factories, and so on) and by the recognition of the most important elements in the case of an emergency such as hospitals, fire stations and civil protection stations. Data about road networks are also required.

The Assessment of the Scenarios in Terms of Spatial Flood Hazard

A pre-flood assessment is required to estimate the intensity of the possible flood event. An estimation of hydraulic characteristics can be obtained from two-dimensional hydraulic models. They provide detailed information on flood hydraulics but require a high quantity of input data and parameters.

From hydraulic simulations the following characteristics of a scenario should be obtained:

- Depth: Height of flood water above ground level at the location site, which indicates flood magnitude and used for estimating direct consequences on buildings and roads;
- Velocity: Flow velocity of the flood wave, which indicates flood magnitude and is used for estimating direct and indirect consequences;
- Total flooded area: Flooded surface at the study area that determines which zones and assets are at risk.

The Analysis of the Physical Vulnerability of the Buildings Affected Directly by the Scenario

Studies of the physical vulnerability of structures from flooding have emerged for a variety of reasons. One priority has been the identification of existing structures that might suffer damage from flooding, primarily aiming to establish where damages will occur or suggest where the public is most vulnerable during flooding. In addition, in recent years focus has moved towards development control, flood-proofing and policies advocating the development of properties in flood risk areas that are more resilient and resistant to the effects of flooding.

Two main approaches have been established to assess the physical vulnerability of structures to flooding: economic damage assessment and a quantification of the structural integrity of buildings. The first of these is arguably the more widely used approach and is essentially a quantification of the expected, or actual, damages to a property or area either through the estimation of a monetary value or through an evaluation of the percentage of the expected loss. This study considers the second of these approaches and focuses on assessing the physical vulnerability of individual structures (or group of structures), and on the estimation of the likelihood of occurrence of physical damage or collapse of a single building.

This information will be incorporated within a more general estimation of vulnerability whereby the likelihood of structural collapse is combined with other information (systemic vulnerability) to provide a comprehensive assessment of flood vulnerability.

This step concentrates on the integrity of individual structures or types of structure to the physical characteristics of the flood hazard; notably the water depth.

In particular, literature studies focus on the failure of structures and the hazard conditions that are likely to cause the collapse or partial collapse of structures.

The methods are fundamentally based on the use of depth-damage relationships that assign a percentage of damage from the resulting water depth during the flood.

Depth-damage curves demonstrate the relationship between the depth of the flood relative to the first finished floor level of buildings and the damage caused to the structures.

To calculate damages, each structure must be assigned to a structure occupancy type (Residential with 1 story, Residential with more than 1 story, Commercial, Industrial, Institutional/ Government, Agricultural, and Transport). For each structure occupancy type, an estimated replacement value and a structure depth-damage must be defined. The depth-damage curves used in this heading are drawn from the report of [12] (Figure 1).

The Analysis of the Vulnerability of the Road Network Affected Directly by the Scenario

In this phase, we estimate the vulnerability of the road network due to the possible effects on those travelling on road networks, specifically concentrating on vehicles, roads and road users instead of evaluating the hazard con-

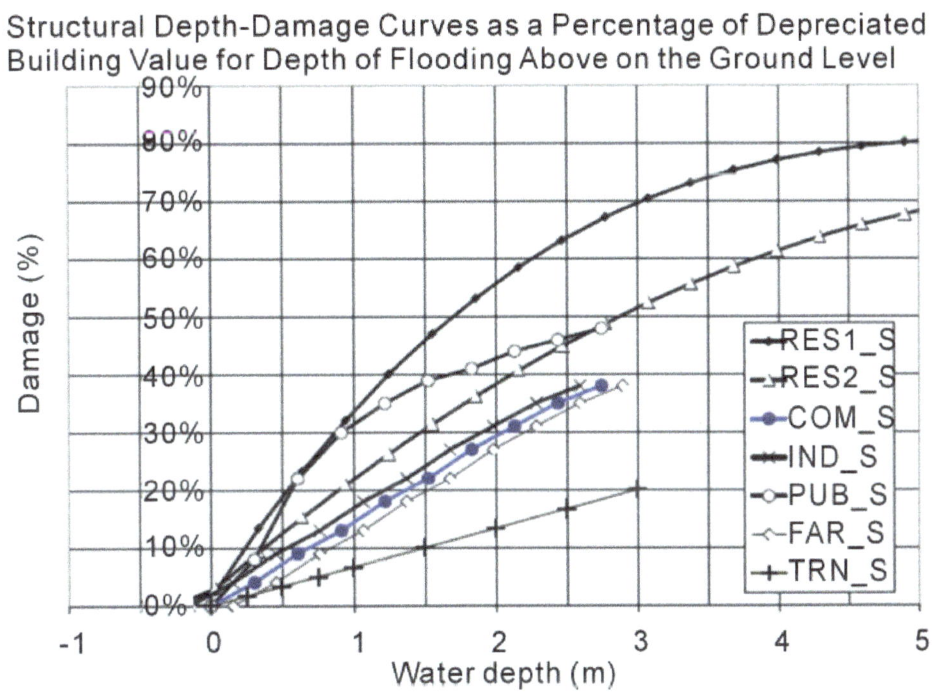

Figure 1. Depth damage curves.

ditions that are likely to cause the collapse or partial collapse of road structures. This choice derived from the difficulty to find structural information about roads and also because the intensity of flood events to cause damage to those travelling on the road is lesser than the one that causes the collapse of the road [13] . Therefore, we could estimate the vulnerability under maximum safety conditions. In addition, the collapse of the structures is greater in the riverbed where the infrastructure is often overcome by the water flow, obstructed by floating materials and showed excavations at the base of the piers of the bridges.

Scientific literature highlights the increase in the number of deaths occurring within, or escaping from motor vehicles [14] [13] .

In this paper, three categories of road hazards are chosen on the basis of these literature studies [15] [16] . A weight concerning critical threshold values of hydraulic instability for idealized vehicles is assigned to each road [17] (Figure 2).

Referring to the envelope curves that have been developed by [17] , with three color zones (i.e. green, yellow, and red), a novel innovative approach has been introduced first as the "Traffic Light of Hydraulic Stability System" (TLHS). Through this innovation, zones of hydraulic stability for each idealized vehicle are easily identified by color with the stable zone in green (left zone in Figure 2), the transition zone in yellow, (central zone) and the unstable zone in red (right zone). Thus, a straight forward envelope curve in Figure 2, in this study, is utilized to evaluate the potential road closures. Actually, all of the vehicles in the red zone of the graph are dragged by the water flow. Therefore, they could block, for example, an emergency vehicle during the rescue actions;, hence, the corresponding road is considered inoperable. The curves are utilized in the study only for the part when incoming flow depths are less than the vehicle height, the low part of the graph in Figure 2. Then, when the incoming flow depth is greater than the vehicle height, the roads are considered always inoperable.

Framework for Dependency Analysis

This framework improves and adapts the approach described by [7] . The evaluation of the dependencies of the elements is due to the inoperability of the road network considering the functionality, in case of emergency, of each element in the system. Inoperability, i.e. the inability of a system to perform its intended functions, analyzes how a given amount of inoperability in one component affects the operability of other components.

This framework analyzes how the presence of dependencies between different infrastructures contributes to the spread of degradation. The approach assumes as atomic entity whose level of operability, (in addition to external causes), depends on the availability of resources supplied by other infrastructures [18].

Dependencies are complex, non-linear, geographically-dispersed clusters of systems and the task to analyze the impact of element outages on the overall system is essential.

The model describes these phenomena mathematically by associating an inoperability level with each infrastructure. The inoperability of the ith infrastructure is expressed by variable x_i between [0,1] where $x_i = 0$

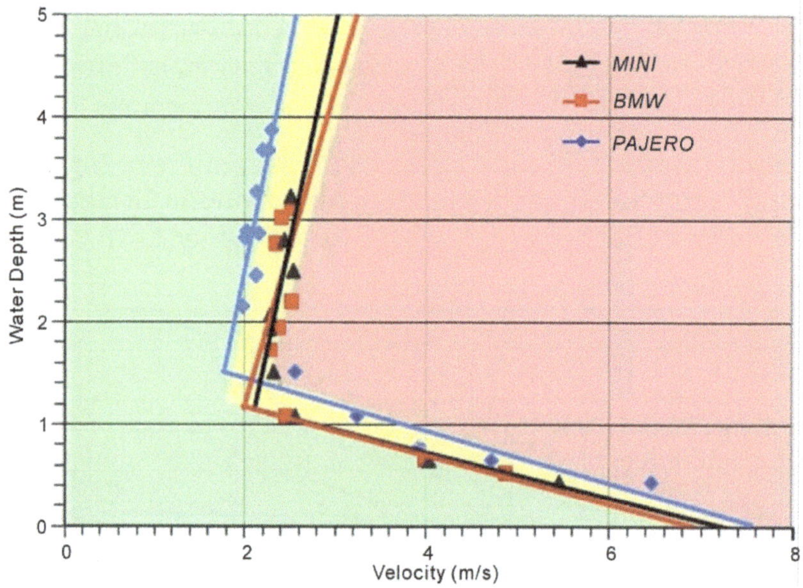

Figure 2. Critical threshold values of hydraulic instability for idealised vehicles (taken from Teo F.Y. 2012).

means that the infrastructure is fully operable and $x_i = 1$ means that the infrastructure is completely inoperable.

$$x_i = (1/n-1)\sum_{j=n} w_{ij} \quad (1)$$

The mutual dependency (interdependency) phenomena are estimated by the following equation [8] :

$$w_{ij} = (1-0.1\alpha_{ij})\frac{\left(1-e^{-\alpha_{ij}x_j^2}\right)}{\left(1-e^{-\alpha_{ij}}\right)}+0.1\alpha_{ij} \quad (2)$$

where x_j is the index of physical vulnerability of the j element. Parameter, that represents the dependence of the ith road network on the jth elements, is calculated through the relation:

$$a_{ij} = -9.9I_{ij} + 10 \quad (3)$$

where I_{ij} is the value of influence that is evaluated by:

$$I_{ij} = (1/n-1)\sum_{j=n} \rho_{ij} \quad (4)$$

where n is the number of element affected by the inoperability of the ith arc of the road network and ρ_{ij} is parameter that evaluates the dependency of the element j on the arc i.

Parameter ρ_{ij} is evaluated through the vulnerability curves taken from the following equation [13] :

$$\rho_{ij} = (1-\alpha)\left(e^{-\alpha\xi_{j+1}^{j(k)}} \Big/ \left(1-e^{-\alpha\xi_{j+1}^{j(\lambda)}}\right)\right) \quad (5)$$

In this study parameter is the percentage elongation of the short path utilized by the element j to reach element j + 1. For example, the percentage elongation of the path utilized by an emergency vehicle to reach a house at flood risk when an arc of the road network, that the emergency vehicle usually drives on when there is not a flood event, was inoperable.

Parameters κ and λ are chosen according to the categories of the elements [7] . In this case, we divided the elements in three categories relative to the element functions for the systems in case of emergency.

In fact, we do not recognize the elements by the physical state but through the role they play in the case of an emergency and for the socio-economic and environmental damage they can cause. For instance, if a hospital is damaged the whole system is affected by an increase in the rescue workload on other forms of assistance.

The risk elements with different physical, social and economic features are set in Categories A, B and C.

The importance of these features moving from Category A to C:

- Category A includes all the most important elements in the case of an emergency such as hospitals, fire stations and civil protection stations. These are all those elements that give assistance when catastrophic events occur. It includes also the major roads;
- Category B includes all the major socio-economic and environmental elements such as factories, which can also deal with dangerous materials, and big shopping centers as well as all other public buildings including Universities, Libraries and Churches. All of these can contain a large number of people and can also be important from a historical, artistic and cultural point of view. It includes also country roads;
- Category C includes private buildings, small business activities and secondary roads.

In the Equation (5), a parameter f is added, as a corrective factor, to consider the population density. The population density is important because it provides a priori information on the number of people who may need to be rescued and on how to act in the case of danger when an appropriate emergency plan can be prepared which foresees the right kind of intervention necessary where the population density is higher.

Evaluation of the Systemic Vulnerability

The last phase is the estimation of the systemic vulnerability of each element by the equation

$$s_i = \max\left(x_i, v_i\right) \tag{6}$$

where v_i is the physical vulnerability and x_i is the influence of the road network on the elements of the territorial systems. The systemic vulnerability index s_i is chosen as cautionary because it highlights the maximum risk.

CASE STUDY

"Ginosa Marina" is a city of the "Puglia Region" located near the mouth of the "Bradano River". The morphological characteristic of the river, i.e. area strongly flat, the great number of field data in the last years and the high resolution DTM built thanks to laser-scan data justify the choice as case study. Moreover, the "Bradano River" is crossed by "S.S. 106 Jonica" main road and the railway "Taranto-Reggio Calabria" that are some of the most important arteries of communication between border regions. All the area near the mouth of the "Bradano River" that can reasonably be considered an area at risk of flooding. This finding is derived primarily from historical information:

- The most important collection of historical information on hydrogeological disasters (landslides and floods) that have occurred across the country in the period 1918-2001, that was carried out as part of the AVI (Affected Italian Areas) by the "National Group for the Defense of Hydrogeological Disasters" (GNDCI) of the "National Research Council" (CNR) on behalf of the "Department of Civil Protection" [19] ;
- The data collected by ARPAB (Regional Agency for Environmental Protection of Basilicata).

These data show that the area at the mouth of the "Bradano River" has been affected in the past by a great number of natural disasters. In this area, there is an urbanized area with a considerable number of people, which usually increases during the summer season, and important roads, such as the "S.S. 106 Jonica", and a railway line that connects different Regions in Southern Italy.

The last event goes back to the 1st March 2011. It was so intense that the authorities had to declare a state of calamity. Hence, we have drawn up, in a GIS environment, the flood intensity maps of the considered urban areas with the hydrological and hydraulic simulation based on 1st March 2011 data.

Data

The hydrological and hydraulic data of the event, which occurred on 1st March 2011, was used to validate the proposed model. This flood event was particularly intense, causing damage to economic activities and residential buildings, that were evacuated, and some provincial and national roads were interrupted by water and mud. It is preferred to validate the model to an event which has actually occurred rather than to a generic simulated event.

A commercial numerical model, "MIKE Flood" by DHI, validated on remote sensing images and land survey in situ, was able to define the areas of the territory directly involved in the flood event, the maximum water depth and the maximum velocity of the water flow [20] .

The maximum discharge of the occurred event can be assimilated to an event with 30 years return time, estimated with the VAPI method [21] , recommended in Italy.

The topological characterization of the territorial system was taken from the ISTAT ("National Institute of Statistic") database coupled with data extrapolated by Remote-Sensing and orthophoto images [20] .

Resolution

The first step was to identify the "Ginosa Marina" territorial network, characterized by 3358 elements: 2274 buildings, (public buildings such as schools, local health centers, hospitals, military and fire stations, industries and so on), 1084 infrastructures, (streets, railways and so on).

"Ginosa Marina" is a small town on the Ionian Sea and is crossed by the "Bradano River". Therefore, there are a great number of farming businesses and small factories. The population density is not so high but in the summer a lot of people visit the touristic town. The economy of the area is based on agriculture and tourism.

The ISTAT data is aggregated at the census areas level, therefore, a statistic method to disaggregate the data at a single building level was developed. The area of the buildings belongs to each single census area which is evaluated and multiplied by the percentage of the number of the floors of the building. Then, the total sum of the areas of the buildings is utilized to calculate the population density in each building. After the evaluation of water depth H and velocity V, and the identification of the total flooded area, see Figures 3 and 4, the elements directly at risk are highlighted Figure 5.

Only 10% of the buildings are at risk because the more populated area of the town is out of the flooded area. The 22% of the roads are directly involved in the events and the main road, "S.S. 106", in the most hazardous area. The health units and the operating unit are far from the flooded area.

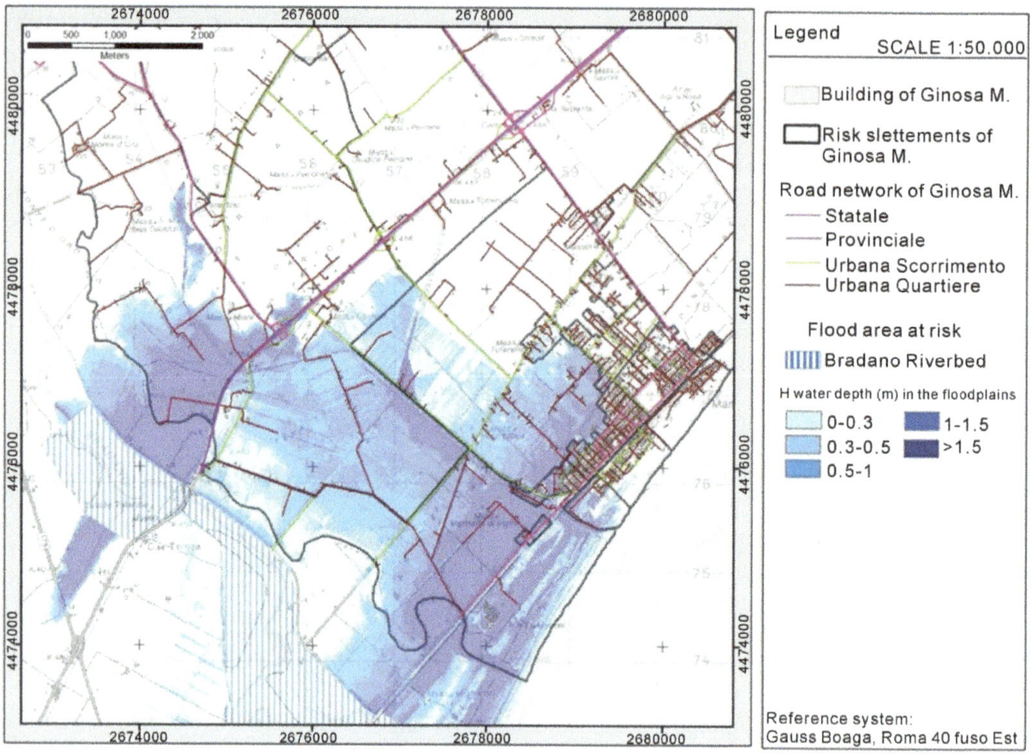

Figure 3. Water depth "H".

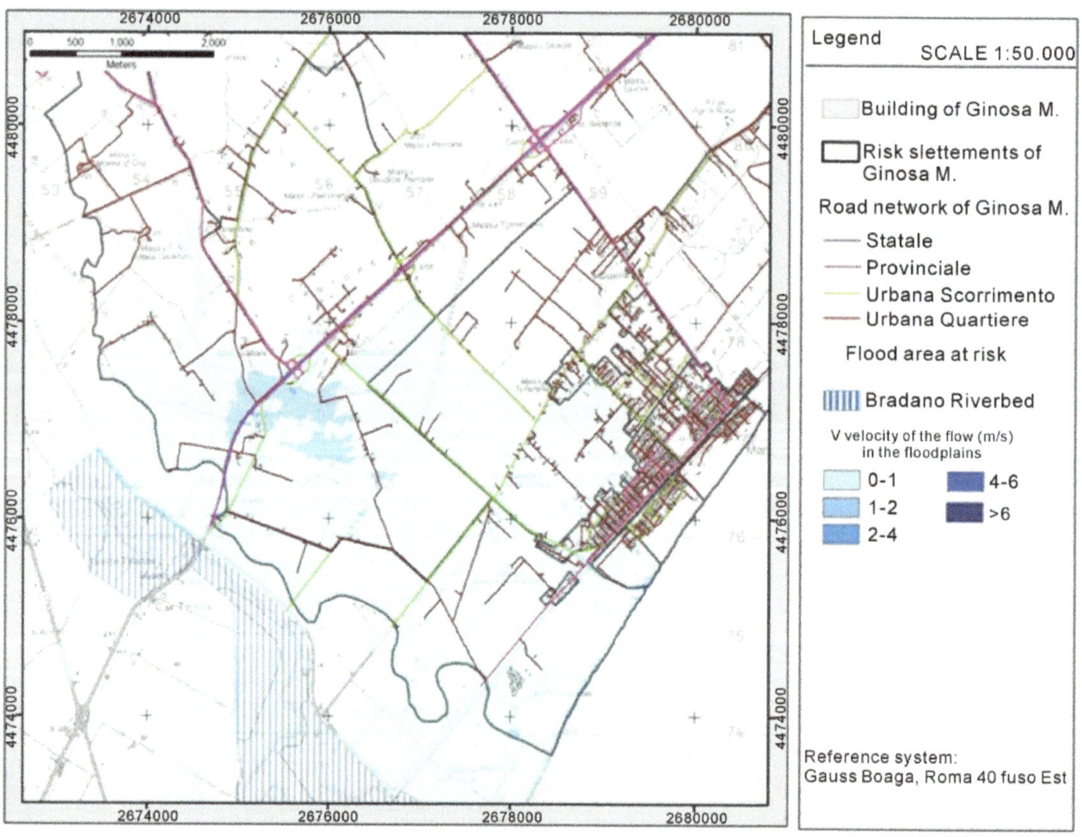

Figure 4. Velocity of the water flow "V".

The flooded area is a level land, therefore, the flow velocity, as you can deduce by the hydraulic data, reported in Figure 4 is on average low. The flow velocity is low in most of the flooded areas but in proximity of the "S.S. 106", i.e. the main road of the system, it is higher.

The water depth is really high in most of the flooded area and, hence, it has influenced the value of the physical vulnerability of the buildings. The physical vulnerability of buildings and road networks are estimated respectively by the curve described in "Heading 3.3" and "Heading 3.4".

The dependency analyses are effectuated considering the elongation path method. In network GIS, computational modeling of an urban network (for example, street network or underground) is based on a graph in which the intersections of linear features are regarded as nodes, and connections between pairs of nodes are represented as edges.

In an open-source GIS, an ad-hoc tool was developed to evaluate the shortest path (pathfinding), to analyze network connectivity (tracing), and to assign portions of a network to a location based.

The result path lengths, i.e. the shortest path with and without the inoperable roads, are estimated, hence, the path elongation is evaluated in percentage.

The inoperable roads are chosen depending on the intensity of the flood: the roads, that belong to the extreme right zone of Figure 2, can be inoperable also during the post-event because the vehicles, that will be on these streets, are dragged by the water flow.

Results

This study has put forward an approach to evaluate the consequences of flood events drawn up by means of the utilization of a territorial model based on a network of influences and an estimation procedure for assessment vulnerability. The significance of such analyses in this particular territory must also be emphasized. The event of

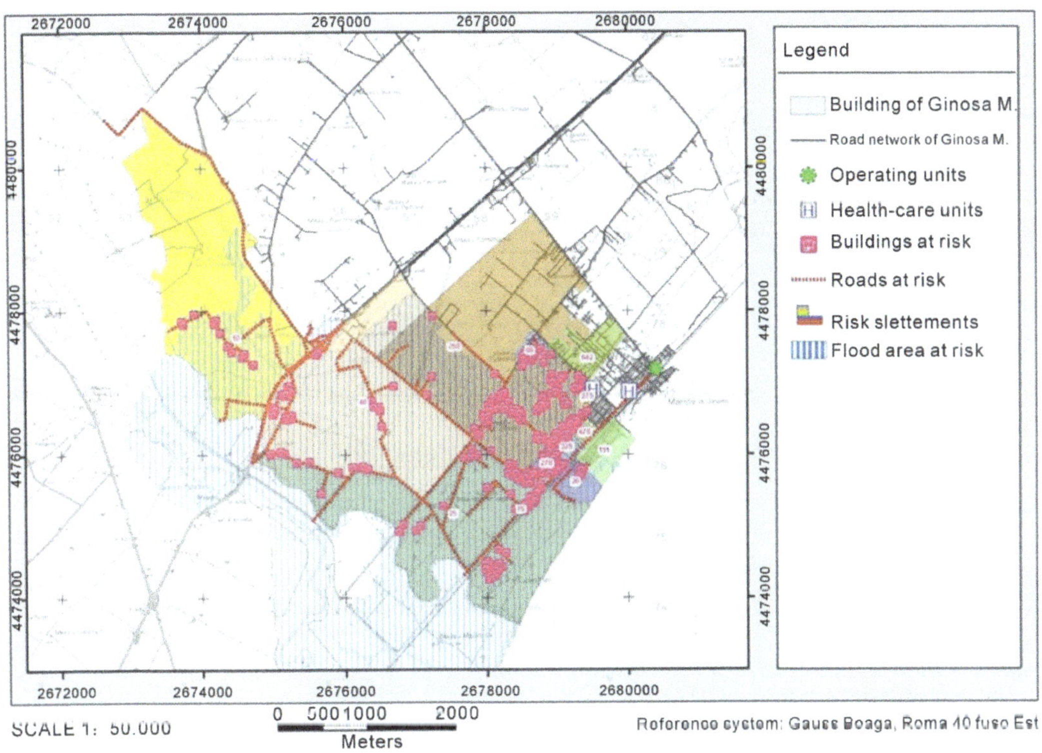

Figure 5 "Ginosa Marina" territorial network overlapped with the scenario to highlight the elements directly involved by the simulated flood event of 1st March.

1st March 2011 has caused serious damage to the main infrastructural systems as well as direct and indirect economic losses on most of the surrounding area. The scenario utilized has highlighted that the most vulnerable buildings with a value of index of systemic vulnerability over 50%, that are 20% of the total elements at risk, are concentrated in the area with the highest water depth and near "Via Lungo Galaso", a street that is almost completely inoperable. Therefore, there are the buildings around "Via Lungo Galaso", in "c/da Marinella", that are almost isolated. Emergency vehicles can reach these buildings in double the time that it usually takes. Most of these buildings are farming businesses, so the economic loss is greater for the system.

During the event of March 2011, the "Galaso Stream", a branch of the "Bradano River", created great damage to the adjacent buildings and roads. The major request for compensation for damage for this event has been made by the people that live in the proximity of this stream.

The validations, performed by comparisons with the case study, show the good reliability of the model, which allows a satisfactory representation of the fragility of the territorial system (Figure 6).

The street which is more affected by systemic vulnerability, index over 0.5, is the "S.S. 106" where the value of water depth and flow velocity are very high (Figure 7). Only 5% of the total roads at risk have an index of systemic vulnerability over 0.5 because most of the streets are urban and extra-urban. Therefore, the streets have a low classification and importance for the system. Another important factor is that the values of flow velocity are on overage low in the area. The "S.S. 106" hat is the most important road in the system, has the highest value of the systemic vulnerability, over 0.7.

The reliability of the proposed approach has been tested on a case study regarding flood risk in the district of "Ginosa Marina" and has led to identify the infrastructural elements that mainly affect the functionality of the territorial system, i.e. the "National Road S.S. 106" and the railway "Taranto-Reggio Calabria". Although, the same conclusion could also have been achieved by the analysis carried out by an expert in the territory, due to the relative simplicity of the territorial system studied. It can be considered as an important result, because it highlights the reliability of the obtained results and, consequently, of the decisions to come.

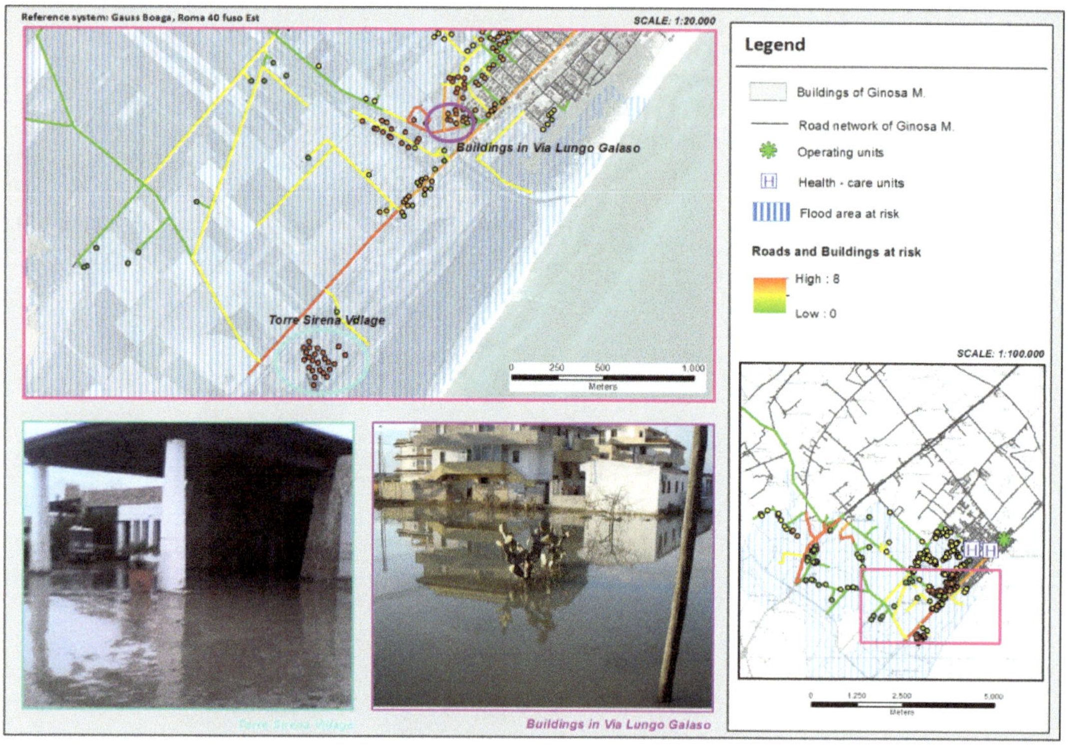

Figure 6.Examples of some critical situation of the real events comparing with the model.

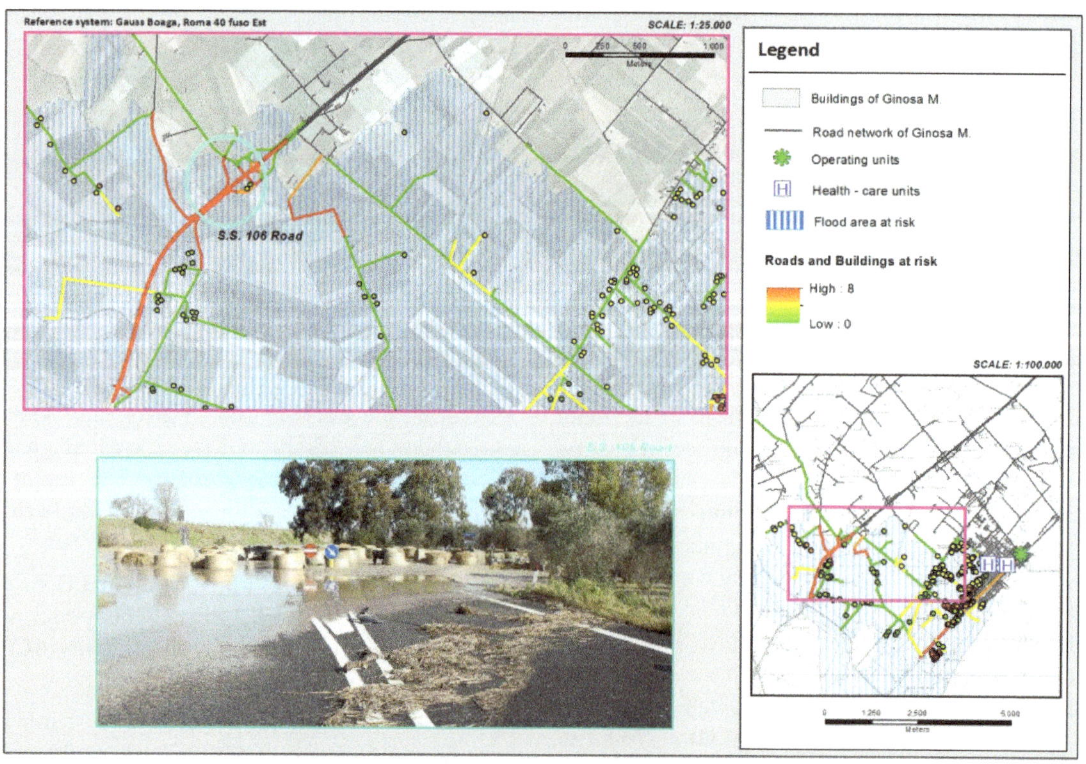

Figure 7. Example of the fragility of the system at the junction of "S.S. 106" due to the 1st March event: comparison with the model results.

CONCLUSIONS

We have proposed a dynamic model that analyzes the system, considering the interrelations of the elements, and evaluates the direct or indirect relationship of vulnerability among the elements as well as the effects that the fragility of a node could produce in the overall functioning of the whole system. It have identified, using spatial analysis, the most critical elements for the system and the most decisive elements in the management of rescuing. The model has been validated on a pilot site, i.e. an urban area in "Puglia Region", Southern Italy, to demonstrate its satisfactory operational degree to provide planners a tool to assess weaknesses in the whole urban system affected by flood. It has provided evidence for the presence of criticality in connections of the urban case study by measuring their amount through quantitative indicators. The indicators, evaluated by the GIS tool, have helped to define a hierarchy among the various structures and infrastructures by identifying those whose operation and efficiency are fundamental to the maintenance of network connectivity; hence, those whose minimal performance needs to be preserved, concurrently with the occurrence of physical damage estimated, in order to guarantee victims assistance and rescue activities.

The systemic approach can offer a great potential to get more insight into potentially large disruptions resulting from hazard impact. Future developments and outcomes will have an increasing effect on flood management policies and urban planning. The integration of systemic evaluation of the vulnerability with combined with economic models may drive many strategic decisions on the perception of contemporary development policies.

REFERENCES

1. Rafiq, L. and Blasch, T. (2012) Disaster Risk and Vulnerability in Pakistan at a District Level. Natural Hazards and Risk, 3, 311-323.http://dx.doi.org/10.1080/19475705.2011.626083
2. Douben, K.J. (2006) Characteristics of River Floods and Flooding: A Global Overview, 1985-2003. Irrigation and Drainage, 55, 9-21. http://dx.doi.org/10.1002/ird.239
3. Jonkman, S.N. (2005) Global Perspectives of Loss of Human Life Caused by Floods. Natural Hazards, 34, 151-175. http://dx.doi.org/10.1007/s11069-004-8891-3
4. Jonkman, S.N., Vrijling, J.K. and Vrouwenvelder, W.M. (2008) Methods for the Estimation of Loss of Life Due to Floods: A Literature Review and a Proposal for a New Method. Natural Hazards, 46, 353-389. http://dx.doi.org/10.1007/s11069-008-9227-5
5. United Nations World Water Assessment Programme (2009) Global Trends in Water-Related Disasters an Insight for Policymakers.
6. Ezell, B.C., Farr, J.V. and Wiese, I. (2000) Infrastructure Risk Analysis Model. Journal of Infrastructure System, 6, 114-117. http://dx.doi.org/10.1061/(ASCE)1076-0342(2000)6:3(114)
7. Pascale, S., Sdao, F. and Sole, A. (2010) A Model for Assessing the Systemic Vulnerability in Landslide Prone Areas. Natural Hazards and Earth System Sciences, 10, 1575-1590.http://dx.doi.org/10.5194/nhess-10-1575-2010
8. Sdao, F., Albano, R., Sivertun, A., Sole, A., Pascale, F. and Giosa, L. (2013) Chapter 19: Model of Systemic Vulnerability Assessment in Urbanized Areas Exposed to Combined Risk of Landslide and Flood. In: Borruso, G., Bertazzon, S., Favretto, A., Murgante, B. and Torre, C.M., Eds., Geographic Information Analysis for Sustainable Development and Economic Planning: New Technologies, IGI Global, 274-294.http://dx.doi.org/10.4018/978-1-4666-1924-1.ch019
9. Tamura, H., Yamamoto, K., Tomiyama, S. and Hatono, I. (2000) Modeling and Analysis of Decision Making Problem for Mitigating Natural Disaster Risks. European Journal of Operational Research, 122, 461-468. http://dx.doi.org/10.1016/S0377-2217(99)00247-7
10. Cantisani, A., Cafarelli, R., Giosa, L., Mileti, C., Sole, A. and Albano, R. (2012) An Easy Hydraulic Monitoring System to Support the Road Management. Le Colloque International des Utilisateurs de SIG.
11. Menoni, S., Molinari, D., Parker, D., Ballio, F. and Tapsell, S. (2012) Assessing Multifaceted Vulnerability and Resilience in Order to Design Risk-Mitigation Strategies. Natural Hazards, 64, 2057-2082. http://dx.doi.org/10.1007/s11069-012-0134-4
12. Department of Water Resources Division of Flood Management (2008) Flood Rapid Assessment Model (F-RAM) Development 2008, State of California, The Resources Agency.
13. Sharif, H.O., Hossain, M., Terrace, J. and Bin-Shafique, S. (2012) Person-Place-Time Analysis of Vehicle Fatalities Caused by Flash Floods in Texas. Geomatics, Natural Hazards and Risk, 3, 311-323. http://dx.doi.org/10.1080/19475705.2011.615343
14. Gruntfest, E. (2000) Flash Floods in the United States. In: Pielke, J.R. & Pielke, S.R., Eds., Storms, London, Routledge, 192-206.
15. Shu, C., Xia, J., Falconer, R.A. and Lin, B. (2011) Incipient Velocity for Partially Submerged Vehicles in Floodwaters, Journal of Hydraulic Research, 49, 709-717.http://dx.doi.org/10.1080/00221686.2011.616318
16. Xia, J., Falconer, R.A., Lin, B. and Tan, G. (2011) Numerical Assessment of Flood Hazard Risk to People and Vehicles in Flash Flood. Environmental Modelling & Software, 26, 987-998. http://dx.doi.org/10.1016/j.envsoft.2011.02.017
17. Teo, F.Y., Xia, J., Falconer, R.A. and Lin, B. (2012) Experimental Study on the Interaction between Vehicles and Floodplain Flows. International Journal River Basin Management, 10, 149-160. http://dx.doi.org/10.1080/15715124.2012.674040
18. Setola, R., De Procellinis, S. and Sforna, M. (2009) Critical Infrastructure Dependency Assessment Using the InputOutput Inoperability Model. International Journal of Critical Infrastructure Protection, 2, 170-178. http://dx.doi.org/10.1016/j.ijcip.2009.09.002
19. CNR IRPI () AVI Project: Inventory of Areas Affected by Landslides and Floods in Italy from 1918 to 2000. http://avi.gndci.cnr.it

20. Sole, A., Mancusi, L., Giosa, L., Cantisani, A., Grippa, A., Albano, R. and Varasano, A. (2012) Effetti Della Vegetazione Nella Valutazione Delle Aree a Rischio Inondazione. XXXIII Convegno Nazionale di Idraulica e Costruzioni Idrauliche, Brescia, 10-15 September 2012.

21. Claps, P. and Fiorentino, M. (2005) Valutazione Delle Piene in Italia, Rapporto di Sintesi per la Regione Basilicata (Bacini del Versante Ionico), GNDCI-CNR. Dipartimento di Ingegneria e Fisica dell'Ambiente, Università della Basilicata, Potenza.

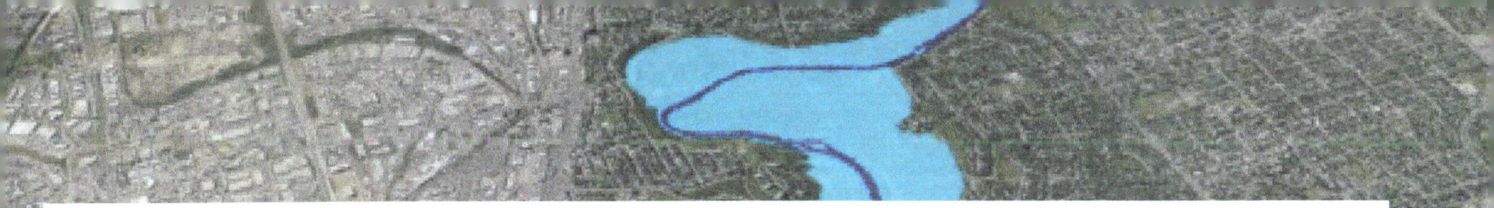

CHAPTER 11

A GIS-BASED MODEL TO ESTIMATE FLOOD CONSEQUENCES AND THE DEGREE OF ACCESSIBILITY AND OPERABILITY OF STRATEGIC EMERGENCY RESPONSE STRUCTURES IN URBAN AREAS

R. Albano[1], A. Sole[1], J. Adamowski[2], and L. Mancusi[3]

[1]School of Engineering, University of Basilicata, Potenza, Italy

[2]Bioresource Department, McGill University, Montreal, Canada

[3]Sustainable Development and Energy Resources Department, Research on Energy Systems SpA, Milano, Italy

ABSTRACT

Efficient decision-making regarding flood risk reduction has become a priority for authorities and stakeholders in many European countries. Risk analysis methods and techniques are a useful tool for evaluating costs and benefits of possible interventions. Within this context, a methodology to estimate flood consequences was developed in this paper that is based on GIS, and integrated with a model that estimates the degree of accessibility and operability of strategic emergency response structures in an urban area. The majority of the currently available approaches do not properly analyse road network connections and dependencies within systems, and as such a loss of roads could cause significant damages and problems to emergency services in cases of flooding. The proposed model is unique in that it provides a maximumimpact estimation of flood consequences on the basis of the operability of the strategic emergency structures in an urban area, their accessibility, and connection within the urban system of a city (i.e. connection between aid centres and buildings at risk), in the emergency phase. The results of a case study in the Puglia region in southern Italy are described to illustrate the practical applications of this newly proposed approach. The main advantage of the proposed approach is that it allows for defining a hierarchy between different infrastructure in the urban area through the identification of particular components whose operation and efficiency are critical for emergency management. This information can be used by decision-makers to prioritize risk reduction interventions in flood emergencies in urban areas, given limited financial resources.

INTRODUCTION

Urban flooding is a serious and growing challenge. Against the backdrop of demographic growth, urbanization trends and climate change, the causes of floods are shifting and their impacts are accelerating (Jha et al., 2012). Between 1975 and 2002, floods due to drainage problems, flash, and river floods accounted for 9 % of all deaths from natural disasters, with about 175 000 fatalities worldwide and more than 2.2 billion people affected (Jonkman et al., 2005). From 2000 to 2006, water-related disasters killed more than 290 000 people, affected more than 1.5 billion people, and inflicted more than USD 422 billion in damage (Adikati and Yoshitani, 2009). In light of this, there has been increased emphasis on new policies for increasing resilience to flooding (Djordjevic et al., 2011), "preparing for floods" (ODPM, 2002), ´ "making space

Citation: Albano, R., Sole, A., Adamowski, J., and Mancusi, L.: A GIS-based model to estimate flood consequences and the degree of accessibility and operability of strategic emergency response structures in urban areas, Nat. Hazards Earth Syst. Sci., 14, 2847-2865, doi:10.5194/nhess-14-2847-2014, 2014.

for water" (Defra, 2004), and "living with risk" (UN/ISDR, 2004). This emphasis reflects in part the perception that a risk management paradigm is more complex than a more traditional standard-based approach as it involves "whole systems" and "whole life" thinking. However, this is its main strength and a prerequisite for more integrated and informed decision-making in the face of flood emergencies (Sayers at al., 2013). For example, in the Netherlands, seeking to provide "room for the river", scientists, policy-makers, and stakeholders have focused their attention on warning and evacuation systems, improvements in maintenance standards, and a decision-making process that re- flects greater attention to economic efficiencies (Sayers at al., 2013). Flood forecasting, warning, emergency management, and other non-structural measures are increasingly being seen as critical for reducing flood consequences. As part of this, there is a need to refine methods to estimate flood risk and consequences, with particular attention on emergency management.

The Hyogo Framework for Action 2005–2015 (ISDR, 2004) highlights the central role of emergency planning in ensuring that a flood event does not become a flood disaster. The internationally accepted and most common flood damage models (FLEMO model (Apel et al., 2009; Vorogushyn et al., 2012); HAZUS-MH (FEMA, 2003; Scawthorn, 2006); Damage Scanner Model (Klijn et al., 2007); Multi-Coloured Manual (Penning-Rowsell et al., 2005)) place economic values on flood risk in order to help planners in the estimation of the benefits of flood protection measures in terms of prevented flood damage. The latter approach does not take into account the dynamic nature of the urban system, with its interconnections and relationships among elements, and hence the performance of strategic structures and infrastructure in the case of emergency. Hence, indirect damages in the field of emergency management are not considered in these currently available consequence-estimation models. For example, the inaccessibility of inundated roads during emergency management activities could cause indirect damage to the operability of strategic structures such as hospitals or fire stations.

Other studies have dealt with specific aspects of emergency management, as well as identification of safest access routes (Dalziell et al., 2011), or evaluations of the number of unassisted people (Taylor et al., 2006). These studies have provided useful contributions to the analysis of road accessibility (Franchlin et al., 2006) and reliability (Lhomme et al., 2013). However, these studies did not consider emergency management of the whole system (i.e. quantification of the contributions of each structure or infrastructure in the maintenance of the performance of the rescue, and also its degree of vulnerability). On the one hand, the latter papers have not estimated the degree of physical damage of road networks and buildings due to natural events. On the other hand, although these papers analysed the accessibility and operability of road networks, they did not consider their typology (main roads, local roads, etc.) or the contribution of strategic structures (hospitals, civil protection centres, etc.) and hotspots (industries, resorts, and hotels) in the system.

Menoni et al. (2002) attempted to evaluate the systemic vulnerability of an urban system by using a model to assess the vulnerability due to lifeline failures (i.e. road system, water system, gas system, power system, etc.) for earthquake events. They proposed a regional-scale model that concentrates on the assessment of the large number of indirect damages to define where to engage in more detailed studies on vulnerability analysis (i.e. the cities and towns most affected by indirect damages evaluated through the model). This study highlighted the need to quantify, through spatial analysis, the contribution of infrastructure (e.g. road networks) and structures (hospitals, industries, schools, etc.) in a city system to support decision-making regarding the type and location of the mitigation interventions.

Pascale et al. (2010) and Sdao et al. (2013) focused on the estimation of dependences within an urban system in the case of floods and/or landslide events by studying the "systemic" vulnerability, in terms of physical damage and functional relationship between operative centres and industries at risk or roads and private buildings at risk, etc., due to landslide or flood events. However, they did not analyse the spatial accessibility and operability relationships within the urban system based on the path connections and analysis, which is very important during the emergency phase of a flood event (i.e. during and immediately after a flood).

The proposed study overcomes the limitations of the approaches and models discussed above by integrating the concepts and methods of the previously mentioned studies, based on an accessibility and reliability analysis of the road network, within a systemic flood impact estimation. The proposed methodology couples the flow approach (Dalziell et al., 2001; Franchlin et al., 2006), based on flow and functionality of paths (i.e. comparison between the flow during normal working conditions and under disruption), with an approach based on topology (Lhomme et al., 2013) that considers structural analysis (i.e. it considers the number of alternative paths to the disruptions of one or several paths). In addition, the impact of road networks and dependencies between hotspots, i.e. buildings at risk (schools, private building, industries, etc.), and strategic structures, i.e. rescue centres (hospitals, fire stations, etc.), are estimated with a spatial analysis approach based on flows and topologies in order to evaluate the indirect impacts to the system during the emergency phase. Finally, the latter accessibility and operability model is integrated with a consequence-estimation model for urban areas based on the main concepts that drive the internationally used flood-consequence-estimation methods that were previously cited in order to evaluate the maximum impact of a chosen flood event in terms of direct and indirect damages only during the emergency phases of a flood event. The proposed model does not aim to estimate the entire wide range of indirect impacts that may have effects on time scales of months and years, i.e. macro-economic effects or longterm barriers to regional development (Merz et al., 2010). Instead, the model focuses on how the impact of a flood hazard on individual elements of strategic infrastructure or single nodes in network systems may influence the system as a whole (Meyer et al., 2013) in the emergency phase of a flood. Hence, the proposed model for consequence estimation in urban areas provides a quantitative evaluation of direct damage, to inform decision-making in terms of loss of life and structural and economic damages, which is useful in order to support an innovative methodology for investigating the relationships of spatial accessibility and functional/operability failure (i.e. the performance to guarantee victim assistance and rescue activities) in a complex urban system during the emergency phase. Concurrently with the occurrence of physical and functional damage to urban areas, the operability of the strategic emergency structures, their accessibility, and connection within the city – or in general the urban area – is an important priority in emergency management.

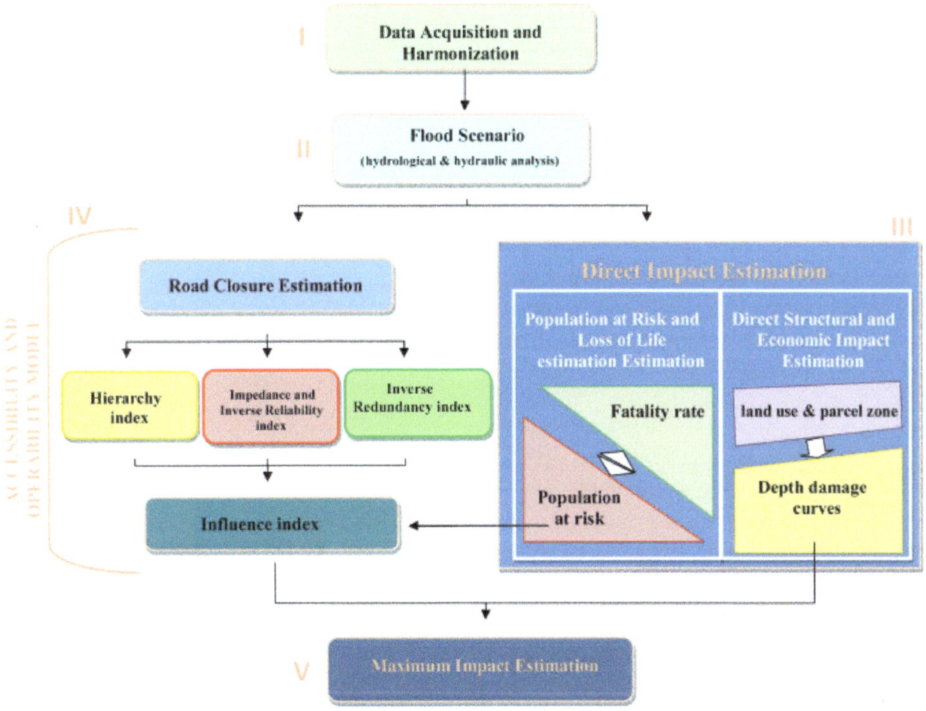

Figure 1. Phases of the proposed methodology.

The present framework, integrated in a geographic information system (GIS), aims to estimate the direct and indirect damage of a flood event in order to understand the strengths and fragilities of a particular urban area. The scope is to de- fine a hierarchy between the various structures (hospitals, fire stations, town halls, schools, industries, etc.) and infrastructure (main roads, secondary and local roads, bridges, etc.) through the identification of those structures/ infrastructure whose operation and effectiveness are critical in emergency management. The proposed model can aid in prioritizing the decisions on flood mitigation strategies that should be planned. This could support the maximization of the benefit of limited investments by selecting the highest-priority ones for emergency service. In Sect. 2, the overall GIS framework is outlined, in Sect. 3 the application and results of the proposed model on a real flood event are described, and overall conclusions are provided in Sect. 4.

OVERALL FRAMEWORK

This section describes the integration of a methodology that estimates the impact on accessibility and operability of strategic emergency response structures within an urban system, and a methodology for flood-consequence estimation in urban areas, with the aim of prioritizing actions for flood consequence reduction (Fig. 1). Sections 2.1 and 2.2 describe the preliminary phases needed for the implementation of the methodology. Section 2.3 summarizes the proposed GIS methodology for the estimation of the consequences for an urban population, which can also be used to estimate the direct structural and economic damages for residential, commercial, and industrial buildings. Section 2.4 describes the proposed approach to explore the dependencies among the structures and infrastructure of a city during the emergency phase of a flood event (i.e. during or immediately after a flood), in terms of the accessibility of flood-prone areas and the operability of road networks for emergency service. Finally, in Sect. 2.5, this latter indirect-consequence estimation is coupled with direct-consequence estimation through a maximum-impact index.

Data acquisition and Harmonization

The level of epistemic uncertainty in estimating potential damage by the model depends on available data (data collection, site visits, etc.). An analysis of the data considers land use distribution, data population census, digital elevation terrain models, and buildings and roads categorized on the basis of the function/typology (main roads, local roads, industries, resorts, hospitals, etc.). Therefore, the proposed approach requires the characterization of the system during the preliminary phases of the scheme in Fig. 1, i.e. phase I: input "data acquisition and harmonization" (data collection, site visits, etc.).

Definition of the Flood Scenario

Phase II ("flood scenario: hydrological analysis and flood scenario evaluation") is concerned with the definition of a flood scenario, or flood scenarios, required to estimate the potential damages and/or in order to determinate the possible flood events. A flood scenario can be identified by a return period, a combination of loads that determine a failure scenario, the result of flood routing, etc. If the proposed model runs several times for different flood scenarios with different return times, it can relate probabilities of each flood event to potential consequences.

However, the evaluation of a flood scenario could be performed via a hydrological analysis, which could be essential for estimating flood probability of a scenario or of more scenarios, coupled with a flood simulation, that should preferably be conducted using a 2-D flood model (e.g. MIKE FLOOD developed by the Danish Hydraulic Institute, Telemac2D developed by the National Hydraulics and Environment Laboratory of the Research and Development Directorate of the French Electricity Board, and CCHE2D developed by the National Center for Computational Hydroscience and the School of Engineering of the University of Mississippi) that is likely to be data intensive but provides more detailed results in terms of velocity and water depth distribution. The latter parameters are essential to estimate the flood severity of the chosen scenario; flood severity is usually assigned using flood depth multiplied by average velocity value (DHS, 2011b).

GIS Direct-Impact Estimation

This phase of the methodology is composed of two parts and provides two main results: the estimation of the loss of life and of the direct economic damages due the flood event.

Population at Risk and Loss of Life Estimation

During urban flooding events, consequences in terms of loss of life can be estimated as the combination of population exposed to the flood, i.e. population at risk and fatality rates (Escuder-Bueno et al., 2012) related to the characteristics of the flood (flood severity), evaluated in phase II. Indeed, the results of flood modelling and data from the population census are used. Geographic analyses are carried out using map algebra techniques implemented in a set of scripts tested and developed using the Python scripting language (http://www.python.org), the open-source Geospatial Data Abstraction Library (GDAL) (http://www.gdal.org), and the NumPy Python module (http://www.numpy.org). To combine multiple maps in map algebra, all data were required to be converted into grid format.

The outputs of the hydrodynamic model are processed to derive the information required for the analysis (flood wave arrival time, peak unit flow rate, etc.). Using GIS scripts, a flood wave arrival time (Twv), i.e. the time of occurrence of the flood wave, grid was obtained. In addition, the two components (x coordinate and y coordinate) of the vector unit flow rate were combined to obtain the maximum peak unit flow rate values ($m^2\ s^{-1}$) (i.e. the flow discharge for each linear metre of cross section). These values, termed parameter DV, proposed by Graham (1999), are representative of the general level of destruction that would be caused by the flooding. The DV values are then categorized, as illustrated in Table 1 based on guidelines of DHS (2011b), widely used in the United States. The values are classified into ranges de- fined as low-, medium-, and high-severity zones that define the rating of the flood severity.

Table 1. Flood severity rating criteria (source: Department of Homeland Security (DHS), 2011b).

Flood severity rating	Rating criteria
Low	DV less than $5\ m^2\ s^{-1}$
Medium	DV equal to or greater than $5\ m^2\ s^{-1}$ and less than $15\ m^2\ s^{-1}$
High	DV equal to or greater than $15\ m^2\ s^{-1}$ combined with rate of rise at least 3 m in 5 min

If the information on population is aggregated at the census area level, it could be hypothesized that it is distributed homogeneously within the vector polygon that represents the census areas. Hence, the vector polygons of the population census block are converted into grid format. By overlaying grid maps of flood with the grid of the population, it was possible to develop a map of population at risk (PAR).

Estimates of loss of life are obtained by multiplying the PAR with the fatality rate (fraction of people at risk projected to die from (severe) flood events). The fatality rates proposed in the SUFRI (Sustainable Strategies of Urban Flood Risk Management with non-structural measures to cope with the residual risk) project (Escuder-Bueno et al., 2012) are adopted in the model because it is based on a literature study and procedures that cover the loss-of-life estimation of historical flood events (e.g. Graham, 1999; DHS, 2011a, b; Penning-Rowsell et al., 2005), and it has been applied with good results in Italy (Escuder-Bueno at al., 2012). Ten categories were established by Escuder-Bueno et al. (2011) to estimate potential loss of life in urban areas in the case of river flooding. This classification of categories from low to high flood severity understanding, from C1 to C10, was developed based on levels of public education on flood risk, warning systems, risk communication, and coordination between emergency agencies and authorities (see Table 2). It defines a certain level of flood severity understanding for each category, linked to fatality rates and based on a compilation of historical data and existing reference values on loss of life (Graham, 1999; Escuder-Bueno et al., 2012). Consequently, different fatality rates are considered for each category (C1 to C10) depending on available warning times (0–24 h) and three flood severity levels described previously (Table 1). The warning time, which is a function of the Twv, at night is defined as a time period 15 min lower than the warning time during the day, such as in Escuder-Bueno et al. (2011). If there is no warning time or data are not available, the available warning time is estimated from the difference between the time of occurrence of the first-notice flow and the firstdamage flow, such as in Escuder-Bueno et al. (2011).

Table 2. Fatality rates in the case of river flooding (Escuder-Bueno et al., 2012).

ID	Warning time (h)	Flood severity		
		High	Medium	Low
C1	0	0.9	0.3	0.02
	0.25	0.9	0.3	0.02
	0.625	0.7	0.08	0.015
	1	0.3	0.06	0.0006
	1.5	0.3	0.0002	0.0002
	24	0.08	0.0002	0.0001
C2	0	0.9	0.3	0.02
	0.25	0.9	0.3	0.02
	0.625	0.675	0.075	0.014
	1	0.3	0.055	0.00055
	1.5	0.3	0.0002	0.0002
	24	0.075	0.0002	0.0001
C3	0	0.9	0.3	0.02
	0.25	0.85	0.2	0.015
	0.625	0.6	0.07	0.012
	1	0.3	0.05	0.0005
	1.5	0.3	0.0002	0.0002
	24	0.075	0.0002	0.0001
C4	0	0.9	0.3	0.02
	0.25	0.75	0.15	0.01
	0.625	0.5	0.04	0.007
	1	0.3	0.03	0.0003
	1.5	0.15	0.0002	0.0002
	24	0.04	0.0002	0.0001
C5	0	0.9	0.3	0.02
	0.25	0.75	0.15	0.01
	0.625	0.5	0.0375	0.0065
	1	0.3	0.0275	0.000275
	1.5	0.15	0.0002	0.0002
	24	0.375	0.0002	0.0001
C6	0	0.9	0.3	0.02
	0.25	0.75	0.15	0.01
	0.625	0.475	0.035	0.006
	1	0.3	0.025	0.00025
	1.5	0.15	0.0002	0.0002
	24	0.035	0.0002	0.0001
C7	0	0.9	0.3	0.02
	0.25	0.65	0.1	0.0075
	0.625	0.4	0.02	0.002
	1	0.3	0.01	0.0002
	1.5	0.1	0.0002	0.0002
	24	0.02	0.0002	0.0001
C8	0	0.9	0.3	0.02
	0.25	0.55	0.06	0.006
	0.625	0.35	0.01	0.0015
	1	0.25	0.005 0	0.0015
	1.5	0.1	0.0002	0.00015
	24	0.01	0.0002	0.0001
C9	0	0.9	0.3	0.02
	0.25	0.55	0.06	0.006
	0.625	0.35	0.008	0.0015
	1	0.2	0.004	0.000125
	1.5	0.01	0.0002	0.0001
	24	0.01	0.0002	0.0001
C10	0	0.9	0.3	0.02
	0.25	0.5	0.03	0.005
	0.625	0.3	0.005	0.001
	1	0.1	0.002	0.0001
	1.5	0.01	0.0002	0.0001
	24	0.01	0.0002	0.0001

The final step for loss-of-life estimation relies on the combination of fatality rates and population at risk to obtain the number of potential fatalities for each flood scenario.

Direct Structural and Economic Impact Estimation

Methods and values of the parameters used in this section are drawn mostly from the report of the Department of Water Resources Division of Flood Management on Flood Rapid Assessment Model Development (F-RAM, 2008). The model is widely used in the evaluation of structural damage because it was evaluated in laboratories and real survey data from recent flood events in the United States. The methods presented in this subsection (phase III of Fig. 1) are based on the use of depth–damage relationships that assign a percentage of damage from the resulting water depth during the flood.

An economic value of assets or land use was established, and economic losses were obtained from the destruction rate (e.g. percentage of damage) within the flooded area. These curves are used for estimating the direct economic damage for residential, commercial, and industrial buildings. The input data consist of maps of land use and parcel zones of the study area.

The curves allow for the estimation of the damage to buildings and their contents and, when applied to different scenarios, allow for an effective comparison of the impact. The extent of damage to buildings and their contents is estimated from the flood depth by the application of a depth– damage curve associated with each occupancy type. Depth damage curves show the relationship between the depth of the flood relative to the first finished floor level of buildings and the damage caused to the structures and contents. Damage is typically expressed as a percentage of depreciated building replacement value. The adopted method measures the content damage directly as a percentage of structure value rather than using a content / structure value ratio, i.e. the ratio between the unitary value of the content and the unitary value of the building structure, rather than formulating some specific content depth–damage curves that could be more difficult to estimate. To calculate damage, each structure must be assigned to a structure occupancy type. For each structure occupancy type an estimated replacement value, a structure depth–damage curve (Fig. 2) and a content depth–damage curve (Fig. 3) relationship must be defined. The depth–damage curves implemented in the model were obtained from the United States Army Corps of Engineers (USACE; Department of Water Resources Division of Flood Management, 2008). The methodology, here presented, could use other depth–damage curves that are more suitable for the area of interest; however, in the present model the USACE curves are implemented since they are suitable with the case study described in the next section, because they are more precautionary than the one proposed by Luino et al. (2003) for Italy. In assigning an occupancy type, taken usually from a city map at micro-scale, to each parcel, we chose values according to those shown in Table 3.

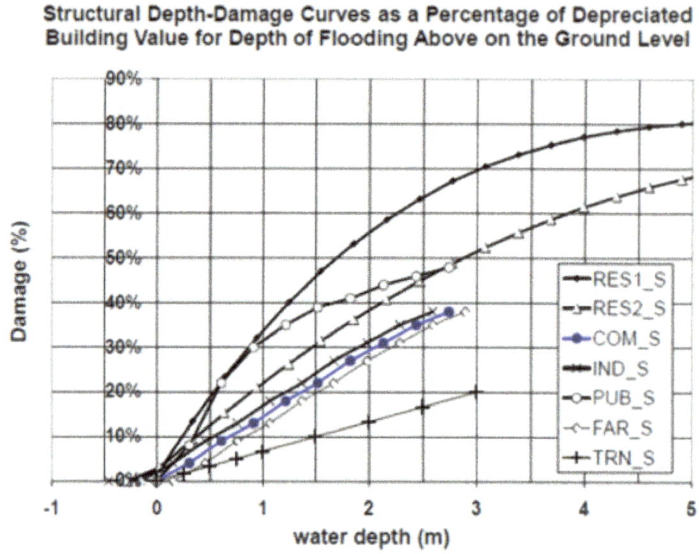

Figure 2. Structural depth–damage curves implemented in the model (source: Department of Water Resources Division of Flood Management, 2008).

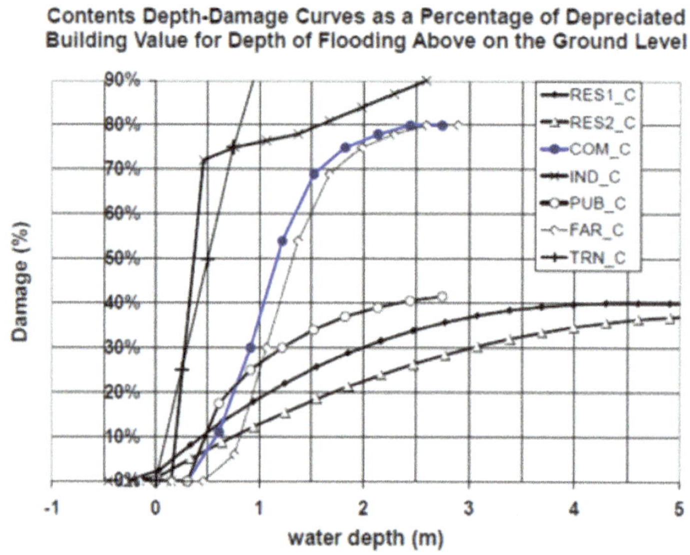

Figure 3. Content depth–damage curves implemented in the model (source: Department of Water Resources Division of Flood Management, 2008).

Table 3. Assigning occupancy type from zoning type.

Zoning type	No. of stories	Occupancy type
Commercial	Any	COM
Industrial/wholesale/ manufacturing	Any	IND
Institutional/government	Any	PUB
Office	1	RES1
Office	2 or more	RES2
Open space/recreation/ agricultural	Any	FAR
Residential	1	RES1
Residential	2 or more	RES2
Transport	Any	TRN

GIS Accessibility and Operability Model for Emergency Management

This section describes how the infrastructural transport dependencies are estimated in the urban area during the emergency phases of a flood event (i.e. the performance of rescue activities taking into account the connections/paths between areas at risk and rescue centres such as hospitals and fire stations). In terms of emergency management, the failure of some part of the transport infrastructure would have the most serious effects on access to specific locations and overall system performance. The road closures due to flood waters, estimated on the basis of velocity and water depth values, could create damages and hence could alter the emergency travel operations from normal conditions. In this context, an analysis of the paths of the emergency travel activities could open the possibility to estimate the operability of the strategic emergency structures and highlight weaknesses (e.g. the most inaccessible area at risk or the strategic connectivity road that are most damaged). We focus on the emergency operations, and not on the evacuation of the people that could have been done in the pre-event phase of the flood event.

Road Closure Estimation

First, it is necessary to estimate road closures due to flood waters in order to estimate the potential inaccessible areas and inoperable roads (phase IV of Fig. 1). The possible road closures due to flood waters or large debris transport are estimated on the basis of literature studies that estimate a weight related to critical threshold values of hydraulic instability for idealized vehicles (Teo et al., 2012). If the vehicles on these streets are dragged by the water flow, the road is inaccessible. The envelope curves developed by Teo et al. (2012) consider three colour zones (i.e. green, yellow, and red), and the hydraulic stability for each idealized vehicle was easily identi- fied by colour. The stable zone is shown in green (left zone), the transition zone in yellow (central zone), and the unstable zone in red (right zone). All vehicles in the red zone of the graph are dragged by the water flow; hence they could block, for example, an emergency vehicle during rescue actions. The curves implemented in the model are used when incoming flow depths are lower than the vehicle height, shown in the lower part of the graph in Fig. 4. When the incoming flow depth is greater than the vehicle height, the roads are considered to be always inaccessible. This choice is justified by the possible presence of emergency vehicles that could work in worse conditions than cars (firefighter trucks, ambulances, small boats, etc.). As such, the methodology, on the one hand, aims to give more importance to closure of roads due to vehicle transport – which is a frequent phenomena in urban areas as highlighted in Albano et al. (2014), Gruntfest (2000), and Gruntfest and Ripps (2000) – and, on the other hand, aims to be precautionary and independent of the type of vehicles available in a specific scenario in the analysis.

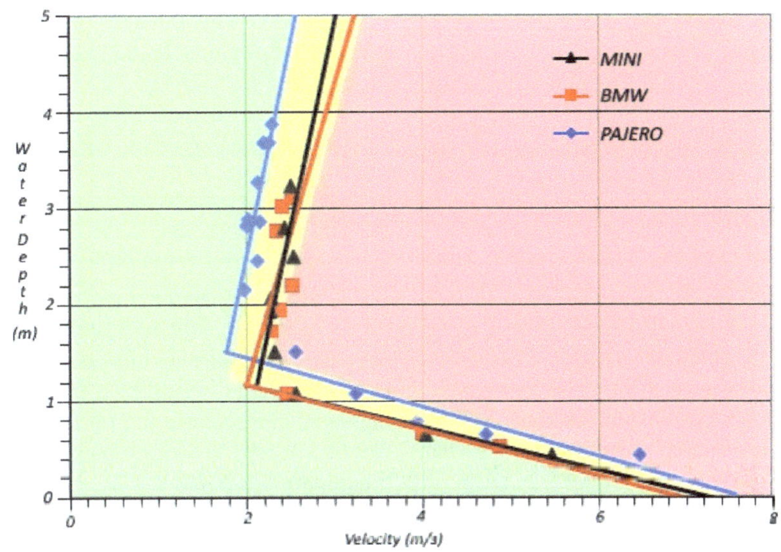

Figure 4. Critical threshold values of hydraulic instability for specific vehicles (taken from Teo et al., 2012).

Accessibility and Operability Analysis of the Urban System

Emergency management systems operate their vehicles in different ways during an emergency such as a flood. For example, they might use local streets in order to take the shortest path to their destination since the lower speed limit of local streets may not apply to those emergency vehicles. As a result, the shortest path will provide them with the shortest time response. In this situation, a road closure due to a flood could alter the path that connects different elements in an urban area, such as the path between a hospital and a damaged school, thereby increasing the distance between them, which would result in a lower level of accessibility. Equation (1) is proposed to estimate the degree of inoperability of the arc ai that is a path (i) within the system, i.e. the inverse (connectivity) reliability index, where the concept of reliability is introduced by Taylor et al. (2006) – see the central part of phase IV of Fig. 1:

$$R_a = \sum_{P_{si}=1}^{k} \left[\frac{\sum_{o=1}^{m} \left[\left(1 - \frac{P_{si,o}}{P_{ei,o}}\right) \cdot \frac{P_{\mathrm{smax},i}}{P_{si,o}} \right]}{\sum_{o=1}^{m} \frac{P_{\mathrm{smax},i}}{P_{si,o}}} \right] \Big/ k, \tag{1}$$

where P_s is the length of the generic standard path, and P_e is the length of the emergency path (i.e. the path that the aid vehicles have to travel due to the flood event). P_{smax} is the value of the longest standard path between all the standard paths that connects the aid centres with the building at risk. A path is defined as "standard" if the latter connects aid centres with buildings at risk in the normal functioning of system connections. These are defined as "emergency" paths if the system is affected by a flood event. Equation (1) is an average of the ratio $1-(P_s/P_e)$ weighted on the ratio P_{smax}/P_s in order to consider the whole accessibility system (i.e. all the shortest paths among the elements at risk and all the emergency centres in the system), normalized on k. The latter represents the number of times that the arch a_i is crossed by the shortest paths (P_s) connecting the relations "origins / destinations", hereafter "o/d", where the origins (o) are the core rescue buildings and the destinations are buildings at risk (i.e. private or public buildings, factories, etc.). If an emergency path does not exist, (i.e. the elements are completely isolated) this means that access to alternative services (such as hospitals and businesses) does not exist. The disruption costs to households, businesses, and communities can therefore be more critical for the whole system.

The inverse reliability index, estimated by Eq. (1), highlights the travel distance reliability of the path. Travel distance reliability considers the probability that a trip between an origin–destination pair (see Fig. 5) can be completed successfully via the shortest distance possible for the normal functioning of system connections; this is represented by the blue line in Fig. 5, and, for the case of a flood event, this is represented by the red line in Fig. 5. The ratio between P_s and P_e is weighted on the basis of the distance between o and d in order to relate this ratio to the urban system network dependencies in the emergency phase; the estimated value for each path is normalized on the basis of the multiple o/d relationships because there can be more than one origin in the system (i.e. core rescue buildings).

Equation (1) is assigned to each arch a_i that composes the shortest path P_{si}, but it is used in a modified version (see Eq. 2) also in order to estimate the degree of inaccessibility of an area that requires rescue (i.e. the impedance index, introduced by Taylor et al. (2006) but here modified in order to consider accessibility in the whole system for emergency service), assigning the estimated value to each building at risk that requires rescue:

Figure 5. Graphical example of the elements (e.g. standard and emergency shortest paths, origin (i.e. first-aid centre) and destinations (i.e. buildings at risk), and node in which there are road closures) involved in Eqs. (1) and (2).

$$I_i = \frac{\sum_{o=1}^{m}\left[\left(1-\frac{P_{si,o}}{P_{ei,o}}\right)\cdot\frac{P_{s\max,i}}{P_{si,o}}\right]}{\sum_{o=1}^{m}\frac{P_{s\max,i}}{P_{si,o}}}. \tag{2}$$

The impedance index in Eq. (2) is used to estimate the impedance of nodes (i.e. buildings at risk), i.e. the remoteness derived from measures that aims to identify the buildings that are more difficult to reach by the emergency services. In Fig. 5, the black building has the highest degree of impedance. The inverse (connectivity) reliability index, instead, in Eq. (1) is useful to highlight the strategic paths that connect the elements of the system. The inverse reliability and impedance index ranges between 0, i.e. no impedance, and 1, the highest value of inverse reliability or impedance, i.e. where the building is completely isolated.

Considering that each shortest path is composed of a number (k) of arches, an index to estimate the strategic importance of single arches is estimated; it is known as the hierarchy index. A network link is critical if loss or substantial degradation of the link significantly diminishes the accessibility of the network or of particular nodes. Therefore, the arches that are involved in a greater number of path connections (i.e. the ones that could be used more often by aid vehicles to reach the flood-prone areas) are the more important arches for maintenance of the emergency management performance.

The hierarchy index, H_a, developed in this study represents the number of paths, P_s, that connect the relations o/d, using the arc a_i:

$$H_a = \sum_{od}\left(k_{ai}\Big/NP_{sod}\right), \tag{3}$$

where k_{ai} is the count, k, of the times that the shortest paths, P_s, used the arch a_i to connect the multiple relations o/d. NP_s is the number of shortest paths, Ps , that connects the multiple relationship o/d. The arch that is more utilized by the shortest paths, i.e. the one with highest k_{aj} (e.g. the one in red in Fig. 6), is of significant importance for the system during emergency management because the performance of emergency services can be affected in a significant way by its inoperability. H_a can range between 0 and 1.

The estimation of the hierarchy index can help to identify the arches most affected by infrastructural relations o/d in order to define a hierarchy between the various infrastructure through the identification of those components in which operation and efficiency are fundamental to the maintenance of network connectivity. Another measure of network performance in flood emergency conditions is the estimation of possible alternatives for each single arch (i.e. the number of outgoing arcs a_i from the arc a_j) in the case of a flood event:

$$IR_a = 1-\left[\left(a_{ijs}-a_{ijE}\right)\Big/a_{ijs}\right], \tag{4}$$

where a_{ijE} is the number of outgoing arcs a_j from the arc a_i that are inoperable due to the flood events, and $a_{ij}S$ is the number of outgoing arcs a_j from the arc a_i in the normal functioning of the system. The redundancy concept was introduced by Lhomme et al. (2013) but here is modified in order to consider the situation before and after the flood event. The inverse redundancy index, which ranges from 0 to 1, suggests the number of potential alternative connections between arch a_j and the others related to that being considered in the emergency phase and, therefore, the number of available and non-available arches, in the case of flooding, that could be utilized by emergency services if the arc a_i is inoperable.

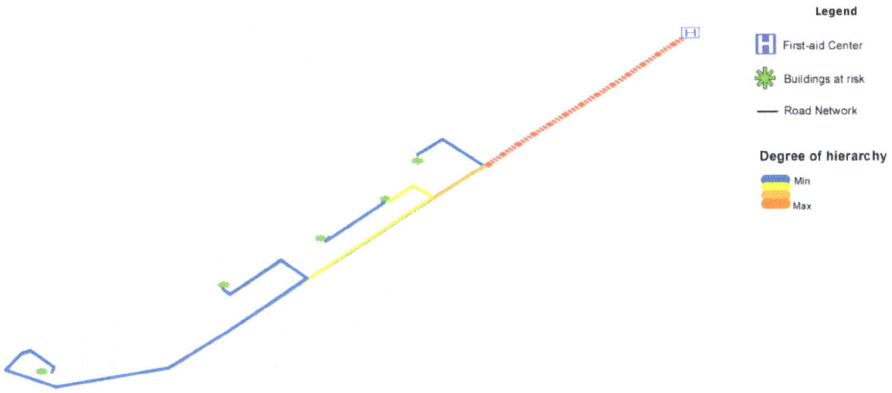

Figure 6. Graphical example of the degree of hierarchy of arches that can be utilized by emergency services during the shortest paths that connect "origin" (e.g. a first-aid centre) with diverse destinations (i.e. the buildings at risk).

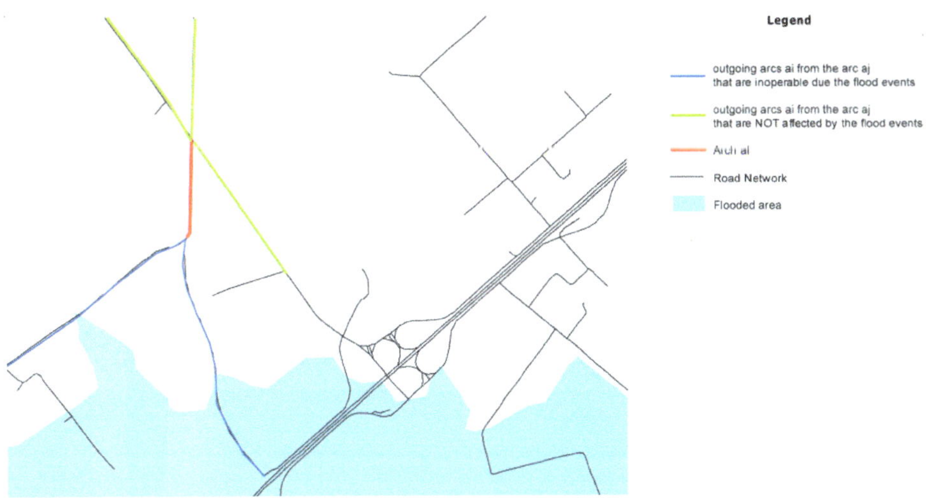

Figure 7. Graphical example of the elements (i.e. arc a_i and its outgoing arcs a_j) involved in Eq. (4).

Figure 7 shows an example of parameters involved in Eq. (4): the red line is the arc a_i, i.e. the arc to which will be assigned the value of inverse redundancy index; in blue arc outgoing arcs a_j from the arc a_i that are inoperable due to the flood events; and in green are the arcs a_j from the arc a_i that are operable even in the case of a flood event. Therefore, the inverse redundancy of arc a_i in the system could be affected by the presence of more arcs a_j that are inoperable due to the flood events. It means that, in the case of inoperability of arc a_i, a great number of outgoing arcs a_i that are also inoperable due the flood event (the blue ones in Fig. 7) will represent a slowing-down in the performance of emergency service that can use less alternatives to the arc a_i during the emergency rescue activities.

Finally, the value of the geometric mean of the product for each arch derived from Eqs. (1), (3), and (4) represents the index of weakness of each arch in the emergency phase:

$$x_i = (\sqrt[3]{R_a \cdot H_a \cdot IR_a}). \tag{3}$$

This value, which coupled the flow and functionality approach with the topology analysis, defines a hierarchy between the various arches through the identification of those arches whose operation and performance are fundamental to the maintenance of network connectivity and accessibility in the whole system during a flood emergency. For the structures, i.e. buildings at risk, only Eq. (2) (i.e. the impedance index) is used in order to estimate the weakness index of structures at risk for each building. The weakness index, which ranges from 0 to 1, is reclassified in order to vary in the interval [1–10].

Finally, an influence index for structures and infrastructure is estimated based upon the typology of each building or road in the system during the emergency response phase. It can be defined by a Gaussian curve corresponding to a mathematical function of an exponential type (Pascale et al., 2010):

$$y_i = a \cdot \frac{e^{-\partial \cdot x_i^{2.2}}}{\left(1 + e^{-\partial \cdot x_i^{2.2}}\right)}, \tag{6}$$

where x_i is the weakness index of each of the elements previously described; a is a constant which takes on a value equal to 2 and is calculated by fixing the boundary conditions ($x_i = 0$ and $y = 0$, where $y = 0$ represents 0 % of vulnerability, equivalent to no loss); and ∂ is a parameter calculated by fixing boundary conditions as follows: $3 < x_i < 6$ and $3 < y < 6$ in a condition of medium to high vulnerability and equal to 0.02 (Pascale et al., 2010). The role of this function is to estimate the degree of influence among the elements of the system, considering the degree of connectivity, accessibility, and the role of each in the system in the emergency phase. It can range between 0 and 1.

Equation (6), as in Pascale et al. (2010), is modified by introducing a correction factor that takes into consideration the population affected by the event, calculated previously in Sect. 2.3.1:

$$y_i = a \cdot \frac{e^{-\partial \cdot x_i^{(2.2+f)}}}{\left(1 + e^{-\partial \cdot x_i^{(2.2+f)}}\right)}. \tag{7}$$

The roads and the buildings at risk located in the census area with higher numbers of population at risk have higher values of the influence index, for the same value of the weakness index and the same functions in the system in the case of an emergency. The influence index takes into account the role of each element in the system in the emergency phase. In this light, the components such as buildings or communication networks were subdivided into Categories A, B, and C. These elements were divided into these categories relative to the element functions in the systems in the case of an emergency. For instance, if a hospital is damaged, the whole system is affected by an increase in the rescue workload for other forms of assistance. The elements at risk with different roles and importance in the emergency management are set in Categories A, B, and C. The importance of these features move from Category A to C in the following manner:

- Category A includes the most important elements in the case of an emergency, such as hospitals, fire stations, and civil protection stations. These are all elements that give assistance when catastrophic events occur. This category also includes main roads.
- Category B includes all the major socio-economic and environmental elements, such as factories, which can also deal with dangerous materials; large shopping centres; and all other public buildings, including universities, libraries, and churches. All of these can contain a large number of people and can be important from a historical, artistic, and cultural perspective. This category also includes secondary roads.
- Category C includes private buildings, small business activities, and local roads.

In this study, it is assumed that ∂ varies in the interval [0.01–0.1], defining in this way the influence index trend on the basis of the different categories previously described; the parameter ∂ assumes a value of 0.06, 0.04, and 0.02, respectively, for Category A, B, and C, such as in Pascale et al. (2010).

Maximum-Impact Estimation

Finally, the direct economic consequence estimation is coupled with the indirect systemic impact in emergency management through a maximum-impact index (i.e. phase V of Fig. 1). The maximum impact of each element within the system is estimated by the equation

$$v_i = \max(y_i, s_i), \tag{8}$$

where s_i is the structural damage, estimated by depth– damage curves as described in the previous subsection (phase III of Fig. 1), and y_i is the influence of the road network on the elements of the territorial systems. The value of the maximum impact, which can vary in the range [0, 1], is the recapitulatory index, and it is also precautionary since it considers the highest value between possible direct and indirect damages. The innovative proposed systemic approach that is integrated in a consequence-estimation model can only increase the value of the damage by taking into account the inoperability of roads or the isolation of buildings due to the flood event.

The model aims to highlight the areas that need priority risk prevention interventions on the basis of accessibility and operability of the structures and infrastructure in the territorial system. In light of this, the estimation of structural damage can be an important indicator of the effectiveness and operability of each element of the system, on the basis of its structural status, and, moreover, the influence index can increase the understanding of decision-making through the estimation of the operability and accessibility in the whole system. The choice of taking the higher value between the direct and indirect consequences provides the possibility of showing how the impact of a flood hazard on the individual elements, i.e. strategic infrastructure or structure, may support decision-making regarding the prioritization of the risk prevention measures, in the emergency phase of a flood, through the estimation of the tendency of a given territorial element to suffer damage of a structural kind and, hence, concerns its intrinsic operability or its influence on the other elements of the system.

CASE STUDY

Ginosa is a city in the Puglia region of Italy, located near the mouth of the Bradano River. The choice of this case study site was justified by the flat morphological characteristics of the river, determined using significant field data collected in recent years as well as the use of high-resolution digital terrain models (DTMs) from laser-scan data. Moreover, the study area includes the mouth of the Bradano River, which is particularly at risk for flooding. This estimation was derived from an analysis of historical data on hydrogeological disasters during the period 1918 to 2000, conducted as part of the "Affected Italian Areas" by the National Research Council (CNR).

As mentioned, analysis of the data shows that the area at the mouth of the Bradano River has been affected in the past by a significant number of natural disasters. The most recent flood event occurred on 1 March 2011. This flood event was deemed so severe that authorities declared a state of emergency. The flood event of 2011 at the mouth of the Bradano River affected the town in the first days of March when the majority of the hotels, resorts, and tourist attractions were essentially closed or empty. Therefore, in the analysis presented in this case study, seasonal variability in tourist numbers was not taken into account because in March there are very few tourists in this area. This flood event was particularly intense, causing damage to economic activities and residential buildings, as well as provincial and national roads which became unusable due to water and mud. The local administration is still in the process of developing both structural and non-structural measures to cope with flood risk in Ginosa, as well as in the neighbouring towns. Regarding this study, it was deemed preferable to validate the model proposed in this study with an event that has actually occurred, rather than a generic simulated event.

Data

Characterization of the urban system of Ginosa

The total population of Ginosa is approximately 22 146 (ISTAT, National Institute of Statistics, 2001), with 32 % comprising children under 14 and adults over 65 years of age. The population data are taken from the Italian Institute of Statistics, which stores all the demographical statistics, also in geographical form, for all of Italy ("Geo demo database", available at www.demo.istat.it). The population is aggregated at the census level scale.

The typical building topology is more than 90 % 1–2-floor cottages (SIT Puglia database, 2011). It should be noted that the ISTAT database and Puglia regional databases were developed at different times, resulting in discrepancies between the data. The discrepancies are related to the different times of the acquisition of the population data (ISTAT, National Institute of Statistics, 2001) and the map of the city which represents buildings and roads, at a scale of 1 : 5000 (SIT Puglia database, 2011). These discrepancies are not believed to affect the final results of the model application.

The principal vulnerable hotspots in the Ginosa territorial system are the two most important throughways. These include the "S.S. 106 Jonica Main Road" and the railway "Taranto–Reggio Calabria". In addition, there is a first-aid unit located in the part of the city closer to the sea as well as diverse operative units that could support rescue activities. Several schools, churches, and banks are also identified in the town. The urban area is mainly composed of residential and agricultural areas but also of key resorts, zootechnical activities, and small and medium enterprises (SMEs). More than 45 % of the workers are employed in the service sector, such as in key resorts and hotels located in the area. Seasonal variability of the demography and tourist numbers could have a significant impact in the flood-consequences analysis.

Hydrological and Hydraulic Characterization of the Simulated Scenario

The flood scenario utilized for the application of the model is a simulated event that has a return time period closer to the real event of 1 March 2011, which occurred in Ginosa, Italy. The maximum discharge of the chosen event can be assimilated to an event with a 30-year return time period, estimated using the VAPI (Valutazioni delle Piene in Italia) method, which is recommended by local authorities (e.g. the Basin Authority of Puglia) in southern Italy (Claps et al., 2005).

Hydraulic simulations of flood scenarios were performed using a 2-D commercial flood model. For this case study, the MIKE FLOOD model was used since it was deemed to be the most appropriate model for this area as highlighted in Sole et al. (2011), who calibrated the model for the study area, using the digital elevation model, which includes cross sections of the river embankment extrapolated from laser scanner data. The friction coefficient of the flooded area was evaluated by the land use map at a scale of 1 : 5000, which is available on the online database of the Puglia region (SIT Puglia database, 2011).

RESULTS

Simulations provided hydraulic characteristics of the chosen flood scenario. Data of water depth, velocity, and wave arrival times were obtained in the urban area.

Due to the flat nature of the flooded zone, the flow velocity was average–low, and the water depth high, in most of the zone (Figs. 8 and 9). Hence, the direct economic damage estimation was performed only on the basis of the water depth parameter. The total flood area was determined to be approximately 30 561 900 m^2 (Table 4).

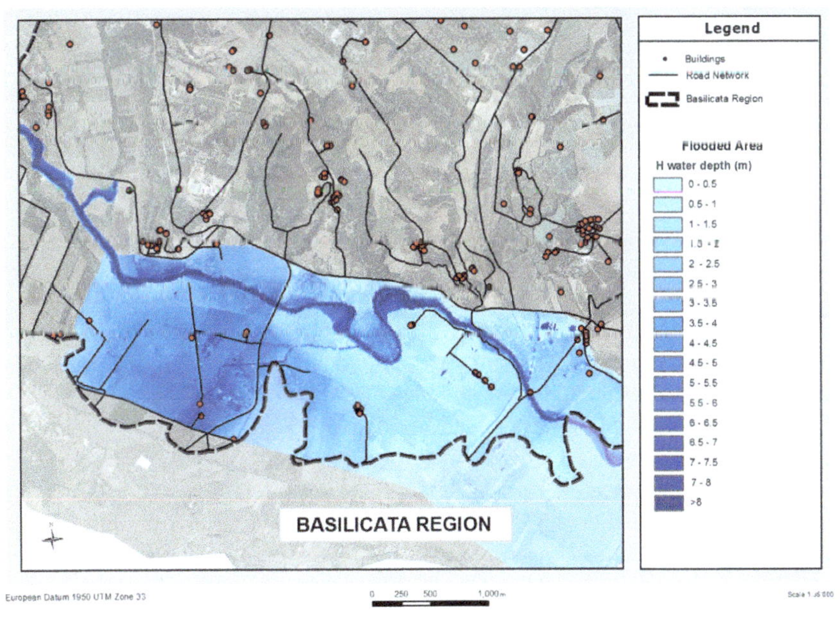

Figure 8. Water depth, H, from hydraulic modelling (upflow).

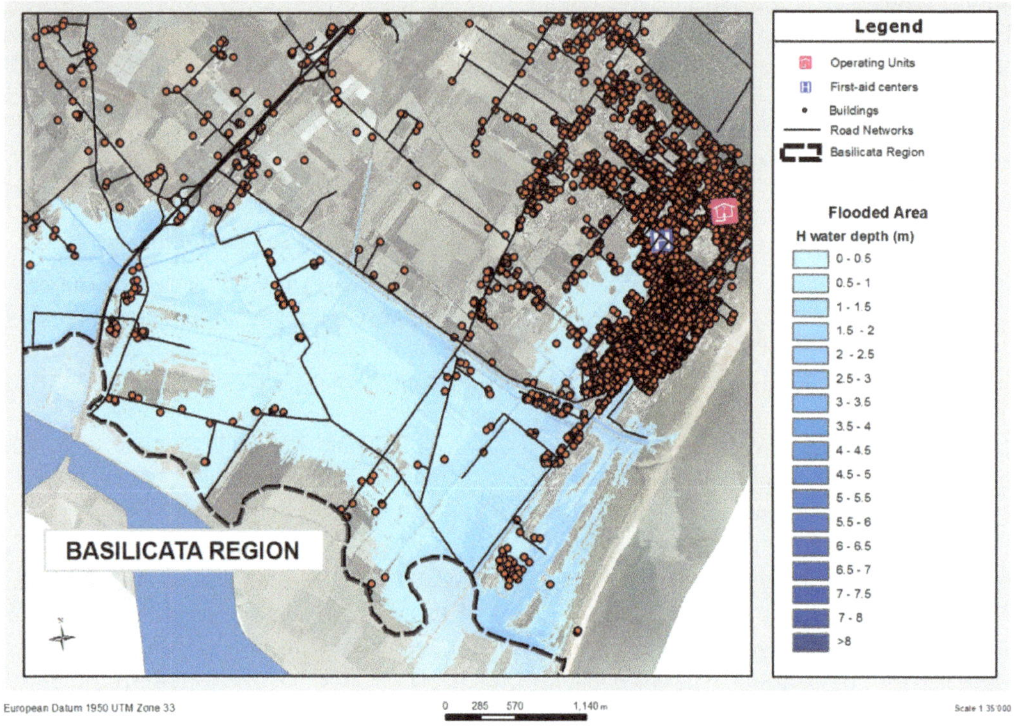

Figure 9. Water depth, H, from hydraulic modelling (downflow).

Table 4. Flooded area for the different categories of water depth, H.

Water depth (m)	Flooded area (m^2)
0.0–0.5	9 707 000
0.5–1.0	7 902 700
1.0–1.5	5 366 700
1.5–2.0	2 692 600
2.0–2.5	1 192 700
2.5–3.0	687 600
3.0–3.5	529 800
3.5–4.0	509 800
4.0–4.5	471 800
4.5–5.0	424 100
5.0–5.5	284 700
5.5–6.0	153 700
6.0–6.5	118 900
6.5–7.0	88 100
7.0–7.5	81 400
7.5–8.0	68 000
>8	282 300

The flood extension maps were able to define the areas of the territory directly affected by the flood event and to incorporate the necessary hydraulic characteristics for the study. Using GIS, flooded areas were identified to estimate the element at risk. Specifically, it was found that less than 10 % of the residential buildings are at risk because the more populated area of the town is located outside the flooded area. However, 30 % of business activities are located in the floodprone area, in particular SMEs and resorts. In the flooded area, 7 % of the population are children or elderly people. A majority of the people at risk are in the downflow area, near the sea. Further, the area characterized by the highest fatality rate estimated by the model, and shown in the area coloured in red in Fig. 10, is the first zone affected by water flow. The comparison between historical data of loss of life in 2000 (AVI project, 2000) and the estimated degree of loss of life (estimated by the model), which is represented in Fig. 10 in categories from low (i.e. the areas coloured in green) to high (i.e. the areas coloured in red), is justified by the fact that during the event of 1 March 2011 there was no loss of life. As such, it is likely important to validate, in a spatial way, the degree of the potential loss of life in the system.

Historical data on loss of life for floods have highlighted that a single flood event in Ginosa prior to the year 2000 resulted in casualties. The largest number of victims was found to be in the area highlighted as most prone to fatalities according to our application shown in Fig. 10. It was assumed that there was minimal warning of flood threats in this zone.

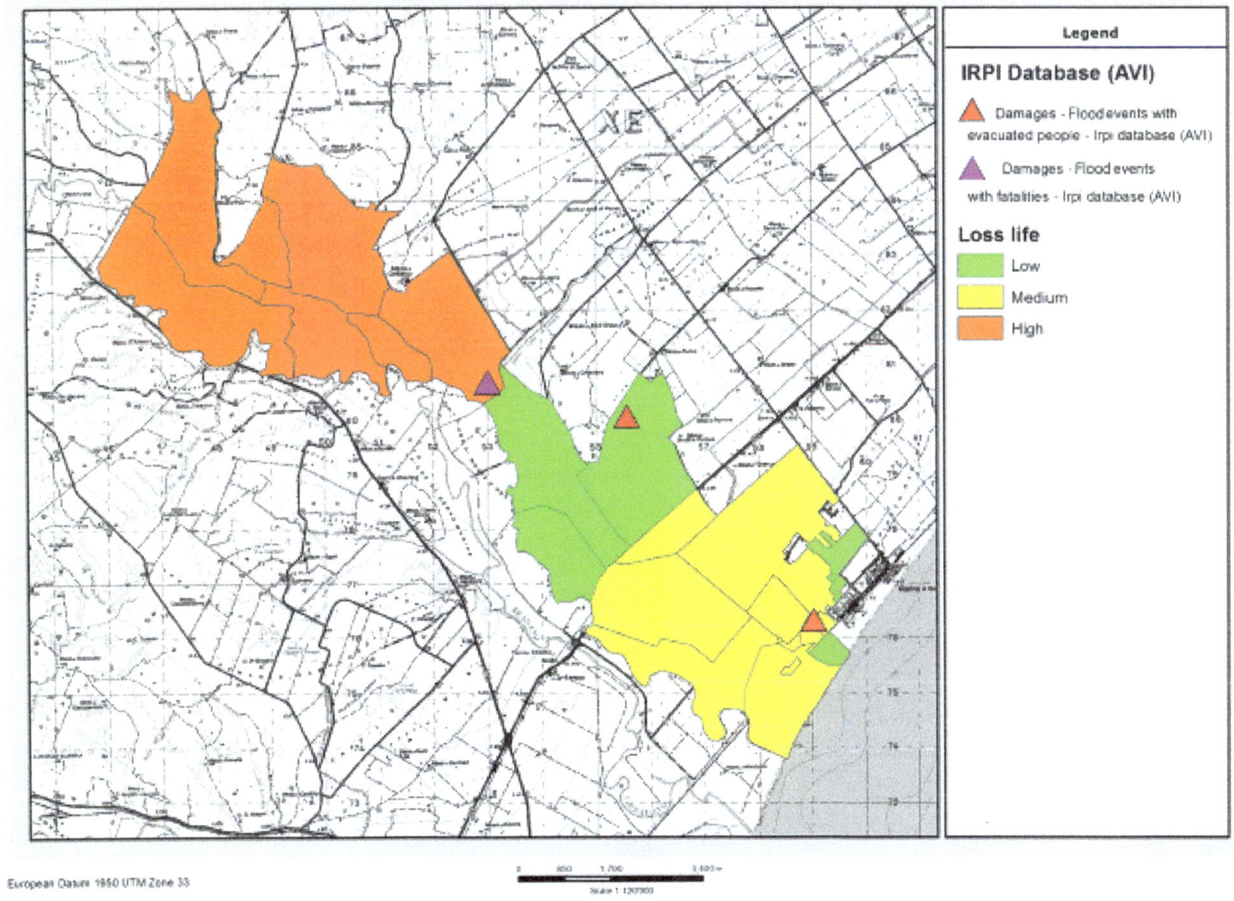

Figure 10. Map of the estimated loss of life divided into categories (low, medium, and high) for the flood event, compared with historical information (AVI project, 2000).

Warning time is defined as the time difference from the firstnotice flow and the first-damage flow. We made the assumption that the first-notice peak corresponded to the time when the first building of Ginosa is affected by the river flow since Ginosa does not have a flood warning system. Additionally, in the literature and on the web there is evidence that there has been no public education on flood risk, risk communication, and recent events have highlighted the lack of coordination between emergency agencies and authorities. Hence, in this case study, the fatality rates considered in the lowest level of flood severity understanding, i.e. category C1 of Table 2, area applied.

The low value of loss of life estimated by the model is addressed by the fact that, even though there is evidence of a lack of a warning system and public education activity, the peak unit flow rate is really low in the area due to the lower flow velocity estimated by the 2-D numerical flood model. The total loss of life estimated by the model corresponds to less than one fatality due to the low population density of the area as well as the low percentage of people at risk. In the event of 1 March 2011, there were no reported fatalities but substantial displacement of populations and damage to infrastructure, farms, and resorts, as highlighted in Table 5, which provides information on the direct economic damage, estimated by the model, considering this chosen flood scenario.

After the 1 March event, the total amount of money requested on the basis of a self-estimation by the citizens of Ginosa to the Italian government for the damages to their properties due to this flood event was around EUR 6 501 741 (source: "Ordinanza ministeriale del 5 luglio 2012 n. 4024"), in comparison to the EUR 4 736 125 estimated by the model as direct economic damages. This discrepancy could be justified by the evidence that the model does not take into consideration the damage caused by pluvial contribution to the flood event (the model simulates only the river flood event). Indeed, the number of buildings affected by the flood estimated in the model is about 63 % of the number of buildings affected by the real event (about 1000 buildings). It should be noted that it is not possible to complete a validation on the other elements (i.e. roads, railways, agricultural areas) involved in the flood event due to a lack of available data from the real event. However, it is possible to make a spatial comparison with photos recorded at 10 observation points throughout the city (Figs. 11, 13 and 14), as was done in this study.

Table 5. Direct economic damage due to the event simulated by the model.

Occup. type	Description	Structural value (€)	Contents' value (€)	Structural damage (€)	Contents' damage (€)
IND	Zoothecnical activities	9 800 000	34 300 000	0	0
IND	SMEs	12 560 000	43 960 000	24 000	84 000
ReS1 and RES2	Residential buildings	452 300 000	226 150 000	1 620 000	752 500
PUB	Public services	7 540 000	15 080 000	0	0
TRN	Main roads	48 516 000	1 940 676	2 528 915	735 294
TRN	Urban roads	145 932 500	5 836 807	6 743 983	2 101 124
TRN	Railways	30 694 000	1 534 700	1 098 666	433 887
COM	Hotels and resorts	19 050 000	38 100 000	928 125	1 327 500
FAR	Agricultural areas	0	5 999 187	0	5 999 187
FAR	Forest areas	0	597 750	0	63 280

Figure 11. Direct damage estimation.

Figure 11 provides a comparison between the model results and several site surveys by or after the events. It gives an overview of the consequences of the event and the potential reliability of the model. The area in which damage potential is greatest and most affected during the flood event is that closest to the river, where residential buildings and a resort are located in "c/da Marinella". Meanwhile, the area on the far end of the riverbed (i.e. "Via Ancona Road") received minimal damage (Fig. 11). During the actual flood, the majority of claims from damage associated with the natural disaster came from residents and proprietors of factories and industries closest to the river. Indeed, one of the most damaged buildings was the "Torre Sirena" resort, which resulted in one of the highest values of the influence index because it has a high impedance index (Eq. 2, Sect. 2.4.2).

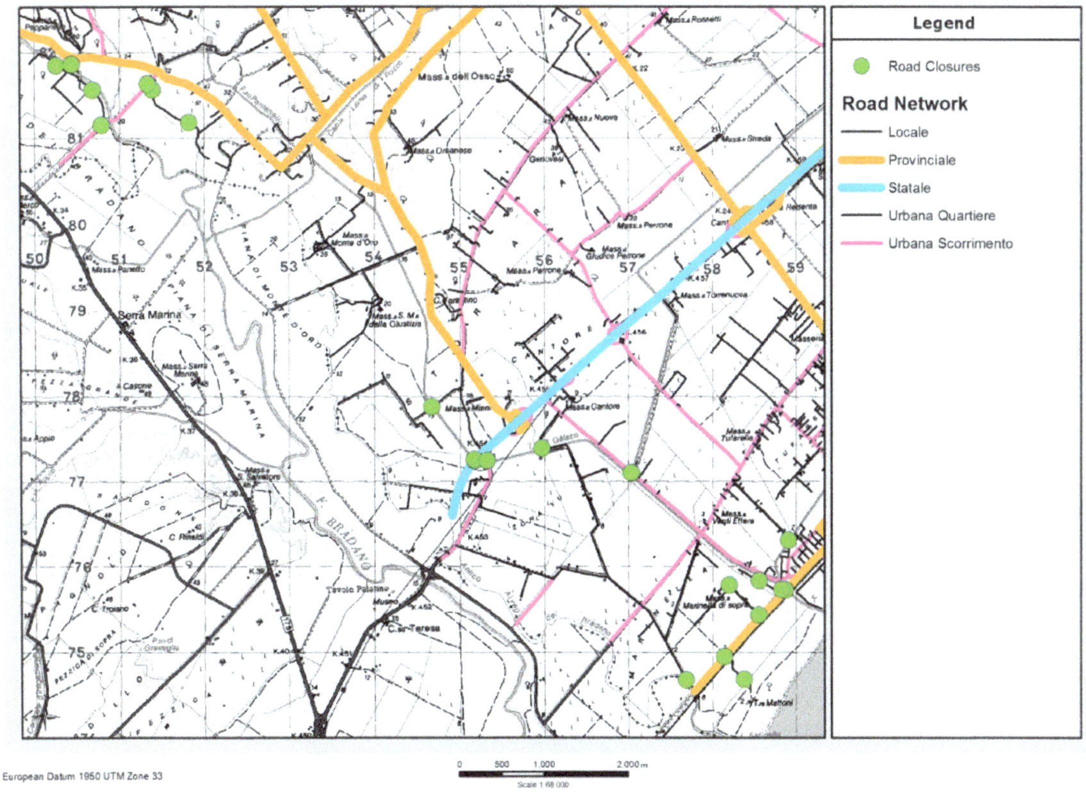

Figure 12. Road closures due the chosen scenario.

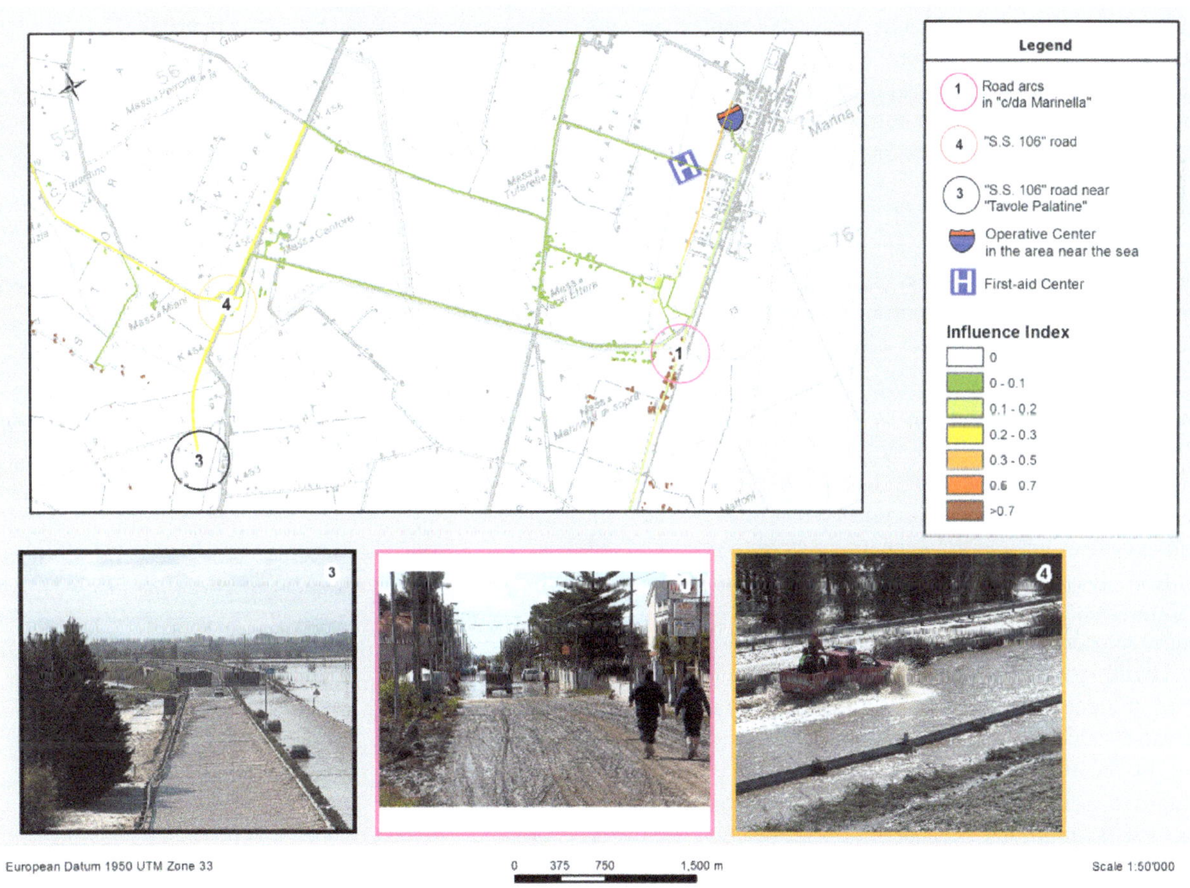

Figure 13. Influence index estimation.

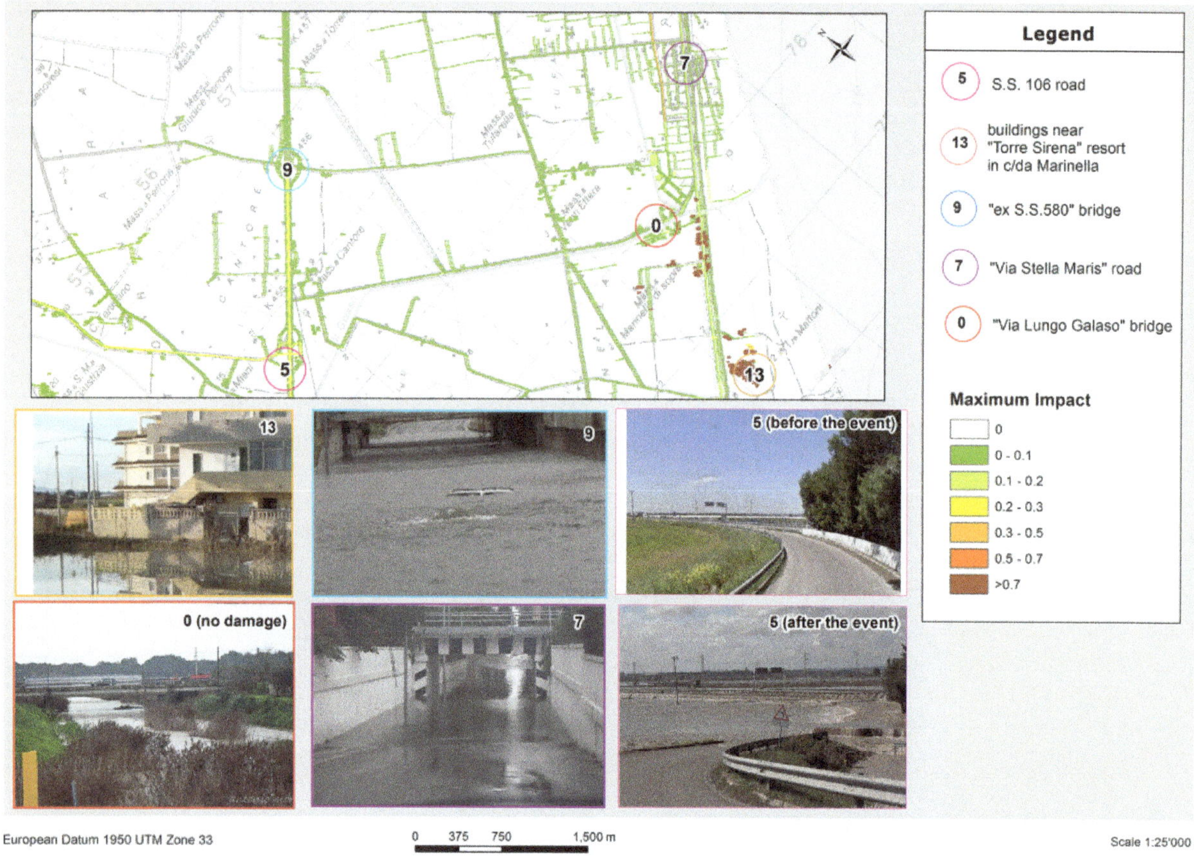

Figure 14. Maximum-impact estimation.

The flood event of 1 March 2011 also caused serious damage to the main infrastructural systems, as well as indirect damage to most of the surrounding area. Indeed, the failure of some parts of the transport infrastructure would have the most serious effects on access to specific locations and overall system performance. Based on the criteria described earlier, the road closures are illustrated in Fig. 12. This estimation allows for the identification of potential inoperable road arches that could affect the whole system during the emergency response activities. Figure 13 outlines the potential fragility in connectivity between emergency centres and the flooded area.

Figure 13 shows the S.S. 106 road has a medium value of the influence index and this is justified by the important function that S.S. 106 has in the system: this road is a highway, i.e. a "Strada Statale" in accordance with the Italian road classification, and it is an important connection between the operative centres located in the central part of the city and the buildings at risk located in the area closer to the sea. Figure 13 also shows that the roads closer to the first-aid centre, i.e. the element represented by the blue rectangle with the white "H", are coloured in orange, and this means that they have a high value of influence index. This is justified because this road has an elevated value of the hierarchy index (Eq. 3 in Sect. 2.4.2). Figure 14 illustrates that the maximum-impact estimation is important to identify hotspots such as the main road, S.S. 106, which is very important because it crosses through the town, dividing it into two parts (e.g. Ginosa Marina located in front of the sea and Ginosa town inland). The neighbouring roads and the main street act as a connection between the area at risk and the middle of the town and beaches. The zone located in c/da Marinella also had a high value for this index because it is almost completely isolated (Fig. 14).

The validations performed by comparisons with the case study illustrate the reliability of the model, which allows for a satisfactory representation of the fragility of the territorial system. It is possible that a similar conclusion could have been obtained simply through expert judgment due to the relative simplicity of the territorial system studied. However, the results we show here can be viewed as important given the reliability of the model adopted and the value of flood emergency management planning.

The proposed model outlined in this paper provides a quantitative estimate of flooding consequences on the basis of direct-impact estimation, i.e. structural and economic loss estimation, and an estimation of loss of life, taking into account the operability of the strategic emergency structures, their accessibility, and connection within the urban area during the emergency phase of a flood. The model can support emergency planning through the definition of a hierarchy among the various structures and infrastructure by identifying those structures and infrastructure whose loss of operability and accessibility could cause vulnerability in the entire system and problems with the performance of rescue activities and victim assistance. In this manner, emergency flood planners can recognize which infrastructure is critical to the maintenance of network connectivity, as well as the structures whose operability and safety are critical during the emergency phase to improve the planning of possible mitigation interventions.

CONCLUSIONS

This paper has presented a new approach to integrate the analysis of an accessibility and operability model for estimation of the strategic elements in the emergency phase associated with a consequence-estimation model during a flood event. The aim is to support decision-making regarding the prioritization of risk prevention measures in order to optimize investments. The innovative aspect of the proposed model is to provide a direct and indirect estimation of flood consequences on the basis of the operability of strategic emergency structures, their accessibility, and connection with the urban system of a city in emergency phases. The accessibility and operability model, illustrated in the GIS model and integrated in the consequence-estimation model, helps to define a hierarchy among the various structures and infrastructure by identifying those structures and infrastructure whose operation and efficiency are fundamental to the maintenance of network connectivity. In this way, the model identifies the structures and infrastructure whose maintenance of performance, in terms of connectivity or operability, could be essential in order to facilitate assistance to victims and rescue activities, and could highlight the areas that need priority interventions. The latter could be extremely useful in cases of limited financial resources. The proposed model was piloted and validated in an urban area of the Puglia region, southern Italy, to demonstrate its operability for providing planners with a tool to identify the hotspots in the urban system affected by floods and to aid in prioritizing interventions.

Future developments of the proposed model could deal with the analysis of risk, implementing in the model the possibility of simulated diverse flood scenarios characterized by probabilities of occurrence, in order to obtain a probability of the maximum impact of the structure and infrastructure within the system. In addition, the estimation of the economic cost of systemic loss during the emergency phase could provide more information on prioritizing risk mitigation measures in terms of cost–benefit analyses of interventions. Finally, the integration of local stakeholders in the development and use of the model could assist authorities to facilitate the quality and fairness of flood risk management. Incorporation of diverse stakeholder views can increase the legitimacy of such processes given the significant uncertainty surrounding the impact of climate change and the dynamics of socio-economic systems.

ACKNOWLEDGEMENTS

We would like to thank three reviewers for their very valuable comments on the first version of this article. Funding for this study was provided, for the contributions of the author belonging to RSE SpA, by the Research Fund for the Italian Electrical System under the contract agreement between RSE SpA and the Italian Ministry of Economic Development–General Directorate for Nuclear Energy, Renewable Energy, and Energy Efficiency, stipulated on 29 July 2009, in compliance with the decree of 11 November 2012.

REFERENCES

1. Adikati, Y. and Yoshitani, J.: United Nations World Water Assessment Programme: Global Trends in Water-Related Disasters an insight for policymakers, United Nations Educational, Scientific and Cultural Organization, UNESCO 2009.
2. Albano, R., Amicarelli, A., Mirauda, D., Agate, G., Sole, A., and Taramasso, A. C.: Experimental validation of a 3-D SPH model for the simulation of a dam-break event involving multiple fixed and mobile bodies, 7th WSEAS International Conference on Engineering Mechanics, Structures, Engineering Geology, Special session "Analysis and modelling of fast-moving flow-like phenomena", Salerno, Italy, June 3–5, 2014.
3. Apel, H., Aronica, G. T., Kreibich, H. and Thieken, A. H.: Flood risk analyses-how detailed do we need to be?, Nat. Hazards, 49, 79–98, 2009.
4. AVI ("Aree Vulnerate Italiane") project, inventory of areas affected by landslides and floods in Italy from 1918 to 2000, available at: http://avi.gndci.cnr.it/, Department of Civil Protection to the National Group for Prevention of Hydrological Hazards (GNDCI) of the National Research Council IRPI ("Istituto Nazionale per la Protezione Idrogeologica"), Italy, 2000.
5. Claps, P. and Fiorentino, M.: Valutazione delle Piene in Italia, Rapporto di sintesi per la regione Basilicata (bacini del versante ionico), GNDCI-CNR. Dipartimento di Ingegneria e Fisica dell'Ambiente, Università della Basilicata-Potenza, 2005.
6. Defra: Making space for water: developing a new government strategy for flood and coastal erosion risk management in England Department for Environment, Food and Rural Affairs, London, 2004.
7. Department of Water Resources Division of Flood Management: Flood Rapid Assessment Model (F-RAM) Development, State of California, The Resources Agency, 2008.
8. Djordjevic, S., Butler, D., Gourbesville, P., Mark, O. and Pasche, ′ E.: New policies to deal with climate change and other drivers impacting on resilience to flooding in urban areas: the CORFU approach, Environ. Sci. Pol., 14, 864–873, doi:10.1016/j.envsci.2011.05.008, 2011.
9. Escuder-Bueno, I., Morales Torres, A., Castillo Rodriguez, J. T., and Momparler, S. P.: 2nd ERA-NET CRUE Research Funding Initiative Flood resilient communities – managing the consequences of flooding Methodology for pluvial and river flooding risk assessment in urban areas to inform decision-making, Report July 2011.
10. Escuder-Bueno, I., Castillo-Rodríguez, J. T., Zechner, S., Jöbstl, C., Perales-Momparler, S., and Petaccia, G.: A quantitative flood risk analysis methodology for urban areas with integration of social research data, Nat. Hazards Earth Syst. Sci., 12, 2843–2863, doi:10.5194/nhess-12-2843-2012, 2012.
11. FEMA (Federal Emergency Management Agency): HAZUSMH Technical Manual. Washington, DC, FEMA, Department of Homeland, Technical Report, 2003.
12. Graham, W. J.: A procedure for estimating loss of life caused by dam failure, DSO-99-06, US Department of Interior, Bureau of Reclamation, 1999.

13. International Strategy for Disaster Reduction (ISDR): United Nation, Living with risk: A global review of disaster reduction initiative, Volume II Annexes, 2004.
14. ISTAT, National Institute of Statistics: i dati geografici delle basi territoriali, available at: http://www.istat.it/it/strumenti/cartografia (last access: October 2013), 2001.
15. Jha, A. K., Bloch, R., and Lamond, J.: Cities and flooding: a guide to integrated urban flood risk management for the 21st century, World Bank Publications, 2012.
16. Klijn, F., Baan, P., De Bruijn, K. M., and Kwadijk, J.: Overstromingsrisico's in Nederland in eenveranderend klimaat: verwachtingen, schattingen en berekeningen voor het project Nederland Later, WL Delft Hydraulics report Q4290.00, 2007 (in Dutch).
17. Lhomme, S., Serre, D., Diab, Y., and Laganier, R.: Analyzing resilience of urban networks: a preliminary step towards more flood resilient cities, Nat. Hazards Earth Syst. Sci., 13, 221–230, doi:10.5194/nhess-13-221-2013, 2013.
18. Luino F., Chiarle M., Nigerelli G., Agangi A., Biddoccu M., Cirio C. G. and Giulietto W: A model for estimating flood damage in Italy, in: Preliminary results, Proceedings of Environmental Economics and Investment Assessment, edited by: Aravossis, K., Karakas, E., Brebbia, C. A., and Kungolos, A., 65–74, 2006.
19. Menoni S., Peergalani F., Boni M. P., and Pertini V.: Lifelines earthquake vulnerability assessment: a systemic approach, Soil Dynam. Earthq. Eng., 22, 9–12, doi:10.1016/S0267- 7261(02)00148-3, 2002.
20. Merz, B., Kreibich, H., Schwarze, R., and Thieken, A.: Review article "Assessment of economic flood damage", Nat. Hazards Earth Syst. Sci., 10, 1697–1724, doi:10.5194/nhess-10-1697- 2010, 2010.
21. Meyer, V., Becker, N., Markantonis, V., Schwarze, R., van den Bergh, J. C. J. M., Bouwer, L. M., Bubeck, P., Ciavola, P., Genovese, E., Green, C., Hallegatte, S., Kreibich, H., Lequeux, Q., Logar, I., Papyrakis, E., Pfurtscheller, C., Poussin, J., Przyluski, V., Thieken, A. H., and Viavattene, C.: Review article: Assessing the costs of natural hazards – state of the art and knowledge gaps, Nat. Hazards Earth Syst. Sci., 13, 1351–1373, doi:10.5194/nhess-13-1351-2013, 2013.
22. ODPM: Preparing for floods, Office of the Deputy Prime Minister, London, 2002.
23. Ordinanza ministeriale del 5 luglio 2012 no. 4024 – Presidenza Consiglio dei Ministri, 2012.
24. Pascale, S., Sdao, F., and Sole, A.: A model for assessing the systemic vulnerability in landslide prone areas, Nat. Hazards Earth Syst. Sci., 10, 1575–1590, doi:10.5194/nhess-10-1575- 2010, 2010.
25. Penning-Rowsell, E., Johnson, C., Tunstall, S., Tapsell, S., Morris, J., Chatterton, J. and Green, C.: The Benefits of Flood and Coastal Risk Management: A Handbook of Assessment. Techniques, London, Middlesex University Press, 2005.
26. Sayers, P., Li, Y., Galloway, G., Penning-Rowsell, E., Shen, F., Wen, K., Chen, Y., and Le Quesne, T.: Flood Risk Management: A Strategic Approach. Paris, UNESCO, 2013.
27. Sdao, F., Albano, R., Sivertun, A., Sole, A., Pascale, F., Giosa, L.: Chapter 19: Model of Systemic Vulnerability Assessment in Urbanized Areas Exposed to Combined Risk of Landslide and Flood, in: Geographic Information Analysis for Sustainable Development and Economic Planning: New Technologies, edited by: Borruso, G., Bertazzon, S., Favretto, A., Murgante, B., Torre, C. M., IGI Global, 274-294, doi:10.4018/978-1-4666- 1924-1.ch019, 2013.
28. SIT PUIGLIA, Banche dati territoriali della Regione Puglia, Geoportal and opendata of Puglia Region, available at: http://www. sit.puglia.it (last access: October 2013), 2011.
29. Taylor, M. A. P., Sekhar, S. V. C., and D'Este, G. M.: Application of Accessibility Based Methods for Vulnerability Analysis of Strategic Road Networks, Netw. Spat. Econ. 6, 267–291, 2006.
30. Vorogushyn, S., Lindenschmidt, K.-E., Kreibich, H., Apel, H., and Merz, B.: Analysis of a detention basin impact on dike failure probabilities and flood risk for a channel-dike-floodplain system along the river Elbe, Germany, J. Hydrol., 436–437, 120–131, 2012.

CHAPTER 12

FLASH FLOOD OCCURRENCE AND RELATION TO THE RAINFALL HAZARD IN A HIGHLY URBANIZED AREA

K. Papagiannaki, K. Lagouvardos, V. Kotroni, and A. Bezes

Institute of Environmental Research and Sustainable Development, National Observatory of Athens, Athens, Greece

ABSTRACT

The paper examines the flash flood events that occurred over a decade in the Attica prefecture, the most urbanized region of Greece, with the aim of identifying triggering rainfall thresholds, as well as assessing the effect of rainfall upon the magnitude of the induced damages. The analysis incorporates rainfall records from a network of 28 surface meteorological stations and information on the spatial distribution of the flash flood events that is derived from the active database of damaging weather events maintained by the atmospheric modelling group of the National Observatory of Athens. The main findings concern firstly the relation between the flash flood impact, as measured by the fire service operations in flooded properties, and precipitation in various time intervals. In the period 2005–2014, 48 damaging flash flood events occurred in the target area and caused more than 3500 fire service operations in flooded properties. Most of the events are associated with maximum accumulated rainfall of more than 20 mm in 24 h and 3 mm in 10 min. However, the flash flood impact intensity, as measured by the number of the fire service operations per event, increases significantly above the levels of 60 mm in 24 h and 10 mm in 10 min. Secondly, graphs of rainfall intensity versus duration are developed for 15 sub-areas of Attica in order to define rainfall intensity thresholds for flood triggering at a more local level. It is shown that conclusions regarding the reliability of the estimated thresholds should take into account the representativity of the rain gauges, which is determined by the local network's density, the gauges' location and record length.

INTRODUCTION

Flooding caused by heavy rainfall is a weather-related hydrological hazard associated with significant economic losses, human fatalities and adverse effects to society and the ecosystem. Even though floods do not appear with the catastrophic power of other natural hazards, such as earthquakes, their overall impact on society and the economy is very serious. In Europe, numerous floods and flash floods occur each year, accounting for 40 % of the total economic damages due to natural hazards for the period 1989–2008 (CRED, 2009). The societal impact of flash floods varies and depends on the hazard intensity and the vulnerability of the affected area as measured by its topographic and demographic features, the quantity and quality of the materials exposed, the anthropogenic interventions and the effectiveness of the prevention measures taken by the local authorities (Tsakiris, 2014). Interestingly, extreme flash floods in the Mediterranean countries tend to be greater in magnitude compared to the

Citation: Papagiannaki, K., Lagouvardos, K., Kotroni, V., and Bezes, A.: Flash flood occurrence and relation to the rainfall hazard in a highly urbanized area, Nat. Hazards Earth Syst. Sci., 15, 1859-1871, doi:10.5194/nhess-15-1859-2015, 2015.

inner continental countries (Gaume et al., 2009), while they occasionally produce catastrophic damages (Llasat et al., 2013; Lastoria et al., 2006). Furthermore, analysis of the spatial distribution of flood events in Mediterranean countries has shown that the urbanized environment increases vulnerability and flood risks (Llasat et al., 2009).

As the risk and consequently the societal impact of the weather hazard is a function of the hazard's magnitude and the local vulnerability (Špitalar et al., 2014), the assessment of damages caused by flash floods in urban areas integrates a variety of factors related to the weather and hydrological hazard and to urban-specific characteristics. Compared to the damages that are caused by riverine or coastal flooding, damages in urban areas are much more dependent on anthropogenic and demographic factors that increase flood risk and are difficult to assess, such as the land development that limits water infiltration and increases run-off (Du et al., 2015; Kandilioti and Makropoulos, 2012) and population density associated with the amount of people and property exposed to risk (Amaro et al., 2010; Llasat et al., 2009). Recently, Barbería et al. (2014) discussed rainfall intensity thresholds in conjunction with regional vulnerability indicators, namely the population density, which is directly linked to the type of urbanization and the land use (Llasat et al., 2009). This study investigates the correlation between several rainfall parameters and the number of requests received by the meteorological services related to insurance claims, which have been already proved to be a good indicator of social impact in the study of Amaro et al. (2010). The urban area discussed is the city of Barcelona, and results for the period 2008–2011 show that there is a statistically significant correlation between short-duration rainfall intensities and the requests related to insurance claims. The high value of this study is that it successfully introduces an alternative measure of the material damages, namely the number of insurance claims, compared to monetary measures such as the compensations paid by the implicated authorities and the insurance companies (Llasat et al., 2008; Lastoria et al., 2006; Jonkman, 2005), which may constitute a better indicator of the impact intensity of the event but are rather hard to find and often cover only part of the overall damage.

As mentioned before, Barbería et al. (2014) also assess the rainfall intensity factor as an indicator of the flash flood hazard. The authors suggest specific thresholds of precipitation and relate them to the impact intensity of the flash flood events. In search of the flood-triggering factors, other studies have also used the accumulated rainfall and/or the rainfall intensity as main indicators of the flash flood occurrence and magnitude. Cannon et al. (2008) and Guzzetti et al. (2008) developed rainfall intensity–duration thresholds for the occurrence of debris flows, landslides and floods. This methodology was later adapted by Diakakis (2012) to examine the role of storm totals, peak and average rainfall intensity, and moisture conditions in flood triggering in northeast Attica, in Greece, for the period 1993–2008. The author found a significant correlation between peak storm intensity and flood triggering and presented rainfall thresholds above which flooding is highly probable. Bracken et al. (2008) studied a semiarid area in south-east Spain and found that floods in the period 1997–2003 have been more highly related to the total rainfall than to the intensity of a storm, although intensity controls the nature and timing of the flood process.

Rainfall thresholds have been also proposed as indicators of early warnings of flash floods. Martina et al. (2006) introduced a methodology for issuing flood warnings by directly comparing the quantitative precipitation forecast to a critical rainfall threshold value, incorporating initial soil moisture conditions as well as expected costs, without the need of real-time meteorological and hydrological forecasting models. In Norbiato et al. (2009) flash flood warning methodologies are based on hydrological models, but they also integrate the flash flood guidance (FFG) method, that is, the rainfall threshold approach on which the US National Weather Service relies for issuing warnings (Georgakakos, 2006). FFG refers to the volume of rain of a given duration necessary to cause minor flooding of small streams (Carpenter et al., 1999). Price et al. (2011a, b) recognized that flash floods are among the costliest natural hazards around the globe and used lightning data to better understand and predict flash floods in the Mediterranean. Over the highly urbanized area of Athens, flash flood forecasting has been explored based on an integrated hydro-meteorological system by Koussis et al. (2003), with promising results.

The present paper contributes to the research on the impact and triggering factors of flash floods in Greece and the Mediterranean, introducing an analysis of the flash flood events that occurred during the 10-year period 2005–2014 in the Attica basin, the most urbanized and densely populated region of the country. The main targets are to detect systematically affected areas within Attica, define and assess flash flood hazard indicators, and discern flash flood patterns and possible rainfall thresholds related to the occurrence of flash floods as well as to the magnitude of the adverse impact. The possibility of using the estimated rainfall thresholds to establish alert systems to warn communities of impending floods is also discussed. For the analysis, the database of high-impact weather events developed by the atmospheric modelling group of the National Observatory of Athens (NOA) was used (Papagiannaki et al., 2013). The database provides information about all the high-impact weather events that have occurred since 2001 and is systematically updated. The number of the fire service operations during flooding events is also incorporated in the analysis and studied in relation to the rainfall parameters, in search of a reliable impact indicator that is systematically and officially recorded. In the context of the present analysis, regional vulnerability to the flash flood hazard is partly addressed by examining the frequency of flooding events in conjunction with the amount of rainfall needed for a place to flood and by discussing the relevance of topographic and vulnerability-related factors, namely the population density, to the frequency of flooding events.

This paper is structured as follows: in Sect. 2 (Data collection), the sources of the high-impact weather events that have occurred in Greece, the meteorological observations and the fire service operations are presented in detail. In Sect. 3 (Methodological issues), the criteria for the specification of the sub-areas studied are described. Then the reasoning behind the use of specific meteorological parameters is explained and the methodology for estimating rainfall thresholds for flood triggering is introduced. In Sect. 4 (Urban flash flood analysis), results that have been derived from the correlation between precipitation parameters and the fire service operations, as well as from the analysis of the rainfall intensity for various accumulated durations, are analysed and discussed. The analysis concerns the entire target area as well as 15 additional sub-areas, thus providing evaluation of the rainfall thresholds for flood triggering at a more local level. Finally, in Sect. 5 (Concluding remarks), main conclusions and future perspectives are given.

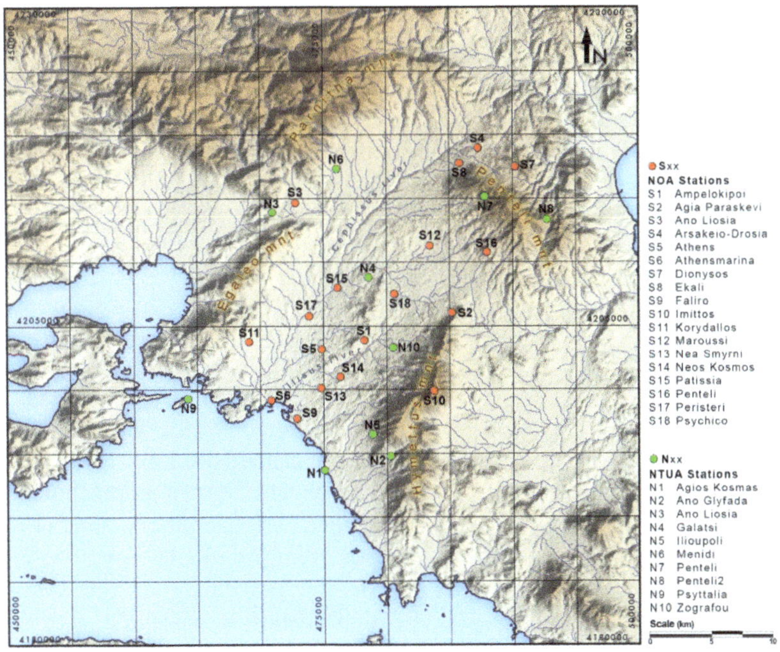

Figure 1. Map of the Greek target area (Athens and suburbs) and locations of the 28 surface meteorological stations. Stations are named by the municipalities they are located in. (Created using QGIS and SRTM-v3 elevation data. Projection GGRS87. Area easting: 450 000–500 000 m. Area northing: 4 180 000–4 230 000 m. Copyright NOA-IERSD.)

DATA COLLECTION

The target area comprises the city of Athens, which is the capital of Greece, and the suburbs, which are subdivided in municipalities. From a geographical point of view, the target area – hereafter called Athens-suburbs – depicted in Fig. 1, is located in the centre of the Attica region, in the central-east of the Greek mainland, bordering the southern part of the Aegean Sea and surrounded by four mountains: Aigaleo and Parnitha in the west-north-west, and Penteli and Hymettus in the east-south-east. Two rivers flow through the Athens plain. The large one, the river Kifissos, is the main water stream of Athens, follows a NNE-SSW flow direction and has produced serious flooding events in the past. Since 2002, it has been systematically monitored to prevent unpredictable over- flows. The second and smaller river, Ilissos, is found in the east part and has been built up to a great extent.

In what concerns the frequency and magnitude of damages caused by weather-related hazards, Attica is by far the most affected region in the country (Papagiannaki et al., 2013; Diakakis et al., 2012), suffering particularly from flash floods, which cause at times serious damages. According to the latest report of the Hellenic Association of Insurance Companies (HAIC, 2015), the 13 most damaging flash flood events that occurred in Greece during the period 1993–2014 caused economic losses of about EUR 93 million, most of which correspond to claims for compensations in Attica. Vulnerability of the Attica prefecture to flash floods is highlighted in the study of Papagiannaki et al. (2013), which introduced the database of high-impact weather events that have occurred in Greece since 2001, followed by the spatial and temporal distribution of the events for the period 2001–2011. A simplified version of the database is available online on the site of the atmospheric modelling group of the National Observatory of Athens (http://www. meteo.gr/weather_cases.asp). A geographical presentation of the recorded weather events is also available online (http:// www.meteo.gr/meteoplus/weatherevents.cfm). The database is systematically updated with every new weather event that generates damages in Greece and includes so far more than 300 cases, which are classified according to their weather and impact intensity. According to the updated database, during the period 2001–2014, 91 flash flood events affected Attica, 48 of which occurred in the Athens-suburbs area, causing buildings, infrastructure and vehicle damages; large-scale transport disruption; flooded road; and in some cases human fatalities. The database, besides the weather and impact classification, also encloses detailed information for each weather event, namely their geographical distribution, the location and type of damages, meteorological data, and detailed references in the press. Thus, it has been used for the purposes of the present analysis in order to define the flash flood events that occurred in the target area and the location of the respective damages.

To address the vulnerability of the specific areas affected by flash floods within the Athens-suburbs region, weather and impact indicators are both incorporated in the present analysis. Meteorological data used in the analysis derive from 28 surface meteorological stations (Fig. 1) spread in the Athens-suburbs region. The atmospheric modelling group of the National Observatory of Athens operates 18 of these stations, while the rest are operated by the Laboratory of Hydrology and Water Resources Management of the National Technical University of Athens (NTUA) and rainfall measurements are freely available to the public. The meteorological stations provide 10 min observations of various meteorological parameters such as temperature, pressure, humidity, wind velocity and direction, rain and rain intensity. Based on the availability of rain observations from the NOA and NTUA meteorological station networks in Athens, the present analysis is limited to the 10-year period 2005–2014. Data on actual financial losses due to flash floods, such as the monetary compensations paid by the implicated authorities and the insurance companies, were not available during this study. Therefore, following the concept of Barbería at al. (2014), the impact of each flash flood event is assessed using the number of the fire service operations that took place due to citizens' calls for flooded properties. Daily data on the operations in flooded properties derive from the Statistical Department of the Hellenic Fire Service and concern the entire Attica region for the period 2003–2013. For the year 2014, the respective data have been found in press articles; thus they may deviate from the official ones, being underestimated only to a small degree. Even though the number of operations during an event concerns the entire region of Attica, the description of the events included in the database is extensive enough to reproduce with high certainty the map with the specific areas affected during each event. The data provided by the fire service are exhaustive, while all the events with more than 10 operations have been included in the NOA database and, thus, are considered in the present analysis. The amount of fire service operations per event derives from the aggregation of the number of operations per day that took place during the days of an event. It should be noted that one operation may concern more than one building; thus the actual number of flooded properties is not known. Judging from the respective press that usually reports the number of citizens' calls, the amount of flooded properties can be as much as 4 times greater than the number of operations in the same event. However, due to the uncertainty in the actual number of properties, this study uses the number of operations which is systematically recorded and therefore highly reliable.

METHODOLOGICAL ISSUES

Specification of Sub-areas

In order to account for the regional characteristics and vulnerability (Barberia et al., 2011), specific sub-areas within the Athens-suburbs region are also studied separately. Methodological issues concern the division of the Athenssuburbs region into sub-areas that can be considered to be adequately covered by a group of specific rain gauges, the specification of the particular areas that were influenced by each flash flood event and the specification of the most representative rain gauges for each flash flood event. Difficulties emerge due to the spatial distribution and the short lifetime of some of the gauges that result in lack of good representation at municipality level. Thus, an optimum exploitation of the existing rain gauges is required in order to match gauges with sub-areas and then with the flash flood events.

The division of the Athens-suburbs region in secondary areas follows geographical criteria and is guided by the location of mountains and river streams, and an important criterion for the determination of each sub-area is the presence of rain gauges located within its perimeter. As Fig. 2 shows, the shaped sub-areas are covered by three to five gauges. The name of each sub-area derives from the respective municipality that covers the major part of it. The formation of the sub-areas based on the above-mentioned criteria rather than the administrative boundaries of the municipalities enables the analysis to be as effective as possible based on a suffi- cient number and spatial distribution of gauges. In what concerns the specification of the areas influenced by each flash flood event, as explained in Sect. 2, the descriptive information included in the NOA database of weather-related events has been used.

Maximum Precipitation and Fire Service Data

In the context of the present analysis, impact indicators are studied in conjunction with hazard parameters. Following the concept of the most recent and relevant articles (Barbería et al., 2014; Diakakis, 2012), precipitation parameters have been used in the analysis, namely the rain records in various time intervals. For each event, the absolute maximum precipitation is calculated based on the observations of the representative rain gauges. Maximum precipitation in different time intervals is then correlated to the number of the fire service operations per event in order to assess the effect of the rainfall intensity on the magnitude of the societal impact. The statistical dependence is measured with Spearman's rank correlation coefficient, ρ, which does not assume normality of data. Also, distributions of the frequency of the operations by month and by rainfall level are presented to complement the discussion. A methodological limitation is the fact that the number of the fire service operations per event, which is given as a total for the entire target area, cannot be distributed to the specific sub-areas affected in each case. Thus, the local impact intensity cannot be measured by the number of operations. Instead, local vulnerability is assessed firstly based on the frequency of damaging flash flood events per area and secondly by the examination of the hazard intensity related to the flood occurrence in each specified area.

Rainfall Thresholds

To estimate possible rain thresholds related to flood triggering in each sub-area, the methodology followed by Diakakis (2012) has been applied. The author developed rainfall intensity–duration thresholds for the occurrence of floods in Marathonas, Greece, and by plotting peak rainfall intensities of various time intervals against their respective durations and then defining rainfall intensity thresholds above which flooding is most likely to occur. Accordingly, in the present study, the graphs of peak rainfall intensities of various time intervals (10, 30 and 60 min; 2, 3, 12 and 24 h) vs. duration have been plotted for each area in order to investigate possible rainfall thresholds above which flooding is most likely to occur. For that, all the time series records of 10 min precipitation have been used. It should be noted that a rainfall event – independently of whether flooding occurred – is considered to start if more than 24 h without rain has proceeded. Also, the maximum accumulated rain in 10, 30 and 60 min, and 2, 3, 12 and 24 h of each rainfall event is calculated as the maximum moving sum, through time steps of 10, 30 and 60 min, and 2, 3, 12 and 24 h respectively.

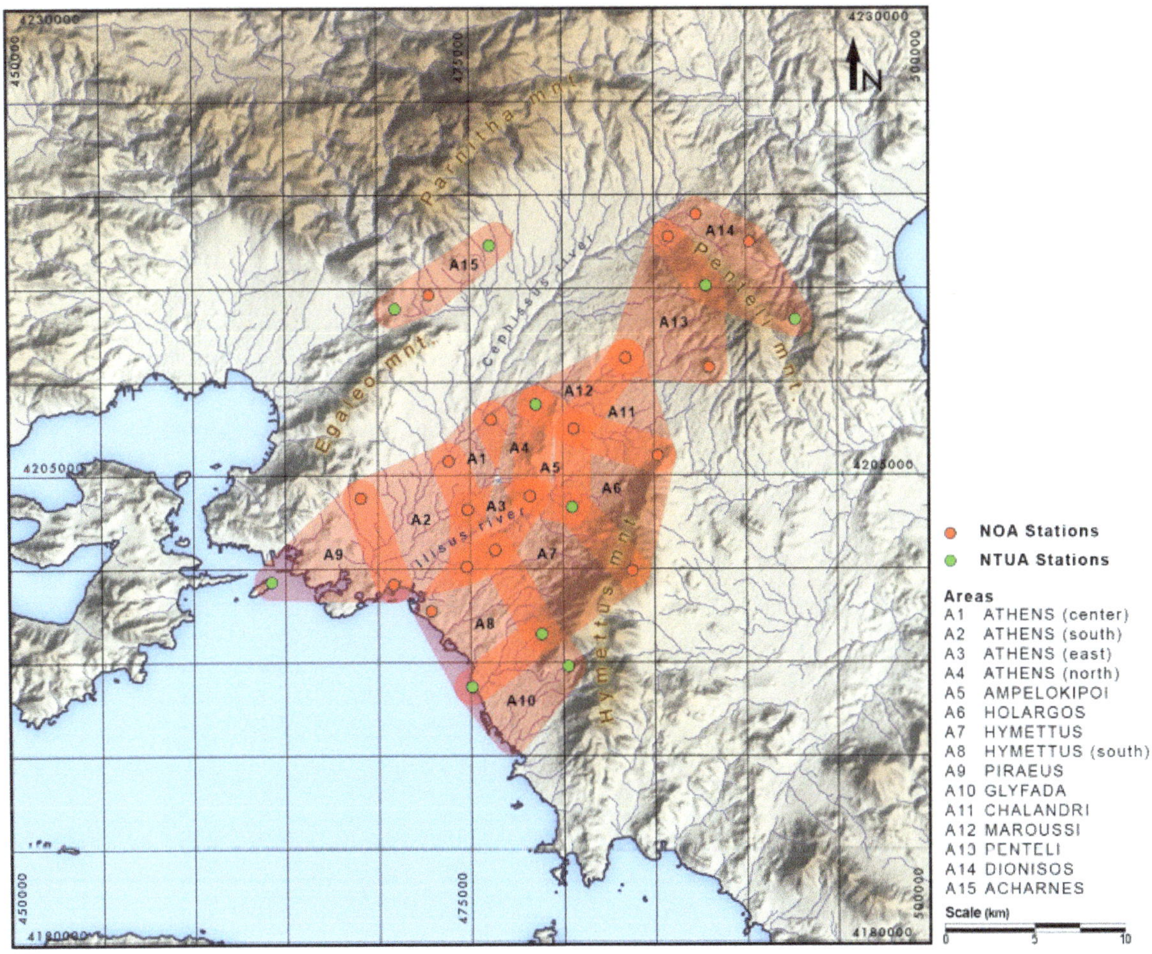

Figure 2. Division of Athens-suburbs region in 15 sub-areas. The representative rain gauges in each sub-area are also shown with red (NOA stations) and green (NTUA stations) bullets. (Created using QGIS and SRTM-v3 elevation data. Projection GGRS87. Area easting: 450 000–50 0000 m. Area northing: 4 180 000–4 230 000 m. Copyright NOA-IERSD.)

URBAN FLASH FLOOD ANALYSIS

During the period 2005–2014, 48 flash flood events occurred in the Athens-suburbs region and caused more than 3500 fire service operations in flooded properties. The maps in Fig. 3a and b highlight the spatial distribution of flash flood events in the Athens-suburbs region and population density respectively. Both maps display the distribution at a municipality level due to the respective availability of population data. Comparison between the two maps leaves no doubt as to the importance of the population distribution to the regional vulnerability. The municipality of Athens, which is the most densely populated area, with around 17 000 inhabitants per km^2, was the most frequently affected area, with 20 events recorded during the study period. High population density is naturally related to intense urbanization, which results in an increase in impervious surface and, consequently, in total run-off volumes. Besides that, the geographical location of Athens also contributes to the increased vulnerability to the rainfall hazard. Athens is located in the centre of the Attica basin, where run-off from the higher sloped areas end.

Relation between Hazard and Impact Indicators

Table 1 includes the areas affected by each flash flood, following the numbering of the specific sub-areas shown in Fig. 2, the number of the fire service operations and the maximum precipitation (in mm) in various time intervals per event. Most of the events affected as many as 6 sub-areas, and only 3 out of the 48 events were actually spread over a large part of the Athens-suburbs region.

The quantile distribution plots of the maximum 24 h accumulated rain (R24) and the maximum 10 min accumulated rain (R10) are displayed in Fig. 4a and b, showing that 50 % of events are related to R24 between 30 and 60 mm and to R10 between 6.5 and 14 mm. Rain in 24 h is more indicative of the ground saturation effect, while R10 is more indicative of the rain intensity effect. There have been only two damaging flash flood events during which less than 20 mm of R24 was recorded and three events with R10 less than 3 mm.

Table 1. Areas affected, number of fire service operations and rainfall (in mm, at various accumulation durations) per flash flood event.

Date	Areas[1]	Operations	R10[2]	R30[2]	R60[2]	R12[2]	R24[2]
03/07/2005	7, 8, 9, 10	35	11.6	25.6	33.4	35.6	35.6
22/09/2005	1, 2, 4, 7, 8, 9	168	13.4	28.8	37.6	53.0	55.2
17/11/2005	7, 8	73	8.0	16.4	19.8	76.8	82.0
24/11/2005	4, 7, 8, 10, 12	245	15.8	44.2	65.4	105.4	122.8
28/12/2005	4, 12	31	4.2	8.4	9.4	15.4	17.2
05/02/2006	1, 7, 8	23	10.6	22.2	29.6	52.0	52.0
11/03/2006	4	11	7.6	22.2	24.8	30.2	35.0
18/09/2006	1, 4, 5, 12	24	11.4	23.6	28.0	28.2	28.2
20/09/2006	1, 3, 4, 11	11	2.2	5.4	7.6	13.0	13.0
21/09/2006	1, 4, 5, 6, 12	55	12.4	29.8	34.0	34.2	34.2
10/10/2006	14	85	13.6	23.2	30.6	82.9	132.9
30/10/2006	11, 12, 13, 15	32	3.0	8.4	15.4	94.2	110.8
24/05/2007	15	130	20.6	41.8	55.6	71.6	72.2
21/10/2007	1, 3, 5	80	11.4	20.4	23.6	46.2	46.4
17/11/2008	2, 9	24	7.0	16.8	25.4	42.6	42.6
12/12/2008	1, 2, 3, 8, 9	27	9.4	15.6	16.6	21.0	23.2
04/07/2009	4, 5, 12	42	9.4	18.8	25.4	28.8	28.8
31/08/2009	15	20	15.2	27.8	28.8	29.0	29.0
25/10/2009	13	25	5.2	8.4	11.4	27.0	33.6
03/11/2009	12	18	4.6	9.6	14.4	20.4	22.4
15/12/2009	15	24	6.6	14.4	19.6	37.4	37.4
03/07/2010	1, 6, 7	25	18.4	33.4	38.4	39.2	39.2
26/07/2010	11, 13	16	11.0	23.8	29.4	45.6	45.6
27/10/2010	1–7, 9, 11, 12	52	16.4	28.4	30.8	36.0	36.0
12/11/2010	6, 7, 8	16	11.0	15.2	16.2	28.0	32.0
03/02/2011	7, 8, 10	210	3.4	9.4	18.0	88.0	99.6
24/02/2011	6, 11	54	2.6	5.6	7.4	29.8	41.0
12/06/2011	1, 2, 3	26	9.6	24.6	32.4	36.4	36.4
14/06/2011	15	37	11.0	22.8	33.8	35.2	35.2
10/10/2011	3	35	2.6	5.8	8.4	23.0	28.8
06/02/2012	1, 2, 3, 7, 9	70	12.2	25.2	39.2	65.2	72.0
18/04/2012	2, 8	10	5.0	10.2	18.6	32.6	32.6
30/11/2012	13, 14	30	20.8	34.8	39.2	54.0	55.2
18/12/2012	1, 2, 8, 9	36	16.0	25.0	29.4	31.0	43.8
29/12/2012	6, 14	118	3.2	4.8	8.4	161.6	186.2
16/01/2013	2, 15	17	6.8	15.4	24.4	39.6	40.0
22/02/2013	1–9, 11, 12, 15	632	15.6	35.0	46.4	95.4	97.6
12/06/2013	11, 12	26	19.6	31.8	42.6	50.4	50.8
11/11/2013	1, 2, 3	90	10.0	23.0	39.0	53.2	53.2
20/11/2013	10	22	4.8	12.4	14.8	27.4	27.4
24/11/2013	1, 2, 3	12	8.0	12.6	19.6	26.8	26.8
25/01/2014	1, 2, 3, 6, 7	100	16.4	34.6	35.6	37.6	42.0
28/01/2014	1, 2, 4	20	10.0	17.2	31.2	56.4	62.2
19/07/2014	3, 7, 8, 10	50	16.6	41.2	51.6	55.4	55.4
07/09/2014	13	25	12.0	31.6	46.6	50.6	50.8
15/09/2014	11, 12	20	15.6	27.6	28.6	28.6	28.8
16/09/2014	1, 7, 14	60	10.2	24.6	26.4	28.4	28.8
24/10/2014	1–4, 9, 12, 15	500	14.0	35.6	43.6	71.8	74.0

[1] For numbering see Fig. 2. [2] Maximum precipitation per event in various time intervals: 10, 30 and 60 min; 12 and 24 h.

To better comprehend the role of the hazard magnitude, Fig. 5a and b display the number of events and operations for various ranges of R24 and R10. Due to the high frequency of events associated with R24 between 30 and 60 mm, most of the fire service operations correspond to this rainfall range (Fig. 5a). However, the impact intensity, measured as the average number of operations per event, is much higher for R24 above 60 mm. On average, 42 operations per event have been recorded for R24 between 30 and 60 mm, while 160 operations per event correspond to R24 between 60 and 90 mm, 291 operations per event to R24 between 90 mm and 120 mm, and 165 operations per event to R24 between 120 and 150 mm. Therefore, the rainfall hazard, when examining the entire Athens-suburbs region, increases significantly above the level of 60 mm for R24. Furthermore, the distribution and impact of flash flood events based on the level of R10 (Fig. 5b) show that the most common R10 records are in the range of 10–15 mm, while the impact intensity increases significantly above 10 mm. Particularly for R10 range 15– 20 mm, the average number of operations per event was over 120 for the examined period.

In what concerns the temporal distribution, Fig. 6 shows that autumn was the season with the highest frequency of flash flood events, as well as with the highest number of fire service operations, followed by winter. Autumn in Greece is actually the season with the highest frequency of rainfall and the highest accumulated rainfall, particularly in what concerns the mainland (Jansa et al., 2014; Lagouvardos and Kotroni, 2007; Kotroni et al., 1999; Lagouvardos et al., 1996). This is in accordance with previous analysis of historical floods in Greece, which concluded that autumn was the season with the richest flood record for the period 1880–2010 (Diakakis et al., 2012). The high impact of February flash floods is mostly due to one event that occurred in 2013 and provoked more than 600 operations. Similarly, one event that occurred in October of 2014 caused 500 operations. Overall, the highest R24 record was 186 mm, observed during an event that caused 118 operations by the fire service and had a quite low R10 (3.2 mm in 10 min). Actually, the correlation between R24 and R10 is weak ($\rho = 0.32$, $p < 0.05$), while most of the high R10 values relate to R24 being less than 70 mm.

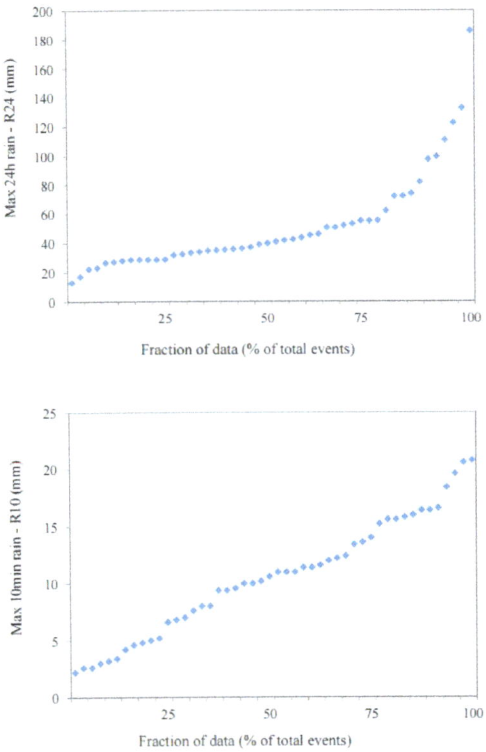

Figure 4. Quantile distribution plots for (a) the maximum 24 h accumulated rain (R24) and (b) the maximum 10 min accumulated rain (R10) observations (in mm).

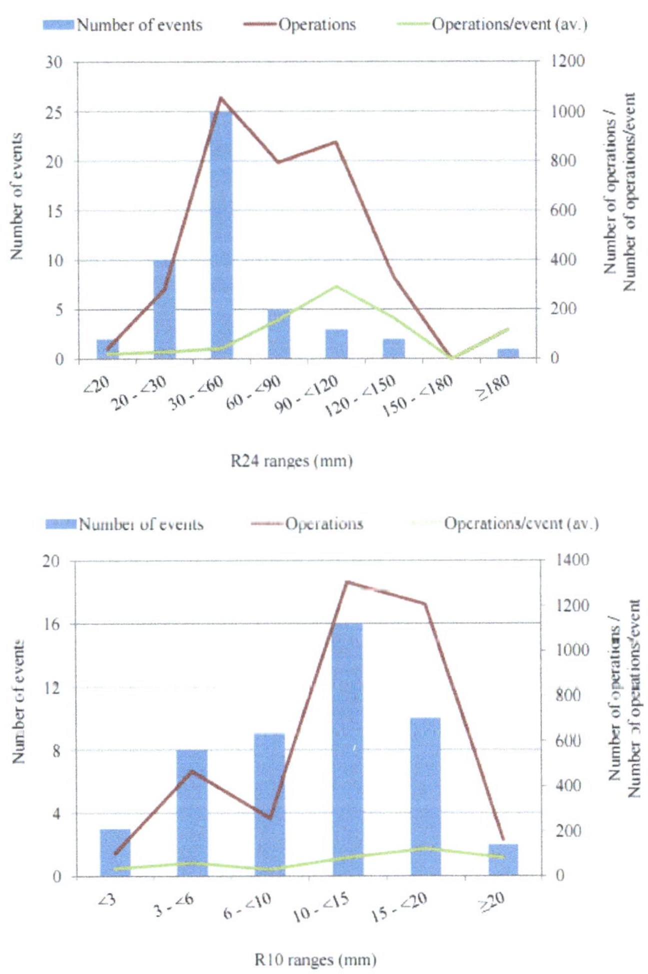

Figure 5. Number of events, fire service operations and average number of operations per event for various ranges of maximum (a) 24 h (R24) and (b) 10 min (R10) accumulated rain (in mm).

Correlation between the number of operations and the maximum rainfall in the various time intervals has been investigated in order to detect the hazard parameter that best relates to the magnitude of the impact. As illustrated in Table 2, the number of operations is strongly correlated with the maximum rainfall in 24 h, R24, ($\rho = 0.61$), while the strength of correlation drops with the decrease in the time interval. The fact that R24 is the precipitation parameter that best represents the magnitude of the impact, at least when considering only the flooded properties, is reasonable since the calculation concerns the entire Athens-suburbs region, which includes, however, areas of unique regional vulnerability that may respond differently to rainfall of the same intensity. These differences, however, tend to become smoother as the accumulation period increases and the various smallscale intensities are normalized.

The significance of the regional vulnerability should be further studied by focusing on the sub-areas of the Athenssuburbs region. However, what is known with certainty is the number of operations per event as well as the areas the event affected, while the exact allocation of the operations is not known. Therefore, to minimize the uncertainty in the estimated results, we attempted to study separately only the city of Athens, focusing on the events that affected the centre, south and east of the municipality of Athens together. This is actually the most populated region in Attica and Greece, exhibiting also the highest frequency of damaging flash flood events, while the network of rain gauges is denser compared to the other areas examined. Some of the events incorporated in the specific analysis have also affected surrounding areas, thus there might be an overestimation of the assigned number of operations. However, when focusing on these neighbouring areas together, the number of operations is strongly correlated with all the durations of accumulated precipitation (ρ 0.69–0.85), the R10 included. Correlation results are included in Table 2. It is obvious that analysis on a more local scale permits better capturing the effect of the volume of rainfall to the magnitude of impact and allows for consideration of short-duration rainfall in the examination of the flash flood impact, as long as a sufficient network of rain gauges exists.

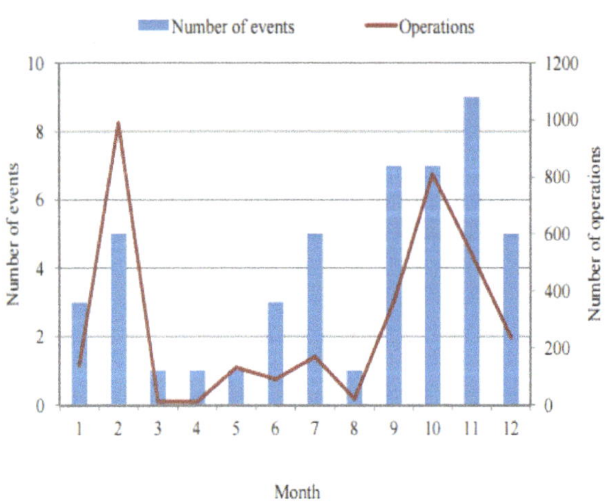

Figure 6. Monthly distribution of the number of flash flood events that occurred in the Athens-suburbs region and fire service operations.

Table 2. Correlation (Spearman's rank coefficient, ρ) between the number of the fire service operations and the maximum rainfall in various time intervals.

Maximum rainfall (time interval)	R24 (24 h)	R12 (12 h)	R60 (60 min)	R30 (30 min)	R10 (10 min)
Operations					
Athens-suburbs	0.61***	0.58***	0.40**	0.37**	0.31*
Athens city	0.85**	0.85**	0.85**	0.83**	0.69*

* $p<0.05$, ** $p<0.01$, *** $p<0.001$.

Rainfall Intensity Thresholds for Flood Triggering, At Regional Level

The results of the previous section raise the question of whether rainfall intensity thresholds can be defined above which flooding is probable or even certain for some areas and of whether short-duration rain intensities can reliably indicate risk of flooding. For that, the entire time series of precipitation records (not only of the flood events) for different accumulation durations have been used and the respective peak rainfall intensities are plotted, following the methodology introduced by Cannon et al. (2008) and Diakakis (2012). Graphs of rainfall intensity versus duration have been developed for each one of the sub-areas specified in Fig. 2 and illustrate the peak rainfall intensities recorded by the most representative rain gauges. Flash flood occurrence is highlighted in red. As expected, rainfall intensity is falls as accumulation duration increases. The goal is to distinguish, if possible, two clear thresholds (denoted by the black lines in Fig. 7) that divide the graph into three parts: (a) the lower one includes peak intensity values that did not lead to flooding throughout the studied period; (b) the middle part includes peak intensities that either did or did not lead to flooding; and (c) the upper part includes peak intensities that always led to flooding.

Results vary considerably depending on the area examined and the local vulnerability, as well as on the density, location and record length of the representative rain gauges. Graphs on which extremely low rainfall intensities are shown to correspond to flood occurrence indicate a poor relationship between the flooding event and one or more of the rain gauges that represent the studied area. One reason for this inconsistency may be the long distance of the gauge from the flooded area; it is even possible for a flash flood to be caused by the run-off of distant intensive rainfall occurring in higher terrain around the affected area.

The distribution of the rainfall intensity values in the short durations varies a lot between areas; as accumulation time increases, peak intensities tend to be more similar. Obviously, areas with high frequency of flash flood events tend to give more explicable outcomes. These are mainly the areas around the centre of Athens (centre, east and south of Athens) and secondarily both Hymettus areas. In most of these cases quite clear thresholds can be defined particularly for the high probability of flooding. These graphs are also consistent regarding the distribution of the flash flood event concentration, which tends to increase as we move to higher rainfall intensity values. Contrariwise, areas such as Holargos and Glyfada are obviously not well represented by the existing rain gauges since their graphs present a wide and inconsistent spread of peak intensities. Low representativity is related either to the small operation record of rain gauges or to the local density of stations, in some cases too low to cover adequately the entire area. It is also possible that some areas demonstrate particular vulnerability in some places that are prone to flooding even for low rainfall intensity. To capture the specific local sensitivities to rainfall, it is obvious that a denser network of rain gauges is required. This applies in the areas of Chalandri and Maroussi, and it is even more necessary for Acharnes, Penteli and Dionisos due to the topographic features of the mountainous areas.

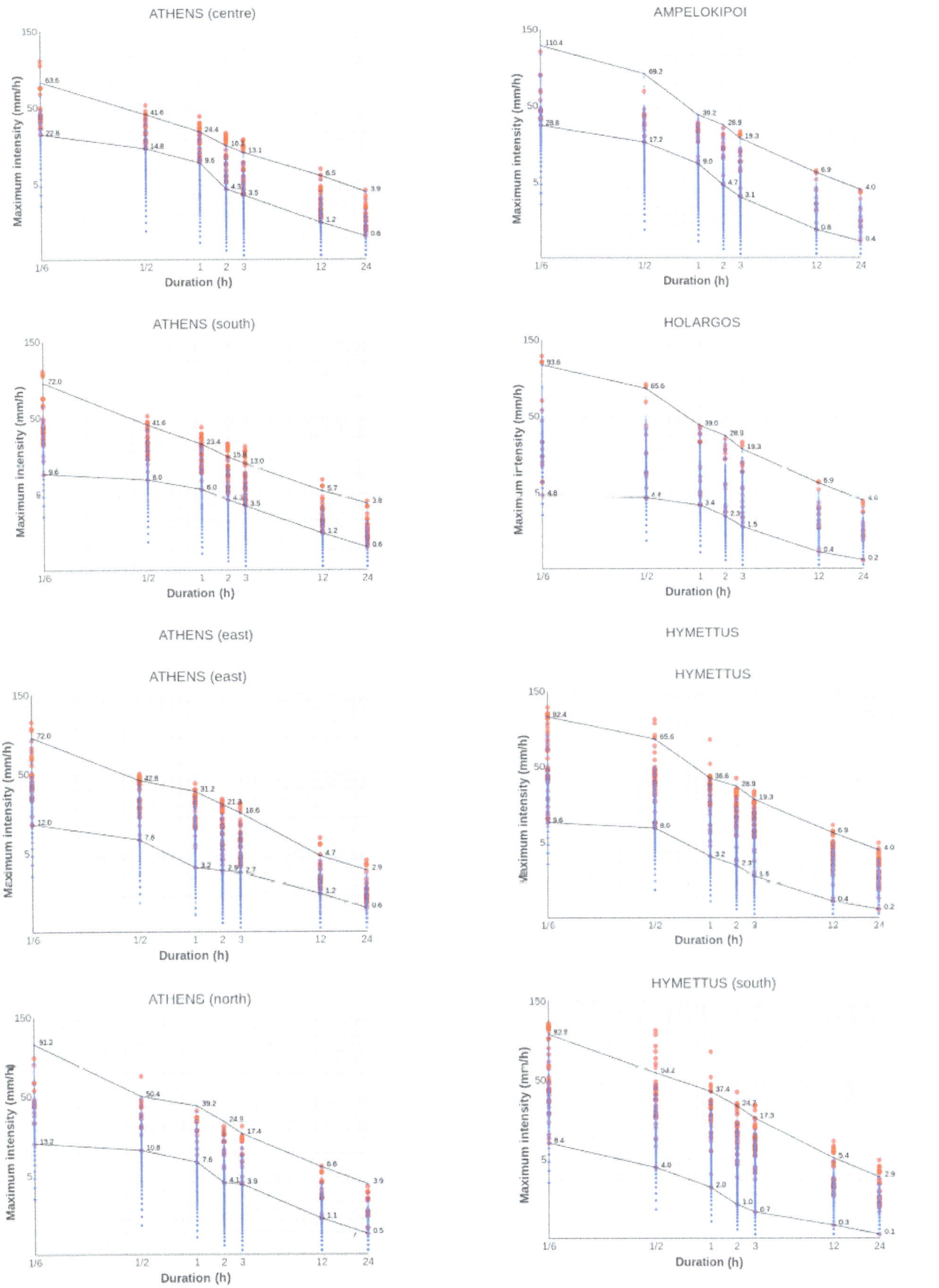

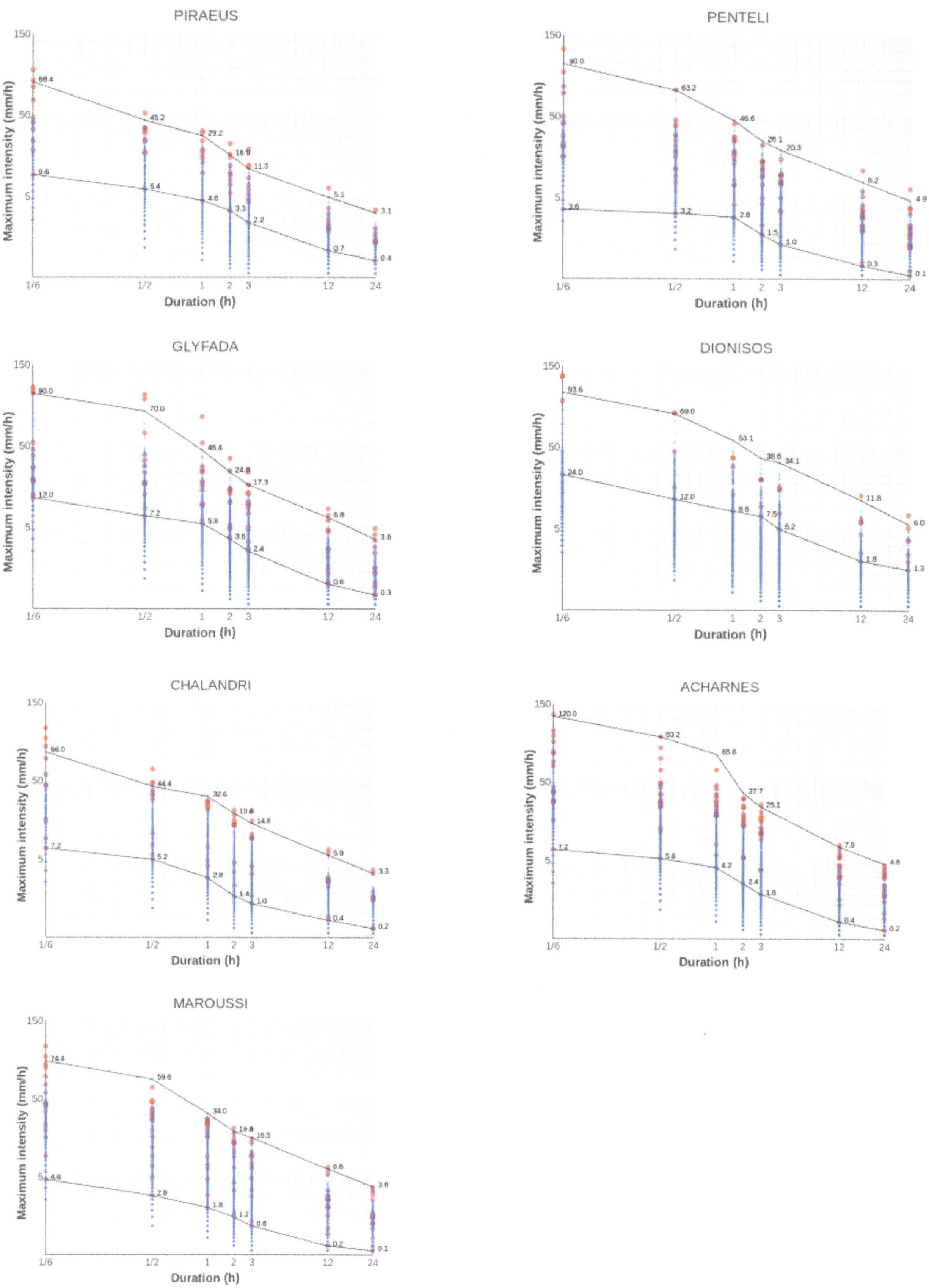

Figure 7. Peak rainfall intensities versus duration graphs, at each sub-area shown in Fig. 2. Peak intensities for each sub-area derive from the rainfall data recorded by all the representative rain gauges shown in Fig. 2. Red dots represent the events that led to reported flooding. All graphs use logarithmic x axis scale.

The areas of Ampelokipoi and Athens centre, opposite of what we would expect for the most intensively urbanized areas, present a high rainfall intensity threshold above which flooding is likely to occur. Specifically, in the area of Ampelokipoi no damaging flash flood occurred for a peak intensity less than 29 mm h^{-1} for 10 min duration or 9 mm h^{-1} for 1 h duration. Even though distribution of population has been found to be relevant to the frequency of flash flood events, it can only partly explain the flash flood thresholds. The threshold value – in other words the hazard intensity needed for a place to flood – may also be related to other factors, such as the drainage system effectiveness and the preparedness of the local authorities (Barrera-Escoda and Llasat, 2015). Therefore, there is not one simple answer to what the key drivers are behind vulnerability of an area. For highly urbanized areas like the centre of Athens there is land cover and population density on the one hand, while on the other hand, as discussed also by Brody et al. (2012), a more coordinated system of flood mitigation infrastructure due to greater resources may exist. Although the centre of Athens is not likely to flood if rainfall intensity is less than 22.8 mm h^{-1} for 10 min duration or 9.6 mm h^{-1} for 1 h duration, it floods for intensities, of 63.3 mm h^{-1} for 10 min duration or 24.4 mm h^{-1} for 1 h duration, that are lower than the respective intensities for which other areas flood, showing a lower resistance of the system. Low threshold values above which flooding is most probable could indicate some limits in the existing infrastructure to cope with the growth of population and materials exposed and, in any case, confirm the necessity for additional protection measures and for the development of an effective warning system.

According to the European Environmental Agency (EEA, 2010), the improvement of forecasting and early-warning systems is seen as the most effective way to mitigate the effects of flash floods. Early warning for flash floods is provided in Greece, based on the weather forecasts, and it is communicated via the media. However, not only the spatial scale of the numerical weather prediction models but also the uncertainty in the quantitative precipitation forecasts do not allow for the risk estimation at local level. For that reason, a dense network of automatic surface meteorological stations providing real-time 10 min accumulated rainfall and intensity observations could constitute a supplementary tool for early warning. Standardization of regional rainfall thresholds in several time intervals could therefore be a very useful tool during a rainfall event. Evidently, what is required in order to produce reliable thresholds on a local level is a denser network of rain gauges as well as long-period data. According to the rain intensity–duration graphs produced in the present study, the rainfall thresholds for flood triggering for the areas around Athens city and Hymettus could be used for the needs of an early-warning system, while for the other areas under examination a denser rain gauge network is required in order to obtain more robust results.

CONCLUDING REMARKS

The objective of this study was to perform an analysis of the flash flood occurrence in an urbanized region, in an effort to identify triggering rainfall thresholds, as well as the effect of the rainfall hazard indicators to the magnitude of the induced impact, considering factors that influence the local vulnerability, such as the population distribution and the geographical features. A major contribution of this paper is that it provides extensive information on the flash flood events of a 10-year period, at the very local level of the most highly urbanized area of Greece, the urban part of the Attica prefecture. This information is considered particularly useful in the context of an integrated approach of the assessment and management of flood risks (EEA, 2010). Moreover, the synthesis of flash flood impact data and complete rainfall records allows the investigation of local rainfall thresholds related to the occurrence of damaging flooding.

Almost all the damaging flash flood events in the target area that occurred during the period 2005–2014 were associated with maximum accumulated rainfall of more than 20 mm in 24 h and 3 mm in 10 min. However, the flash flood impact intensity, as measured by the number of fire service operations per event, increases significantly above the level of 60 mm in 24 h and 10 mm in 10 min. It is interesting that Barbería at al. (2014), in their study on the role of factors related to the social impact of heavy-rain events in the city of Barcelona, examined the maximum 24 h rainfall paying special attention to the rainfall between 40 and 60 mm and the rainfall exceeding 60 mm, especially for the areas with high population density. These thresholds have been found to be significant in what concerns the magnitude of the flash flood impact in an urban environment. They also found good correlation of the maximum rainfall recorded in short time intervals (10 and 30 min) with the impact indicators. In the present study, rainfall in short time intervals has been proved to be a good indicator of the induced impact only when the analysis is performed on a more local scale and provided that a sufficient network of rain gauges exists in the examined area.

The most robust results in what concerns the rainfall intensity thresholds have been produced for the centre of Athens, the area that suffers most from damaging flash flood events. The respective rain gauges have been found to be representative of the area, and the estimated thresholds indicate that the city centre manages to resist very low rainfall intensities, but its resistance has a certain limit above which flooding seems to be the most probable scenario. It is in the authors' future plans to evaluate in practice the rainfall thresholds estimated and improve their reliability by the increase in the precipitation data series, as well as the expansion of the meteorological network. Consideration of the local particularities led to conclusions regarding the rainfall hazard at a local level and the implication of urbanization for the coping capacity. The analysis showed that reliability of these thresholds depends a lot on the representativity of the existing rain gauge network in terms of density, location and record length. The ongoing expansion of the network operated by the National Observatory of Athens ensures the future collection of long-term data, which will enhance the accuracy of defined rainfall thresholds associated with flooding and will permit further assessment of changes in the communities' exposure and resiliency.

ACKNOWLEDGEMENTS

This research has been partly financed by the "Development Proposals of Research Entities – KRIPIS" framework, which is funded by National Programme "Competitiveness and Entrepreneurship", Action: "THESPIA – Development of synergistic and integrated methods and tools for monitoring, management and forecasting of Environmental parameters and pressures". The authors acknowledge the Hellenic Fire Service for the provision of statistical data on the operations made due to harmful weather events. Finally the Laboratory of Hydrology and Water Resources Management of the National Technical University of Athens is acknowledged for the supply of precipitation measurements from 10 stations in the study area.

REFERENCES

1. Amaro, J., Gayà, M., Aran, M., and Llasat, M. C.: Preliminary results of the Social Impact Research Group of MEDEX: the request database (2000–2002) of two Meteorological Services, Nat. Hazards Earth Syst. Sci., 10, 2643–2652, doi:10.5194/nhess-10-2643-2010, 2010.
2. Barberia, L., Amaro, J., Aran, M., and Llasat, M. C.: Social impact analysis of two heavy rain events in Catalonia: 14 and 15 July 2001 and 3 April 2002, Plinius Conference on Mediterranean Storms, Disasters and Climate Change: knowing how to adapt, Savona, Italy, 7–9 September 2011, Programme Book, contribution 13–57, 2011.
3. Barrera-Escoda, A. and Llasat, M. C.: Evolving flood patterns in a Mediterranean region (1301–2012) and climatic factors – the case of Catalonia, Hydrol. Earth Syst. Sci., 19, 465–483, doi:10.5194/hess-19-465-2015, 2015.
4. Bracken, L. J., Cox, N. J., and Shannon, J.: The relationship between rainfall inputs and flood generation in south–east Spain, Hydrol. Process., 22, 683–696, doi:10.1002/hyp.6641, 2008.
5. Cannon, S. H., Gartner, J. E., Wilson, R. C., Bowers, J. C., and Laber, J. L.: Storm rainfall conditions for floods and debris flows from recently burned areas in southwestern Colorado and southern California, Geomorphology, 96, 250–269, doi:10.1016/j.geomorph.2007.03.019, 2008.
6. Carpenter, T., Sperfslage, J., Georgakakos, K., Sweeney, T., and Fread, D.: National threshold runoff estimation utilizing GIS in support of operational flash flood warning systems, J. Hydrol., 224, 21–44, 1999
7. CRED: Disaster data: A balanced perspective, CRED Crunch, Centre for Research on the Epidemiology of Disasters, Issue No. 17, Brussels, 2009.
8. Diakakis, M.: Rainfall thresholds for flood triggering. The case of Marathonas in Greece, Nat. Hazards, 60, 789–800, doi:10.1007/s11069-011-9904-7, 2012.
9. Du, S., Shi, P., Van Rompaey, A. and Wen, J.: Quantifying the impact of impervious surface location on flood peak discharge in urban areas, Nat. Hazards, 76, 1457–1471, doi:10.1007/s11069- 014-1463-2, 2015.
10. EEA: Mapping the impacts of natural hazards and technological accidents in Europe, European Environmental Agency, Technical report No. 13/2010, Copenhagen, doi:10.2800/62638, 2010.
11. Gaume, E., Bain, V., Bernardara, P., Newinger, O., and Barbuc, M.: A compilation of data on European flash floods, J. Hydrol., 367, 70–78, 2009.
12. Georgakakos, K.: Analytical results for operational flash flood guidance, J. Hydrol., 317, 81–103, doi:10.1016/j.jhydrol.2005.05.009, 2006.
13. Guzzetti, F., Peruccacci, S., Rossi, M., and Stark, C. P.: The rainfall intensity–duration control of shallow landslides and debris flows: an update, Landslides, 5, 3–17, doi:10.1007/s10346-007-0112-1, 2008.
14. Jansa, A., Alpert, P., Arbogast, P., Buzzi, A., Ivancan-Picek, B., Kotroni, V., Llasat, M. C., Ramis, C., Richard, E., Romero, R., and Speranza, A.: MEDEX: a general overview, Nat. Hazards Earth Syst. Sci., 14, 1965–1984, doi:10.5194/nhess-14-1965- 2014, 2014.
15. Kandilioti, G. and Makropoulos, C.: Preliminary flood risk assessment: the case of Athens, Nat. Hazards, 61, 441–468, doi:10.1007/s11069-011-9930-5, 2012.
16. Kotroni, V., Lagouvardos, K., Kallos, G., and Ziakopoulos, D.: Severe flooding over central and southern greece associated with pre-cold frontal orographic lifting, Q. J. Roy. Meteor. Soc., 125, 967–991, doi:10.1002/qj.49712555511, 1999.
17. Koussis, A., Lagouvardos, K., Mazi, K., Kotroni, V., Sitzmann, D., Lang, J., Zaiss, H., Buzzi, A., and Malguzzi, P: Flood Forecasts for Urban Basin with Integrated Hydro-Meteorological Model, J. Hydrol. Eng., 8, 1–11, 2003.
18. Lagouvardos, K. and Kotroni, V.: TRMM and Lightning Observations of a Low-Pressure System over the Eastern Mediterranean, B. Am. Meteorol. Soc., 88, 1363–1367, doi:10.1175/bams-88-9- 1363, 2007.
19. Lagouvardos, K., Kotroni, V., Dobricic, S., Nickovic, S., and Kallos, G.: The storm of October 21–22, 1994, over Greece: Observations and model results, J. Geophys. Res.-Atmos., 101, 26217– 26226, doi:10.1029/96jd01385, 1996.
20. Lastoria, B., Simonetti, M. R., Casaioli, M., Mariani, S., and Monacelli, G.: Socio-economic impacts of major floods in Italy from 1951 to 2003, Adv. Geosci., 7, 223–229, 2006, http://www.adv-geosci.net/7/223/2006/.
21. Llasat, M. C., Llasat-Botija, M., Barnolas, M., López, L., and Altava-Ortiz, V.: An analysis of the evolution of hydrometeorological extremes in newspapers: the case of Catalonia, 1982–2006, Nat. Hazards Earth Syst. Sci., 9, 1201–1212, doi:10.5194/nhess-9-1201-2009, 2009.
22. Llasat, M. C., Llasat-Botija, M., Petrucci, O., Pasqua, A. A., Rosselló, J., Vinet, F., and Boissier, L.: Towards a database on societal impact of Mediterranean floods within the framework of the HYMEX project, Nat. Hazards Earth Syst. Sci., 13, 1337– 1350, doi:10.5194/nhess-13-1337-2013, 2013.
23. Martina, M. L. V., Todini, E., and Libralon, A.: A Bayesian decision approach to rainfall thresholds based flood warning, Hydrol. Earth Syst. Sci., 10, 413–426, doi:10.5194/hess-10-413-2006, 2006.
24. Norbiato, D., Borga, M., and Dinale, R.: Flash flood warning in ungauged basins by use of the flash flood guidance and model-based runoff thresholds, Meteorol. Appl., 16, 65–75, doi:10.1002/met.126, 2009.

25. Papagiannaki, K., Lagouvardos, K., and Kotroni, V.: A database of high-impact weather events in Greece: a descriptive impact analysis for the period 2001–2011, Nat. Hazards Earth Syst. Sci., 13, 727–736, doi:10.5194/nhess-13-727-2013, 2013.

26. Price, C., Yair, Y., Mugnai, A., Lagouvardos, K., Llasat, M. C., Michaelides, S., Dayan, U., Dietrich, S., Di Paola, F., Galanti, E., Garrote, L., Harats, N., Katsanos, D., Kohn, M., Kotroni, V., Llasat-Botija, M., Lynn, B., Mediero, L., Morin, E., Nicolaides, K., Rozalis, S., Savvidou, K., and Ziv, B.: Using Lightning Data to Better Understand and Predict Flash Floods in the Mediterranean, Surv. Geophys., 32, 733–751, doi:10.1007/s10712-011- 9146-y, 2011a.

27. Price, C., Yair, Y., Mugnai, A., Lagouvardos, K., Llasat, M. C., Michaelides, S., Dayan, U., Dietrich, S., Galanti, E., Garrote, L., Harats, N., Katsanos, D., Kohn, M., Kotroni, V., Llasat-Botija, M., Lynn, B., Mediero, L., Morin, E., Nicolaides, K., Rozalis, S., Savvidou, K., and Ziv, B.: The FLASH Project: using lightning data to better understand and predict flash floods, Environ. Sci. Policy, 14, 898–911, doi:10.1016/j.envsci.2011.03.004, 2011b.

28. Špitalar, M., Gourley, J. J., Lutoff, C., Kirstetter, P.-E., Brilly, M., and Carr, N.: Analysis of flash flood parameters and human impacts in the US from 2006 to 2012, J. Hydrol., 519, 863–870, doi:10.1016/j.jhydrol.2014.07.004, 2014.

29. Tsakiris, G.: Flood risk assessment: concepts, modelling, applications, Nat. Hazards Earth Syst. Sci., 14, 1361–1369, doi:10.5194/nhess-14-1361-2014, 2014.

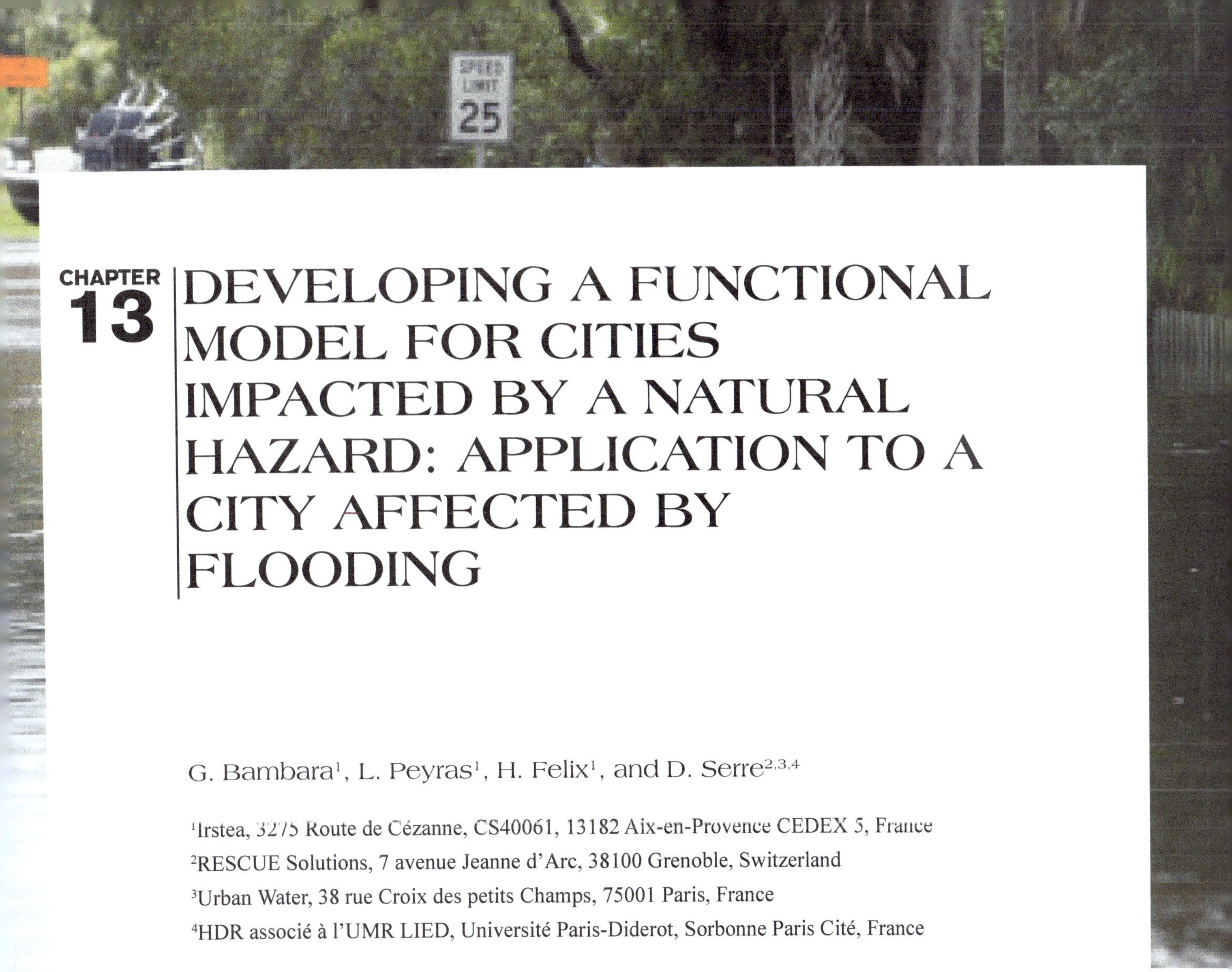

CHAPTER 13

DEVELOPING A FUNCTIONAL MODEL FOR CITIES IMPACTED BY A NATURAL HAZARD: APPLICATION TO A CITY AFFECTED BY FLOODING

G. Bambara[1], L. Peyras[1], H. Felix[1], and D. Serre[2,3,4]

[1]Irstea, 3275 Route de Cézanne, CS40061, 13182 Aix-en-Provence CEDEX 5, France

[2]RESCUE Solutions, 7 avenue Jeanne d'Arc, 38100 Grenoble, Switzerland

[3]Urban Water, 38 rue Croix des petits Champs, 75001 Paris, France

[4]HDR associé à l'UMR LIED, Université Paris-Diderot, Sorbonne Paris Cité, France

ABSTRACT

The experience feedback on a crisis that hit a city is frequently used as a "recollection" tool. To capitalize information about an experience feedback from the cities that have been affected by a natural hazard, the authors propose in this study a functional model to model scenarios of city crises. In this model, the city, considered as a complex system, was modelled using a functional analysis method. Based on such modelling, two risk analysis methods (Failure Mode and Effect Analysis and Event Tree Method) were deployed and adjusted. Lastly, a qualitative reasoning model was used for the scenario modelling of the urban crisis. By functional modelling performed on components of the cities, the objective of this model is to replicate the behaviour of a city affected by a crisis, highlighting the sequences of failure and non-failure modes that have operated during the crisis. This model constitutes a means of understanding the functional behaviour in a crisis of cities and capitalization of the experience feedback of the cities affected by crisis. Such functional modelling was deployed in a case study.

INTRODUCTION

For the past decades, growing urbanization and industrialization has given rise to a sharp increase in harmful disastrous events, thereby demonstrating the city's higher exposure to major risks, e.g. flood risk (Ashley et al., 2007). In addition, climate change will lead to a substantial escalation and increase in frequency of a number of natural hazards (Muller, 2007). Cities, particularly vulnerable, need to acquire new tools to guard against future disasters and improve their resilience when facing such hazards. The RESILIS project – Technical urban systems governance for a resilient city – funded by the French National Research Agency, addresses such issues and aims at improving urban system resilience in the face of natural or anthropogenic hazard-driven crisis. All works described herein were conducted as part of the RESILIS project and focus on developing a functional model for cities.

Improving the resilience of crisis-stricken cities is a key objective for crisis management stakeholders in the city. Such an improvement cannot succeed without a better understanding of how urban systems behave when hit by a crisis. For that reason, performing a detailed review of urban crisis experience

Citation: Bambara, G., Peyras, L., Felix, H., and Serre, D.: Developing a functional model for cities impacted by a natural hazard: application to a city affected by flooding, Nat. Hazards Earth Syst. Sci., 15, 603-615, doi:10.5194/nhess-15-603-2015, 2015.

feedback is essential (Komendantova et al., 2014). The experience feedback can be defined as a process of collecting information about an event in order to trace the detailed evolution of the event and all actions taken to address them. This is usually done within an organization (enterprise, local authorities, state services, etc.) when there is a deviation from normal system operation (Wybo and Van Wassenhove, 2009). Such experience feedback will help identify failures occurring during a dreaded event, analyse the different scenarios and draw meaningful conclusions. Based on such analyses, the city should be able to improve its protection, prevention and crisis management performance, in the event that a similar event were to impact the considered urban system again. However, in order to learn from urban crisis experience feedback, a methodology must be clearly set up so as to formalize the experience feedback, provide a model for the impacted city's failures, and understand the failure processes as well as the sequences and causal relationships thereof. This is not a negligible issue since urban systems appear particularly complex due to their many constituent subsystems, and the multiple organizations and governance (Lhomme et al., 2013; Toubin et al., 2012). The main scientific issue pertains to developing a model that might reproduce the complex operation and deficiencies of urban systems. An underlying issue relates to modelling the experience feedback on crisis-stricken urban systems. The aim is to analyse urban crisis case studies, understand the failure processes thereof and view them within a homogeneous model. The objective of this research is to develop a model that will allow modelling the crisis-stricken cities functional behaviour. Reference work searches did not reveal many models for urban system modelling existing in the technical literature, except for research undertaken by teams working in urban engineering and more particularly in the field of urban resilience (Campanella, 2006; Serre et al., 2012). Resilience is generally included in the concept of response to sudden impact like natural disasters (Alberti et al., 2003; Pickett et al., 2004). The concept was initially a term of ecology (Holling, 1973) and has evolved significantly in the past 30 years (Folke, 2006; Gallopín, 2006). Recognition has increased across disciplines that human and ecological systems are interlinked and that their resilience relates to the functioning and interaction of the systems rather than to the stability of their components or the ability to maintain or return to some equilibrium state (Klein et al., 2003). The management of urban resilience focuses on reducing the extent of the damage, speeding recovery and adapting to unexpected threats, in addition to protecting against known threats (Linkov et al., 2014). Thus, considering the city as a system, urban resilience can be defined (according to Campanella, 2006) as "the capacity of a city to face devastating event reducing damage at minimum". Concerning the natural hazards – particularly for the flood hazard – resilience concepts are comprised of individual preventive and emergency measures at building scale and a land use policy to adapt building activities to floods (Pasche and Geisler, 2005). The resilience concept has been applied to flood risk management by adopting a system approach (De Bruijn et al., 2005). The system approach tends to overcome the segmentations and level of complexity. Indeed, a system approach proposes a common language for different disciplines and can be considered a good way to study complex systems (Batty, 2009). The research work of Lhomme et al. (2011) is based on this approach and he defines urban resilience as "the ability of a city to absorb disturbance and recover its functions after disturbance". He used a system model for studying cities and for modelling their functions. The model seeks to assess the resilience level of urban networks. Indeed, urban networks play an important part in crises and not always in positive ways – for instance, the reliability and rapid restoration of the electric grid is particularly necessary to support the needs of the population within a disaster area effectively (Winkler et al., 2010).

In this study, in representing the functional behaviour of a city affected by a natural hazard, we are interested in all urban system components including technical networks. Our model uses functional analysis methods in order to understand how the systems work (Baroth et al., 2013), risk analysis methods designed for modelling complex systems and qualitative reasoning models for the representation of failure sequences in the form of causal graphs (Forbus, 1996). After reviewing past experience feedback, the goal is to suggest a functional methodology that will allow us to reconstruct the failure scenarios that occurred in the crisis-stricken cities. We have applied this model to urban crisis case studies, to model the cities impacted by natural hazard, by the successions of failure and non-failure modes that took place during crisis. Two advantages arise through this model. Firstly, it is possible to have knowledge of satisfactory functioning and dysfunctions at the component level of a city, when hit by a natural hazard. Secondly, it is possible to capitalize on the experience feedback from urban crises in functional form and benefit from this experience feedback for better management and an improvement of the components of a city. These advantages help improve our understanding of urban system failures. This is an innovative research project as it uses functional analysis, risk analysis and qualitative reasoning models and applies them to urban engineering. Adjusting such methods will result in developing a functional model to outline the behaviour of a crisis-stricken city, and showing how helpful it can be for analysing urban crisis experience feedback. This aspect is described in Sect. 2. In Sect. 3, the model is applied to the case study of an urban crisis caused by flash floods. The city selected as a case study is the city of Nîmes (France), which was struck by major runoff flooding in 1988. Such application will show how much can be learnt from such a model about the weak points to be improved in an urban system. Section 4 specifically address the case study. Based on the different suggested models, the urban system's strengths and potential improvements are described.

DEVELOPING A FUNCTIONAL MODEL FOR CRISIS-STRICKEN CITIES

Our goal is to develop a functional model that will help to better understand the functional behaviour of a city faced with an external hazard, to analyse its failures, reconstruct the failure scenarios, determine which components are more efficient and which are most critical, in order to take the benefit of the experience feedback of crisis and then to improve the city component's resilience in any subsequent similar situation.

Functional Model Development Procedure for Crisis-Stricken Cities

A city is a complex system with many a constituent subsystems and multiple interactions between those subsystems and the city's outer environment (Zevenbergen et al., 2010; Serre et al., 2011). Consequently, risk analysis methods used in industry to identify risks and model complex system functions are considered relevant for modelling and analysing how a crisis-stricken city operates (Baroth et al., 2013; Modarres, 1993). Also, qualitative reasoning models are deemed relevant for representing urban crisis (Forbus, 1996). Such methods have been used in the past as part of civil engineering projects (Peyras et al., 2012; Serre et al., 2008). They have also been applied to urban engineering with regard to technical networks (Serre et al., 2011).

Our procedure for building a functional model that will outline the functional behaviour of a crisis-stricken city can be summarized in four steps: (i) a functional analysis: the urban system and the subsystems' functions are determined; (ii) failure mode and effect analysis (FMEA): information on failures occurring in an urban system due to a dreaded event are capitalized on from experience feedback; (iii) event-tree method (ETM): for each component, a representation of potential scenarios is suggested, including the scenario resulting from experience feedback, as a sequence of failure modes; (iv) a qualitative reasoning model with causal graph modelling: a dynamic and gradual model of failure is suggested to describe the urban failure scenario from experience feedback.

Developing a Functional Model for Crisis-Stricken Cities

Functional Analysis Applied To Cities

The functional analysis consists of two stages: the external functional analysis, including analysing the system within its environment, and internal functional analysis, including a structural analysis of the system, i.e. identifying its various constituent subsystems, and exploring the subsystems' functions (Baroth et al., 2013). Many functional analysis methods are suitable for industry, services or organizations. It was decided to rely on one method deriving from value analysis: the APTE method as successfully used and applied to civil engineering systems (Peyras et al., 2006). In functional analysis, two categories of functions are considered to reflect the actions of a system: main and technical functions. The main functions convey the purpose of a system's action. Technical functions model how the system reacts to stresses set by the external environment. Functional analysis also uses functional block diagrams (FBDs) to represent the system and external environment (Baroth et al., 2013).

The external functional analysis aims at defining the system to be reviewed, its limits and interacting external environments (Baroth et al., 2013). When applied to cities, the system is regarded as consisting of the urban system delineated by its geographic and administrative boundaries. Figure 1 shows the FBD of a generic urban system. This FBD makes it possible to show how the urban system interacts with external environment – including neighbouring cities, rural areas, outer technical networks (power, water, telecommunications, transport networks, etc.), environmental factors (meteorology, seismic situation, etc.) and governance. These elements were defined by different experts in urban engineering during work sessions. From this FBD, we deduce the main and technical functions of a generic urban system. The internal functional analysis is performed for a typical urban system. It is suggested (by experts in urban engineering) that four large categories of subsystems should be differentiated: technical networks, housing, businesses and public infrastructure. The population is implicitly included in each of these four large categories of subsystems. Such categories, referred to as grade 1 categories, may be detailed as grade 2 or grade 3 categories. Table 1 shows how the city is structurally subdivided into different subsystem categories. The main and technical functions have been taken from the FBDs.

As an example, the FBD for the Drinking Water Supply (DWS) technical network is shown in Fig. 2. From the FBD, the DWS technical network, main and technical functions may be differentiated and highlighted. The main functions are to "provide of drinking water for the housing, businesses and public infrastructures" and to "protect the city in case of fire". Technical functions are to "resist to mechanical loadings" and to "resist to failures from other technical networks interrelated with DWS technical network". All the results from the internal functional analysis are compiled in a functional analysis table detailing the main and technical functions of a city's constituent subsystems. Table 2 provides a sample of such functional analysis table for the DWS technical network subsystem. For an analysis of other urban subsystem functions, see the RESILIS project (www.resilis.fr). In this paper, we focus on five essential hand or technical functions required for the system to work properly when a dreaded event impacts a city:

- the function "provide and/or drain off flows" for "water" technical network,
- the function "provide flows" for "power" technical network (power flows, natural gas flows, district heating flows and "telecommunications" technical network (communication flows: broadband, landline telephone or mobile telephone services)
- the function "meet the needs for travel" for "transport" technical network (for road and railway transport),
- the function "ensure": "ensure accommodation" for housing, "ensure their economic and social duties are performed" for businesses and "ensure and serve a public interest purpose" for public infrastructures,
- the function "resist to mechanical loadings" for all subsystems.

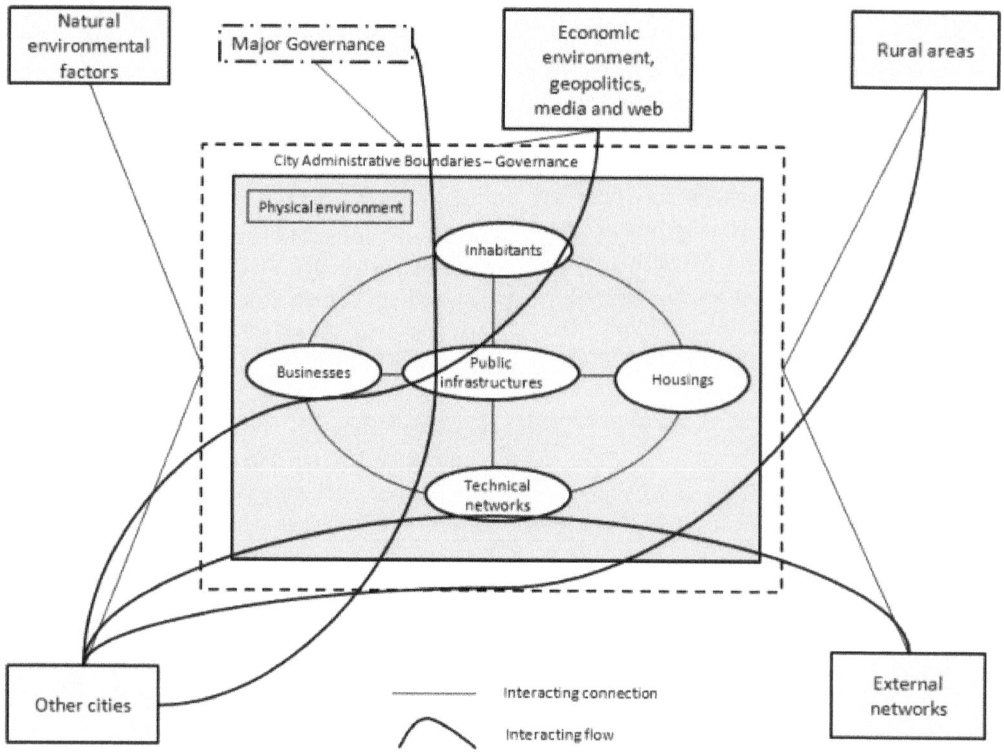

Figure 1. Functional block diagram for urban system.

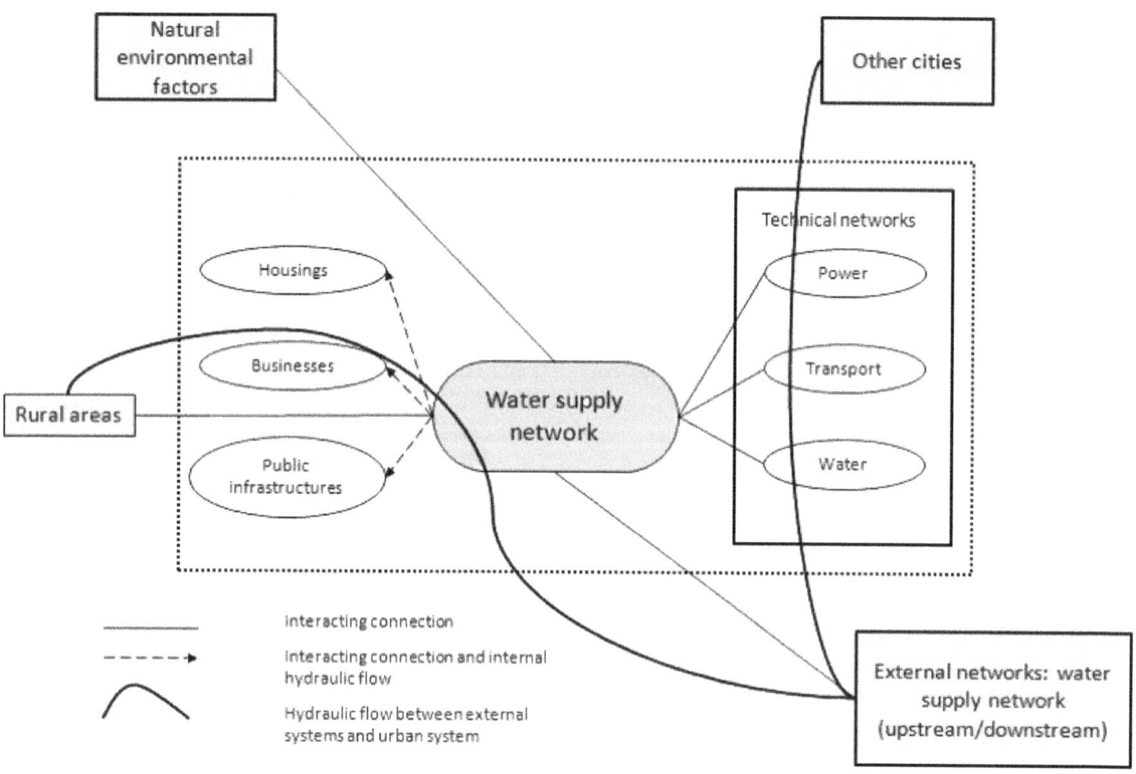

Figure 2. Functional block diagram for drinking water supply technical network.

Finally, the functional analysis drawn up for a typical city may subsequently be applied to a specific city for a case study through adjustments and quick streamlining. Therefore, this is a generic functional analysis that may apply to most urban systems.

FMEA applied to cities

FMEA is one of the most frequently used risk analysis methods in industry. FMEA is an inductive method for analysing the potential failures in a system (Modarres, 1993). Each component is consistently and consecutively considered and their failure modes (lost or impaired function) and effects

are analysed. FMEA analysis results are outlined in tables specifically designed for the system type of interest. A FMEA procedure will include a system functional analysis then a systematic research for failure modes and effects thereof (Baroth et al., 2013; Modarres, 1993).

Table 1. City structural breakdown.

#	Grade 1 category	#	Grade 2 category	#	Grade 3 category
1.	Technical networks	1.1	Water	1.1.1	Drinking water supply
				1.1.2	Sewage and storm water
		1.2	Power	1.2.1	Electric network
				1.2.2	Natural gas network
				1.2.3	District heating system
		1.3	Telecommunications	1.3.1	Broadband
				1.3.2	Landline telephone services
				1.3.3	Mobile telephone services
		1.4	Transport	1.4.1	Road
				1.4.2	Railway
2.	Housing	2.1	Mobile		
		2.2	Individual		
		2.3	Group		
3.	Businesses	3.1	Services		
		3.2	Manufacturing industry		
		3.3	Stores		
4.	Public infrastructures	4.1	Decision-making crisis management infrastructure		Urban system administration (city hall)
		4.2	Operational crisis management infrastructure		Police, security, emergency response services...
		4.3	Infrastructure dedicated to vulnerable inhabitants		Health (hospitals, retirement homes...)
					Education (childcare centres, primary schools, secondary schools, high schools, universities)
					Specific accommodation (hostels, prisons...)
		4.4	Other public services		Welfare benefits
					Housing allowances
					Business support
					Record keeping
					Legal and judicial matters...

Table 2. Functional analysis table for drinking water supply technical networks.

No. and subsystems	Main functions	Technical functions
1.1.1 Water technical network: "drinking water supply"	To provide flows:	To resist mechanical loadings (multi-hazards)
	To provide drinking water supplies to housing, businesses and public infrastructure	To resist (piping) seismic loadings
	To protect in case of fire	To resist external contamination: to ensure water drinkability
		To ensure undisrupted service,
		To resist power outage (by providing a backup generator for instance)
		To last a long time

In order to be applied to urban engineering and serve to model the failure scenarios during crisis, the FMEA method had to undergo a number of adjustments. To that end, and to show the causal connections between phenomena and failure modes during an urban crisis, each parameter in the FMEA table and how they are achieved is precisely defined: failure mode, cause, effect. The suggested FMEA analysis includes the following

- subsystems;
- subsystem main and technical functions, provided by the functional analysis;
- subsystem failure modes, matching a function's failures and/or damages;
- possible causes of failure, originating from either the strength and/or location of the given external mechanical loading, or the failure of a subsystem interacting with the analysed subsystem;
- possible effects of failure

For instance, the DWS technical network subsystem's technical function is to "resist to mechanical loadings", hence the failure mode thereof: "did not resist to mechanical loadings". A possible cause of such failure is "damaged DWS pipes". Consequently, the effect of such failure is that the population (housing, businesses and public infrastructures) no longer have access to drinking water. This example is shown in Table 3 as a FMEA table. Unlike in industry, where FMEA is used as a tool to identify a system's failure modes, the FMEA is used here to identify, based on an experience feedback, the failure modes, causes and effects of crisis-stricken urban systems. At the end of the FMEA, the city's subsystems failure modes, causes and effects are known in connection with an urban crisis as detailed in the experience feedback.

ETM Applied to Cities

Based on the identification of failure modes, the failure scenario modelling methods are used to construct the sequences of failure modes that may lead to system failure. There are different methods for modelling risk analysis, one of the main ones being the ETM. The sequence of events on the tree is deductive or inductive (Zwingelstein, 1995), from the initiating event to the final events (Baroth et al., 2013; Modarres, 1993). The tree is chronologically developed according to how each modelled subsystem behaves. As part of our study on urban systems, the ETM will be developed after the FMEA method has been applied. Indeed, failure modes that occurred during an urban crisis are determined. The ETM will then help model the chronological sequence of effects as identified during this urban crisis. For instance, we consider, based on fictitious experience feedback, an urban system hit by severe flooding, for which the successive sequence of impacted subsystem failure modes is to be modelled. More specifically, three impacted urban subsystems are in focus: the electrical the power technical network, the DWS technical network, and housing. Failure of the functions "provide electric power" and "resist to mechanical loadings" for the electric power technical network lead to a defaulting technical function ("resist to power outage") for the DWS technical network. Thus, the failure of these technical functions results in the failure of the function "ensure suitable accommodation for population" for the housing subsystem. Ultimately, the flooding scenario for such urban crisis consists in the sequence of three failure modes: (i) failure of the electric power technical network that was flooded, (ii) failure of the DWS technical network due to power outage, and (iii) disruption of the electric power and DWS services for housing. This scenario, i.e. the sequence of failure modes described above, is modelled in Fig. 3.

In the end, the ETM allows us to produce a sequence of the subsystem failure modes according to their functional and chronological occurrence. Thus, an event tree consists of modelling an urban crisis as a functional scenario with sequenced urban subsystem failure modes.

Modelling Crisis-Stricken Cities through Qualitative Reasoning Models

The dynamic quality-based models resulting from qualitative reasoning models will help develop dynamic models in order to reproduce the systems behaviours from their causal relationships: any event (e.g. a P phenomenon occurring in an urban system, see Fig. 4) impacting a variable (e.g. an urban subsystem's F function, see Fig. 4) will spread to other variables (other phenomena and functions in city) having a causal connection therewith, through causal paths (Forbus, 1996). Propagation may be simulated using quality-based and dynamic transfer functions with time as a parameter.

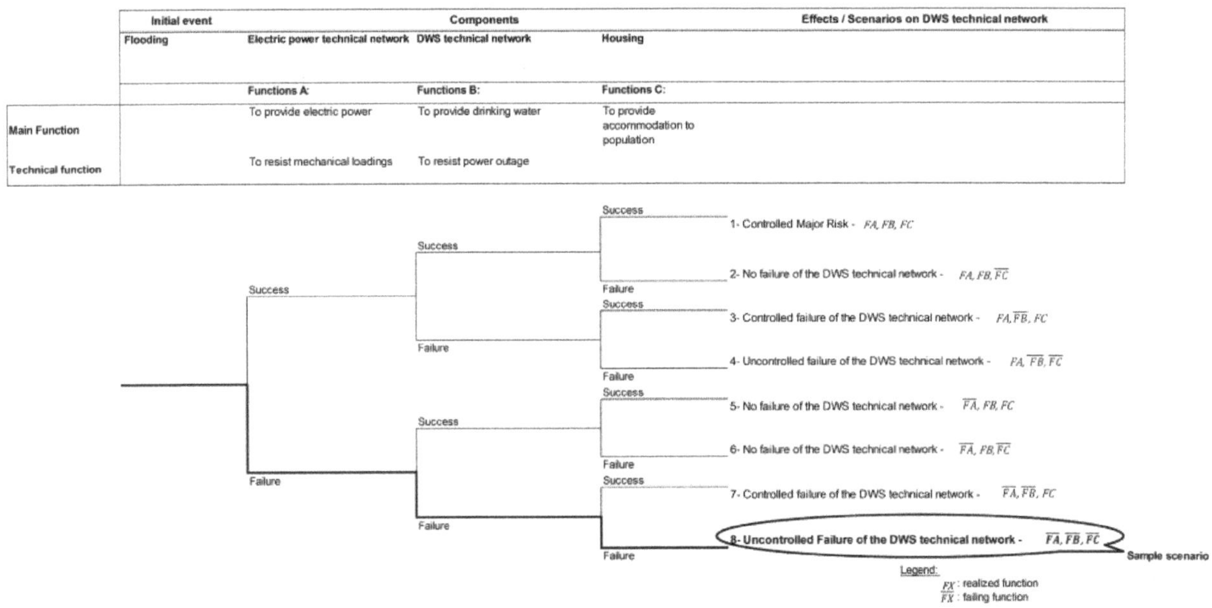

Figure 3. Sample of ETM: application to drinking water supply technical network.

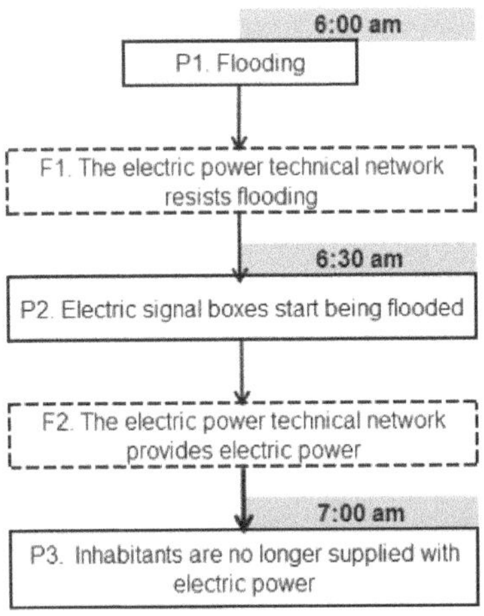

Figure 4. Sample of a directed graph dynamic causal model application, for the electric power technical network.

Of all dynamic quality-based models, the directed graph dynamic causal model is the most relevant for modelling an urban crisis scenario based on given experience feedback. Indeed, it will help show the dynamic and gradual change in impaired functions and subsystems over time. Also, redundancies may thereby be incorporated into the scenarios (introducing multiple identical phenomena in the scenario model) in order to take loop and non-linear scenarios into account. Figure 4 gives an example of a directed graph dynamic causal model considering the experience feedback on a flooding event impacting an urban system and, more specifically, the impact of such an event on the electric power technical network subsystem.

APPLYING THE FUNCTIONAL MODEL FOR FLOOD-STRICKEN CITIES

The functional model is applied to a case study pertaining to the city of Nîmes (France) after the October 1988 flooding. The event is first described and subsequently modelled following the above procedure: functional analysis, FMEA, ETM and causal model. Finally, the urban crisis is analysed based on the model results in Sect. 4.

Event Description

The information description is based on reference work searches that helped draw up an experience feedback analysis grid of the 1988 runoff flooding in Nîmes as part of the RESILIS project. The pieces of information are derived from a variety of essays, media and books. For instance, the main reference (Fabre et al., 1994) described the event and the 1988 flooding hazard impact on Nîmes and its subsystems.

A runoff flooding occurred on 3 October 1988 within Nîmes catchment area. This rain and storm event took place between 04:00 and 12:00 UTC. This produced a sheet of water of between 300 to 420 mm, corresponding to an estimated 100-year return period, which caused the runoff flooding, especially as it had rained heavily for several days before 3 October 1988. This previous rainfall caused ground saturation, hence the substantial increase in runoff.

Table 4. Subsystems impacted by flooding event in Nîmes (1988).

Classification (#)	Impacted subsystems
1.	Electric power technical network
2.	Drinking water supply technical network
3.	Road transport technical network
4.	Landline telecommunications technical network
5.	Local radio telecommunications technical network
6.	Housing
7.	Businesses
8.	Decision-making crisis management public infrastructures
9.	Operational crisis management public infrastructures
10.	Public infrastructure dedicated to vulnerable inhabitants

The natural event was greatly exacerbated by the city's urban structures. Indeed, the intermittent stream beds located upstream the city of Nîmes had been narrowed by road infrastructure and housing. The intermittent streams were directed downstream of the city centre through pipes that were either undersized or cluttered up by technical networks. And the outlets of these intermittent streams down the city were blocked by road or railway embankments. Therefore, Nîmes and its population ended up being trapped by water. The flooding event in Nîmes had disastrous outcomes. Regrettably, there were several casualties, 30 houses were destroyed and about 2000 had their fittings damaged, causing 45 000 disaster victims, of which 200 had to be rehoused. Between 450 and 700 businesses were also damaged. The crisis management public infrastructure buildings were flooded (command post for organization of civil security response in the prefecture, fire and rescue service department, departmental fire and rescue operational centre, emergency ambulance service, police station), as well as the city's utility services (electric power, telecommunications, DWS, road transport). The estimated overall disaster cost was FF 3.3 billion. Nîmes was not prepared to respond to such a large-scale crisis: there was no prevention plan except for the weather report, no hydraulic protection works or structures capable of responding to such a severe flooding event, no flood forecast system since there is no large watercourse flowing through the city, no delineation regulation applied to areas liable to flooding and no specific crisis management procedure. The city of Nîmes had to deal with an unexpected and large-scale event without any specific protection or adaptation resources available.

Functional analysis of 1988 flooded Nîmes

Identifying the Subsystems Impacted

The subsystems of Nîmes impacted during flooding were determined based on the experience feedback (see Sect. 3.1). Table 4 shows all 10 impacted subsystems. Based on such subsystems and by reviewing the event description contained in the experience feedback on the 1988 flooding in Nîmes, the functional model was applied.

Identifying the Main Functions of Subsystems Impacted

By analysing the experience feedback in Nîmes and considering the functions identified during an urban system generic functional analysis (Sect. 2.2.1), the essential main functions impacted during an urban crisis were determined. The first function is "provide drinking water" which is ensured by the water technical network: due to the flooding event which began at 06:45 UTC on 3 October 1988, the DWS technical network was cut off as of 09:30 UTC, affecting 50 % of the city. Thus the impacted function during crisis was not ensured across the system

The second function is "ensure communication and information" (to provide telecommunications flows): this function was ensured in Nîmes by two technical subsystems: the landline telecommunications technical network and the local radio telecommunications technical network. Four telephone exchange offices were flooded due to runoff flooding. Such flooding resulted in 60 000 users having their telephone communications cut off. In addition, the police telecommunications network was also cut off, thereby hindering communication in crisis management and emergency response operations. The "ensure information communication" function was indeed impacted by this event, although the second local radio telecommunications technical network subsystem remained in running order. The third function is "meet the needs for travel": this function, ensured by the road transport technical network, was impacted by the flooding event due to the damaged road system. Indeed, over 30 km of roads were destroyed, resulting in restricted access to housing for crisis management stakeholders.

The fourth function is "ensure accommodation and shelter for the population": this function is ensured by all constructed entities in the city, especially during urban crisis. A large number of constructed entities were impacted by the flooding event, resulting for the population in loss of their homes, their jobs and access to public services for a variable length of time. The main technical function involved in urban crisis is "resistance to flooding".

FMEA Applied To Subsystems Impacted By the Event

Once all functions impacted by the urban crisis were identified, the FMEA helped highlight the failure modes, causes and effects thereof on subsystems impacted by such urban crisis. FMEA is applied only to subsystems involved in the crisis event, as determined by the functional analysis (see Sect. 3.2.1). A FMEA table was used for modelling the failure modes, causes and effects of the subsystems (10 subsystems) impacted by the flooding in Nîmes. The subsystem main and technical functions were analysed in the FMEA table. From such functions, the failure modes, causes and effects were determined based on information provided by the experience feedback of the Nîmes crisis. Information contained in this experience feedback were subsequently organized and classified in the FMEA table. Table 5 provides sample results for the "DWS technical network". This table indicates that the function "provide drinking water for the housing, businesses and public infrastructures" of the DWS technical network has the following failure mode: "does not provide drinking water to 50 % of the city". This failure mode has the following cause: "interruption drinking water for 50 % of the city of Nîmes" and has the following effects: "50 % population, including housing, businesses and public infrastructure, no longer have access to drinking water due to their geographical location".

Event-Tree Method Applied to Nîmes Crises

The ETM is used to model the potential failure scenarios of October 1988 in Nîmes, in linking the different failure mode sequences. The initiating event of the ETM to be considered is the runoff flooding. As described in Sect. 3.2.1, the flooding event in Nîmes impacted 10 subsystems. The 10 subsystems require that the urban crisis should be modelled with regard to four essential main functions: (i) "provide drinking water", carried out by the DWS technical network, (ii) "ensure communication and information", carried out by the telecommunications technical networks (landline telephone and local radio services), (iii) "meet the needs for travel", carried out by the road transport network, (iv) "ensure accommodation and shelter" for the population, carried out by constructed entities. A model of the failures occurring in Nîmes was subsequently sought for each of these four essential main functions using the ETM and the procedure outlined in Sect. 2.2.3.

We present in the paper the modelling of the failure of the main function "provide drinking water" (Fig. 5). Event-tree modelling of the "DWS technical network failure" involves the following subsystems: electric power technical network, DWS technical network, decision-making crisis management public infrastructures and operational crisis management public infrastructures. In Fig. 5 we detail the scenario that occurred during the Nîmes crisis in October 1988. The ETM makes it possible to highlight the different potential scenarios, in a flood crisis-stricken city, induced by various sequences of failure modes, for each one of four urban subsystem essential main functions.

Modelling the City Of Nîmes during the Runoff Flooding Of 1988 Using the Directed Graph Dynamic Causal Model

The directed graph dynamic causal model helps to show the dynamic and gradual change in impaired functions and subsystems over time. It demonstrates the impact of an initiating event (P as phenomenon) on one or more variables (F as function), and how it would spread to other variables (other phenomena and functions in the city) having a causal connection thereto through causal paths. With regard to the experience feedback for Nîmes, a directed graph dynamic causal model may be developed considering the sequence of subsystem failures as suggested in the event trees (see Sect. 3.4). In this respect, this model provides a dynamic perspective on the scenario occurring in Nîmes from 3 to 4 October 1988. In Fig. 6, the DWS technical network section demonstrates that, faced with the runoff flooding phenomenon that occurred on Monday 3 October 1988, the function "provide drinking water" proved poorly efficient. However, the crisis management public infrastructures managed to handle the technical network failure. The population were supplied with substitute drinking water as soon as the following morning

(Tuesday 4 October 1988). Therefore, the directed graph dynamic causal model shows the global scenario derived from the experience feedback on the considered case study, including each relevant subsystem-specific scenario. The scenario is then more specifically defined by describing all phenomena at stake as a result of the failure modes. Unlike the ETM, the dynamic causal model does not highlight the different potential scenarios resulting from an initiating event. For that reason, a scenario's criticality may not be assessed against a whole set of potential scenarios.

DISCUSSION

Discussion on Functional Modelling Of the Nîmes Crisis

In the example of the case study, the flooding of Nîmes in October 1988, we chose to analyse the DWS technical network failure (see Fig. 5). The different scenarios, obtained with the ETM method, show how the function "provides drinking water" may be affected. These different scenarios may result in the failure or non-failure of subsystems' functions that are involved in potential failures of the DWS technical network. Scenario #5 (see Fig. 5), of all possible scenarios, is the one derived from the experience feedback on the flooding's in Nîmes 1988. Indeed, during the crisis a cut-off of DWS technical network occurred in 50 % of the city. This cut-off is due to the DWS technical network contamination on 3 October 1988 at 09:30 UTC. However, the crisis management public infrastructures ensured their functions with respect to such failure. The function of component "decision making crisis management infrastructure" that is "to decide on implementing substitute of drinking water" was performed with success. Substitute drinking water was quickly supplied to the population on 4 October 1988, as shown in scenario #5 of the event-tree through the sequence of failure modes regarding: "resist a contamination" and "ensure undisrupted service" technical functions for the DWS technical networks; "decide on implementing substitute of drinking water" principal function for decision-making crisis management public infrastructures; "ensure substitute of drinking water" principal function for operational crisis management public infrastructures.

Table 5. FMEA table for Nîmes during the 1988 runoff flooding event: sample of drinking water supply technical networks.

#1	Subsystems	Functions		Failure modes	Potential causes of failure	Failure effects
		Main	Technical			
2.	Drinking water supply technical network	To provide drinking water supply to housing, businesses and public infrastructure		Does not provide drinking water (50 % of the city)	50 % of Nîmes drinking water supply technical network is cut off	50 % inhabitants, including housing, businesses and public infrastructures, have no longer access to drinking water due to their geographic location
			To resist (piping) mechanical loadings (water pressures)		Damaged technical network	
		To protect in case of fire				
			To resist external contamination: to ensure water drinkability	Did not resist external contamination: non-drinkable water	Flooded network = contaminated water over 50 % of the city	
			To ensure undisrupted service	Did not ensure undisrupted service	Electric power technical network was cut off	
			To resist power outage (by providing a backup generator for instance)			
			To last a long time			

In light of the different scenarios shown in Fig. 5, it may be inferred that the most critical scenarios are related to defaulting crisis management public infrastructures. As a matter of fact, a situation will be deemed under control when crisis management public infrastructures are able to respond, by carrying out their functions, to other urban subsystem failures. For example, a scenario more critical that scenario #5 is scenario #8 which corresponds in experience feedback to Monday 3 October 1988. Indeed, no decision or emergency response operation was considered on this day to respond to the DWS technical network.

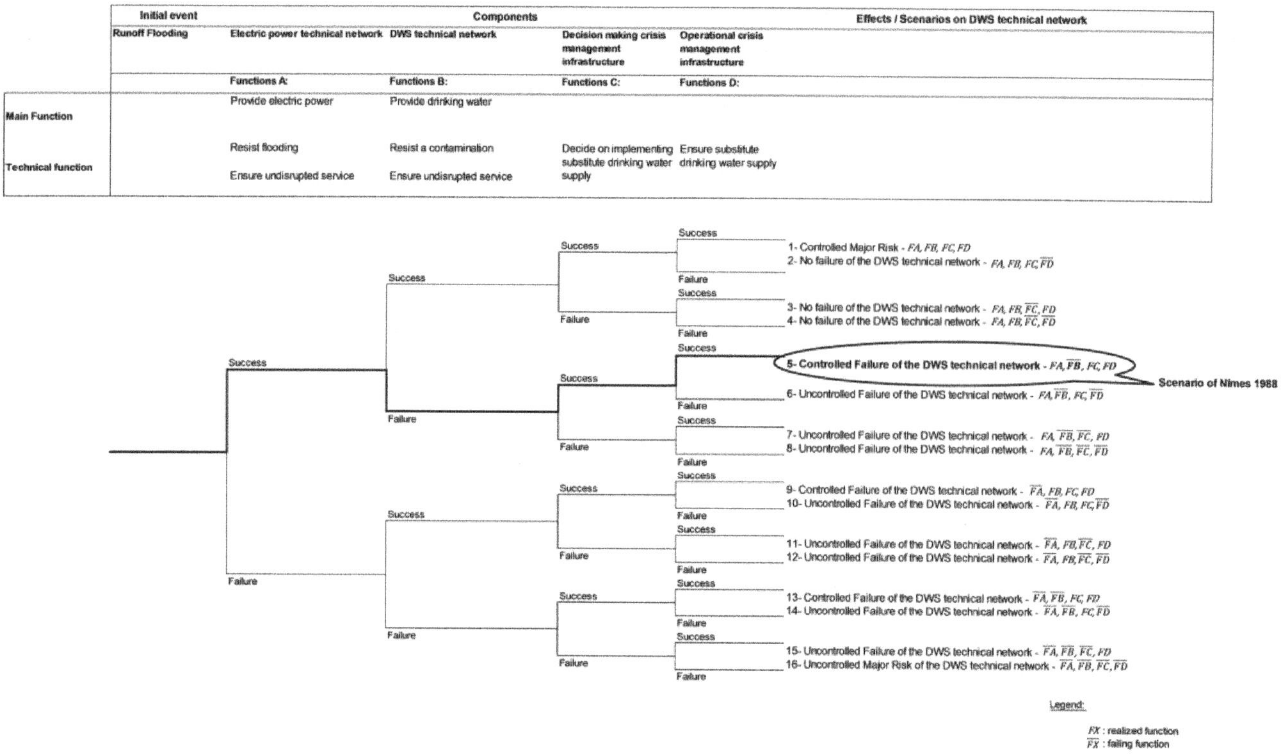

Figure 5. Event tree representing the drinking water supply technical network failure during the 1988 flooding in Nîmes.

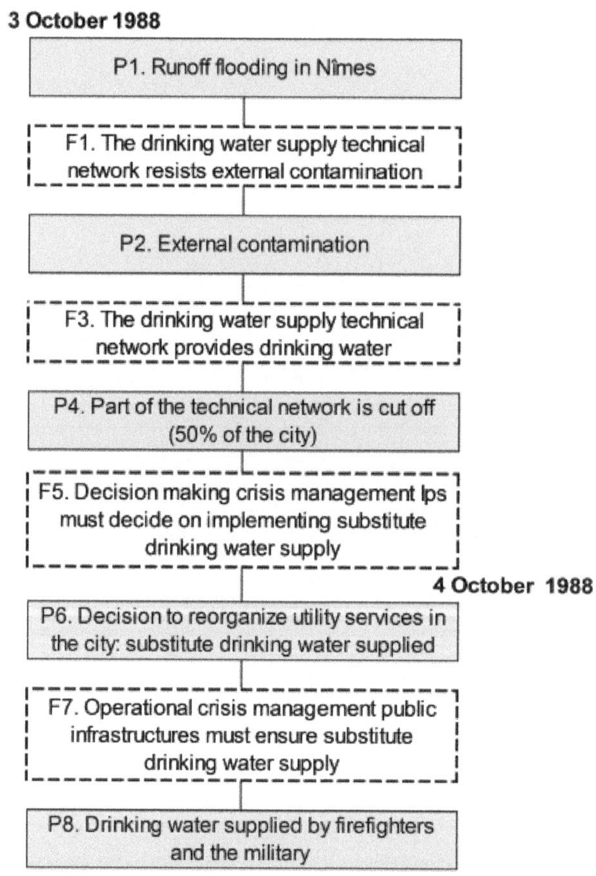

Figure 6. Representation of directed graph dynamic causal model for the 1988 crisis scenario in Nîmes, for drinking water supply technical networks.

The city of Nîmes can be viewed as exhibiting intermediate behaviour in regard to their defaulting DWS technical network on a time limit of 2 days, by calling on their crisis management public infrastructures for action. The city's resilience with regard to the defaulting "provide drinking water" function would most likely be improved with a faster response from crisis management public infrastructures (response in 1 day). Indeed, the principal function "decide on implementing substitute of drinking water" for decisionmaking crisis management infrastructures and the principal function "ensure substitute of drinking water" for operational crisis management infrastructures should have been carried on Monday 3 October 1988 instead of Tuesday 4 October 1988. A better resilience in Nîmes would also imply securing the DWS technical network against the flooding phenomenon in order to reduce the length of technical network cut-off. The ETM model proposed is a generic event tree for urban crisis affecting any one of the considered essential main functions, and may therefore apply to a large range of cities that were or might be subjected to this type of crisis. For each event tree, the procedure helped to determine how the city behaved in the event of a crisis, and such behaviour was identified and modelled by one potential scenario.

Utilization, Advantages and Limitations of the Functional Model

The functional model developed in this study has a dual purpose: capitalize on the failures of the city affected by a natural hazard and analyse behaviour. We present here the application of our model to the city of Nîmes, which was impacted by a flood in October 1988. Based on information from the experience feedback established following this event, we capitalized the failures of the functions of the various subsystems that have been mentioned in this experience feedback. Failures of the functions of affected subsystems, their causes and their effects are compiled into a FMEA table. The behaviour of each subsystem is then analysed by the ETM. The dependencies and the functional relationships between the subsystems are taken into consideration chronologically. Thus, the functional model can scan potential scenarios of the function failures and their effects for each subsystem. A directed graph dynamic causal model highlights the scenario that occurred during the crisis, by a logical chain of phenomena and functions for each subsystem mobilized during the event.

The novelty in the use of these methods is that their use is based solely on experience feedback. They have been adapted to this problematic, such as FMEA does not aim to find the set of failure modes, their causes and effects possible but retains only the information from the experience feedback. The application of our model can be performed on a city that has been impacted by various natural hazards (flood, earthquake, landslide. . .) and for which an experience feedback has been realized, including a complete register of failures subsystems and actions undertaken by the actors during the crisis management. In the framework of the RESILIS project, we have applied our model to four other cases of cities impacted by natural and technological hazards: the city of Annecy (France) impacted by an earthquake in July 1996; the city of Arles (France) impacted by flooding in 2003; the city of Toulouse (France) impacted by the explosion of factory AZF "Grande Paroisse" in 2001; and the city of Voûltesur-Rhône (France) impacted by the derailment of a train carrying hydrocarbons. For these four cases, the functional model has facilitated the analysis of positive and/or negative behaviour of the city facing a catastrophic event (Toubin et al., 2013).

This model based on risk analysis methods helps to improve the formalization of an experience feedback. It may be considered as a first step towards the analysis of the actions to be implemented with the aim to improve the resilience of a city impacted by a hazard. This model is intended for municipalities and local authorities wishing to use simple tools to reorganize procedures for prevention and protection, and the risk management for crises that may arise. Our functional model is limited primarily by the fact that its effectiveness depends on the availability and quality of information from the experience feedback of the city crisis. Second, this model is considered as a first step in the analysis of the crisis in order to improve the resilience of the city studied. It must be supplemented by other tools. Nevertheless, it can give indications of the subsystems of the city that are potentially the most critical and where there is urgent need to implement further study.

CONCLUSIONS

The functional model developed with a view to evaluate the behaviour a crisis-stricken city comes down to five basic steps: (i) to perform a "generic" functional analysis for an urban system. Such analysis includes, on the one hand, an external functional analysis showing all items outside the urban system as well as the principal and technical functions thereof and, on the other hand, an internal functional analysis defining each subsystem's main and their functions. The "generic" functional analysis underlines the essential functions required by the system to work properly; (ii) to have an experience feedback on an urban crisis; (iii) to analyse failure modes and effects using the FMEA method based on the experience feedback on the city of interest; (iv) to model different potential failure scenarios for the different potential sequences of failure modes, based on the initiating event and using ETM; (v) to represent the dynamic and gradual change in failed functions and subsystems as described in the experience feedback using a directed graph dynamic causal model. Such representation will also offer a global perspective on the scenario that occurred during the urban crisis of interest. This urban crisis experience feedback analysis tool helps modelling the functional behaviour of a crisis-stricken city. The purpose of this model is to help the various stakeholders of a city impacted by a natural hazard to better prepare for a similar hazard occurring in the future. Also, applying a crisisstricken city's functional model to experience feedback will allow the lessons learned from such modelling to be extended to other cities liable to similar hazards. Most lessons learned will stem from the ETM where trees are, in this tool, generic trees representing an urban crisis that impact a specific main function and involve specific subsystems. These may therefore apply to a large range of cities that were or might be subjected to this type of crisis, without further extension to other subsystems. It should also be noted that this functional modelling tool for crisis-stricken cities applies to any types of hazards, should they be natural or technological. Lastly, the functional modelling tool for crisis-stricken cities is limited, as it would show the city's weak points only based on previous experience feedback. As a matter of fact, this tool may not be used to implement proactive crisis management procedures based on hazards the city has never experienced before.

Acknowledgements

This research is part of the Project RESILIS, led by Egis with EIVP as scientific coordinator (www.resilis.fr) and funded by the French National Research Agency (ANR Sustainable Cities 2009).

REFERENCES

1. Alberti, M., Marzluff, J. M., Schulenberger, E., Bradley, G., Ryan, C., and Zumbrunnen, C.: Integrating humans into ecology: opportunities and challenges for studying urban ecosystems, Biosciencess, 53, 1169–1179, 2003.
2. Ashley, R., Blanksby, J., Chapman, J., and Zhou, A.: Towards integrated approaches to reduced flood risk in urban areas, in: Advances in Urban Flood Management, edited by: Ashley, R., Gravin, S., Pasche, E., Vassilopoulous, A., and Zevenbergen, C., Taylor & Francis Group, London, UK, 415–432, 2007.
3. Batty, M.: Cities as complex systems: scaling, interaction, networks, dynamics and urban morphologies, in: Encyclopedia of Complexity and Systems Science, edited by: Meyers, R. A., Springer, New York, 1041–1071, 2009.
4. De Bruijn, K. M., Van Beek, E., and TU Delft: Delft University of Technology: Resilience and flood risk management: a systems approach applied to lowland rivers, Delft University Press, Delft, 11 October 2005.
5. Campanella, T. J.: Urban resilience and the recovery of New Orleans, J. Am. Plann. Assoc., 72, 141–146, 2006.
6. Fabre, G., Lequeux, J.-M., and Scheurer, P.: Hydraulique pluviale à Nîmes après le 3 octobre 1988, Houille Blanche, 8, 27–31, 1994.
7. Folke, C.: Resilience: the emergence of a perspective for social– ecological systems analyses, Global Environ. Chang., 16, 253– 267, 2006.
8. Forbus, K. D.: Qualitative reasoning, in: CRC Handbook of Computer Science, Draft, Qualitative reasoning group, the Institute for the Learning Sciences, Northwestern University, USA, 715– 733, 1996.
9. Gallopín, G. C.: Linkages between vulnerability, resilience, and adaptive capacity, Global Environ. Change, 16, 293–303, 2006.
10. Holling, C.: Resilience and stability of ecological systems, Annu. Rev. Ecol. Syst., 4, 1–23, 1973.
11. Klein, R. J. T., Nicholls, R. J., and Thomalla, F.: Resilience to natural hazards: how useful is this concept?, Global Environ. Change B, 5, 35–45, 2003.
12. Komendantova, N., Mrzyglocki, R., Mignan, A., Khazai, B., Wenzel, F., Patt, A., and Fleming, K.: Multi-hazard and multi-risk decision support tools as a part of participatory risk governance: feedback from civil protection stakeholders, International Journal of Disaster Risk Reduction, 8, 50–67, 2014.
13. Lhomme, S., Serre, D., Diab, Y., and Laganier, R.: A methodology to produce interdependent networks disturbance scenarios, in: Vulnerability, Uncertainty, and Risk: Analysis, Modeling, and Management, Proceeding of the ICVRAM 2011 and ISUMA 2011 Conferences, 11–13 April 2011, Hyattsville, Maryland, USA, 724–731, 2011.
14. Lhomme, S., Serre, D., Diab, Y., and Laganier, R.: Analyzing resilience of urban networks: a preliminary step towards more flood resilient cities, Nat. Hazards Earth Syst. Sci., 13, 221–230, doi:10.5194/nhess-13-221-2013, 2013.
15. Linkov I., Bridges T., Creutzig F., Decker J., Fox-Lent C., Kröger W., Lambert J. H., Levermann A., Montreuil B., Nathwani J., Nyer R., Renn O., Scharte B., Scheffler A., Schreurs M., and Thiel-Clemen T.: Changing the resilience paradigm, Nat. Clim. Chang., 4, 407–409, 2014.
16. Modarres, M.: What every engineer should know about Reliability and Risk Analysis, Marcel Dekker, New York, USA, 1993.
17. Muller, M.: Adapting to climate change water management for urban resilience, Environ. Urban., 19, 99–113, 2007.
18. Pasche, E. and Geisler, T. R.: New strategies of damage reduction in urban areas proned to flood, in: Urban flood management, Taylor & Francis Group, London, 101–117, 2005.
19. Peyras, L., Royet, P., and Boissier, D.: Dam ageing diagnosis and risk analysis: development of methods to support expert judgment, Can. Geotech. J., 43, 169–186, 2006.
20. Peyras, L., Carvajal, C., Felix, H., Bacconne,t C., Royet, P., Becue, J. P., Boissier, D.: Probability-based assessment of dam safety using combined risk analysis and reliability methods – application to hazards studies, Eur. J. Environ. Civ. Eng., 16, 795–817, doi:10.1080/19648189.2012.672200, 2012.
21. Pickett, S. T. A., Cadenasso, M. L., and Grove, J. M.: Resilience cities: meaning, models, and metaphors for integrating the ecological, socio-economic, and planning realms, Landscape Urban Plan., 69, 369–384, 2004.
22. Serre, D., Peyras, L., Tourment, R., and Diab, Y.: Levee performance assessment methods integrated in a GIS to support planning maintenance actions, J. Infrastruct. Syst., 14, 201–213, 2008.
23. Serre, D., Lhomme, S., Heilemann, K., Hafskjold, L., Tagg, A., Walliman, N., and Diab, Y.: Assessing vulnerability to floods of the built environment – integrating urban networks and buildings, in: Vulnerability, Uncertainty, and Risk, American Society of Civil Engineers, Hyattsville, Maryland, USA, 746–753, 2011.
24. Toubin, M., Serre, D., Diab, Y., and Laganier, R.: Brief communication "An auto-diagnosis tool to highlight interdependencies between urban technical networks", Nat. Hazards Earth Syst. Sci., 12, 2219–2224, doi:10.5194/nhess-12-2219-2012, 2012.
25. Toubin, M., Raymond, M., and Arnaud, J. P., coord.: RESILIS – Guide pratique de la résilience urbaine, Pont et chaussée, presse des Ponts, France, 2013.
26. Wybo, J. L. and Van Wassenhove, W.: Retour d'expérience et maîtrise des risques-Pratiques et méthodes de mise en œuvre, Technique & Documentation Lavoisier, France, 2009.
27. Winkler, J., Dueñas-Osorio, L., Stein, R., and

Subramanian, D.: Performance assessment of topologically diverse power systems subjected to hurricane events, Reliab. Eng. Syst. Safe., 95, 323– 336, 2010.

28. Zevenbergen, C., Cashman, A., Evelpidou, N., Pasche, E., Garvin, S., and Ashley, R.: Urban flood management, CRC Press, London, 2010.

29. Zwingelstein G.: Diagnostic des défaillances – Théorie et pratique pour les systèmes industriels, Hermes, Paris, 1995.

CHAPTER 14

FLOODING AND FLOOD RISK REDUCTION IN NIGERIA: CARDINAL GAPS

Nkwunonwo UC[1,2], Whitworth Malcolm[2] and Baily Brian[3]

[1]Department of Geoinformatics and Surveying, University of Nigeria, Enugu Campus, Nigeria

[2]School of Earth and Environmental Sciences, University of Portsmouth, UK

[3]Department of Geography, University of Portsmouth, UK

ABSTRACT

This study reflects on critical issue relating to flooding in Nigeria such as causes, impacts and remedies. Flooding which arguably has been more damaging for Nigeria has worsened recently due to a number of possible factors including rapid population growth, urbanization, poor urban planning and climate change especially in increased frequency and intensity of rainfall. Attempts to tackle the hazard in Nigeria appear to be limited by lack of flood data and other remote causes which are yet to be identified. In view of this background, the present study reviews the widespread flooding in Nigeria and efforts to tackle it. Over the period 1985 to 2014, flooding in Nigeria has affected more than 11 million lives with a total of 1100 deaths and property damage exceeding US$17 billion. Lagos state has experienced most of the floods while more frequent floods are recorded in Niger, Adamawa, Oyo, Kano and Jigawa states possibly due to the influence of rivers Niger, Benue, Ogun and Hadeja. It is argued that more robust and scientific approaches to flood risk reduction such as: flood modeling and vulnerability assessment are lacking. To align the focus of flood risk reduction in Nigeria with the objectives of such a task in more developed countries (such as the United States, United Kingdom and the Netherlands) which is among three fundamental issues to be addressed in Nigeria, the present study makes pivotal recommendations.

INTRODUCTION

Arguably, the rate of flooding occurrence in recent times has been unprecedented. With 70 million people globally exposed to flooding every year, and more than 800 million living in flood prone areas [1], climate change with more frequent and severe rainfall events, sea level rise, rapid population growth and urbanization, the rate of development on floodplains, the level of awareness of flood risk and the ineffectiveness of efforts towards tackling flooding in many places are factors of concern within the global [2,3]. The economic consequences of flooding reported in the last two decades amount to tens of billions of US dollars [4]. Over 3700 flood disasters globally are recorded in the EM-DAT database, covering the period 1985 to 2014 [5]. These events were responsible for hundreds of thousands of deaths mainly in Asia (most notably China, Thailand and Bangladesh) and adversely affected billions of people mostly through homelessness, spread of diseases, physical injuries, mortality (mainly through drowning) and psychological conditions mainly depression, anxiety and

Citation: Nkwunonwo UC, Malcolm W, Brian B (2015) Flooding and Flood Risk Reduction in Nigeria: Cardinal Gaps. J Geogr Nat Disast 5:136. doi: 10.4172/2167-0587.1000136.

posttraumatic stress [6-10]. In the US, 32.9% of the total natural disasters in 2012 were hydrological with floods accounting for the most part, affecting more than 9 million people and causing about US$ 0.58 billion worth of damage [5]. The same source shows, for that year, more than US$4.7 billion worth of damage recorded for Europe, and about US$0.83 billion and US$19.3 billion damage for Africa and Asia respectively resulting from flooding. Four different floods that hit cities in UK in 2012 caused a total loss of $2.9 billion, with hundreds of people who were affected [4]. In many African countries for example Nigeria, flooding has impoverished hundreds of thousands of people through displacement from homes and loss of tangible properties [11].

In Nigeria, flooding and means of addressing its challenges are critical issues [12]. Evidently, the country has experienced devastating floods which affected millions people and caused fiscal losses amounting to billions of US dollars [13]. These hazards were generally linked to poor urban planning and climate change especially in increased frequency and intensity of rainfall [11,14-16]. The impacts of floods in Nigeria include mortality, physical injuries, widespread infection and vector-borne diseases, social disorders, homelessness, food insecurity, economic losses (mainly through destruction of farmlands, social and urban infrastructure) and economic disruption (most notably in oil exploration in the Niger delta, traffic congestion in many cities in Nigeria, disruption in telecommunication and power supply) [17-19]. In 2012, Nigeria experienced the worst flooding in more than 40 years as a result of heavy rainfall that lasted for many days. The incidence affected 32 states with 24 considered severely affected [13]. The floods extended from July to October that year and affected 7.7 million people with more than 2 million others reckoned as internally displaced (IDPs). More than 5000 people were physically injured along with over 5900 houses which were destroyed.

Historically, flooding in Nigeria which dates back to the early 1950' are fluvial, coastal and pluvial in nature and have been a major cause of concern for rural areas and cities within the country [20,21]. Fluvial and coastal flooding both of which affected mainly coastal environments were influenced by seasonal interruption of major rivers and water overtopping their natural and artificial defences and overflowing areas not typically submerged [22]. Fluvial floods account for the majority of the flood threats experienced in locations along the plains adjoining major rivers in the country, including rivers Niger, Benue and Hadeja. The states in Nigeria mostly affected by such floods are Adamawa, Kano, Niger, Jigawa, Kaduna, Cross River and Kebbi [23,24]. The worst fluvial flood in Nigeria was the Kaduna flood disaster of 2006 which affected hundreds of thousands of human lives with economic loss worth millions of US dollars [25,26]. Coastal floods in Nigeria affect the low-lying areas in the southern part of the country (comprising for examples Lagos, Oyo, Ondo, Akwa-Ibom and Bayelsa states). The impacts of such floods have been severe due to the number of human populations exposed following the attractions of coastal areas for economic and social reasons [27]. Nigeria is globally ranked with the top 20 countries whose present population and future scenarios in the 2070s (including climate change and socio-economic factors) are exposed to coastal flooding [28].

Pluvial floods usually occur annually during rainy seasons (between July and October) and affect mainly the urban areas in Nigeria. Such floods which are arguably unprecedented in recent times are caused by more frequent and severe rainfall which overwhelms the efficiency of drainage systems and soil infiltration capacity [29,30]. The significance of urban areas in the economic and political development of Nigeria is generally acknowledged [31]. However, urbanization is a critical anthropogenic influence on climate change and hydrological cycle in the country given that much impervious surfaces increase surface water runoffs and reduce soil infiltration capacity [32-35]. Along with urbanization is the rapid population increase in many Nigerian cities which is also a global concern within the context of flooding in urban areas [36]. It is estimated that more than half of the world's population has been residing in cities since the last 6 years and by 2030 the number of people living in urban areas (with urban areas of developing countries (DCs) accounting for the most part) will grow to 5 billion (i.e. 60% of the world's population) [37-39]. Regrettably, a major challenge with rapid population growth and urbanization in Nigeria which also seems to influence the risk of flooding in the country both presently and in the future (if not addressed) has been poor urban planning (in particular inadequate drainage system and the range of poorly serviced urban utilities) [14]. The authors argue that urbanization in Nigeria has not to be accompanied by corresponding strategies to support humanitarian needs and anthropogenic activities. Apparently, this concern has not received adequate attention in the literature, especially with regards to the implications of future urban scenarios on environmental sustainability. Moreover, the reality of widespread flooding in Nigeria coupled with the notion that floods are inevitable phenomena which can never be fully constrained within the natural environment appears to overwhelm efforts towards finding a solution. Thus, it is rather easier to feel some developing attitudes of anticipating increasing human vulnerability within the country as the years succeed each other, and the authors perceive that "there seems to be fear on the horizon".

However, various levels of government, the community and other stake holders have been active with measures to tackle flooding in Nigeria [40]. These measures have been criticized as ad-hoc, nongeneralizable and not well established [12]. In the light of 'best practices' in flood risk reduction and 'lessons learned' from other countries' experiences of flooding, it can be argued that such stake holders' efforts are at best limited most probably due to lack of quality data, which among other things are needed to systematically tackle flooding, poor perception of flooding among the general public, lack of funds and improved technology as well as poor political will power. Best practices in flood risk reduction ideally is based on "living with floods and not fighting them" idea, which dominates key environmental risk research themes (for examples: Disaster Risk Reduction (DRR) and Climate Change Adaptation (CCA)) and integrates structural and nonstructural measures to reduce the impacts of flooding on social systems and to achieve key requirements in risk management which are prevention/ mitigation, preparedness, emergency response, recovery and lessons learned [41-47].

Arguably, given the growing number of flood victims and the constrained sustainable development caused by flooding within the country, researches into solutions to widespread flooding in Nigeria are weak towards flood risk reduction. Recent studies focus attention on the impacts of flooding and its remedies in local communities, geopolitical regions and various states within the country [27,48-52]. It is argued that a more comprehensive understanding of the widespread flooding in Nigeria requires wide ranging discussions and accounts which few studies presented [12,20,53-58]. The influence of climate change and anthropogenic activities, poor urban planning and environmental management on the widespread flooding in Nigerian cities has been extensively discussed [14,50,54,59]. While the lack of definite measures and capacity to radically address the challenges of flooding within the country has been arguably overwhelming, concerted efforts in the form of environmental and infrastructural planning, policy directives, social responses, physical intervention and enhanced public enlightenment programmes, vital to tackling the prevailing flood hazard in the country have been considered [20,21,24]. Other vital tools needed to address flooding in Nigeria are community based early warning systems [56], humanitarian aids from government and private sectors [54,60] and appropriate level of preparedness and capacity building [57]. The need for science and technology to embrace environmental education in Nigeria is posited [58]. In [27,48,61,62], flood hazard mapping and assessment of vulnerabilities of lives and properties to flooding are important procedures factors which play key roles towards building community resilience to flooding. Few studies considered modeling of flood by means of hydrodynamic, GIS, cellular automata and statistical techniques [63-72]. The importance of reinforcing present strength and capacities of all agencies, including local communities within Nigeria to deal with flood hazard situations underlined [12]. The study investigated existing institutional approaches that deal with flood challenges in Nigeria with a view to more efficient, robust and satisfactory flood intervention strategies in the country.

question: "what is the remedy to the recurrent flooding in Nigeria?" remains unanswered. The lack of flood data and other ancillary data which is a major setback towards tackling flooding in Nigeria was not addressed. Attention has solely rested on general knowledge of the causes, impacts and remedies of flooding, suggesting that the general view of the situation in these studies has been lop-sided and sloppy. The need for more scientific approaches such a flood modeling which drives flood risk management in more developed countries was not highlighted. A general critique, which should provide a nuanced understanding of the strengths and limitations of present efforts to addressing the threats of flooding in the country, is lacking and gaps between increasing flood occurrences and vulnerabilities of local communities were not identified. It is well known that the success of flood risk reduction depends to a large extent on knowledge-based decision, robust institutional framework and flood risk communication [3,18] but these factors are missing in Nigeria and where they exist they are poorly addressed. Knowledge-based decision uses available information relating to flooding to draw conclusions on possible strategies to be adopted for tackling flooding. Flood risk communication aims at creating awareness of flooding and its impacts in stake holders and the general public. Institutional framework that tackles flooding in Nigeria refers to government response procedures which include policies, regulations, guidelines and agencies engaged in planning and managing flood emergency conditions or in helping victims to cope with and recover speedily from extreme flooding events [12]. Many flood risk based studies argue that these requirements are fundamental to information relating to flood hazard and its consequences and may be sourced from public opinions, expert knowledge, research findings and flood risk/hazard maps of an area [73-75]. It has also been argued that various flood modelling approaches are pivotal components of flood risk reduction because they are capable of quick, continuous and routine simulation of flood data (most notably flood water depth, inundation extent and duration as well as water flow velocity) required for flood hazard / risk assessment [30,76,77].

Despite the lack of these critical data for tackling flooding in Nigeria, flood modelling is poorly carried out [69]. Although the acquisition of these datasets (particularly the flood risk/hazard maps) have been a daunting task for Nigeria as they require critical understanding of the drivers of flood hazard/risk. However, it is important to acknowledge the specific roles flood modelling can play in this regard and in general towards tackling flooding in Nigeria. Over the recent past and in many flood management policies such as the EU commission directive on flood, the United States flood control policy, national flood insurance program (NFIP) and other regionally-based flood risk management policies, the relevance and provision of flood data highlight the significance of flood modeling [30,45]. For Nigeria, besides the general roles of flood modelling, the technique can assist in provision of critical information for strategic planning and effective flood risk management within the country. It will equally drive an improved understanding of flood phenomena and prompt improvements into more robust flood management approaches such as flood forecasting, flood early warning system and flood damage estimation.

This need for flood modelling was emphasized by the DG of Nigerian Hydrological Services Agency [78] in a recent mission statement:

"...in view of flooding in Nigeria, governments at all levels should create awareness on the need for communities to relocate to safer terrain. Moreover, while the current trends in climate variations prevail, the need to develop flood modeling and early warning systems cannot be overemphasized... There is also need to carry out a comprehensive flood hazard mapping for all areas considered at risk of flooding in the country..."

It is argued that whilst this statement recognizes flooding situation in Nigeria and the relevance of flood data for flood risk reduction, it laments the dearth of flood data for the country and the peculiar challenges that arise from utilizing commercial flood modelling codes to simulate such data. Moreover, the fact that Nigeria is one of the most populated countries of the world with population size estimated at over 170 million people compounds the situation [79]. The theory that future population growth will drive future flood risk highlights the importance of urgency in finding means of preparing human population in the country to adapt to floods. It will be a welcome development for Nigeria to implement flood modelling since within European Union framework, such a technique is operationalized towards flood hazard/risk mapping of the constituting States [73]. In the Netherlands and the US, flood modeling, among other roles, supports investigation into estimation of damage caused by flooding [80- 83]. Many Asian countries, notably China, Vietnam and Bangladesh although having 'not too well' established flood management systems and methodologies utilize flood modeling methodologies for flood risk assessment and mitigation [38,84].

Thus, to assist human populations towards adapting to flooding and to support stake holders' efforts towards tackling flooding in Nigeria, the present study is an attempt to bridge several gaps in flood risk reduction in Nigeria. The authors synthesized and analyzed relevant literature and available historical flood data spanning the period 1985 to 2014, to identify these gaps and to make relevant recommendations towards building flood resilient communities and more effective remedies to the threats of flooding. It is argued that promoting flood modelling in Nigeria will underpin current efforts at tackling the hazard.

This study in general and the recommendations in particular are driven by three key issues, which are: firstly, to demonstrate the roles more robust and scientific techniques such as flood modeling can play in flood risk reduction within the context of Nigeria, secondly, to align the focus of flood risk reduction in Nigeria with the objectives of such a task in more developed countries such as the US, the Netherlands and United Kingdom, and finally, to promote flood risk awareness in the general public as well as to delineate safer terrains for relocation of human populations during flooding in Nigeria.

The methodology and data for the research are discussed in section 2 while the study area is described in section 3. Section 4, the result, discusses efforts towards tackling flooding in Nigeria and cardinal gaps which have been identified to undermine such efforts. Recommendations relevant to strengthening present efforts of tackling flooding in the study area are presented in section 5. Section 6 gives a general conclusion of the paper.

METHOD AND DATA

A search process to identify the body of literature relevant to flooding and efforts towards addressing its threats in Nigeria was undertaken. Combination of terms such as "flooding and management in Nigeria", "flooding and human health in Nigeria", "flooding and modeling in Nigeria" and "flooding and climate change in Nigeria" was applicable to the search. Overall, 429 publications were identified of which 17 focused on the causes of flooding in Nigeria, 132 addressed the impacts, 181 discussed the remedies, 54 looked at climate change issues, 14 discussed public perception of flooding while 31 addressed urban management and planning (Table 1). The scientific quality of these papers was assessed based on the publishing journal. This is consistent with academic standard and regulations. Although locally published articles provided most of the information to establish the case in the present study, however, the greater weight was given to articles published by Elsevier, Science Direct, Taylor and Francis, Wiley and sons, ASCE, Nature, Sage, Springer and Copernicus publishers and on International conferences. The data that provided much of the evidence regarding the prevalence of flooding in Nigeria was sourced mainly from EM-DAT database and Nigerian ministry of Environment. These findings are fundamental to discussions presented in this study.

Table 1: The results of literature search of issues relating to flooding in Nigeria. From these results, whilst remedies have received ample attention, improving public perception of flooding is poorly addressed. This is a reason for concern knowing that the success of present approaches of flood risk reduction depends on public participation which is driven by public perception of flooding.

S. No.	Issues relating to flooding in Nigeria	Literature search
1.	Flooding causes – causes	17
2.	Flooding – impacts	132
3.	Climate change issues	54
4.	Mitigation activities	181
5.	Public perception	14
6.	Urban planning and management	31

Description of the Study Area

Nigeria, a sub-Saharan West African country, is on the Gulf of Guinea, east of the Greenwich and north of the equator. The country, made up of 36 states including the federal capital territory (FCT), Abuja, lies between latitudes 4° and 14°N, and longitudes 2° and 15°E, with a total land area of 923,768 km^2 (Figure 1), and borders with Republics of Benin and Niger, Chad, and Cameroon. It maintains a large expanse of coastline, over 853 km in magnitude, with hydrological features which includes the rivers Niger and Benue, both of which confluence at Lokoja, and flows further southwards passing through the Niger Delta to empty into the Atlantic ocean.

The 2006 census confirmed over 140 million people in Nigeria, but this population has grown steadily, and is presently estimated at more than 170 million people, making the country the seventh most populous country in the world [12,85]. According to United Nations projections, Nigeria is one of the eight countries expected to account collectively for half of the total population increase in the world from 2005-2050, and will by 2100, record a population amounting between 505 million and 1.03 billion people [86].

RESULT AND DISCUSSION

First and foremost, the present study argues that the impacts of flooding in Nigeria continue to trigger concerns for food security, vulnerability of the general public and local communities within the country, humanitarian needs and services, primary health delivery, environmental management, solid waste management, urban development, professionalism in journalism practice and the dynamism or lethargy of Nigerian democracy and political system [12,21,27,60,87- 89]. Whilst investigating these factors vis-à-vis flood risk reduction in Nigeria is vital, key features of flooding which influence its level of impacts in the country include flood water depth, inundation extent and duration as well as velocity of flood flow. Respectively, flood width, height, annual frequency and duration in Nigeria can measure up to 700 m, 11 m, 10 and 25 days respectively [53]. It is shown from EMDAT database that most floods in Nigeria lasted up to 79 days. Thus based on these characteristics, the dangers posed to human lives and properties by flooding in Nigeria can be appreciated (Figure 2).

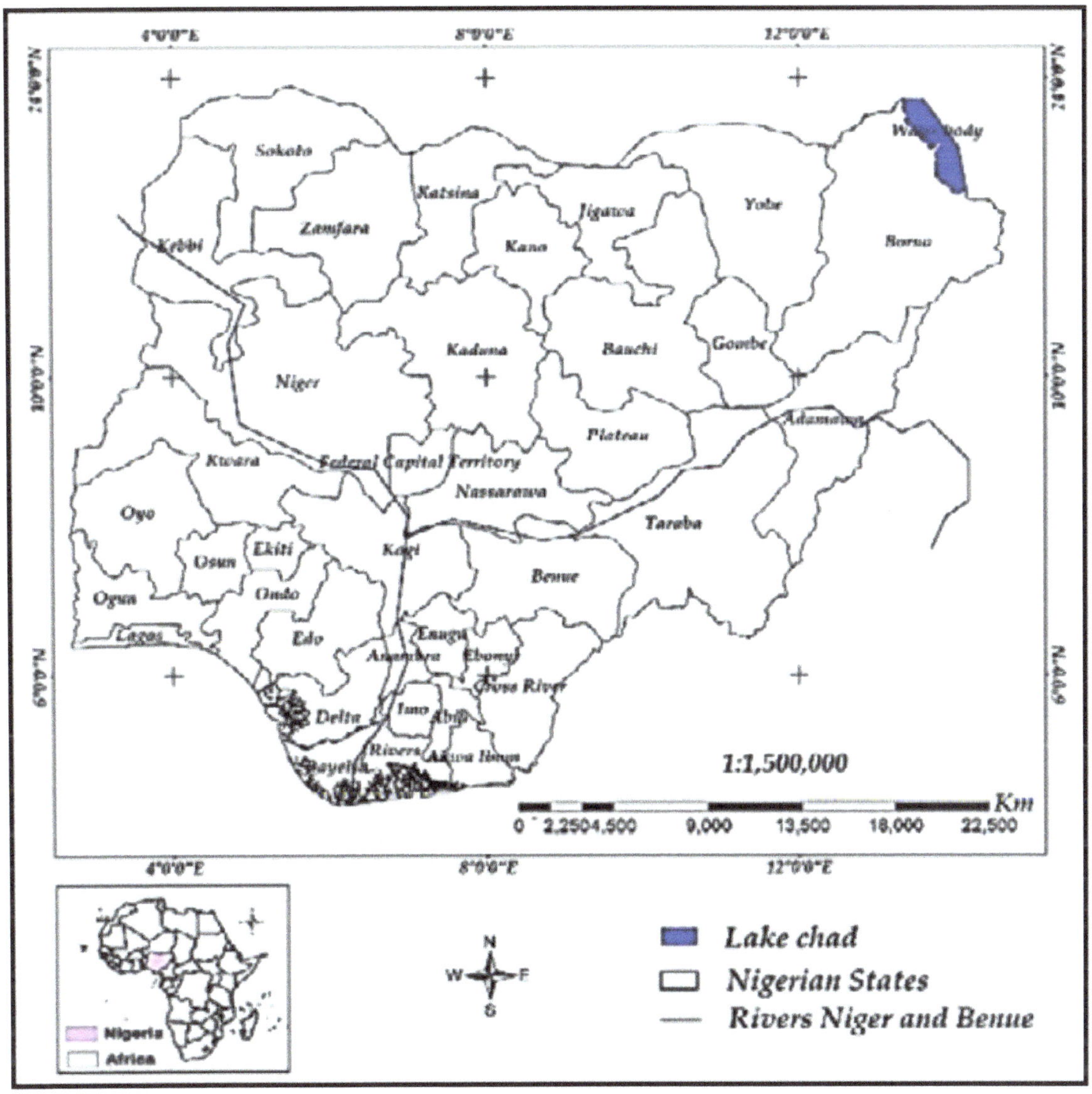

Figure 1: Map of Nigeria showing the 36 states and Rivers Niger and Benue. Inset is Africa showing Nigeria's location. Source: Drafted by the authors

Figure 2: Evidence of flooding impacts in Nigeria: (a) Major public access road affected by flood water (b) street lanes submerged by flood water, (c) community center submerged by flooding, and (d) residential house submerged by flooding. Source: Authors' field work.

Paradoxically, relevant flood data which reveal widespread flood threats in Nigeria and how the situation differs from other places are lacking. Arguably, such dearth of flood data not only constrains investigations into gaining a better understanding of the spatial and temporal distribution of flooding across the states of Nigeria, but it also undermines efforts towards tackling the hazard. Although there is a large body of literature that highlights the gravity of flooding situation in the country, inconsistency of flood narratives in Nigeria has been overwhelming [59,90]. Undoubtedly, this inconsistency of flood narratives will undermine the general flood awareness which underpins more proactive ways of tackling the hazard in Nigeria. However, based on flood data from EM-DAT, CRED and Dartmouth Flood Observatory (DFO) databases, present investigations (which extend studies by [12,24,25,54,91]) relating to flood threats in Nigeria have been made.

Based on data on historical flooding in Nigeria from 1985 till 2014, the present study reveals that flooding is a common phenomenon for various states in Nigeria. Although Lagos state appears to have experienced most of the floods in the country, more frequent flooding are recorded in Niger, Adamawa, Kano, Oyo and Jigawa states. Whilst Lagos state flooding can be attributed to urban growth and coastal influence among other key factors, the influence of rivers such as Niger, Benue, Ogun and Hadeja may account for the rest of the states with more frequent floods (Figures 3 and 4).

To understand where Nigeria stands in global and regional perspective since 1985, the 2012 flooding in Nigeria reputed as the worst in 40 years is compared with the most devastating floods in the world between 1985 and 2014. The result of this comparison shows that within the period under consideration, Nigeria may be ranked as third in the world, following peoples Republic of China and Soviet Union (Table 2). In Africa, the study shows that the country ranks topmost (Table 3), overtaking Mozambique and Algeria in terms of economic loss. This reality highlights the need for more proactive efforts towards tackling flooding in the country [92].

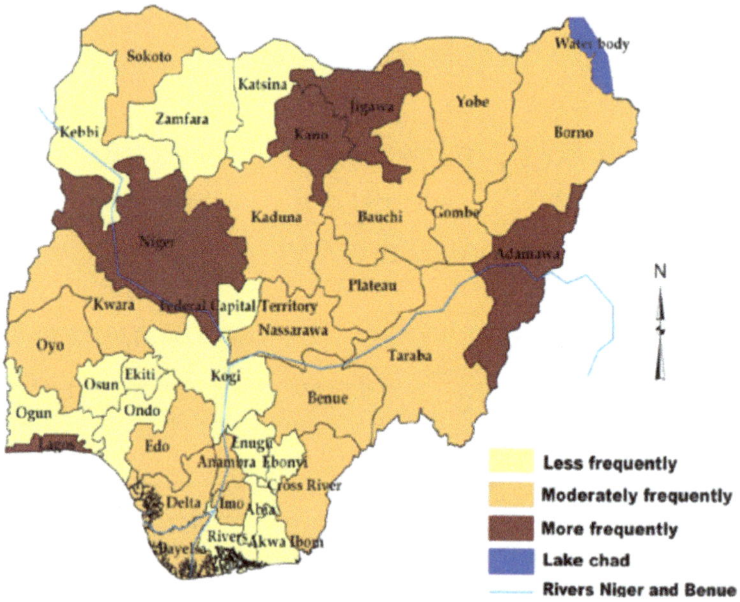

Figure 3: Spatial distribution of flooding across the states of Nigeria from 1985 to 2014. Here it is shown that Lagos, under coastal influence along with under states under the influence of Rivers Niger, Benue, Hadeja and Ogun have experienced more frequent floods in the country. Source: Authors' analysis of historical flood data from 1985 to 2014

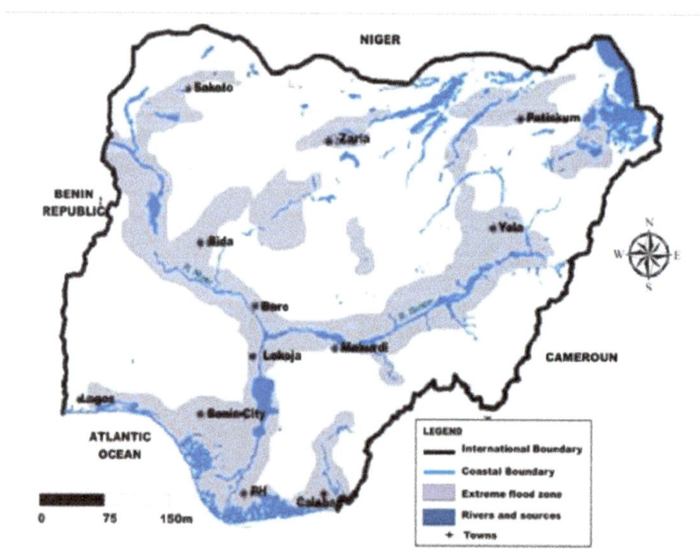

Figure 4: Spatial distribution of areas affected by extreme floods in Nigeria between 2000 and 2012. Source: Federal Ministry of Environment [92].

Apart from China which presently reputes as the most flood prone country in the world, characterized by recurrent perennial floods due to among other things, the influence of population growth and mainly the River Yangtze [93]. The fact that other countries with known extreme flooding experience (for examples: Netherlands, the US, Brazil, United Kingdom and many other European countries) may be ranked below Nigeria in terms of devastating floods within the period considered suggest among other things the significance of flood risk reduction measures which are presently in place in those countries.

The Netherlands with more than half of the country at or below sea level experienced a severe flood in 1953 that devastated majority of the nation's economic and human infrastructure. The estimated impact of the flood was 1835 deaths and 1 billion Dutch guilders (US$ 558 million). That flood challenged various stake holders, particularly the general public and local communities and Dutch government towards more effective strategies of mitigating the threats of flooding. The reduced impacts of flooding in the country in recent times validates these efforts. Similarly, recent floods in the US have lesser impacts compared to the flood of 1972 which caused 238 deaths, 357 injuries and destroyed about 1335 homes with estimated fiscal loss of over 800 million US$. In the UK, although there have been devastating recent floods, the 1947 floods were considered the worst in recent history with overall impact estimated at about £4.5 million (USD$ 6.81 million) at current value, with millions of displaced human populations, farm animals and agricultural products [94]. For Brazil, the floods of 2010 have reduced impacts compared to the floods of 1967 which was considered the deadliest in the country's history, claiming about 610 lives with economic loss amounting to about US$ 1.2.

Table 2: 20 most devastating floods in the world between 1985 and 2014. Source: Complied by authors

S.No.	Country	Year	Mortality	Economic loss (billion US$)	Affected Human Population (millions)
1.	Soviet Union	1991	N/A	60	0.04
2.	China	1998	3656	20	240
3.	Nigeria	2012	363	16.9	7.7
4.	Korea (P Dem Rep)	1998	405	15	5.2
5.	United States	1993	50	15	N/A
6.	Russia	2002	167	0.9	0.3
7.	Italy	1994	N/A	6	N/A
8.	Poland	1997	N/A	4.3	0.24
9	United Kingdom	2007	30000	3.2	N/A
10.	Ecuador	2008	N/A	1	0.289
11.	Bangladesh	1998	3836	2	30
12.	Netherland	1998	N/A	5.3	0.02
13.	Argentina	1985	N/A	1.3	1
14.	Vietnam	1999	702	0.78	1.7
15.	Philippines	2012	1268	2.19	0.45
16.	Pakistan	1993	3084	1	6.184
17.	Afghanistan	1992	3000	N/A	N/A
18.	Colombia	2011	450	1.2	4.98
19.	Nepal	1993	1048	N/A	0.53
20.	Brazil	2008	246+	0.75	1.5

Table 3: 10 most devastating floods in Africa since 1985. Source: Complied by authors

S/No	Country	Year	Mortality	Economic loss (billion US$)	Affected Human Population (millions)
1.	Nigeria	2012	363	16.9	7.7
2.	Algeria	2001	921	0.3	0.045
3.	Mozambique	2000	800	0.419	4.5
4.	Ethiopia	2006	705	0.32	N/A
5.	South Africa	1987	506	0.765	0.065
6.	Burkina-Faso	2009	N/A	0.15	0.15
7.	Malawi	1991	472	0.024	0.15
8.	Kenya	2006	114	N/A	0.723
9	Egypt	1994	600	0.14	0.16
10.	Sudan	2007	N/A	0.3	0.56

Efforts towards tackling flooding in Nigeria

The threats of flooding in Nigeria have been tackled through a number of specific and general actions including physical intervention, legislation and policy formulation, creation of awareness of flooding, urban renewal and development, engineering structures such as dams, bridges and drainage systems, relocation of human populations during flooding and assisting flood victims with basic humanitarian needs [40,78,95]. To simplify discussion regarding these efforts, the present study simplifies and groups these actions into institutional approach, actions by local communities and the general public, action by humanitarian agencies and actions by research institutions and the media.

Institutional Approach

Institutional approach towards addressing the threats of flooding in Nigeria dates back to the early 1960's with the establishment of federal and state ministries of works [96]. However, the increasing frequency and severity of floods across the country prompted the establishing of the Federal Environmental Protection Agency (FEPA) as a unit in the Federal Ministry of Works and Housing in 1988 [26] and the Federal Ministry of Environment (FME) in 1999 [92]. Among other things, the key roles of FME towards flooding risk reduction in Nigeria is to assess the flooding potentials as well as design, determine, develop and/ or authorize the development of appropriate flood reduction measures for the country [92].

With the FME comes various ministries and agencies for tackling flooding in Nigeria which include: Federal Emergency Management Agency (FEMA), National Emergency Management Agency (NEMA), State Emergency Management Agency (SEMA), Local Emergency Management Agency (LEMA), National Orientation Agency (NOA), National Commission for Refugees (NCR), National Environmental Standards and Regulations Enforcement Agency (NESREA) which by 2009 Nigerian Acts supersedes the FEPA, Nigerian Meteorological Agency (NIMET), Nigeria Hydrological Services Agency (NIHSA), NEST (Nigeria Environmental Study/Action Team) and Building Nigeria's Response to Climate Change (BNRCC).

NEMA is basically a coordinating body for disaster management in Nigeria. Actions towards addressing the threats of flooding which the agency coordinates include but not limited to policy formulation, leasing with and assessing the state of preparedness of all other relevant agencies, data collation from relevant agencies, education of the general public on flooding and interaction with SEMA towards the distribution of relief materials to disaster victims within states and local government areas. Recently, a memorandum of understanding was signed with NESREA and NOA to intensify efforts towards flood risk management in Nigeria [13].

The protection and development of the environment, a measure that helps keep check flooding, is the key role of NESREA, the agency which enforces all environmental laws, guidelines, policies, standards and regulations in Nigeria, as well as enforcing compliance with provisions of international agreements, protocols, conventions and treaties on the environment to which Nigeria is a signatory

Whilst the provision of humanitarian needs such as shelter, clothing and floods for internally displaced persons is anchored by NCR, the importance of local communities being aware of flooding and actively participating in discussions and decisions which might increase their resilience and adaptability to the hazard highlights in the roles of NOA, which re-orientates and keeps Nigerians informed about ways of taking part in issue that affects them. The poor perception of flooding in the country should be the concern of this agency.

NIHSA provides reliable and high quality hydrological and hydrogeological data on a continuous basis for the for the purposes of assessing the status and trends of the nation's water resources including its location in time and space, extent, dependability, quality and the possibilities of its utilization and control. Since 2013, the agency has been creating awareness of flooding through the "flood outlook" initiative. Other activities of NIHSA include; provision professional advice to various levels of government in Nigeria on all aspects of hydrology, collaborates with NIMET to issue flood forecast and contributes towards creating awareness of flooding among local communities. NIMET furnishes the country with weather report, and other meteorological information, issues alerts and early warning and forecast on impending flood disasters within the country.

Issues relating to flood insurance are coordinated by FEMA, the agency which makes federally funded insurance protection policy available for property owners in Nigeria. Policies relating to assisting flood victim at state and local government levels are coordinated by SEMA and LEMA. As climate change is complicit with other factors that influence flooding in Nigeria, BNRCC's key role is to collaborate with other agencies to promote the capacities of the generality of human populations within the country to cope with all effects of climate change. NEST undertakes continuous research required for enhancing decisions and robust measures towards addressing flooding in Nigeria [97,98].

Action by Local Communities and the General Public

The peculiar attitude of Nigerians towards issues they probably have no solution to is to find a possible natural means to adapt. Although such attitude has cost many lives and properties, however, it is arguably a significant potential for Nigerians and has severally favored them in emergency situations. Families in Nigeria co-habit and this offers a comparative advantage in the event of flood disasters. In many flooding incidences in Nigerian cities, the general public has often offered assistance to victims, assisted in evacuation of those displaced and in protecting property from further damage. Many IDPs easily find shelter and other humanitarian needs from families and friends while awaiting intervention by authorities. Although anti-social behavior, such as looting and sexual harassment of some of the internally displaced victims often arise, however the civilized attitudes of the general public, which may be comparable to those in the developed world such as the US following the hurricane Katrina of 2005 [99] has been commendable.

Action by Humanitarian Agencies

Humanitarian response to flooding in Nigeria dates back to the early 1980s following the Ogunpa, Sokoto-Rima, Banguada flood disasters [100]. Almost in all cases of flooding in Nigeria victims have received humanitarian supports with most notably the International Federation of Red Cross (IFRC), United Nations, World Bank, Foreign countries including UK, the United States, China, Japan, France as well as religious organizations including the Catholic, Anglican and Pentecostal churches and missionary societies. Although this study has not provided an empirical evidence of supports received so far in view of flooding in Nigeria, recent records by OCHA [60] indicate that such supports have been considerable. The 2012 flooding saw humanitarian response amounting to over US$70 million [60]. These supports which seem to indicate solidarity for Nigeria as a whole and the people affected by flooding within the country in particular have been evidently considerable.

Action by Research and the Media

Arguably, much of what is known about flooding in Nigeria can be attributed to efforts by research and media. In the literature, undoubtedly, considerable attention has been given to flooding in Nigeria. Presently there are more than five hundred publications indexed in the Google scholar that relates to flooding and means of tackling it in Nigeria. In addition, the country has a number of countrywide research-based groups such as the NEST (Nigeria Environmental Study/Action Team), BNRCC (Building Nigeria's Response to Climate Change) and university-based research groups that focus on flooding within Nigeria. Similarly, the Nigerian media has been given the credit of providing qualitative information regarding the widespread flooding in Nigeria [59].

GAPS IN FLOOD RISK REDUCTION IN NIGERIA

The fundamental idea of flood risk reduction is to build the resilience of human populations to flooding [44]. In the US, UK and Netherland, it can be argued that this idea is underpinned by much profound measures for tackling flooding. For Nigeria, there are potential efforts for tackling flooding but in the light of best practices in flood risk reduction, there are gaps and limitations which the present study identifies and discusses as follows.

Poor Attention to Flood Modelling and Assessment of Vulnerability to Flooding

The crucial gap in flood risk reduction in Nigeria is poor attention to flood modelling and assessment of vulnerability to flooding. The present study has shown that flood risk reduction under the idea of living with floods integrates structural and nonstructural approaches which are underpinned by flood modelling. Flood data is fundamental to tackling flooding in Nigeria, but the dearth of such data is overwhelming. Similarly, assessment of vulnerability to flooding aides the understanding of variations in human sensitivities, exposures and lack of capacities to cope with floods. These approaches are poorly considered in efforts at tackling flooding in Nigerian.

Weak Institutional Framework

Despite the seeming organized nature of Nigeria's institutional approach to flood risk reduction, the approach has been acutely flawed [12,24,97]. Although key roles within the approach has consisted of facilitating the evacuation of victims affected by floods and providing them with urgent humanitarian needs, the level of dissatisfaction and agitations from large numbers of the flood victims, especially the IDPs, queries the effectiveness of institutional approach in Nigeria. Although it is unjustifiable to claim that the weakness of this institutional approach in Nigeria probably leads to more frequent flooding in the country, however, it can be argued that whilst that such measures have not improved the country with regards to the idea of "living with floods" is clearly acknowledged [27,35,98].

Arguably, institutional framework with a complex chain of action which characterizes Nigeria is not ideal for a country with urgent needs to address the threats of flooding. The widespread flooding in Nigeria requires less complex framework with reduced chain of action and increased authority and responsibilities towards flooding and ways of managing its threats, as is the case in the United States, UK and Netherlands where a centralized authority such as FEMA, Environment Agency (EA) and Rijkswaterstaat respectively operates within the framework of institutional approach with more abundant resources towards addressing the challenges of flood risk.

Poor Human Responsibility

The vulnerabilities of local communities to flooding in Nigeria may indicate among other factors an overwhelming level of irresponsibility towards flooding and ways of addressing its challenges. Although a number of actions by the people towards flooding is commendable, some actions which characterize a cross section of Nigerian such as the failure to comply with environmental laws and regulations and to adhere to weather warnings and alerts are possible situations where lack of responsibilities of the general public is highlighted [53]. Poor perception of flooding among local communities is a major issue which underscores the current activities of NOA within Nigerian institutional framework [18,62]. The indifference towards research questionnaires and surveys most likely constrain research effort at tackling flooding in Nigeria. In Nigeria there seems to be a gap between efforts by the government and activities by the people such that it appears difficult to know the peoples' responsibilities towards addressing the challenges of flooding. The government claims the people are sabotaging its efforts. Whilst this claim is hard to substantiate, the attitude of a considerable population of Nigeria towards environmental management, regulations and rules in the country shows irresponsibility towards the threats of flooding.

Lack of Detailed Plan

We argue that the lack of detailed plan and strategy for disbursing funds and inaccurate information relating to those who have been affected by flooding most probably undermine humanitarian support in Nigeria. The disbursement of these funds and relief materials, although unjustifiable to argue that it highlights the level of corruption and financial mismanagement in Nigeria has been argued to indicate the weakness of institutional framework within the country [12]. Humanitarian actions in Nigeria are generally for post-disaster and emergency situations suggesting some limitation based on what can be achieved through financial support. Given that most local communities in Nigeria consist of poor human populations, it is arguably a topbottom approach to concentrate humanitarian supports in the event of crises rather than prioritizing improving the living condition of these people which will not only ultimately reduce their chances of being vulnerable to flooding and assist in minimizing financial mismanagement, but also it will boost the credence of humanitarian supports towards natural disasters in general and flooding in particular in Nigeria.

Insufficient Research and Skewed Focus of Local Education and Publications

Considering what has been done globally in research (mostly in China, India, UK, the US, Bangladesh and Netherlands) the specific situation of Nigeria regarding research in terms of flooding is arguably weak. The study of science and technology in Nigeria are yet to embrace environmental education and this has been identified [58]. Whilst flooding and climate change are passive subjects in education curriculum of studies in Nigerian schools, current key issues in flood research such as flood modeling, uncertainty analyses, early warning systems and flood forecasting, vulnerability assessment and climate change models are lacking in the country. From the literature search conducted in this study, it is clear that perception of flooding in Nigeria has only received little attention, and this seems to undermine possible efforts as assessing the level of perception and public awareness of flooding in Nigeria. A number of researches relating to flooding in Nigeria seem to recycle issues that are well known such as causes and impacts of flooding. More scientific investigations are few probably due to lack of funds and the indifference of political leaders.

Poor Media Reporting

The inconsistency in media reporting in Nigeria which was identified in [90] highlights some disconnect between the media and agencies tackling flooding in the country particularly the NEMA. The standard of media reporting in Nigeria following recent flood incidences has been very poor. This is compounded by the irregular power supply in the country which overwhelms the chances to take advantage of information gadgets such as radio and TV. A large number of the population seems to rely on Facebook pages, which also has its own challenges based on internet accessibility. This situation seems to have caused a general lack of confidence in the media generally such that a growing parlance among researchers in recent times regarding flooding is "of what relevance is what the media have to say?" Whilst this situation undermines the professionalism in Nigerian media industries in particular, a key question that lingers with regards to widespread flooding in the country is "how would the media industry contribute to addressing the challenges of flooding Nigeria?"

It is possible political factors and market competition to a large extent influence media reporting, but it also appears the importance of piloting the perception and awareness of flooding in the general public and local communities as well as sensitizing them on issues relating to flood alerts and early warning has been compromised.

RECOMMENDATIONS

Based on the findings of this study, lessons learned from other countries' experiences of flooding and "best practices" in flood risk reduction [100,101], the authors argue that the following key issues are fundamental to the success of efforts towards addressing the challenges of flooding in Nigeria. Firstly, the understanding and demonstration of the roles more scientific approaches such as flood modeling, can play in flood risk reduction within the context of Nigeria. Secondly, the need to align the focus of flood risk reduction in Nigeria to the objectives of such task in more developed countries such as the US, UK and the Netherlands, and thirdly, the need to promote awareness of flooding among local communities and the general public and to delineate more suitable locations for relocation of human populations during flooding events. For this reasons, the study makes the following recommendations:

- Following the European Union framework, which requires all constituting States to prepare flood hazard/ risk maps [102], a strong legislation that requires each state of Nigeria to produce a flood hazard/ risk map, is recommended for Nigeria. This will to a large extent strengthen existing institutional framework and stimulate increased responsibility towards flood risk reduction among the states in the country.
- The Environment Agency in United Kingdom is adequately empowered to coordinate, among other things, research which drives current approaches of addressing the challenges of flooding [83]. Thus, it is recommended that Nigeria's NEST be reviewed in terms of its research activities, how relevant they have been and how they can be further strengthened to promote more scientific researches towards ways of addressing the challenges of floods and other environmental challenges in Nigeria.
- The importance of a centralized institution having all the powers and resources to manage the challenges of flooding highlights the success of flood risk reduction in the Netherlands, the US and UK [103]. The roles of NEMA in Nigeria are acknowledged, but improvement is required. Lack of adequate resources especially accurate data clearly undermines the effectiveness of the agency in disaster management. Thus, we recommend improvement in resources allocation to NEMA, but also acquisition of data should be the agencies top priority.
- Hurricane Katrina, among other major disasters in the US showed the unique roles of security personnel in managing emergencies and securing victims of disasters. In Nigeria, the victims of flooding live in apprehension of theft and sexual abuse. These dimensions which if addressed can restrict the impacts of flooding on human populations, are often overlooked by security personnel in the course of duty. For this reason, we recommend training for rescue and evacuation operations during for the civil defense corps and other military and paramilitary personnel, including the police, immigration and road safety commission.
- • Flood risk reduction under the "living with floods" idea is multi-disciplinary indicating that various industries can assist in reducing the impacts of flooding. In UK, evidences of collaboration from various companies

and institutions towards addressing flood challenges are undisputable [83,101,103]. Thus, the need for multinationals and banking industries in Nigeria to sponsor research and promote sustainable development within Nigerian cities, as well as augment humanitarian supports to improve the living standards of local communities whilst reducing their vulnerabilities and building their resilience is should not be ignored.

- Flood insurance is a non-structural approach which many property owners have benefitted from in developed countries following flood disasters. To support the roles of flood insurance in Nigeria, it is recommended that the role of FEMA in this regard should be extended to states and whilst encouraging insurance companies to commence sensitization exercises for properties owners to take positive step in this direction.
- Enforcement of environmental standard and laws is often a key factor towards containing adverse effects of climate change including flooding. Indiscriminate waste disposal, production of Carbon monoxide, construction along flood plain, felling of trees and indiscriminate car parking, among other anthropogenic activities which influence flooding in Nigeria are illegal. In view of addressing these matters, NESREA should embark on arrest, prosecution and proportionately fine Nigerian who violate these rules.
- The flooding of 1953 is historic in the Netherlands, which with its half at or below sea level is reputed as the top in the world in terms of flood management [81]. Such activities in the Netherlands described as imperfect but pragmatic is built on a strong commitment to resist any attempt of a repeat of history. The people are committed, and so is the government implying that collective efforts underlie success towards addressing the challenges of flooding. The old English adage "God created the world, but the Dutch created Netherlands" is often used to highlight the commitment and responsibilities of people in the Netherlands towards flooding and its challenges. It is claimed that flood defense in the Netherlands cost each person a few hundred Euros each year and the people rarely flinched at the responsibility [81]. The high level of adherence to regulations and rules shown by British citizens is highlighted in the conservation of nature and high environmental standards within the country. Such positive attitude is also exhibited towards weather reports, disaster warning and alerts informing a significant level preparedness which appears to influence reduced damage following flooding event. Against this background, Nigerians need to known that the role of protecting the country is a collective one. Thus a change of attitude towards flooding its management in which the people should participate in matters relating to flooding which most largely affects their lives. This can be done by asking relevant questions, seeking to know and willing to adapt to individual actions which can potentially influence flood risk reduction within the country. Individuals in politics should ensure that laws which underlie the enforcement of environmental standards and regulations are made. Equally, the general public and local communities in Nigeria should support research through positive and accurate responses to questionnaire and surveys.
- Arguably, it appears research globally is proportional to success towards tackling flooding. From a routine Google scholar search, literature relating to flooding in Nigeria appears insignificant compared to those of the US, Netherlands, UK, etc. This is a strong pointer and indicator that more research is required for the country. Equally, the universities and research agencies should be empowered to include in their curriculum studies and programmes tailored towards improving flood awareness and management. Flood alert and flood early warning systems should be introduced and utilized at all levels of flood disaster management. More research should be directed towards developing bespoke hydrologic and hydraulic flood models for simulating flood hazard and other hydrological parameters in Nigeria.

The government should play more active parts in this regards. Research is capital intensive and many Universities and research agencies are constrained by this factor. Indeed, many universities have benefitted from the Tertiary Education Intervention (TETFUND). However, due to limited funds available, accessibility by a number of academics has been poor. Thus we recommend improvement on funding and acquisition of high accuracy input data like LiDAR (Light Detection and Ranging) and SAR (Synthetic Aperture Radar) for flood modeling within the country.

CONCLUSION

Flooding has been devastating for Nigeria with large scale threats on people, critical infrastructure and economic activities. The hazard which has been generally linked to climate change and poor urban planning has received the attention of government at all levels, local communities, humanitarian agencies and research. However, efforts so far at tackling the hazard seem limited mostly due to lack of data relating to flooding and other factors which are yet to be identified. Although flood hazard is widespread in the country, its knowledge in the wider population is poor.

To attempt towards solving the challenges of flooding in Nigeria, the present study reviews flooding and ongoing efforts at tackling the hazard in the country over a period 1985 till 2014, the present study reveals considerable impacts of flooding, arguing that such impacts have not been accompanied by corresponding efforts to address them. A number of limitations and gaps which needs to be addressed were identified with those ongoing efforts and they include poor attention to flood modelling and assessment of vulnerability to flooding, weak institutional framework, poor human responsibility, etc.

It is argued that flood modeling and vulnerability assessment, have both been relevant to all aspects of flood risk reduction [39,104], as they provide information about flooding and also serves as the bases for evaluating the susceptibilities of social systems to the hazard. Efforts towards addressing the challenges of flooding in most places for example the US and Europe include these more robust operations to achieve the three-tier risk management cycle, which includes mitigation/ prevention, preparedness, response and recovery. Unfortunately, the implementation of these operations in

Nigeria has been inadequate. Across the world, there are many predictions of the worsened threat of flooding in the future, which in turn is stimulating research into finding solutions. There seems to be some increased interest in preparing for future uncertainties with regards to exposure and vulnerabilities of social systems flood risks. For example in Europe, the EU adopted a directive in 2007 which requires all member States to carry out flood hazard and risk mapping in their territory, in order to support the transition from traditional flood defense strategies to a flood risk management approach at the basin scale [76]. In the US, despite the high level of flood events, based on available records, it can be argued that the country seems to have contained the degree of impacts on social systems at least for the present [4]. This may be as a result of various on-going initiatives especially in flood modeling (for example: HEC-1, HEC-RAS, and HEC-HMS). Given the proliferation of models existing at present for simulating flood hazard along with recent development in remote sensing and computer technology, it is now possible to undertake flood forecasting and flood early warning systems in many places [105].

However, looking at these developments vis-à-vis Nigeria, one wonders and asks "how prepared are the government and the people in this present and future fight against flooding?" If perhaps the situation with Nigeria is that of unpreparedness and limitations with regards to addressing the threats of flooding, them much apprehension should abound bearing in mind that the country has one of the fastest growing urban areas and population density in the world [9], and knowing that the risks of flooding will capitalize on those factors leaves more to be desired. It is not productive to attribute blame as to why flooding occurs, yet responsibility is required to mitigate the extent of damage and disruptions to economic activities. It is in view of this responsibility, which is often defined by awareness, and positive action that the present study has made a number of recommendations hoping that Nigeria will join her counterparts in various parts of the world to become more proactive in finding lasting solutions to the threats of flooding.

Acknowledgements

This paper results from a PhD research funded by the UNN/TETFUND. Some grant was equally provided by the Surveyors Council of Nigeria (SURCON). Most of the analyzed flood data were sourced from the center for research in epidemiology of disasters (CRED/EM-DAT), Nigeria's ministry of environment and Nigeria Environmental Study Action/Team (NEST). A number of previous studies which were consulted and the anonymous reviewers of this paper are acknowledged.

REFERENCES

1. Peduzzi P, Dao H, Herold C, Mouton F (2009) Assessing global exposure and vulnerability towards natural hazards: the disaster risk index, Natural. Hazards Earth System. Science 9: 1149-1159.
2. http://www.art.man.ac.uk/planning/cure
3. Raaijmakers R, Krywkow J, van der Veen A (2008) Flood risk perceptions and spatial multi-criteria analysis: an exploratory research for hazard mitigation. Natural hazards46: 307-322.
4. Guha-Sapir D, HoyoisPh, Below R (2013) Annual Disaster Statistical Review 2012: The Numbers and Trends. Brussels: CRED.
5. http://www.emdat.be/
6. Ahern M, Kovats RS, Wilkinson P, FewR, Matthies F (2005) Global health impacts of floods: epidemiologic evidence. Epidemiologic reviews27: 36-46.
7. Tapsell SM, Tunstall SM (2008) "I wish I'd never heard of Banbury": The relationship between 'place' and the health impacts from flooding. Health & place14: 133-154.
8. Hunter PR (2003) Climate change and waterborne and vector-borne disease. Journal of Applied Microbiology94: 37-46.
9. Tapsell SM, Penning-Rowsell EC, Tunstall SM, Wilson TL (2002) Vulnerability to flooding: health and social dimensions. Philosophical Transactions of the Royal Society of London. Series A: Mathematical. Physical and Engineering Sciences360: 1511-1525.
10. Tunstall S, Tapsell S, Green C, FloydP, George C (2006) The health effects of flooding: social research results from England and Wales. J Water Health4: 365-380.
11. ActionAid (2006) Climate change, urban flooding and the rights of the urban poor in Africa: Key findings from six African cities. London: Action Aid International.
12. Obeta CM (2014) Institutional Approach to Flood Disaster Management in Nigeria: Need for a Preparedness Plan. British Journal of Applied Science & Technology 4: 4575-4590.
13. NEMA (Nigeerina Emergency Manageent Agency) (2013) Report on flood disasters in Nigeria. Abuja: Government Press.
14. Adeloye AJ, Rustum R (2011) Lagos (Nigeria) flooding and influence of urban planning. Urban Design and Planning 164 (DP3): 175-187.
15. Cline WR (2007) Global Warming and Agriculture. Centre for Global Development, Peterson Institute for International Economics, Washington, DC, USA.
16. www.infoo@climatechange.org
17. Ogunbodede EF, Sunmola RA (2014) Flooding and traffic management in Akure (Nigeria) metropolitan environment. International Journal of Innovation and Scientific Research7: 121-130.
18. Ologunorisa TE, Adeyemo A (2005) Public perception of flood hazard in the Niger Delta, Nigeria. Environmentalist25: 39-45.
19. Fadairo G, Ganiyu SA (2010) Effects of flooding on the built environment in Akure, Nigeria. In West Africa built environment research (water) conference. pp. 281.
20. Bashir O, Oludare H, Johnson O, Aloysius B (2012) Floods of fury in Nigerian cities. Journal of Sustainable Development 5: 69-79.
21. Douglas I, AlamK, Maghenda M, Mcdonnell Y, McLean L, et al. (2008) Unjust waters: climate change, flooding and the urban poor in Africa. Environment and Urbanization20: 187-205.

22. Akintola EO (1994) Flooding phenomenon. Ibadan Region. Ibadan: Rex Charles Publication, pp.244-255.
23. Iloje NP (2004) A New Geography of Nigeria. Lagos, Longman publishers.
24. Agbola BS, Ajayi O. Taiwo OJ, Wahab BW (2012) The August 2011 flood in Ibadan, Nigeria: anthropogenic causes and consequences. International Journal of Disaster Risk Science 3: 207-217.
25. Adebayo AA, Oruonye ED (2013) An Assessment of the effects of the 2012 Floods in Taraba State, Nigeria. Paper delivered at the Annual National Conference, organized by the Association Hydrological Science at University of Agriculture, Abeokuta, Ogun, state. Nigeria 13-18.
26. Obeta MC (2009) Extreme river flood events in Nigeria: A geographical perspective of Nigerian. Journal of Geography and the Environment 1: 170-179.
27. Adelekan O (2010) Vulnerability of poor urban coastal communities to flooding in Lagos, Nigeria. Environment and Urbanization 22: 433-450.
28. Nicholls R, Hanson JS, Herweijer C, Patmore N, Hallegatte S, et al. (2008) Ranking of Port cities with high exposure and vulnerabilities to climate extremes. OECD Environment Working Papers, No.1. University of Southampton, United Kingdom: OECD Publishers.
29. Houston D, Werritty A, Bassett D, Geddes A, Hoolachan A, et al. (2011) Pluvial (rain-related) flooding in urban areas: the invincible hazard. York, UK: Joseph Rowntree Foundation.
30. Merz B, Thieken AH, Gocht M (2007) Flood risk mapping at the local scale: concepts and challenges. In Flood risk management in Europe. Springer Netherlands.
31. Ujoh MF, Isa D, Olarewaju O (2009) Understanding urban sprawl in the Federal Capital City, Abuja: Towards sustainable urbanization in Nigeria. Journal of Geography and Regional Planning 3: 106-113.
32. Mujumdar PP (2001) Flood wave propagation: The St. Venant equations. Resonance 6: 66-73.
33. Shuster WD, Bonta J, Thurston H, Warnemuende E, Smith DR (2005) Impacts of impervious surface on watershed hydrology: a review. Urban Water Journal 2: 263-275.
34. Kalnay E, Cai M (2003) Impact of urbanization and land-use change on climate. Nature 423: 528-531.
35. Seto KC, Shepherd JM (2009) Global urban land-use trends and climate impacts. Current Opinion in Environmental Sustainability 1: 89-95.
36. UNSTTPUNDA (United Nations System Task Team on the Post-2015 UN Development Agenda). (2012) Disaster and Resilience. United Nations. New York.
37. UN (United Nations) (2007) World Population Prospects: the 2006 revision - executive summary. Department of Economic and Social Affairs, population division, United Nations. New York.
38. Huong HTL, Pathirana A (2011) Urbanization and climate change impacts on future urban flood risk in Can Tho city, Vietnam. Hydrology and Earth System Sciences Discussions 8: 10781-10824.
39. Li Y, Wang C (2009) Impacts of urbanization on surface runoff of the dardenne creek watershed, St. Charles county, Missouri. Physical Geography 30: 556-573.
40. Olorunfemi FB (2011) Managing flood disasters under a changing climate: lessons from Nigeria and South Africa.NISER Research Seminar Series, NISER, Ibadan.
41. BalbiS, Giupponi C, Gain A, Mojtahed V, Gallina V, et al. (2012)A Conceptual Framework for Comprehensive Assessment of Risk Prevention Measures: The KULTURisk Framework (KR-FWK). Available at SSRN 2184193.
42. Di Baldassarre G, Uhlenbrook S (2012) Is the current flood of data enough? Atreatise on research needs for the improvement of flood modelling. Hydrological Processes 26: 153-158.
43. DEFRA (Department for Environment Food and Rural Affairs) (2013). Desktop review of 2D hydraulic modelling packages. Environmental Agency. Bristol.
44. http://www.unisdr.org/files/657_lwr1.pdf
45. EC (European Commission) (2004) Flood risk management - Flood prevention, protection and mitigation. Communication from the Commission to the Council, the European Parliament, the European Economic and Social Committee and the Committee of the Regions.
46. ZhuD, Zhou N, Jiang S (2011) Research overview of runoff model for urban rainwater. Journal of Water Resources and Water Engineering 22: 132-137 (in Chinese).
47. http://scholarship.kentlaw.iit.edu/cgi/viewcontent.cgi?article=1772&context=fac_schol
48. Ologunorisa TE (2004) An assessment of flood vulnerability zones in the Niger delta,Nigeria. International Journal of Environmental Studies 61: 31-38.
49. Ojigi ML, Abdulkadir FI, Aderoju MO (2013) Geospatial Mapping and Analysis of the 2012 Flood Disaster in Central Parts of Nigeria. 8th National GIS Symposium. Dammam. Saudi Arabia 1067-1077.
50. Aderogba KC, Martins OM, Oderinde S, Afelumo T (2012) Challenges of Poor Drainage Systems and Floods in Lagos Metropolis, Nigeria. International Journal of Social Science and Education 2: 412-427.
51. Ogwuche JA, Abah IA (2014) Assessment of flood disaster vulnerability for flood insurance programme in part of Makurdi floodplain, Benue state, Nigeria. Donnish Journal of Ecology and the Natural Environment 1: 1-5.
52. Nwilo PC, Olayinka DN, Adzandeh AE (2012) Flood modelling and vulnerability assessment of settlements in the Adamawa state floodplain using GIS and cellular framework. Global Journal of Human Social Science 12: 10-20.
53. Aderogba KA (2012a) Qualitative studies of recent flood and sustainable growth and development of cities and towns in Nigeria. International Journal of academic research in Economics and management science 1: 11-25.
54. Adeoye NO, Ayanlade A, Babatimehin O (2009) Climate

change and menace of floods in Nigerian cities: Socio-economic implications. Advances in Natural and Applied Sciences 3: 369-377.

55. Ali D, Hamidu S (2014) Environmental hazard: climate change and flooding, the impact on the built environment in Nigeria. Journal of Environmental Sciences and Resources Management 6: 136-144.
56. Agbonkhese O, Agbonkhese EG, Aka EO, Joe-Abaya J, Ocholi M, et al. (2014) Flood Menace in Nigeria: impacts, remedial and management strategies. Civil and Environmental Research 6: 32-40.
57. Adedeji OH, Odufuwa BO, Adebayo OH (2012) Building capabilities for flood disaster and hazard preparedness and risk reduction in Nigeria: need for spatial planning and land management. Journal of Sustainable Development in Africa 14: 45-58.
58. Terungwa UC, Torkwase IC (2013) Current issues in flood disaster: challenges and implications for science and technology to enhance environmental education. Academic Journal of Interdisciplinary Studies 2: 61-65.
59. Aderogba KA (2012b) Global warming and challenges of floods in Lagos metropolis, Nigeria. Academic Research International 2: 448-468.
60. http://www.unocha.org/
61. Adeaga O (2008) Flood Hazard Mapping and Risk Management in Parts of Lagos N. E.
62. Ajibade I, McBean G, Bezner-Kerr R (2013) Urban flooding in Lagos, Nigeria: Patterns of vulnerability and resilience among women. Global Environmental Change 23: 1714-1725.
63. Pinho J, Ferreira R, Vieira L, Schwanenberg D (2014) Comparison between two hydrodynamic models for flooding simulations at river Lima basin. Water Resources Management 1-14.
64. Thompson JR, Hollis GE (1995) Hydrological modelling and the sustainable development of the Hadejia-Nguru wetlands, Nigeria. Hydrological Sciences Journal40: 97-116.
65. Adewale PO, Sangodoyin AY, Adamowski J (2010) Flood routing in the Ogunpa River in Nigeria using HEC-RAS. Journal of Environmental Hydrology 18.
66. Onwuteaka J (2014) GIS Modeling of Flooding Exposure in Nigerian Coastal Areas from Sea Level Rise. Journal of Environment and Earth Science4: 81-94.
67. Nwilo PC, Olayinka DN, Adzandeh AE (2012) Flood Modelling and Vulnerability Assessment of Settlements in the Adamawa State Floodplain Using GIS and Cellular Framework Approach. Global Journal of Human Social Science.
68. van de Sande B, Lansen J, Hoyng C (2012) Sensitivity of Coastal Flood Risk Assessments to Digital Elevation Models. Water4: 568-579.
69. Nkwunonwo UC, Whitworth M, Baily B, Inkpen R (2014) The development of a simplified model for urban flood risk mitigation in developing countries. In Vulnerability, Uncertainty, and Risk@ Quantification, Mitigation, and Management. ASCE.
70. Wolski P, Savenije HHG, Murray-Hudson M, Gumbricht T (2006) Modelling of the flooding in the Okavango Delta, Botswana, using a hybrid reservoir-GIS model. Journal of Hydrology331: 58-72.
71. Garba H, Ismail A, Ibrahim A, Ahmed S, Faustinus B (2013) Hydrological Modeling of the impact of climate change on a tropical perennial river flooding. Research Inventy: International Journal of Engineering and Science. 3: 30-35.
72. Adegbola AA, Jolayemi JK (2012) Historical rainfall-runoff modeling of river Ogunpa, Ibadan, Nigeria. Indian Journal of Science and Technology5: 2725-2728.
73. Sayers P, Li Y, Galloway G, Penning-Rowsell E, Shen F, et al. (2013) Flood risk management: a strategic approach. Paris, UNESCO.
74. Sayers PB, Hall JW, Meadowcroft IC (2002) Towards risk-based flood hazard management in the UK. Proceedings of ICE Civil Engineering 150: 36-42.
75. Smith K (2013) Environmental hazards: assessing risk and reducing disaster sixth edition. Routledge. London.
76. deMoel H, van Alphen J, Aerts JC (2009) Flood maps in Europe – methods, availability and use. Natural Hazards and Earth System Sciences 9: 289-301.
77. de Almeida GA, Bates P, Freer JE, Souvignet M (2012) Improving the stability of a simple formulation of the shallow water equations for 2-D flood modeling. Water Resources Research 48: 1-14.
78. http://www.nihydro.gov.ng/wp-content/uploads/2012/08/AMENDED-REPORT-OF-The-Nigerian-Hydrological-Services-Agency.pdf
79. World Bank (2013) The world Bank: Working for a world free of poverty, Population (Total). Washington, DC: World Bank Group.
80. Jonkman SN, Bočkarjova M, Kok M, Bernardini P (2008) Integrated hydrodynamic and economic modelling of flood damage in the Netherlands. Ecological Economics66: 77-90.
81. Vis M, Klijn F, De Bruijn KM, Van Buuren M (2003) Resilience strategies for flood risk management in the Netherlands. International journal of river basin management1: 33-40.
82. Neal J, Schumann G, Fewtrell T, Budimir M, Bates P, et al. (2011) Evaluating a new LISFLOOD-FP formulation with data from the summer 2007 floods in Tewkesbury, UK. Journal of Flood Risk Management4: 88-95
83. EA (Environment Agency) (2010) Working with natural processes to manage flood and coastal erosion risk. London, Environment Agency.
84. Renyi L, Nan L (2002) Flood area and damage estimation in Zhejiang, China. Journal of environmental management66: 1-8.
85. National Population Commission (NPC) (2007) 2006 Population and Housing Census: National and State Population and Housing Tables: Priority Tables I-IV. FCT, Abuja: Federal Government of Nigeria.
86. United Nations (2004) World Population to 2300. New York: United Nations Department of Economic and Social Affairs, Population Division.

87. Clement AR (2013) An application of Geographic Information System in mapping flood risk zones in a north central city in Nigeria. African Journal of Environmental Science and Technology7: 365-371.

88. Uzochukwu CE, Ekwugha UP, Marion NE (2014) Media coverage of the environment in Nigeria: issues and prospects. International Journal of Social Sciences and Humanities Review4: 1-10.

89. Ochuko MO (2014) Solid waste management in Obantoko area of Abeokuta, Nigeria. Journal of Emerging Trends in Engineering and Applied Sciences5: 111-115.

90. http://www.e-ir.info/2013/02/04/the-inconsistency-of-the-flood-narrative-in-nigeria/

91. http://www.fig.net/pub/fig2011/papers/ts06j/ts06j_etuonovbe_5002.pdf

92. FME (Federal Ministry of the Environment) (2012) Bulletin on Ecological Disasters, Abuja, Nigeria. FME.

93. Zhang Q, Liu C, Xu CY, Xu Y, Jiang T (2006) Observed trends of annual maximum water level and stream flow during past 130 years in the Yangtze River basin, China. Journal of Hydrology324: 255-265.

94. EA (Environment Agency). (1997) Tyne Estuary Project 1996. Report of the Tyne Estuary Group. Newcastle.

95. Odunuga S (2008) Urban land use change and the flooding in Ashimowu watershed, Lagos, Nigeria. University of Lagos, Nigeria.

96. Ibitoye M (2007) The need for planning of peri-urban growth in south western Nigeria: the surveyors' perspective. Symposium organized by the Nigerian Institution of Surveyors (NIS). University of Lagos, Lagos.

97. Kolawole OM, Olayemi AB, Ajayi KT (2011) Managing flood in Nigerian cities: Risk analysis and adaptation options–Ilorin city as a case study. Archives of Applied Science Research3: 17-24.

98. Akintola FO, Ikwuyatum GO (2012) Issues in sustainable flood management in Nigeria. In Ivbijaro FA, Akintola FO. Editors; Sustainable Environmental Management in Nigeria. Ibadan, Oxford University Publishers.

99. Wilson S, Temple B, Milliron M, Vazquez C, Packard M, et al. (2007)The lack of disaster preparedness by the public and its effect on communities. The Internet Journal of Rescue and Disaster Medicine. 7.

100. Anih SC (2004) Effective survival measures against natural hazards in settled area. In L. Mbah, U. Uchegbu, & L. Muoghalu, Management of environmental problems and hazards in Nigeria. England: Ashgate publishing limited.

101. Water UK (2008) Lesson Learned from Summer Floods 2007-Phase 1 report-Emergency Response.Review Group on Flooding.

102. EA (Environment Agency) (2003) Risk Assessment for Flood & Coastal Defence for Strategic Planning R&D Technical Report W5B-030/TR1, Authors Sayers PB, Hall JW, Meadowcroft, IC, Dawson, R, Deakin R, Rosu C.

103. Pitt M (2008) Lessons from the 2007 floods. London: Pitt Review.

104. Fewtrell L, Kay D (2008) An attempt to quantify the health impacts of flooding in the UK using an urban case study. Public Health122: 446-451.

105. Bates PD, Dawson RJ, Hall WJ, Horritt MS, Nichollos RJ, et al. (2005) Simple two-dimensional numerical modelling of coastal flooding and example applications. Coastal Engineering 52:793-810.

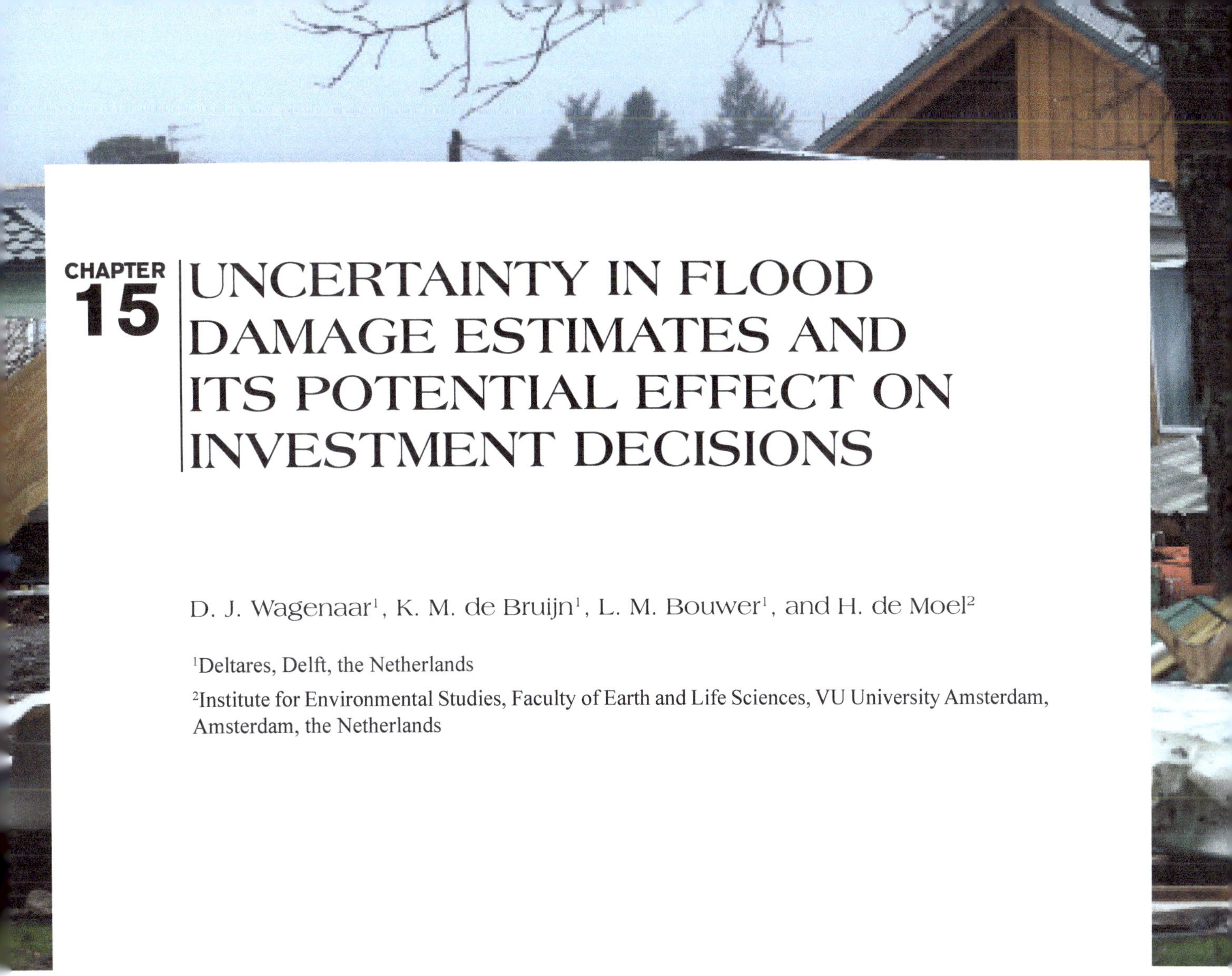

CHAPTER 15

UNCERTAINTY IN FLOOD DAMAGE ESTIMATES AND ITS POTENTIAL EFFECT ON INVESTMENT DECISIONS

D. J. Wagenaar[1], K. M. de Bruijn[1], L. M. Bouwer[1], and H. de Moel[2]

[1]Deltares, Delft, the Netherlands

[2]Institute for Environmental Studies, Faculty of Earth and Life Sciences, VU University Amsterdam, Amsterdam, the Netherlands

ABSTRACT

This paper addresses the large differences that are found between damage estimates of different flood damage models. It explains how implicit assumptions in flood damage functions and maximum damages can have large effects on flood damage estimates. This explanation is then used to quantify the uncertainty in the damage estimates with a Monte Carlo analysis. The Monte Carlo analysis uses a damage function library with 272 functions from seven different flood damage models. The paper shows that the resulting uncertainties in estimated damages are in the order of magnitude of a factor of 2 to 5. The uncertainty is typically larger for flood events with small water depths and for smaller flood events. The implications of the uncertainty in damage estimates for flood risk management are illustrated by a case study in which the economic optimal investment strategy for a dike segment in the Netherlands is determined. The case study shows that the uncertainty in flood damage estimates can lead to significant over- or under-investments.

INTRODUCTION

Flood damage assessment is an essential aspect of flood risk management (Merz et al., 2010). It is used for supporting policy analysis and flood insurance. In the Netherlands flood damage estimates are used, for example, to determine economic optimal protection standards for flood defenses (van der Most et al., 2014), prioritize investments (Jongejan and Maaskant, 2013) or to compare the impact of different flood risk management strategies (Kind et al., 2014). The most commonly used method for flood damage assessment is the unit loss method (De Bruijn, 2005). This method assesses the damage for each unit separately. This assessment is based on a maximum damage per object and a damage function. A damage function describes the relationship between a flood characteristic (most often water depth) and the fraction of the economic loss that occurs to the object that is damaged. There are many different flood damage models all based on the unit loss method (e.g., HIS-SSM for the Netherlands (Kok et al., 2005), Multi-Coloured Manual in the UK (Penning-Rowsell et al., 2005), HAZUS in the USA (Scawthorn et al., 2006) and FLEMO in Germany, Thieken et al., 2008; Kreibich et al., 2010). These models differ for good reasons. Each model is specifically derived for a specific country, region and/or flood type and tailored to characteristics of the flooding and objects in the considered region (Cammerer et al., 2013).

Citation: Wagenaar, D. J., de Bruijn, K. M., Bouwer, L. M., and de Moel, H.: Uncertainty in flood damage estimates and its potential effect on investment decisions, Nat. Hazards Earth Syst. Sci., 16, 1-14, doi:10.5194/nhess-16-1-2016, 2016.

When these different models are applied to one and the same event, they will yield significantly different results (De Moel and Aerts, 2011; Jongman et al., 2012; Chatterton et al., 2014). Jongman et al. (2012) compared the damage outcomes of seven different flood damage models with the recorded flood damages from events in the UK and Germany. The difference between the smallest and largest estimate/recording was a factor of 5 for the German event and a factor of 10 for the event in the UK. Chatterton et al. (2014) compared two different damage assessments for a region in the UK. The damage estimates differed by about a factor of 5 to 6 for both residential and commercial damages.

These large differences in outcomes, for events for which the different models all should be applicable, indicate that flood damage estimation is prone to large uncertainties and thus that potentially large errors can occur when flood damage models are applied. The uncertainties and potential errors in damage estimates affect decision-making based on those damage estimates. A quantification of the uncertainty in the damage estimates can help to get an insight in the potential error that can occur in a decision based on the flood damage estimate and may improve the decision-making process. USACE (1992) and Peterman and Anderson (1999) both showed that taking ranges of uncertainty into account can lead to different decisions than using single value estimates. Furthermore, uncertainty quantification is useful to expose key focus points for the improvement of flood damage estimation methods. To reduce uncertainties, additional effort may be needed in researching the flood damages or in collecting data on damaged objects during floods.

Previous uncertainty analysis did not show a common understanding of the size or cause of uncertainties, which also indicates that further research is needed. Generally, uncertainty in flood damage assessment is quantified with forward uncertainty propagation methods which use Monte Carlo simulations (Merz et al., 2004; Egorova et al., 2008, Apel et al., 2008; De Moel et al., 2012). The results of Egorova et al. (2008) indicate moderate uncertainties, which is in contrast with the large differences between flood damage models that were found by De Moel and Aerts (2011), Jongman et al. (2012) and Chatterton et al. (2014). This paper provides a method to get a robust estimate of the uncertainty in damage estimates based on an analysis of the cause of the large differences between the various existing damage models. The method is illustrated with clear hypothetical examples and then applied to a case to show its use in decision-making for protection standards. The method makes use of a damage function library with 272 damage functions from seven different flood damage models. The paper focuses on direct material damage. Indirect damages including damages due to business interruption are not considered here, since their analysis requires different methods. The paper starts with a qualitative analysis of the uncertainty found in flood damage models. This qualitative analysis is the basis for the assumptions made in a Monte Carlo analysis which is used to quantify uncertainty. Next, the Monte Carlo analysis is described and discussed in detail. Finally, the Monte Carlo analysis is applied to a case study in the Netherlands and the resulting uncertainties in damage estimates and the effects on flood risk management decisions are discussed.

QUALITATIVE UNCERTAINTY ANALYSIS

This section first provides a detailed description of unit loss flood damage models and then places all elements of such models into a framework for uncertainty classification in order to generate a detailed qualitative understanding of the uncertainty sources and effects and their correlation. The understanding is applied in Sect. 3 to enable a quantitative uncertainty analysis.

The Unit Loss Method for Flood Damage Assessment

The unit loss method uses relationships between flood characteristics and damages to a unit. The unit loss method consists of four elements: the maximum damage s_i for each category, the flood characteristics (such as water depth d) at all locations j , the damage functions (f (d)) for all categories which determine the damage fraction and the number of objects affected n. Damage of an area is assessed as the sum of all damage categories i for all grid cells n by the following formula (Egorova et al., 2008).

$$\text{Damage} = \sum_{i=1}^{m} s_i \sum_{j=1}^{n} f_{ij}\left(d_j\right) n_{ij} \tag{1}$$

Potentially relevant flood characteristics are the maximum water depth, flood duration, flow velocity, pollution, warning time and other possible aspects of the flood. Often, only the water depth is used in flood damage modeling, occasionally supplemented by one or two other parameters. The uncertainties in the flood characteristics which are used as input for the damage estimation are not part of this paper

Damage is usually calculated for categories such as houses, industries, commercial companies, roads and agriculture. These object categories differ in maximum damage and flood damage functions. The object and flood characteristics are linked by damage functions which give the fraction of the maximum damage which occurs as a function of the flood intensity. The damage fraction is then multiplied by the maximum damage to get the damage. (Some methods such as the one in the Multi-coloured Manual (Penning-Rowsell et al., 2005) use absolute damage functions which relate the flood intensity directly to the damage and not to a fraction of the maximum damage.) The maximum damage can be defined in different ways. In this analysis we define the maximum damage as the expected damage corresponding with an extreme water depth. This means that the damage function will reach the value of one/unity for the most extreme water depths and it means that the maximum damage already holds information about what part of the total value of the object or unit is susceptible to flood damage. The maximum damage does not include value which is not, or unlikely to be, susceptible to floods, such as the value of the land surface, the costs of building foundations or the value present on high floors in buildings that are unlikely to collapse. Not all damage methods use this definition. Some include more items in the maximum damage and apply damage functions which never reach the value of one if part of that value is on average not susceptible to flooding.

When comparing different models, the definitions of maximum damage and damage functions first need to be aligned to make a fair comparison. In this paper we discuss flood damage models, which we define as a set of maximum damages, damage functions, object data and their relationships with which a damage estimate for a flooding in a certain area can be made.

Types of Uncertainty

In the uncertainty analysis in flood damage assessment two types of uncertainty are distinguished: aleatory and epistemic uncertainty (Merz and Thieken, 2009). Aleatory uncertainty is related to the variability or heterogeneity within a population which can be expressed by statistic parameters such as the mean, variance and skewness. This uncertainty is introduced by using average data: we use the maximum damage value of an average residence, although we know that some houses will suffer more, and other will suffer less damage. In small flood events which only affect a few houses, these few houses may differ significantly from the "average house" and therefore the damage estimate for these houses is uncertain. In large flood events which affect many houses, it is likely that deviations from the mean damage cancel out. This means that for large floods this type of uncertainty is of lesser importance.

Aleatory uncertainty by using averages can sometimes be reduced by applying more differentiation, e.g., the uncertainty within the maximum damage of a residence is reduced by using more differentiation in house types. The variation in maximum damage per house type would then be less than if all houses together were considered as one category "houses". Epistemic uncertainty is the lack of understanding of a system and can in theory be reduced by further study or by collecting more or better data. In other words, also the average damage itself is not certain. For flood damage assessments, data are only available for a small number of events and those events often differ significantly from each other. This variation between events is still poorly understood and is therefore related to epistemic uncertainty. The epistemic uncertainty as stated above is not reduced when many objects are flooded. Therefore, it is the dominant uncertainty type for large flood events. This type of uncertainty is especially relevant when a damage module developed for one area is applied to another area. In such a case, for example, the maximum damage values within the model related to houses may not be valid for the types of houses in the area under consideration. It would therefore be good not to mix up maximum damages and damage functions from different areas. However, given the scarcity of data and flood damage models, this leaves many modelers with the difficult choice between damage functions and maximum damages based on recorded data from another area or a local estimate.

Uncertainty in the Unit Loss Method

Uncertainty in the unit loss method consists of uncertainty in object data, in maximum damage figures and in the damage functions. Table 1 shows an overview of the uncertainties present in these aspects of the unit loss method.

Uncertainty In Object/Land Use Data

Uncertainties are found in the quantity of objects and their precise location. The precise location of objects is important, since the flood hazard characteristics (e.g., depth, or flow speed) may differ substantially from location to location. Each object should be linked to the hazard value present at the location of the object. The effect of the uncertainty in the precise location of objects on damage estimates is smaller for more homogenous hazards. For example, in a deep, flat polder, the exact location of an object is not important because the water depth is approximately the same everywhere. When a hazard becomes more heterogeneous the uncertainty in the exact location becomes more relevant.

Geographical location data uncertainties are especially significant in areas which flood frequently but have small water depths because, in this type of area, a small error in the location or height of an object can make the difference between an object getting wet frequently or very rarely. Furthermore, valuable objects susceptible to flood damage are unlikely to be placed in a location that floods frequently, so it is much more likely to count too many wet objects than to count too few. Therefore, damage estimates may be very wrong if the approach used was too coarse to discover local elevations or the exact locations of objects. For example, the Dutch standard damage model HIS-SSM estimated EUR 100 million damage for an event in an unprotected area that in reality had only caused about EUR 30 000 in damages (price level 2012) (Slager et al., 2013). Such errors are, however, unlikely when objects are not elevated on purpose, placed on safe locations or protected in other ways. Without this local protection of some objects, the damage will be overestimated, and for others, it will be underestimated. If many objects are affected, these errors compensate each other which reduces the uncertainty in the total damage. Use of high-resolution elevation information is also very useful to reduce this uncertainty (Koivumäki et al., 2010).

Uncertainty in the quantity of objects can be caused by errors in data, or by using data sources that are inappropriate for the intended application. This uncertainty depends on the quality of the data set that is used. De Moel and Aerts (2011) illustrated that this type of uncertainty may be small as they showed that different types of land use maps for the same area only have a small impact on the resulting damage estimate. In the uncertainty quantification for this paper, the uncertainty in the geographical location data is neglected.

Table 1. Overview of the uncertainties in flood damage modeling

Element	Uncertainty	Type	Expected significance	Included in analysis
Object data	Quantity	Both	Depends on input data, expected to be often in-significant	No
	Location	Both	Depends on area, often insignificant	No
Maximum damage	Value of the object	Mostly aleatory	Varies	Yes
	Susceptible to flood damage	Mostly epistemic	Significant	Yes
Damage function	Parameter representation	Both	Significant	Yes
	Knowledge/data about damage	Epistemic	Significant	Yes

In addition, objects can be represented in different ways and each way can cause different uncertainties. A company office, for example, can be represented by either the floor space in the office, the footprint area of the office building, an area larger than the office building on a rough land use map or the number of jobs within the office. All these object representations correlate in some way with value present in the building but the indicator will not precisely correspond with the value of the building. The uncertainty this causes is aleatory, because if the indicator overestimates the value at one point it will underestimate it somewhere else, assuming that the maximum damage is a good average.

Uncertainty in the Maximum Damage Figure

The uncertainty in the maximum damage figure can be divided into two parts: the uncertainty in the value of the object and in the part of that value that is susceptible to flood damage. There are generally two ways to obtain the maximum damage for a flood damage model: deriving this from economic data or by looking at synthetic (hypothetical) buildings.

Economic data typically provide a total value per sector of all physical assets in the economy. To obtain a maximum damage figure per unit, this total value can be divided by the number of units within that sector. Next, the part of this total value which is susceptible to flooding must be identified. The strength of this method is that the mean object value will be accurate. However, uncertainty is still present in the part of the object that is susceptible to flood damage. A similar method is to use average construction costs and to correct this for the fraction that is actually susceptible to flooding. Alternatively, the maximum damage of a category can be obtained by defining a hypothetical average company or object, and assessing the damage of all parts/aspects within that hypothetical company. The strength of this method is that the damage function and the maximum damage are well connected. Furthermore, the part of the value that is susceptible to flood damage is determined in a systematic way. The disadvantage of this approach is that epistemic uncertainty is introduced in the value of the object as the assumptions about this may be wrong.

Because generally a lot of good data are available on the value of objects, the uncertainty in the value of objects is considered to be aleatory. The uncertainty in the part of the value which is susceptible to flooding is epistemic, because a small amount of data or knowledge about that is available.

Uncertainty in Damage Functions

Damage functions can be obtained in two ways: by analyzing data on observed damage to objects and flood characteristics in past flood events, or by defining hypothetical average objects and assessing their damage corresponding to different flood intensities. A combination of both approaches may also be used. Flood damage data are rarely collected in a systematic way (Thieken et al., 2005) and are not always available for research. When available, data are often limited to a single or a few events. These events are often not representative of other types of floods or other countries or areas. Cultural or geographical differences can cause the use of different building or interior materials between regions and events, making one data set not applicable to other areas. Another problem is that data are often limited to certain ranges of a flood parameter. For example, data may be only available for low water depths or the flood that was the source of the data may have coincided with a storm. In such cases the data cannot be used for events with larger water depths or no storm.

In general, transferring data from one event to another is error-prone. This makes it very difficult to apply knowledge derived from one event to another. Even if the data are applied to the same area as the data were taken from, problems may arise. Different flood events in the same area may lead to very different damage due to different human responses. For example, the same area in the Netherlands flooded in 1993 and 1995 with approximately the same water levels. The second time, the damage to housing content was about 80 % less (Wind et al., 1999). Also, the damage due to the Rhine floods of 1995 was less than half of the damage that occurred in 1993, as a result of precaution measures taken by households (Bubeck et al., 2012). This shows the sensitivity of flood damage to factors other than water depth. These other factors (in this case, flood experience) are often neglected in the recordings. This example shows that a data set based on a small number of events does not capture all possible variable values. Synthetic damage functions solve many of the problems of having too few empirical data on actual damages. In this method, a hypothetical building is defined and flood damage is assessed for each building part. The hypothetical building should be representative of an average building in the area. If it is not, or if the damage estimates for the different building parts are not right, the damage function is inaccurate. Damage data can also be combined with expert knowledge. Probably the most common method to create a damage model is by picking and choosing damage functions from other models based on an analysis of which existing damage function best represents the area considered. Or, the average between different functions could be used as a damage function. The challenge with this combined method is to understand the background assumptions between the models that are brought together or compared. For example, a common challenge may be that the maximum damage definitions do not match.

The ideal case is to combine the best of the two methods. The damage data available should be used to calibrate a synthetic model. This limits the possibility that large errors are made in the interpretation of the damage data (e.g., wrong definition of the maximum damage), by forcing the modeler to think about the processes. Furthermore, it gives the modeler the freedom to diverge from the observed data in situations that do not match any of the recorded events. A common problem in constructing damage functions is that it is difficult to include the large number of parameters that may influence the flood damage. The parameters that are not used are implicitly considered. Each flood damage model based on a limited number of parameters is therefore making assumptions about the effect of the nonexplicitly considered parameters. Those unconsidered parameters have been very significant in a subset of flood events. For example, in the 2002 Elbe floods, contamination was critical (Thieken et al., 2005), in the Meuse floods, flood experience was critical (Wind et al., 1999) and in the 1945 floods in the Wieringermeer polder in the Netherlands, the waves in the flood water were critical (Duiser, 1982). This last example is complicated by a study of Roos (2003) who showed that the findings of the 1945 Wieringermeer polder flood are not valid for modern buildings; so the construction year/type of building can in some cases also be a critical parameter. Other possibly significant parameters are, for example, building style, flow velocity, flood duration, warning time and preparation. Parameters that are not used can have a correlation with parameters that are used. For example, the water depth is correlated with the flood duration for floods in the Netherlands (Duiser, 1982; Wagenaar, 2012). Because of this correlation, the uncertainty caused by not knowing the flood duration is limited in the Netherlands. This relationship between two parameters may, however, be completely different for other types of floods (e.g., flash floods). A generally applicable flood damage model therefore still needs all parameters. Table 1 splits the uncertainties involved in flood damage functions into two groups: those related to parameter representation (using fewer parameters than theoretically necessary to describe the damage processes) and those related to a lack of knowledge about the damage processes. The uncertainties related to a lack of knowledge are epistemic, while the group of uncertainties related to the use of fewer parameters than necessary is aleatory, because this group of uncertainties would remain even with perfect knowledge. In Sect. 3 the analysis of the epistemic and aleatory uncertainty components is used to assess the uncertainty in the outcome of a single damage calculation. The epistemic uncertainty will be estimated by using the difference between damage functions from entirely different flood damage models (Sect. 3.1.2). The aleatory uncertainty will be considered by looking at the variation within one flood damage model (e.g., the difference between the low and the high estimate of the same flood damage model).

METHODOLOGY

Overview of the Method

This section proposes a method for quantitative uncertainty analyses using a Monte Carlo analysis. The qualitative uncertainty analysis discussed in the previous section is used to estimate the uncertainty in the inputs and the correlations between the different input parameters. The general assumption behind this uncertainty analysis is that no good local damage functions are available and that the modeler therefore does not know which damage functions to choose. For areas for which good local damage functions are available, the approach discussed in this section may overestimate the uncertainty.

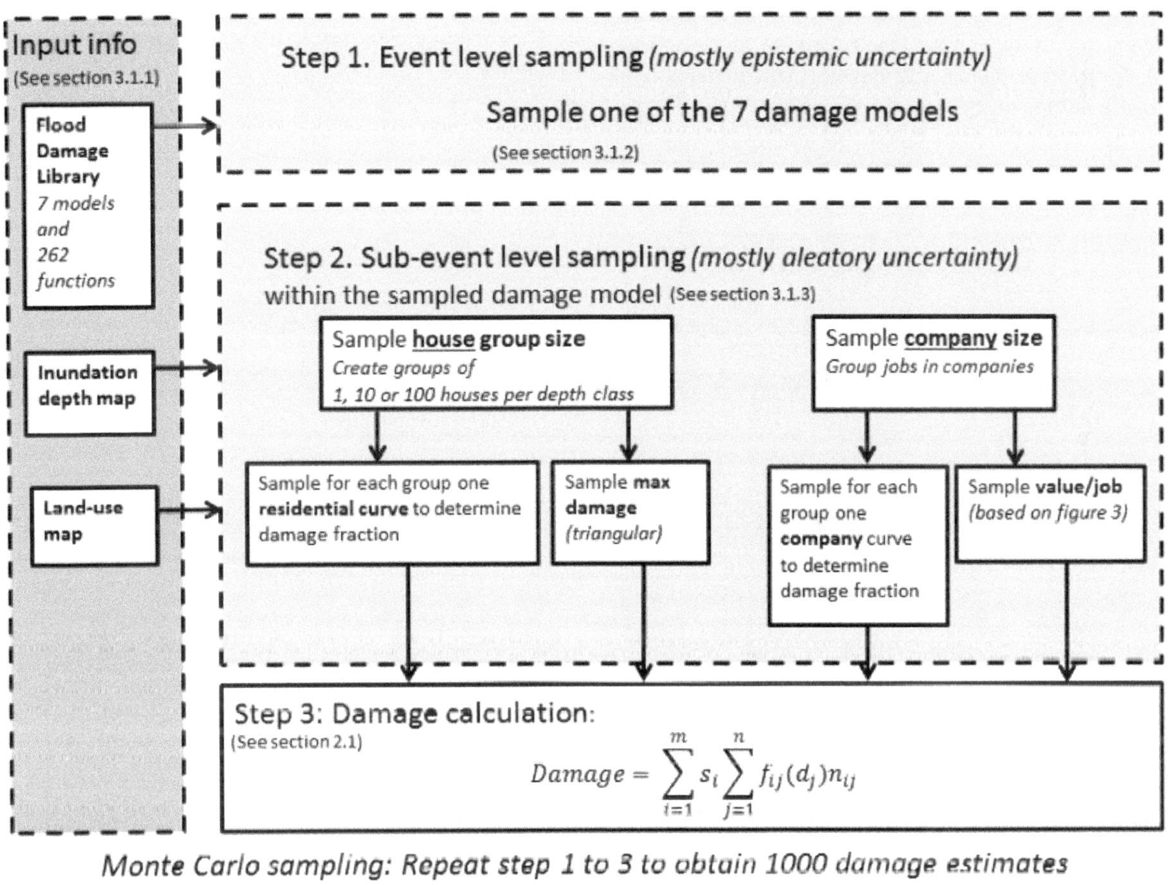

Figure 1. Overview of the different sample steps undertaken in the Monte Carlo analysis.

The damage analysis in this paper is limited to two damage categories: houses and companies, as they are represented in the different flood damage models on which this research is based. These were selected since all damage models contain functions to assess them. This is not the case for other damage types. In flood damage models that do consider other damage categories, houses and companies usually make up the majority of the direct damage. Both damage categories are divided into damage to buildings and damage to content. Many individual flood damage models provide several more detailed subcategories for these basic categories. Our approach may therefore lead to slightly larger uncertainties than are present in such models that are more detailed. A crucial aspect of the Monte Carlo analysis of uncertainties is the correlation amongst the uncertainty of the different input parameters, such as the maximum damage of houses. If for example, the maximum damage of house X is overestimated, the maximum damage of house Y may also be overestimated. The parameters that are homogenous within one event, but vary between events will have a strongly correlated uncertainty value e.g., if damage depends on warning time and on one particular event, the warning time is unusually short; this would more or less be the case for all houses which are affected in such an event. Such aspects are therefore sampled for the entire area at once. Other parameters vary between neighborhoods, or from place to place, such as for example, the building type. These need to be sampled on a smaller level then the entire area at once. Sampling will therefore be done at two different levels: for the entire event and on a more detailed sub-event level.

Figure 1 gives an overview of the calculations process which is repeated 10 000 times. This results in 10 000 different damage estimates which together make up the distribution of possible damages.

Input information

Flood Damage Library

A damage function library was constructed containing 262 different damage functions from seven different flood damage models. Damage functions from flood damage models were included if they were made for developed countries and available to the author at the time of the study (2013–2014). These functions were the basis for the damage fraction and the measure of susceptibility to flooding. The damage fraction was sampled by picking damage functions from a flood damage model. These functions were individually all scaled to 1 to ensure that the same maximum damage definition is applied everywhere.

Table 2 gives an overview of the models included in the damage function library. The Tebodin model only has damage functions for companies and the Billah (2007) model only has damage functions for houses. Since both models were made for the Netherlands and the damage functions were constructed using similar techniques, these two models have therefore been merged into one flood damage model. Figure 2 shows the average damage functions for the different flood damage models. In this picture the damage functions are not scaled up to 1. In this library the definition of the zero height point is the ground level (elevation in the digital elevation model) rather than the ground floor level. Some flood damage models use the floor level as the zero point and combine this with vertical elevation data of the ground floor. For this paper no damage functions with significant damage below the zero point were used. In the few functions used with damage below the zero point, the zero point was shifted to the point where the first damage occurred to make them comparable with other functions.

Table 2. Overview of the damage models from which damage functions are included in the damage function library

Model	Description
HIS-SSM	The HIS-SSM is a standard Dutch flood damage model (Kok et al., 2005). It is based on several earlier Dutch flood damage studies (Duiser, 1982; Briene et al., 2002). The functions are based on expert estimates combined with data from the 1953 flood in Zeeland and the 1945 flood of the Wieringermeer polder.
HAZUS-MH	The HAZUS-MH is an American disaster impact model with a flood module. This model was created by the federal government agency FEMA. It is described in FEMA (2009) and Scrawthorn et al. (2006). HAZUS provides a large set of American flood damage functions. A subset from the HAZUS library was used in the library presented here. The functions taken for houses were based on American insurance data and the functions for companies are based on expert judgment from the USACE.
MCM	The Multi-Coloured Manual (MCM) is a British flood damage model. For the library presented in this document, the version by Penning-Roswell (2005) was used. The MCM is based on a systematic expert judgment approach, in which a hypothetical building is split up into smaller parts, with each part being evaluated separately. The model has a large number of functions for different types of company buildings.
FLEMO	FLEMO is a German flood damage model based on data from the Elbe floods of 2002. The functions were derived from FLEMOps for houses (Thieken et al., 2008) and FLEMOcs for companies (Kreibich et al., 2010). The functions include a low and a high estimate.
Rhine Atlas	This second German model is based on expert judgement, taking data from an earlier German damage database (HOWAS) into account. More information about these functions is available in Jongman et al. (2012).
Tebodin	This is a Dutch study, based on a detailed, systematic and well-documented expert judgement approach. This study only provides damage functions for industry. It is detailed; it provides functions for many different industrial types and it has separate functions for areas protected by flood defences and for unprotected areas (Snuverink et al., 1998; Sluijs et al., 2000).
Billah, 2007	This is a research project in which the systematic expert judgment approach as used in MCM was applied to Dutch houses.

Land Use Maps

A flood damage model needs input about the number of houses and jobs affected. For the case study, the number of houses was taken from the geographical database BAG. This database was made by the Dutch Cadaster, Land Registry and Mapping Agency. For the number of jobs, the background data of HIS-SSM were used (Kok et al., 2005).

Step 1: event-level sampling (epistemic uncertainty)

The sampling on the event level is done by sampling a flood damage model (e.g., HAZUS or MCM) and using that throughout the damage calculation. This sampled model will be applied to all categories and will be used as a source for the damage functions and for the measure of susceptibility to flooding of the maximum damages. The advantage of this is that a realistic combination of inputs will be sampled. This procedure prevents, on average, a higher damage from being sampled for small water depths than for large water depths or functions with different implicit assumptions from being merged.

Step 2: sub-event-level sampling (aleatory uncertainty)

Group size and Dependency

For the sub-event-level sampling, uncertainty values are sampled for small groups of houses or for a company. Houses and jobs are grouped because also in reality, similar houses are often built near each other and a company is also expected to be relatively homogenous in damage per job. It is therefore not realistic to sample all houses and all individual jobs in an area completely independent from each other. By sampling in small groups of houses/jobs, total dependency within the group and total independency between the groups is assumed.

The way in which the area is grouped determines the dependency for this aleatory uncertainty. This buildup of groups is therefore also sampled again for each Monte Carlo simulation. This sampling should therefore be seen as a sampling of the dependencies between the damage of different objects. For houses the area is split into groups of 1, 10 or 100 houses for every simulation. Houses are only grouped together when they have a similar water depth. This is done to keep the calculation simple but also because similar water depths typically occur in locations that are geographical close. Furthermore, the group sizes are so small that for medium- or larger-sized events, this assumption has no influence on the results. For companies, the jobs are grouped per company.

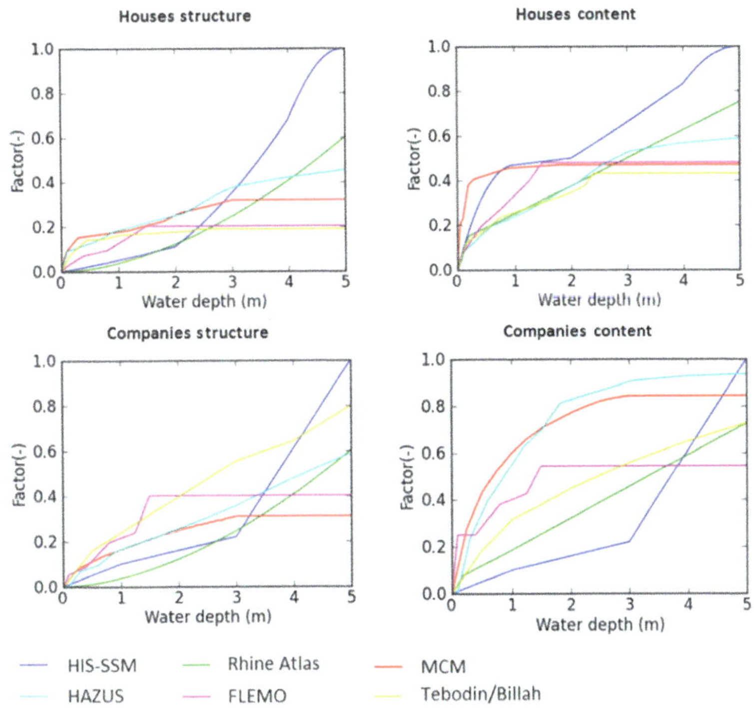

Figure 2. Average damage functions for the different flood damage models in the damage function library. Blue: HIS-SSM, green: Rhine Atlas, red: MCM, light blue: FLEMO, pink: HAZUS, yellow: Billah (2007) for houses and Tebodin for companies.

Damage Functions

Within each group, every house/job receives the same damage fraction and maximum object value. Sampling for the damage fraction is done based on the set of damage functions within the flood damage model sampled in step 1. For example, if the flood damage model sampled has three damage functions for houses, for each group, one of the three damage functions is randomly used.

Maximum Damage

The values are based on De Bruijn et al. (2014) and Gauderis (2012). De Bruijn et al. (2014) estimated the structural value of a house at EUR 125 000. The minimum and maximum from the triangular distribution are roughly estimated at ± EUR 75 000 for structural damage. For content damage De Bruijn et al. (2014) estimated a maximum damage of EUR 70 000, for which here also a triangular distribution is assumed, at ± EUR 50 000. These maximum damages all use the price level 2011. These assumptions lead to a symmetric probability distribution, while it is probably in reality positively skewed. This is neglected in this study because it is difficult to estimate and the impact on the uncertainty is expected to be very small. The maximum damage for companies is assessed in this study as the maximum damage per job times the number of jobs per company. Gauderis (2012) estimated material value per job for 62 different categories of companies. These estimates are taken together to produce a distribution of the physical value of a company per job. Because not all company categories are equally common, the values were weighted in the distribution based on their quantity in the Netherlands. This more complex method was used in order to correctly incorporate the skewness and because the data from Gauderis (2012) make it possible to do this, which was not the case for houses. The results are shown in Fig. 3. These values include both the structure and the content. Assumptions from Gauderis (2012) on the part of the maximum damage that belongs to the structure and the part that belongs to the content were adopted.

MONTE CARLO ANALYSIS BEHAVIOR: TRIAL OF THE METHOD ON HYPOTHETICAL FLOOD MAPS

To gain understanding of the Monte Carlo analysis behavior, the analysis was tested on hypothetical flood depth maps, one with small water depths (< 0.5 m), one with medium water depths (0.5–2 m) and one with large water depths (2–3 m). These had average water depths of 0.35, 1.25 and 2.5 m. These maps were used for calculations with 150 and 15 000 houses and jobs; thus in total, six different trials were carried out and the resulting uncertainty values were compared. The uncertainties in the damage estimates are expressed with the coefficient of variation. This is the standard deviation of the damage divided by the mean of the damage. It has no unit and is therefore independent of the size of the flood event. This makes it a good measure to compare the uncertainties in different areas. Figure 4 shows the results of this hypothetical analysis. It stands out that both a smaller water depth and a smaller area increase the uncertainty significantly. This is because at small water depths, the different flood damage models differ significantly more from each other than at large water depths. This indicates that the uncertainty in damage estimates for events like, for example, small regional levee failures, is much larger than the uncertainty in damage estimates for large-scale floods with large water depths.

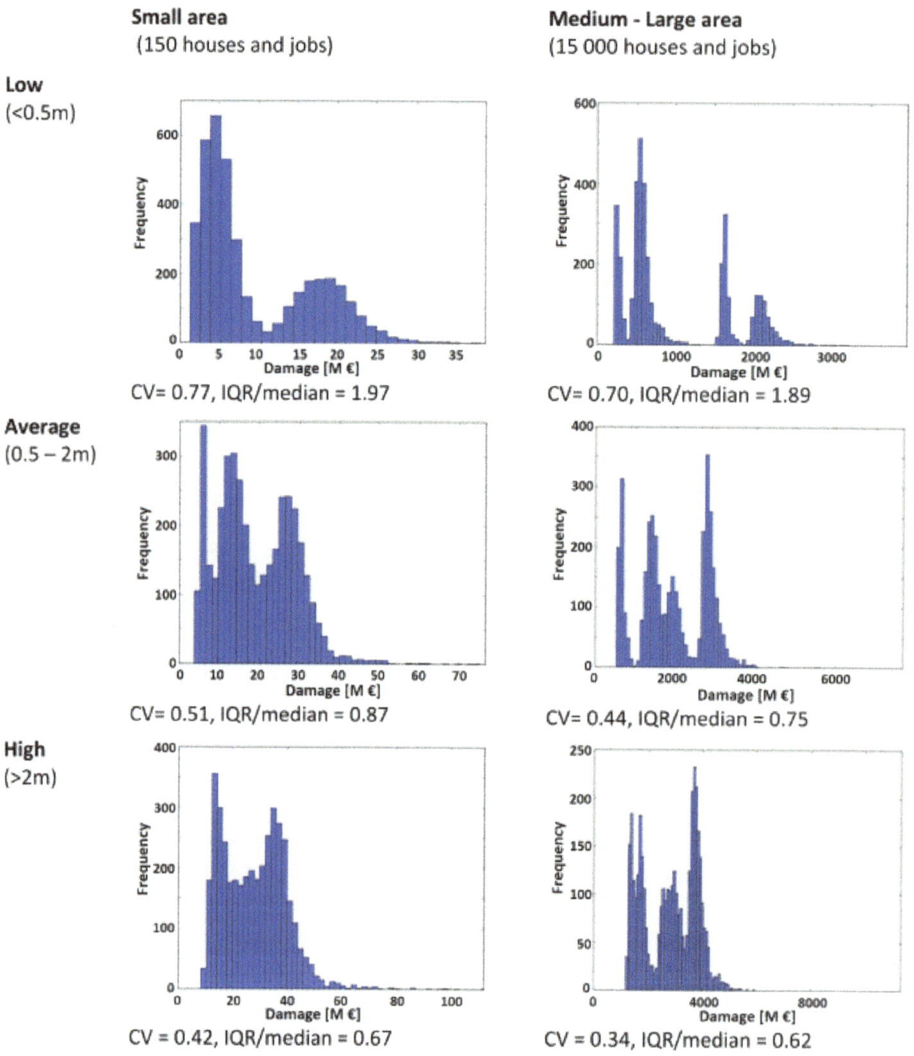

Figure 4. Results of the Monte Carlo simulations applied on synthetic flood maps shown as frequency distributions, coefficient of variations (CV) and interquartile range (IQR)/median for different types of hypothetical areas based on 4000 samples. The CV and IQR/median are both simplified (imperfect) representations of the dispersion in a single number; they show a similar view of the dispersion for the different test runs. The x axis in this figure is equal to 4 times the mean damage.

Another observation is that the distribution of the damage for small events looks very different from the distributions of the damage of large events. The main reason for this is that for large events, the aleatory uncertainty in the flood damages can be reduced significantly by the law of large numbers, but not the epistemic uncertainty. Epistemic uncertainty is therefore the significant uncertainty for larger events. The frequency distributions therefore then show clearly separate peaks related to the damage functions of the separate flood damage models. It is difficult to determine for what event size the variation between the flood damage models (epistemic uncertainty) becomes more important than the variation within the flood damage models (aleatory uncertainty). For the uncertainty model created in this paper this point is somewhere between 100 and 3000 houses, plus jobs. This critical size depends on the dependencies between individual objects. These dependencies determine how fast the law of large numbers will reduce the aleatory uncertainty. For this paper, this was estimated by sampling in groups instead of sampling individual objects. The size of these groups therefore determines when the epistemic uncertainty becomes dominant. These group sizes were based on a rough estimate in this paper and more research should be done for better results.

CASE STUDY

A case study is done in the Betuwe area, Tieler-en Culemborgerwaarden (dike ring 43) in the Netherlands, to show the effect of uncertainty in the flood damage estimation on investment decisions for flood risk management. Dike ring 43 is located between Rhine branches in the Netherlands. In the west, the area is closed with a high dike (border to next dike ring area). The area slopes down to the west. The difference in height between the eastern and western part is about 10 m. The Monte Carlo analysis is applied to a water depth map resulting from a simulated dike breach (VNK, 2014) along the Rhine River, near Bemmel in the Netherlands (see Fig. 5 for its location). This dike section is about 26 km long. Bemmel is situated in the eastern upstream part of the Betuwe area. When the dike breaches, water flows through the Betuwe area to the west where after about 70 km, it is stopped by the western embankment. The maximum water depths due to this dike breach vary from less than 50 cm in the east to over 5 m in the west. In this dike breach scenario a total area of 626 km² is inundated. This area contains several small towns and villages, with a total population of around 300 000 people. The large flood extent, the large number of affected residences and companies and the large variation in water depths are expected to have a reducing effect on the aleatory uncertainty in the total damage of the dike ring area. The damage assessed for this flood scenario was EUR 16 billion (price level 2011) with a standard deviation of EUR 5.6 billion based on 10 000 simulations. The resulting damage outcomes are shown in Fig. 6. The peaks in Fig. 6 are related to the damage models and illustrates the large differences between the different damage models (two damage models overlap).

Figure 5. Map of the case study area. The green line is the dike segment under investigation; the cross indicates the location of the breach scenario which was used.

Table 3. Optimal investment strategy given different damage estimates. The flood protection standard is a return period based on the direct method described in Kind (2011), note that the actual return period of the investment strategy differs per year. Price level 2011 is used.

Damage (EUR in million)	Extra height for the investment (cm)	Flood protection standard (year)	Present value investment cost (EUR in million)	Present value EAD (EUR in million)	Present value total cost (investment cost + EAD) (EUR in million)
8000	70	25 000	109	11	121
12 000	77	40 000	114	11	126
15 000	80	50 000	117	11	128
18 000	82	58 000	118	12	130
25 000	88	83 000	122	12	134

The results in Fig. 6 are used to find the economic optimum flood protection standard and investment strategy for the dike segment from an economic viewpoint. The optimum flood protection standard and investment strategy is calculated using a simplified version of the approach of Kind (2013). Kind (2013) assesses which investment strategy (set of dike improvements at different moments) has the smallest total cost for the future and hence is the economic optimum. This total cost consists of all the discounted expected annual damage (EAD) considering future changes and the discounted future investments in the dike. The expected annual damage depends on the flood probability and the flood damage given to a dike breach (as calculated in this case study). Long-term economic growth forecasts are used to increase the damage for every year in the future. Furthermore, a correction factor was used to take indirect damage into account. The flood probability depends on the quality of the dike which again depends on the investments carried out. This flood probability is calculated for each year into the future based on the current status of the dike, investments up to that point, consolidation of the dike and climate change predictions. Investments in the dike are simplified as height increases and the cost of these investments is based on fixed and variable cost for the dike segment considered. A height increase is converted into a flood probability reduction based on a parameter that describes the height increase necessary to decrease the flood probability by a factor of 10. This parameter and all other parameters used for determining the optimal investment strategy (except for the flood damage) were taken from the WV21 project (Kind, 2011).

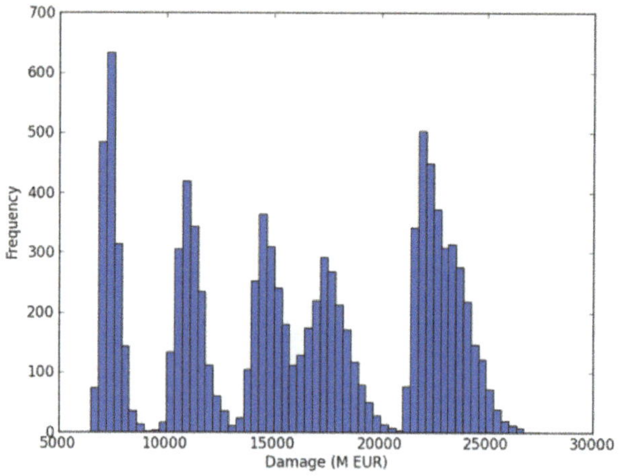

Figure 6. Frequency distribution of the damage in the case study area (based on 10 000 Monte Carlo simulations). The five peaks represent the six models used in this study, with two models overlapping. The damage has the price level of 2011.

In this paper we assess the effects of uncertainty in damage estimates on the economic optimal flood protection standard and the total investment costs. We do that by determining the investment strategy for five different damage estimates. The first four estimates relate to the first four peaks. For the highest damage estimate, the 98 % percentile of the damage outcomes was used. The analysis in this paper focuses on the first investment made. In all five alternatives this investment is done in 2015. The second investment is in all alternatives planned about 75 years later and a third investment is suggested about 50 years after the second one (around the end of the time span considered). The total investment costs are mainly determined by the first investment, because the weight of later investments is very small due to the use of the net present value, which gives future costs and benefits a much lower weight than current costs and benefits. The calculations assumed a discount rate of 5.5 % (based on WV21; Kind, 2011).

The results in Table 3 show that the optimal investment strategy is at first glance not very sensitive to the precise damage estimate. The difference between the five alternatives in required dike heightening is only 18 cm (88 – 70 cm). This small difference is partly explained by the strong sensitivity of the flood probability to the precise height of the dike. The dike segment in this case study becomes 10 times safer by raising it only 34 cm. If the flood probability were less sensitive to height changes, the differences in dike height between a low and a high damage estimate could be much larger. If the dike were increased by 1 m to reduce the flood probability by a factor of 10, the difference between the top and lower damage estimate would be 47 cm. If the flood damage applied in the cost benefit analysis differed from the flood damage that would actually occur, a suboptimal investment strategy would be applied. Table 4 shows the costs of using a wrong damage estimate. It gives the unnecessary cost made by assuming a certain damage for different real damage values. This cost varies in this case study between EUR 0 and 12 million and is on average about EUR 2 million, which is 1.4 % of the total cost and for this case study about EUR 75 000 km^{-1}. The maximum error is 9 % of the total cost and for this case study EUR 500 000 km^{-1} (price level 2011). This case study illustrates how the Monte Carlo analysis may be used to assess the uncertainty in damage assessments, and how the effect of this uncertainty on investment costs may be determined. In the case study here, the effect is small. However, if we take into account the fact that in the Netherlands we have about 3000 km of embankments and that EUR 12 million might be unnecessarily spent per 26 km, the total amount of money spent unnecessarily may then be large. It is also likely that in cases with lower flood probability standards, or with smaller flood events, the effects of this uncertainty are much larger. A striking observation in the results of Table 4 is that the costs of overestimating the damage are significantly lower than the costs of underestimating the damage. The difference in costs is on average a factor of 2 (see Table 4). This can be explained by the nonlinear relationship between the flood probability reduction and the investment costs. The flood probability can be reduced a lot with a small extra investment; thus when too little is invested, the EAD increases faster than the investment cost decreases. This implies that under uncertainty, it would be economically efficient to add a safety factor to avoid investing too little.

DISCUSSION

This paper discusses a new method for the quantification of uncertainties and applied this method in a case study. The case study is a good illustration of the method and its use, but the calculated uncertainty, the damage frequency distribution and the effect of uncertainty on investment decisions may not be representative of all situations, first of all, because several damage-determining aspects were neglected in the case study. The damage is assumed to consist only of damage to buildings and companies. The quantification of other damage categories, such as affected persons and fatalities may also be relevant and can be taken into account in a CBA (cost benefit analysis). Another simplification is that the entire cost benefit analysis in the case study is based on only one flood scenario, at one breach location and at one water level (at the design water level of the dike). A more precise way would have been to include multiple breach locations and water levels. These effects are however assumed to be negligible for the conclusions of this paper, because in most cases, the uncertainty for the different damage estimates will be similar and highly correlated (because it is for the same area).

Secondly, the results may not be representative of all situations because the exact location and number of peaks in the damage frequency distribution depend on the input damage models in the uncertainty analysis. The set of seven damage models used does not cover all possible damage models. If an extra flood damage model were added to the damage function library, an entire new peak could appear. The frequency distributions of the outcomes must therefore be considered as an example of what a frequency distribution could look like and how far the peaks are approximately apart from each other. It is impossible to make a real frequency distribution because the major uncertainties are epistemic uncertainties. Epistemic uncertainties are by definition not understood and can therefore not be represented by a frequency distribution (Helton and Oberkampf, 2004). However, there are alternative concepts to describe uncertainty, such as imprecise probabilities or Bayesian statistics, that deal with this problem (Reichert and Omlin, 1997; Zadeh, 2005).

Thirdly, the cost of a wrong estimate, which was estimated for the case at about 1 % and at a maximum of 10 % of the total cost, may also be different for other cases. It depends, amongst others, on the cost required to reduce the failure probability by a factor of 10, on the damage itself and on the uncertainty in the damage interaction (which will be larger for small areas and areas with little flood water depths). Finally, the uncertainty in the damage estimate was, in this case study, directly linked to an error in the investment strategy. However, in the determination of the optimal investment strategy, not only uncertainties in the damage estimate, but also in other components play an important role. Uncertainty in the costs of dike strengthening, in the discount rate, in the future economic growth, in the flood pattern and so on, all add to the uncertainty in the optimal investment strategy. These uncertainties may partly compensate each other, but can also aggregate each other. Their relative importance differs per case, depending on local characteristics (De Moel et al., 2014).

Table 4. Cost of a damage estimate error for this dike segment in EUR million (price level 2011).

Damage estimate reality	Damage estimate for calculation investment strategy 8000	12 000	15 000	18 000	25 000
8000	0	1.1	2.1	2.7	5.5
12 000	1.3	0	0.2	0.5	2.3
15 000	3.2	0.3	0	0	1.2
18 000	5.2	0.9	0.2	0	0.7
25 000	12.2	3.8	1.8	0.9	0

We tried to combine information from different damage models to get a better quantification of uncertainties in damage outcomes. This can only be done when the damage models may all be applicable to the flood scenario which is being modeled. Whether flood models are equally applicable is sometimes difficult to establish. Metadata of the source of the damage models are not always available and sometimes information on the event on which the model is based is also lacking. This makes it difficult to compare damage models and to understand why they have different estimates for the same flood patterns. Relevant metadata on parameters which may be obvious for a certain event, but vary from event to event are needed. Examples of such parameters are, for example, flood experience of the population, building style, flood duration and contamination of the flood water. Metadata for flood damage functions should give clear instructions about the type of events for which damage functions are applicable and for what events they are not. This could lead to a classification of different flood types with their own damage functions. This would first lead to a better transferability of damage functions and maximum damages and could eventually lead to generally applicable flood damage models.

Vogel et al. (2012) and Schröter et al. (2014) also carried out detailed uncertainty analyses for flood damage assessment. The most obvious difference with these studies is that this paper uses a Monte Carlo approach, while they use fundamentally different methods. However, the more interesting difference is that this paper assumes a situation where no good local data are available and that little is known about the expected conditions during the potential flood (apart from the maximum water depth). Therefore, this paper used relatively simple data from many different countries and flood types as input for the uncertainty analysis, while these other papers used relatively complex data, only from Germany. The strength of this approach therefore is that it has a wider coverage of the spectrum of possible flood damage. The disadvantage of this approach is that it is not applicable when a good local flood damage model is available, based on many data.

CONCLUSION

Uncertainties in flood damage estimates can be large. This study showed uncertainties of an order of magnitude of 2– 5. This uncertainty is mainly caused by a lack of knowledge. Most flood damage models are based on data resulting from a small number of events. Because flooding can occur in many different ways (water depths, contamination, flow velocities, flood durations, etc.) and in many different types of areas (building types, flood experience of the local population) any model will miss considerable parts of the spectrum of possible options. Data from one event therefore are often not transferrable to other areas or events. Since only data representative of the event under consideration can be used, few data are available and hence large uncertainties are introduced in flood damage modeling. This study introduced a method to quantify these uncertainties using a set of damage models which have all been applied in the past to river floods (not flash floods) or storm surges in developed countries. To quantify the uncertainty, a distinction was made between epistemic and aleatory uncertainties. Epistemic uncertainties are introduced by a lack of knowledge about the spectrum of possible flood events and areas in which they could occur. The size of this spectrum was estimated for this study by using the difference between flood damage models. Aleatory uncertainties are introduced by local variations between objects and circumstances. These uncertainties were estimated for this study with the variations within different flood damage models.

These aleatory uncertainties are large for small flood events and much smaller for large flood events, affecting many objects. Epistemic uncertainties are not smaller for large areas, since they are not related to deviations of single objects from the average object for which the damage functions were derived. Epistemic uncertainties can only be reduced with new knowledge. The resulting Monte Carlo analysis therefore shows larger uncertainties for small areas. However, at a certain event size, the epistemic uncertainties become dominant. These uncertainties in flood damage modeling can potentially have a significant effect on investment decisions. In this study a case study was carried out to calculate the economic optimal investment strategy for a dike segment. This case study showed that uncertainties in damage estimates can lead to suboptimal investment decisions. In the worst case scenario (maximum error in damage estimate), the difference between the total cost (remaining risks and investment cost) may be as high as EUR 500 000 km^{-1} dike (price level 2011). The expected difference between the optimal and suboptimal investment strategy was, however, significantly lower (EUR 75 000 km^{-1} dike). These findings need to be verified with further research in other areas.

The paper provides a good first approach for uncertainty quantification in damage estimates and shows how this approach can be used to improve investment decisions. Further research including other areas and more flood events is recommended to develop the approach further.

Acknowledgements

We would like to thank William Lehman, one anonymous reviewer and the editor Bruce D. Malamud for their constructive feedback and helpful comments which helped to improve this paper.

REFERENCES

1. Apel, H., Merz, B., and Thieken, H.: Quantification of uncertainties in flood risk assessments, International Journal of River Basin Management, 6, 149–162, 2008.
2. Briene, M., Koppert, S., Koopman, A., and Verkennis, A.: Financiele onderbouwing kengetallen hoogwaterschade, NEI, Ministerie van Verkeer en Waterstaat, Rijkswaterstaat DWW, 2002.
3. Bubeck, P., Botzen, W. J. W., Kreibich, H., and Aerts, J. C. J. H.: Long-term development and effectiveness of private flood mitigation measures: an analysis for the German part of the river Rhine, Nat. Hazards Earth Syst. Sci., 12, 3507–3518, doi:10.5194/nhess-12-3507-2012, 2012.
4. Cammerer, H., Thieken, A. H., and Lammel, J.: Adaptability and transferability of flood loss functions in residential areas, Nat. Hazards Earth Syst. Sci., 13, 3063–3081, doi:10.5194/nhess-13- 3063-2013, 2013.
5. Chatterton, J., Penning-Rowsell, E., and Priest, S.: The Many Uncertainties in Flood Loss Assessments, in: Applied Uncertainty Analysis for Flood Risk Management, Imperial College Press, London, UK, 2014.
6. De Bruijn, K. M.: Resilience and flood risk management: A system approach applied to lowland rivers, PhD Thesis – Delft University of Technology, Delft, the Netherlands, 2005.
7. De Bruijn, K. M., Wagenaar, D. J., Slager, K., De Bel, M., and Burzel, A.: Proposal for the new flood damage assessment method: SSM2015, Deltares, 2014.
8. De Moel, H. and Aerts, J. C. J. H.: Effect of uncertainty in land use, damage models and inundation depth on flood damage estimates, Nat. Hazards, 58, 407–425, 2011.
9. De Moel, H., Asselman, N. E. M., and Aerts, J. C. J. H.: Uncertainty and sensitivity analysis of coastal flood damage estimates in the west of the Netherlands, Nat. Hazards Earth Syst. Sci., 12, 1045– 1058, doi:10.5194/nhess-12-1045-2012, 2012.
10. De Moel, H., Bouwer, L. M., and Aerts, J. C. J. H.: Uncertainty and sensitivity of flood risk calculations for a dike ring in the Netherlands, Sci. Total Environ., 473–474, 224–234, 2014.
11. Duiser, J. A.: Een verkennend onderzoek naar methoden ter bepaling van inundatieschade bij dijkdoorbraak, 82-0644 TNO, Delft, the Netherlands, 1982.
12. Egorova, R., van Noortwijk, J., and Holterman, S.: Uncertainty in flood damage estimation, International Journal River Basin Management, 6, 139–148, 2008.
13. FEMA: HAZUS-MH MR4 Technical Manual, chapter 5, 5-1–5-31, 2009.
14. Gauderius, J.: Verbetering schadebepaling buitendijkse bedrijven Rijnmond-Drechtsteden. Discussiememo waarderingsprincipes schade bedrijven, RebelGroup, Antwerp, Belgium, 2012.
15. Helton, J. C. and Oberkampf, W. L.: Alternative representations of epistemic uncertainty, Reliab. Eng. Syst. Safe, 85, 1–10, 2004.
16. Jongejan, R. and Maaskant, B.: The use of quantitative risk analysis for prioritizing flood risk management actions in the Netherlands, CDA 2013 Annual Conference, 5–10 October 2013, 2013.
17. Jongman, B., Kreibich, H., Apel, H., Barredo, J. I., Bates, P. D., Feyen, L., Gericke, A., Neal, J., Aerts, J. C. J. H., and Ward, P. J.: Comparative flood damage model assessment: towards a European approach, Nat. Hazards Earth Syst. Sci., 12, 3733–3752, doi:10.5194/nhess-12-3733-2012, 2012.
18. Kind, J.: Maatschappelijke kosten-batenanalyse Waterveiligheid 21e eeuw, Deltares – 1204144, 2011.
19. Kind, J., Vos, R., Tijssen, A., Jeuken A., and Slootjes, N.: Towards the Evaluation of Adaptive Flood Risk Management Strategies for the Rhine estuary – Drechtsteden, 6th International Conference on Flood Management (ICFM6), 16–18 September 2014, Sao Paulo, Brazil, 2014.
20. Kind, J. M.: Economically efficient flood protection standards for the Netherlands, Journal of Flood Risk Management, 7, 103–117, 2013.
21. Koivumäki, L., Alho, P., Lotsari, E., Käyhkö, A., Saari, A., and Hyyppä, H. Uncertainties in flood risk mapping: a case study on estimating building damages for a river flood in Finland, Journal of Flood Risk Management, 3, 166–183, 2010
22. Kok, M., Huizinga, H. J., Vrouwenvelder, A. C. W. M., and van den Braak, W. E. W: Standaardmethode 2005, Schade en Slachtoffers als gevolg van overstroming. HKV, TNObouw, Rijkswaterstaat DWW, 2005.
23. Kreibich, H., Seifert, I., Merz, B., and Thieken, A.: Development of FLEMOcs – a new model for the estimation of flood losses in the commercial sector, Hydrolog. Sci. J., 55, 1302–1313, 2010.
24. Merz, B., Kreibich, H., Schwarze, R., and Thieken, A.: Review article "Assessment of economic flood damage", Nat. Hazards Earth Syst. Sci., 10, 1697–1724, doi:10.5194/nhess-10-1697- 2010, 2010.
25. Penning-Rowsell, E. C., Johnson, C., and Tunstall, S.: The bene- fits of Flood and Coastal Risk Management: A Manual of Assessment Techniques, Middlesex University Press, London, UK, 2005.
26. Peterman, R. M. and Anderson, J. L.: Decision analysis: A method for taking uncertainties into account in risk-based decision making, Hum. Ecol. Risk. Assess. 5, 231–244, 1999
27. Reichert, P. and Omlin, M.: On the usefulness of overparameterized ecological models, Ecol. Model., 95, 289–299. doi:10.1016/S0304-3800(96)00043-9, 1997.
28. Roos, W.: Damage to buildings, Delft Cluster, DC1-233-9 TNO, 2003.

29. Scawthorn, C., Flores, P., Blais, N., Seligson, H., Tate, E., Chang, S., Mifflin, E., Thomas, W., Murphy, J., Jones, C., and Lawrence, M.: HAZUS-MH flood loss estimation methodology II. Damage and loss assessment, Natural Hazards Review, 7, 72–81, 2006.

30. Schröter, K., Kreibich, H., Vogel, K., Riggelsen, C., Scherbaum, F., and Merz, B.: How useful are complex flood damage models?, Water Resour. Res., 50, 3378–3395, doi:10.1002/2013WR014396, 2014.

31. Slager, K., De Bruijn, A., Burzel, K. M., Bouwer, L. M., and Wagenaar, D. J.: Verbeteringen gevolgenbepaling van overstromingen in buitendijkse gebieden Rijnmond-Drechtsteden, Deltares 1208261-000, 2013.

32. Sluijs, L., Snuverink, M., van den Berg, K., and Wiertz, A.: Schadecurves industrie ten gevolge van overstromingen, Tebodin. Rijkswaterstaat DWW, 2000.

33. Snuverink, M., Van Den Berg, K., and Sluijs, L.: Schade bij inundatie van buitendijkse industrie, Tebodin, Rijkswaterstaat DWW, 1998.

34. Thieken, A. H., Müller, M., Kreibich, H., and Merz, B.: Flood damage and influencing factors: New insights from the August 2002 flood in Germany, Water Resour. Res., 41, 1–16, doi:10.1029/2005WR004177, 2005.

35. Thieken, A. H., Olschewski, A., Kreibich, H., Kobsch, S., and Merz, B.: Development and evaluation of FLEMOps – A new flood loss esimation model for the private sector, WIT Trans. Ecol. Envir., 118, 315–324, 2008.

36. USACE: Guidelines for risk and uncertainty analysis in water resources planning, Institute for Water Resources, Fort Belvoir, VA, USA, 92-R-1, 1992.

37. van der Most, H., Tanczos, I., De Bruijn, K. M., and Wagenaar, D. J.: New, Risk-Based standards for flood protection in the Netherlands, 6th International Conference on Flood Management (ICFM6), 19–21 September 2014, Sao Paulo, Brazil, 2014.

38. Vogel, K., Rigg, C., Kreibich, H., Merz, B., and Scherbaum, F. Flood Damage and Influencing Factors: A Bayesian Network Perspective, in: Proceedings of the 6th European Workshop on Probabilistic Graphical Models (PGM 2012), Presented at the 6th European Workshop on Probabilisit Graphical Models, 19– 21 September 2012, Granada, Spain, 347–354, 2012.

39. Wagenaar, D. J.: The significance of flood duration for flood damage assessment, Master Thesis Delft University of Technology – Deltares, the Netherlands, 2012.

40. Wind, H. G., Nierop, T. M., de Blois, C. J., and de Kok, L.: Analysis of flood damages from the 1993 and 1995 Meuse floods, Water Resour. Res., 35, 3459–3465, 1999.

41. Zadeh, L. A.: Toward a generalized theory of uncertainty (GTU) – an outline, Inform. Sciences, 172, 1–40, 2005.

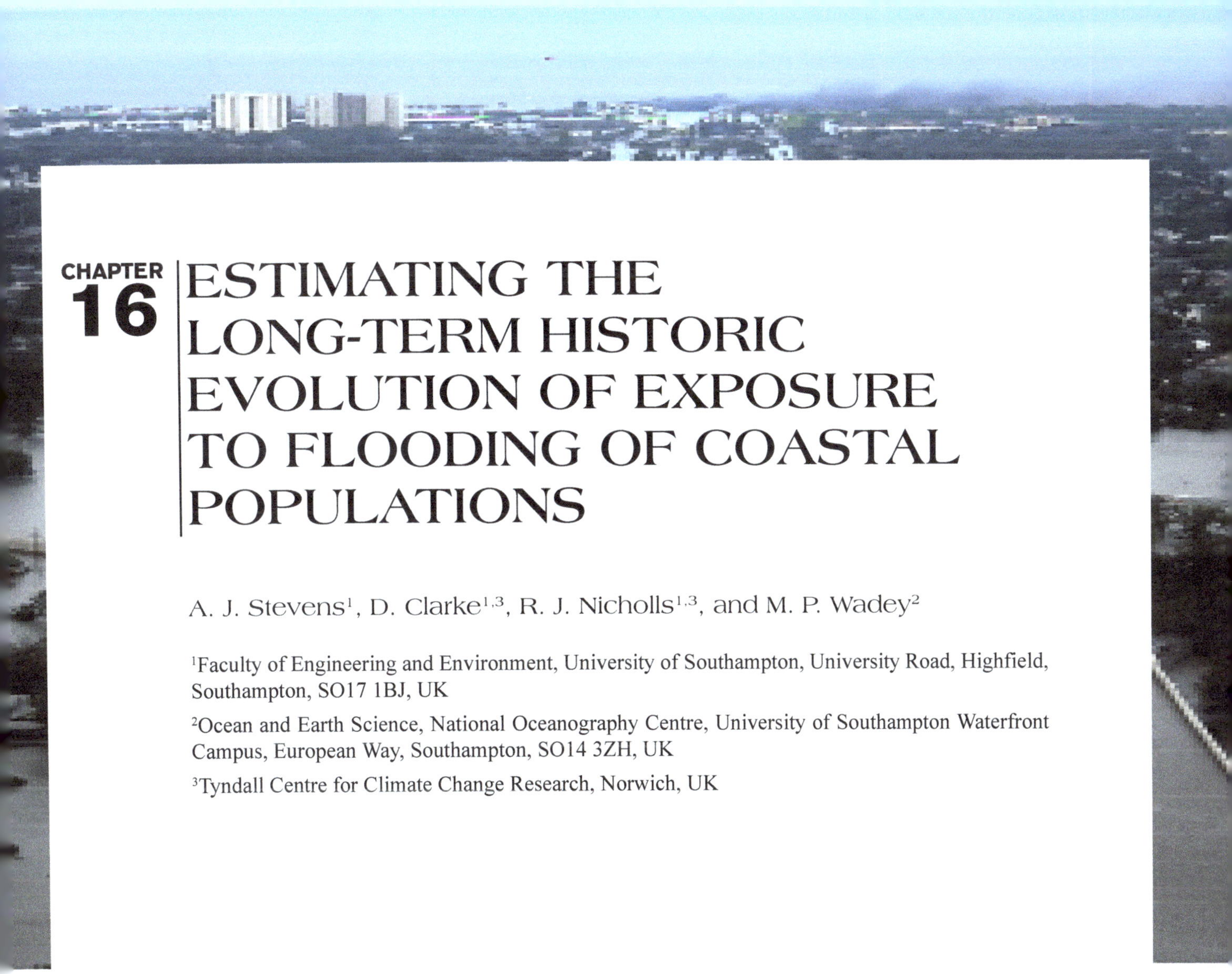

CHAPTER 16

ESTIMATING THE LONG-TERM HISTORIC EVOLUTION OF EXPOSURE TO FLOODING OF COASTAL POPULATIONS

A. J. Stevens[1], D. Clarke[1,3], R. J. Nicholls[1,3], and M. P. Wadey[2]

[1]Faculty of Engineering and Environment, University of Southampton, University Road, Highfield, Southampton, SO17 1BJ, UK

[2]Ocean and Earth Science, National Oceanography Centre, University of Southampton Waterfront Campus, European Way, Southampton, SO14 3ZH, UK

[3]Tyndall Centre for Climate Change Research, Norwich, UK

ABSTRACT

Coastal managers face the task of assessing and managing flood risk. This requires knowledge of the area of land, the number of people, properties and other infrastructure potentially affected by floods. Such analyses are usually static; i.e. they only consider a snapshot of the current situation. This misses the opportunity to learn about the role of key drivers of historical changes in flood risk, such as development and population rise in the coastal flood plain, as well as sea-level rise. In this paper, we develop and apply a method to analyse the temporal evolution of residential population exposure to coastal flooding. It uses readily available data in a GIS environment. We examine how population and sea-level change have modified exposure over two centuries in two neighbouring coastal sites: Portsea and Hayling Islands on the UK south coast. The analysis shows that flood exposure changes as a result of increases in population, changes in coastal population density and sea level rise. The results indicate that to date, population change is the dominant driver of the increase in exposure to flooding in the study sites, but climate change may outweigh this in the future. A full analysis of changing flood risk is not possible as data on historic defences and wider vulnerability are not available. Hence, the historic evolution of flood exposure is as close as we can get to a historic evolution of flood risk. The method is applicable anywhere that suitable floodplain geometry, sea level and population data sets are available and could be widely applied, and will help inform coastal managers of the time evolution in coastal flood drivers.

INTRODUCTION

One tenth of the world's population live in the low elevation coastal zone (Lichter et al., 2011), or are exposed as temporary residents due to coastal tourism and industry (Kron, 2008). More than 200 million people are estimated to be at risk of flooding from extreme sea levels caused by storms (Nicholls, 2010). Hence there is an urgent need for coastal managers to understand coastal flood risk, the drivers of the risk and how the drivers change over time. Drivers of flood risk include population exposed to flooding, frequency of extreme events and the effectiveness of any flood defences and of any other adaptation. All of these drivers can change over time so a full analysis should include an evaluation of how these drivers evolve both historically and into the future (via scenario analysis). While there are many future analyses of flooding, historic analyses are less common, which misses important empirical

Citation: Stevens, A., Clarke, D., Nicholls, R., & Wadey, M. (2015). Estimating the long-term historic evolution of exposure to flooding of coastal populations. Nat. Hazards Earth Syst. Sci., 15(6), 1215-1229. http://dx.doi.org/10.5194/nhess-15-1215-2015.

insights on what has happened. Flood risk can be assessed in a framework which considers the interacting elements of the SPRC (Source–Pathway– Receptor–Consequence) model (Holdgate, 1979) or more recently the "flood system" concept (Evans et al., 2004; Narayan et al., 2014; Sayers et al., 2002). Methods to assess exposure to coastal floods have focused on understanding the sources (e.g. extreme sea levels (Haigh et al., 2010; Batstone et al., 2013) and waves (Wolf et al., 2011; Chini and Stansby, 2012)) or pathways (e.g. simulations of defence failure and inundation via event-based approaches (Wadey et al., 2012, 2013) and flood risk assessment (Gouldby et al., 2008; Dawson et al., 2009)). These studies can include the effects of anticipated sea level rise (SLR) which changes the probability of extreme events (Church et al., 2013; Wahl et al., 2013; Haigh et al., 2011). Coastal flood risk is bound to change in time because sea level is rising (IPCC, 2013) and more people are living closer to the sea (Nicholls, 1995; Small and Nicholls, 2003). However, previous studies have not looked at the detailed historic time evolution of this risk. Population assessments have only been considered in time-aggregated analyses such as Foresight (Evans et al., 2004).

Table 1. Summary of required data and sources.

Data	Source
Population size and distribution	Census data (10 year time steps)
Urban/residential extent	Historic maps digitised in GIS (~ 20 year time steps)
Flood extent	Inundation model (after Wadey et al., 2012)

Receptors and consequences have usually been incorporated into risk assessments by evaluations of economic consequences in the form of expected annual damages (PenningRowsell et al., 2005, 2013). Tools to model human responses and risk to life have been demonstrated via agent-based models (e.g. Dawson et al., 2011) and empirical methods (e.g. Jonkman et al., 2008; Wallingford et al., 2006). In this paper, flood risk is considered as the interplay between the probability of a given event occurring, the people and property exposed to the flood event and the vulnerability of those at risk, as defined in earlier work (e.g. Samuels et al., 2009; Blaikie et al., 1994; Gwilliam et al., 2006; Kron, 2005; Fielding, 2007; UNDRO, 1982; United Nations and Birkmann, 2006; USACE et al., 2011).

Probability is included in the source component of the SPRC and it is commonly expressed as a return period (e.g. this work considers the 1 in 200 year flood event – an event that would be expected to occur, on average, once every 200 years, or more formally have a likelihood of occurrence of 0.5 per cent in a single year). This return period was chosen as it is a typical design standard for coastal defences and so is a critical threshold to assess. Exposure describes the area flooded (pathways of the SPRC) and the people/property within this area (receptors) (Narayan et al., 2014). Vulnerability links the receptors and consequence terms of the SPRC and determines the expected damages for given flood characteristics (e.g. in Fig. 1 a house with a raised floor level is less vulnerable, and thus expected damages would be reduced). In this paper, the change in the "exposure" component of flood risk is evaluated (i.e. we do not account for changes in vulnerability or attempt to evaluate the time-evolving cost of damage caused by flooding). We assume that no defences are present. This reflects that we do not have historic data on defences and beach state and these factors are probably not amenable to historic analysis. In this paper we present a method for assessing the historic exposure of coastal residential populations, and how this has evolved over approximately 200 years (since 1800) for two UK case study sites. The analysis will enable us to determine the key drivers of changes in risk of flooding in the coastal environment. A study site is chosen that represents typical areas of the well-developed UK coast that have already undergone assessments of plausible changes in sea levels and inundation, and has good data sets on population density, coastal floodplain elevations and historic sea levels. Quantifying the number and spatial location of people in the floodplain is vital for effective flood risk management in relation to evacuation planning. It is important to note that the approach in this paper focuses on the population exposure rather than the financial cost of flooding. The paper is structured into the following sections: Sect. 2, an introduction to the case study region; Sect. 3, methodology including Sect. 3.1, model outline and data sets used, Sect. 3.2, population distribution model, Sect. 3.3, flood inundation model and Sect. 3.4, exposure model (see Appendix A for modelling assumptions); Sect. 4, analysis/results of the exposed population calculations; Sect. 5, discussion and Sect. 6, conclusions and recommendations for future research.

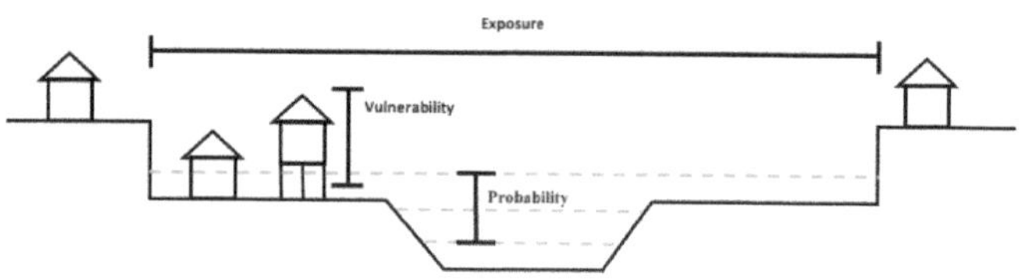

Figure 1. Cross section of a floodplain showing the components of risk.

CASE STUDY SITE

The study site (Fig. 2) is based in the densely populated region of the UK along the Solent estuary which includes the cities of Southampton and Portsmouth. The coastline spans approximately 55 km "as the crow flies" from Hurst Spit in the west to Selsey Bill in the east but it is heavily indented. The Solent region topography, population and land use is representative of many developed coastal areas, with approximately 25 000 properties on land exposed to a 1 in 200 year coastal flood (NFDC, 2010). Portsmouth has the UK's highest population density outside of London, and is a major site where properties are at risk of coastal flooding (RIBA and ICE, 2008). The Solent region faces many of the typical global development pressures on the coast: high population density, a strategic trade location (road and sea transport routes) and tourist/environmental attractions (NFDC, 2010). Some parts of the coastline (notably Portsea Island) have hard engineered sea defences, whereas other sections use softer approaches such as beach nourishment (e.g. Hayling Island). These defences are managed whilst sea levels have been rising, increasing the probability of extreme sea level and flood events (Haigh et al., 2011; Wadey et al., 2013).

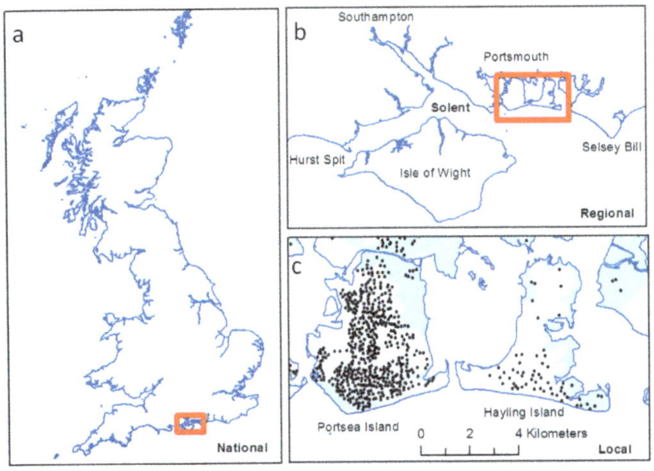

Figure 2. (a) and (b) Location of Portsea and Hayling Islands. (c) Centroid points for population data assigned to the 2011 UK national Census and the Environment Agency's 1 in 200 year indicative floodplain map (IFM, shaded blue) (Centroid points are Crown copyright/ database right 2013. An Ordnance Survey/EDINA supplied service. IFM is ©Environment Agency copyright and database rights 2015.).

There is already a substantial flood history and presentday threat: a study assessing the history of extreme sea levels and media accounts of floods identified 40 flood events in Portsmouth between 1960 and 2005 (Ruocco et al., 2011). On 10 March 2008 a storm surge, high tide and waves in the English channel led to significant coastal flooding in the Solent area (Wadey et al., 2013). The storms and high tides of the 2013–2014 winter caused a number of coastal flood events (Wadey et al., 2015). The study area has been zoned for flood "risk" by the UK Environment Agency for a 1 : 200 year extreme event assuming that no flood defences are present (Fig. 2c). In this study we continue to use the worst case undefended scenario in consistency with current management practices.

This case study tests the developed concept that is transferable to other densely populated coastal regions with appropriate data.

METHODOLOGY

Outline and Data Sets

In this study we are evaluating the evolution of exposure (as a proxy for risk), measured as the number of people within the indicative undefended coastal floodplain, for a 1 in 200 year flood event, given population change, residential development and sea level rise. A detailed digital elevation model of the floodplain was developed by Wadey et al. (2012). Sea level data are available for the study area for 1960 to 2008 (Haigh et al., 2011). Population data are available from the UK Census for Portsea and Hayling from 1801–2011 at 10 year time steps. Historic maps are available at roughly 20 year time steps (1870s, 1890s, 1910s, 1930s, 1960s, 1970s, 1990s and 2010s). From 1870–1990 the maps are at a scale of 1 : 10 560. For the 2010s map a scale of 1: 2000 is available. Data required and sources are summarised in Table 1. The methodology used in this study is shown in Fig. 3, and details of how the population is located and the flood extent generated are presented in the following subsection. We use known population data from the UK Census, locate the population spatially using historic maps and then identify the number of people exposed to flood risk in the 1 in 200 year floodplain. This process is repeated every 10 years between 1801 and 2011. Exposure is evaluated in a time step of 10 years to match the time step of the census data. Data sets for the physical system (sea levels, tidal curve and land elevations) are combined in a floodplain extent model. This gives the extent of the floodplain at different stages of time (e.g. accounting for changes in sea level, and excluding defences). The changes in historic shoreline position are not accounted for as part of this study.

The socio-economic data sets (population, historic maps) are combined in a population distribution model. This gives the spatial distribution of the population at each time step. For simplicity the extent of the housing development is assumed to be constant between the historic map years, as interpolation of housing development between map dates is difficult and unlikely to provide additional knowledge or understanding.

Population Distribution Model

Population Count

Demographic data from the UK Census were used to reconstruct the spatial population distribution at the study site since 1801 at 10 year intervals (Hampshire County Council, 2001; Registrar General for England and Wales, 1971; Of- fice of Population Censuses and Surveys, 1981, 1991; ONS, 2001, 2011). These data were used within the model to identify the coastal population at risk of flooding (Fig. 3).

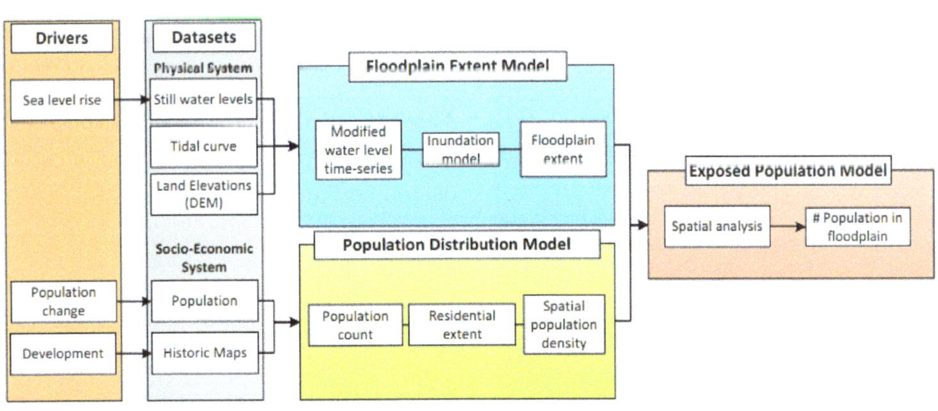

Figure 3. Methodology for evaluating changes to flood exposure.

Prior to 1971 the aggregate population for Portsea and Hayling Islands are used (shown as "non-spatial" data in Fig. 4), because the location of the population was not recorded. Some interpolation was necessary for the Hayling population (see Appendix A). For census years 1971–2011 spatial census data are available as centroid points. Centroid points (Fig. 2c) represent the population within a census output area. Output areas (OAs) are the lowest geographical level at which census estimates are provided. The output areas are designed to have between 40–125 households, with a minimum population of 100. Census data from the 2001 and 2011 censuses at the output area level were used, (OAs 504 and 522 within Portsea Island, respectively). For 1971, 1981 and 1991, data were extracted at the enumeration district level (312, 314 and 303 EDs, respectively – these censuses pre-date output area levels). Enumeration districts are less well defined, containing between 45–940 people for the 1971 Portsea and Hayling data, for example.

Historic Residential Extent

Maps of Portsea and Hayling Islands between 1870 and 2012 were used to identify the level of development and which areas were populated. Urban areas were digitised to create a residential mask in ARC GIS (geographical information system) and these were used to distribute the population count from the census data into the populated areas and to constrain population to residential areas (see Appendix A). The digitised residential areas are seen in Fig. 5. Development has increased on both islands between 1870 and 2012. On Portsea, early residential development (1870s) was centred near the dockyards area to the west of the island with small pockets of residential development elsewhere. The centre and east of the island began to be developed between the 1890s and 1910s and by 1930, the island was largely developed. Major developments since the 1930s include Anchorage park to the north-east of the island (seen in the 1990s map and expanded in the 2010s map), and developments in the Eastney area in the south-east corner of the island (seen from 1960 onwards). Hayling was sparsely developed from the 1870s through to the 1910s. In the 1930s development increased, mostly in the south of the island. As for Portsea, the pattern in the 1930s is similar to that of the modern day, although unlike Portsea, the population has grown more than 4 times larger. For instance, noticeable development did occur in the Eastoke peninsula (south-east corner of the island) seen in the 1960s through to the 2010s map. Portsea Island remains more developed than Hayling throughout the record.

Spatial Population Density

The Census data provided a population count and a centroid point to locate the population in each output area (OA) or enumeration district (ED) (see Fig. 2c). Surface Builder™ was used to distribute the population spatially (Martin, 1989). This model creates a raster grid with population density in each cell calculated as a function of the distance from each population centroid (see Fig. 6a, b and Appendix A). A raster grid is used as it offers ease of integration with other data sources (e.g. the raster flood maps) (Martin et al., 2011). Complications arose because census areas have changed over time (i.e. are different for each census) and the different geographies between censuses make longitudinal studies problematic (Langford, 2007; Martin et al., 2002). A solution is to use interpolation techniques to transform the population data to a common set of zones (Langford, 2007). For small spatial areas, such as output areas and enumeration districts, remodelling of the data to an underlying surface-based representation may prove the only alternative (Martin et al., 2002). In this study, the census population centroid data were aggregated to raster grid cells of size 50 m by 50 m using the SurfaceBuilder™ program. This grid-based method provides a consistent method of assessing the relationship between social vulnerability and exposure to flooding, as opposed to simpler methods based on census output areas (Martin, 1989; Thrush et al., 2005).

Floodplain Extent Model

As already noted, sea defences are excluded due to lack of data. An analysis of the effectiveness of coastal flood defences is beyond the scope of this paper. The lack of historic data on flood defences makes a temporal study of risk evolution including defences time unfeasible. Our aim is to assess the worst case scenario.

To determine the floodplain extent, we used a combined hydraulic model (LISFLOOD FP) (Bates et al., 2010) and digital elevation model (DEM) (Wadey et al., 2012) for a range of flood simulations by return period assuming no sea defences. LISFLOOD FP is an inertial formulation of the shallow water equations (Bates et al., 2010). It has been used to simulate coastal flood events (Smith et al., 2012; Quinn et al., 2014), including within the Solent (Wadey et al., 2012) where the model has been validated (Wadey et al., 2013). Floodplain flows are treated using a "storage cell" approach and implemented for a raster grid to allow an approximation to a two-dimensional (2-D) movement of the flood wave. A continuity equation is solved linking flow into a cell and its change in volume, and a momentum equation for each direction where flow between cells is calculated. With good quality topographic data, this model can produce similar results to full 2-D formulations of the shallow water equations (for sub-critical gradually varied flows only). The model is run for a single tidal cycle. This model has been shown to identify properties exposed to flooding in the Portsmouth case study with a vertical accuracy of approximately ±10 cm. The model application by Wadey et al. (2012) was modified in this application for historic simulations of flooding by adjusting the still water level boundary condition. Sea level rise was based on the estimates of Haigh et al. (2011) at Portsmouth from 1960 to 2008 and extrapolated back to 1801 (1.21 mm yr^{-1} ± 1 s.e.).

Exposure Model (Number of People at Risk)

The population layer and flood extent layer are combined to determine the exposed population in the floodplain (Fig. 6). The exposed population in each grid cell is summed to give a total exposed population for that time step. The process was repeated for each census year to assess the evolution of exposure of the coastal population.

RESULTS: CHANGES IN POPULATION EXPOSED TO FLOODING AND IT'S DRIVERS

The temporal evolution of exposure in Portsea and Hayling

are shown in Figs. 7 and 8. The error bars show the variability in calculated exposure due to uncertainty in the estimates of sea level, population size and distribution (for a breakdown of the uncertainty, see "Sources of uncertainty" in the Appendix). Three rates of sea level rise were used; the mean value for the Portsmouth tide gauge of 1.22 mm yr^{-1} (Haigh et al., 2011), and ± one standard deviation of this value (0.94 and 1.48 mm yr^{-1}, respectively).

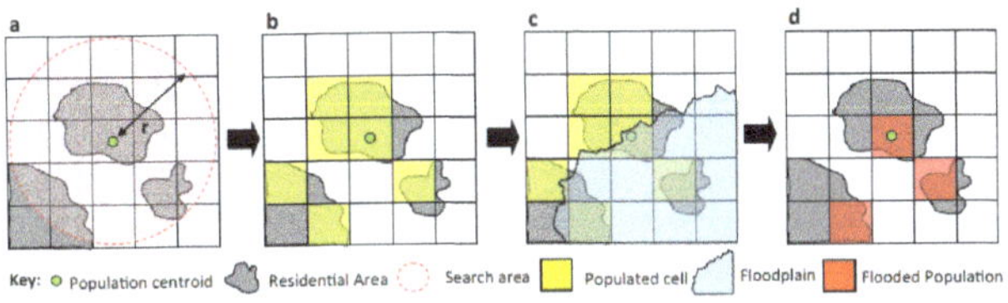

Figure 6. Method to calculate exposed population: (a, b) population is spread from centroid points to a raster grid according to specified search area (see Appendix A), (c) floodplain is overlain and (d) exposed population calculated.

Between 1801 and 2011, the exposed population in Portsea has increased from approximately 1500 people in 1801 to 19 800 in 2011. This represents a greater than 10-fold increase in exposure. Figure 7 shows the temporal evolution; there is a slow rise 1800–1850, a faster rise 1850–1930. Exposure then levels off and falls slightly 1940–1970, followed by a further rise 1980–2011. The curve follows the same pattern as the island's total population (Fig. 4). In Hayling there was only a very small population (< 100 people) exposed to flooding prior to 1921 and this result is consistent across all sea level rates applied (Fig. 8). From 1921 to 2011 there is an almost 15-fold increase in population exposed to flooding over this period – rising from 120 in 1921 to 1759 in 2011. There are two periods with significant increases in exposure: 1951–1961 and 1971–1981.

To determine the relative importance of sea-level rise and population change as the drivers of flood risk, the exposed populations are re-calculated for two scenarios:

- sea levels do not change from the extrapolated 1801 level, and population rises;
- population in 1801 remains static and sea level rises at the mean rate of 1.22 mm yr^{-1}.

The results are shown in Fig. 9. The differences between the two curves in each plot indicate the relative contribution to exposure caused by sea level rise and population change. For Portsea, sea level rise between 1801 and 2011 results in an increase in flood exposure to the 1801 population from 2200 to 4000 (i.e. +1800 people, 82 %), whereas population change over the same period with a static 1801 sea level accounts for +7600 people exposed to flooding (i.e. 2200– 9800, 345 %). In Hayling, the equivalent figures are 50 to 50 (+0, i.e. no change in exposure due to sea level), but for population change the exposure rises from 50 people in 1801 to 1080 people in 2011 (i.e. +1030 people, 2060 %).

This demonstrates that population change has been a more important driver of flood risk than sea level rise in both Portsea and Hayling. Indeed at Portsea, population change is 5 times more important in changing flood risk over this period, which in Hayling, in relative terms, has been even more dominant, even though absolute figures are lower. This analysis was repeated for a range of return period water levels including 1 in 1, 1 in 5, 1 in 10, 1 in 50, 1 in 100 and 1 in 1000 year levels. All of the results show the same trend (albeit for Hayling; there is no exposure for the low return period storms): only the 1 in 200 year results were included in the paper to provide a succinct analysis. This analysis used the mean change in estimated sea level; there is some uncertainty in the actual sea levels as shown in the error bars in Figs. 7 and 8. This uncertainty may account for a variation in calculated exposure of up to 1000 people in 1981. There is no easy way to assess the accuracy of the population data, but the data are the best available and it is a legal requirement for all UK residents to register in the Census.

DISCUSSION: OVERVIEW AND APPLICABILITY TO OTHER SITES

This research builds upon that of Foresight (Evans et al., 2004) and Smith (2015) with its strength being in its transferability to other sites. The methodology described here could be applied to any coastal site where adequate spatial data sets (land use, elevations, population) and sea level data are available.

A national analysis of flood risk is possible using this approach, taking advantage of the modern day data collection systems available in many countries. To demonstrate this, a snapshot national analysis was carried out for the present day flood exposure in England and Wales. We used the present day Environment Agency Indicative floodplain map for both river and coastal flooding plus Census data for 2011. There are some limitations in this approach, for example the floodplain map includes both fluvial and marine flood extents. The algorithm took less than 1 h to run. The calculated exposure to the 1 in 200 year flood event (without sea defences) was 4.8 million people, which is within 10 % of the figure of 5.2 million quoted by the National Flood Forum (NFF, 2015). This quick analysis gives credence to the methodology, however, for a full national scale analysis, a more detailed population data set and DEM model would be necessary. To reduce data processing times, analysis could be restricted to only those areas known to be at risk of flooding and it is estimated that a national scale study could be completed in a few months.

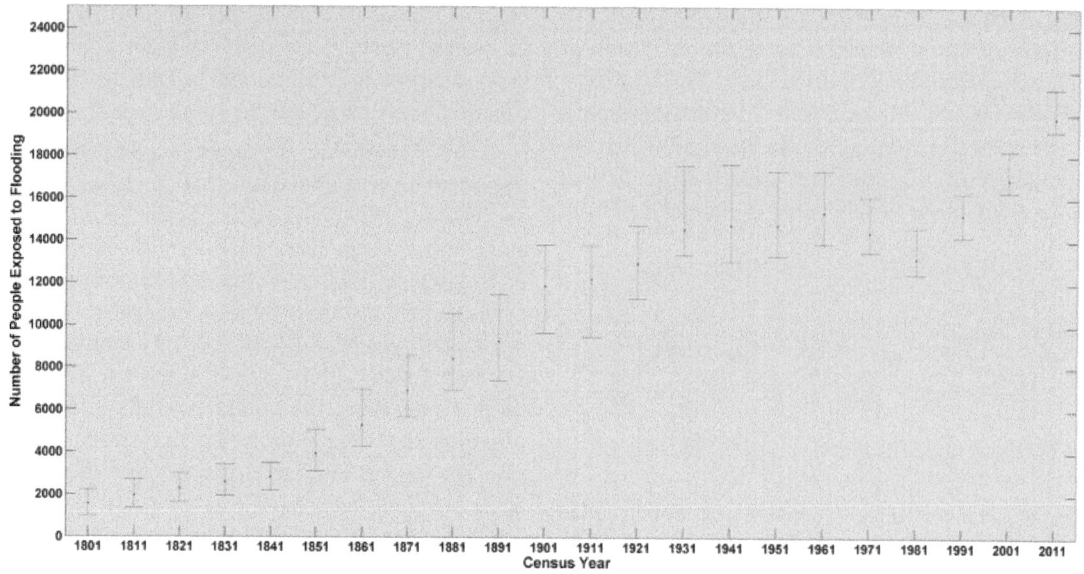

Figure 7. Estimated number of people exposed to flooding in Portsea (1 in 200 year recurrence interval, no defences). Error bars represent uncertainty in estimated rate of sea level change, population distribution and population size.

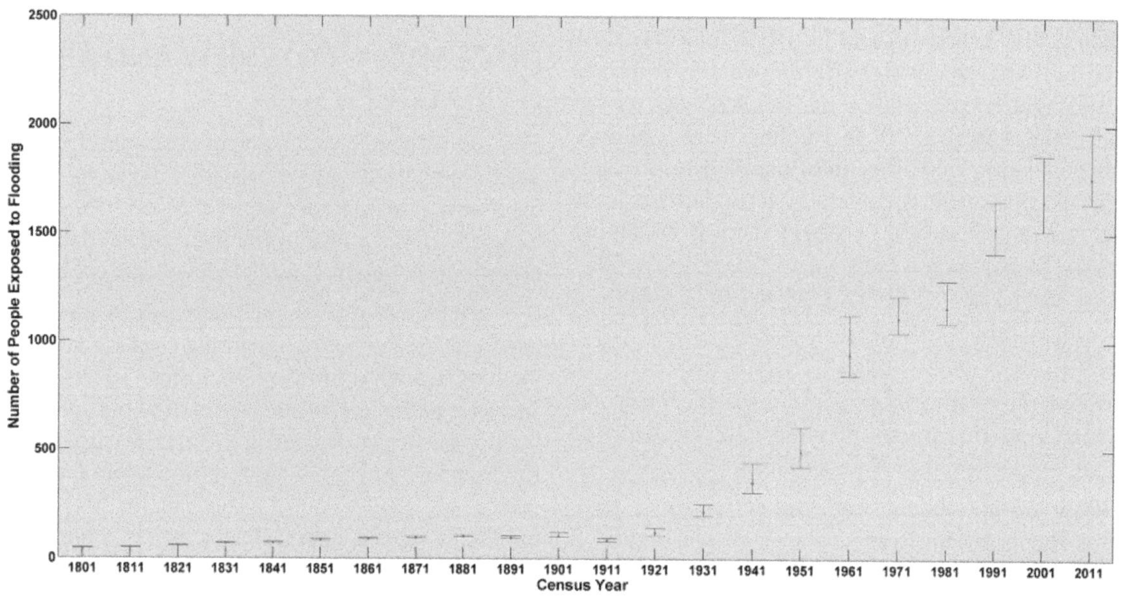

Figure 8. Estimated number of people exposed to flooding in Hayling (1 in 200 year recurrence interval, no defences). Error bars represent uncertainty in estimated rate of sea level change, population distribution and population size.

For an historical analysis users would need access to population data and indicative floodplain maps at regular intervals. The 10 year time step used in this study was chosen on the basis of the UK Census timings and some interpolation was necessary between the spatial data obtained from maps published at irregular time steps. However, the large time step (10 years) may hide changes in coastal population over shorter timescales because urban development can be rapid and significant areas of new coastal settlements can be constructed in less than 5 years This highlights the need for regular high quality data collection on both physical variables (land elevations, sea levels) and socio-economic variables (population size and density, residential extent). The methodology can be developed to look explicitly at attributing flood risk to the underlying drivers. Applying the methodology to different case studies will test whether the attribution of flood risk is consistent across a nation or whether regional differences exist. Over the last 200 years, population has increased across the UK, leading to increased encroachment of development and a higher population density upon floodplains so we would expect a similar pattern to that seen in the present case study. Only in lowlying areas where development/population rise has remained static would observed sea level rise have played a more significant role than that of population change. We suggest that this is more likely to be the case in the future as cities such as Portsmouth reach “saturation point” in their development. The existence of exceptions could be tested by repeating the method across the whole country; we propose this as necessary future work.

The evolution of the effectiveness of flood defences is an area for further study as when combined with exposure, it allows estimate of changing flood risk. However this presents significant challenges for historical analyses, for example, we found that information on flood defences at Portsea before 1990 is poorly recorded. This is likely to be the general case and hence while we may estimate historic exposure back to 1800, we cannot similarly estimate flood risk. This emphasises the importance of documenting defences and vulnerability characteristics over time, such as seen in the UK's Strategic Regional Coastal Monitoring Programme (e.g. see http://www.channelcoast.org/).

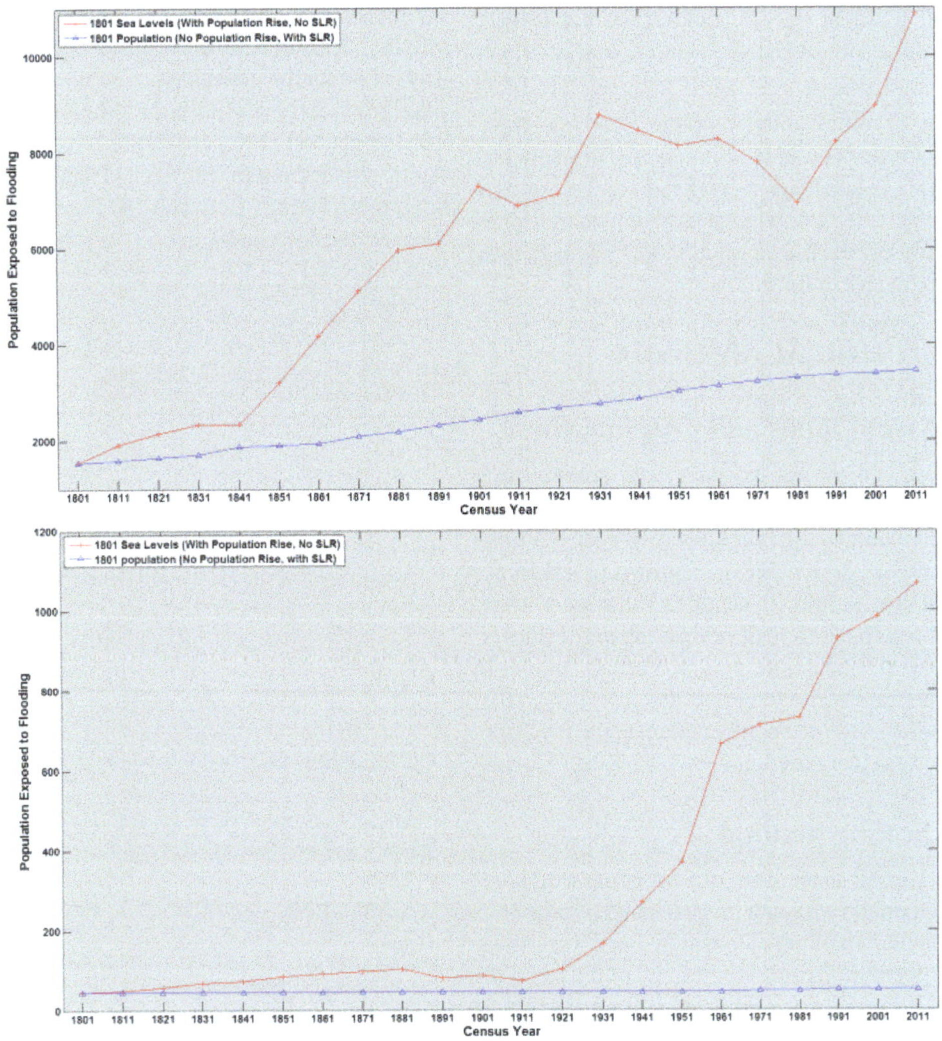

Figure 9. Estimated number of people exposed to flooding (1 in 200 year recurrence interval, no defences) in Portsea (above) and Hayling (below) for no change in sea levels since 1801 (red line) and no change in population since 1801 (blue line).

CONCLUSIONS

This paper has identified and filled a gap in our knowledge of the drivers of risk of coastal flooding, and how this exposure has developed over time. This has implications for the current assessment of coastal flood events, and also for future planning decisions.

In the Solent case study, population change has been shown to be the most significant driver of flood exposure from 1801 to the present time. Observed sea-level rise has a lesser but still significant effect on flood exposure estimates, especially over long timescales (100+ years).The rate of sea level rise is expected to increase, and rising sea levels are likely to have a larger effect on exposure in the future. Furthermore, for Small Island communities, such as Portsea and Hayling, the area available for development may become a limiting factor in the future, causing a shift in drivers that increase the exposure of the population to flood risk towards sea level rise. The estimated exposure to flooding shows that large numbers of people are potentially at risk (18 000 in Portsea for a 1 : 200 event), but they are currently mostly protected by sea defences constructed to a present day 1 : 200 event, with a GBP 44 million defence improvement programme recently announced (Dredging Today, 2015). This paper further demonstrates that assuming a stationary system (for example, assuming the urban extent is static, that population does not change, or that sea levels do not change) is likely to lead to inaccurate estimates of flood exposure and thus flood risk.

A limitation of this work is the inherent unpredictability of future changes in population dynamics across the UK. Agent-based approaches have been used to predict development and population change (such as developed by Fontaine, 2010). Coupling the method presented in this paper with such approaches will develop insights on these processes.

The approach developed here agrees with an independent, national scale assessment of exposure. The methodology can be applied to other areas of the UK, or elsewhere, where population, urban extent and sea level data exists. Attribution of local flood exposure and risk will depend on relative sea level and morphology/hydrology and population dynamics. National studies have shown development in flood risk areas in the UK is increasing, in some cases at a higher rate than development outside of the floodplain (ASC, 2011). Hence, exposure to coastal flooding due to socio-economic drivers seems likely to continue, following the historic trends shown here.

A combination of novel methodologies such as those developed in this paper, and continued collection of high quality data sets on floodplain geometry, sea level and population will contribute towards increased knowledge and understanding in this field. This will aid coastal managers as they prepare to face the challenges of an uncertain future.

APPENDIX A: TECHNICAL APPENDIX

A1 Modelling Assumptions and Considerations

The assumptions used in the methodology are summarised in Table A1. The temporal resolution of the available demographic data constrained the time step to 10 years. Whilst this time step may miss shorter term changes (i.e. seasonal/ yearly variations in hydrology), it captures the longer term dynamics of population change and development, and sea level rise which occurs over a long time period. Further, the high spatial resolution and quality of the census data used gives the study greater reliability than if supplementary data (perhaps with a smaller time step) were used.

A2 Population Scaling Method

This data 1971–2011 exists in the form of population weighted centroid points. Each point represents a census output area and contains the total population of the output area. For census data pre-1971 aggregate population counts for the city of Portsmouth (scaled to represent population within Portsea Island) and for Hayling Island were used. Scaling the total counts in this way deals with the problem of changing geographies through time (e.g. changing administrative boundaries). The populations were scaled using aggregate population counts for the city of Portsmouth for census years 1801–1961 and the modelled counts (spatial populations from centroid points) for census years 1971–2011 (Eq. A1).

$$\mathrm{Pop}_{\mathrm{scaled}_i} = \mathrm{Pop}_{\mathrm{total}_i} \times \frac{\sum_{n=1971}^{2011} \frac{n_{\mathrm{modelled}}}{n_{\mathrm{total}}}}{n_{\mathrm{years}}}, \qquad \text{(A1)}$$

where: $\mathrm{Pop}_{\mathrm{scaled}_i}$ = is the scaled population used within the model at time step i; $\mathrm{Pop}_{\mathrm{total}_i}$ = is the total population for Portsmouth from the census data at time step i; n_{modelled} = is the modelled population used in the spatial census study (1971–2011); n_{total} = the total population for Portsmouth from the census data (1971–2011); n_{years} = is the number of years where spatial data exists (which is = 5 for the case study).

Figure 4 summarises our reconstruction of the population in Portsea and Hayling; which for the former rose from 39 000 in 1841 to a peak of 194 000 in 1931. The population then falls to a low of 134 000 in 1981 before rising again to 164 000 in 2011. The modelled populations from 1801–1961 were from scaled population counts, and 1971–2011 from spatial census data. Historic census data for Hayling parish (which covers the spatial area of Hayling Island) extend to 1801. However, it is not complete due to changing administrative boundaries during the 19th and 20th centuries. Therefore the population counts for missing census years were interpolated. The population in Hayling rose steadily from just under 600 in 1801 to 4000 in 1941. Population continued to increase at a higher rate until the maximum of 17 400 in 2011. Modelled populations in 1801–1851, and 1881–1931 are formed from raw counts from census data, with values in 1861–1871 and 1941–1961 interpolated from these counts. Between 1971 and 2011 spatial census data for Hayling were used.

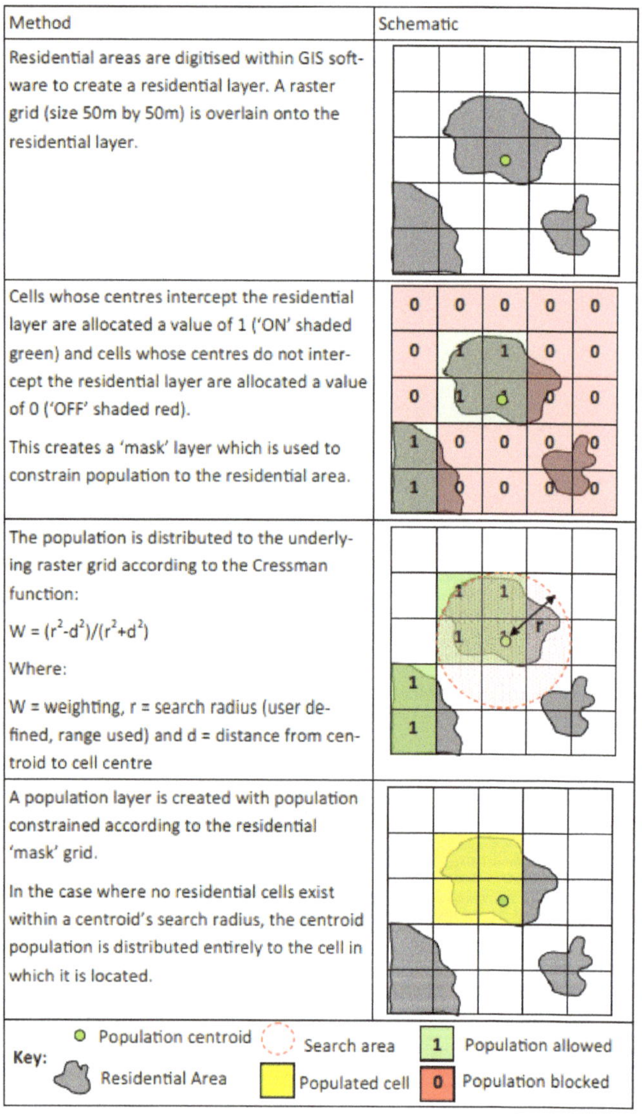

Figure A1. Population spreading method used in this study. See Martin (1989) and Bracken and Martin (1989) for further information on the centroid distribution method.

A3 Residential Layer Method

Maps (sourced from Digimap®, University of Edinburgh) for the 1870s, 1890s, 1910s, 1930s, 1960s, 1970s, 1990s and 2012 are summarised in Table A2. Developed areas were hand-digitised to create a residential layer of where population is situated. This allowed population to be spread more realistically. Non-residential features such as schools, hospitals and industrial units (e.g. the Portsmouth Dockyard) were removed from the residential layer in order to increase the accuracy of the population spreading. Use of a residential layer addresses the problem of differing census geographies by constraining population to the area developed for each time step. The time between publication of the maps used averages 20 years between 1870–2011, which is typical of spatial planning timescales and so a reasonable assumption. Assuming static development over a 70 year time period (1801– 1871) is more uncertain, however the low level of development seen in 1871 does limit the effect of this assumption. Analysis from 1801 is therefore included in the analysis but with the caveat that we are less certain of the results over this time frame.

The vector residential layer was converted to a 50 m raster mask for compatibility with Surface Builder™. A 50 m resolution includes adjacent roads in residential masks. However, the spatial resolution of census data makes higher resolution (e.g. 10 m grid cells) unrealistic. This layer was used as a mask within SurfaceBuilder™ which prevented the program placing population into areas that should not be populated.

A4 Population Spreading Method

The methodology used within SurfaceBuilder™ is shown in Fig. 8. A range of search radii were used in order to account for uncertainty due to this method. The search radii limits the distance from each centroid that the population can be distributed.

A5 Sources of Uncertainty

There is uncertainty inherent in the estimated sea level, and the number and spatial location of the population. The uncertainty in rate of sea level applied was quantified by modelling for three different rates; the mean change from Haigh et al. (2011) and ± one standard deviation from this. Uncertainty in the population estimates are harder to quantify. The measured undercount in the 2001 census was calculated as 6 % (ONS, 2012a). There is a smaller potential for overcount which was estimated as 2 % for the 2011 Census (ONS, 2012b). These uncertainties are accounted for in census population counts, however for older censuses the adjustments may not have been performed and so as a conservative estimate we assume a potential uncertainty of +6 and −2 % in the population estimate (i.e. potential 6 % undercount, 2 % overcount). The spatial location of the population is sensitive to the search radii used when distributing the population from the centroid points. The uncertainty in population location was quantified by testing a variety of search radii. The relative contributions of these three sources of uncertainty are shown in Fig. A2.

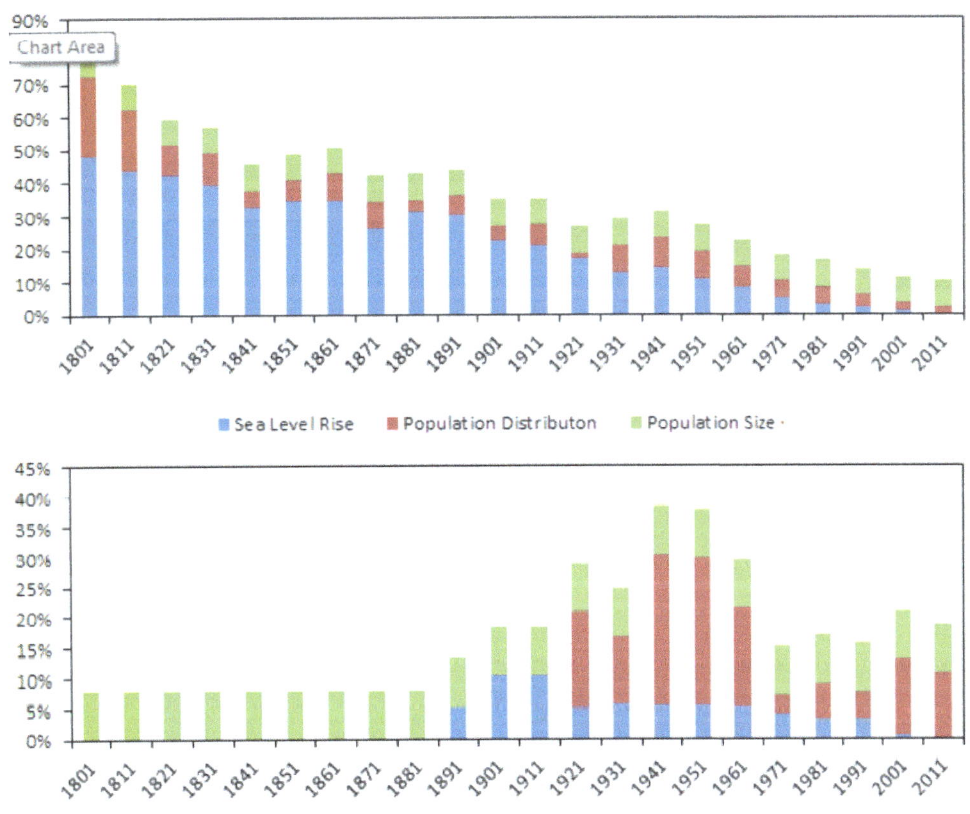

Figure A2. Quantification of the sources of uncertainty within the methodology for Portsea (above) and Hayling (below).

For Portsea, the uncertainty in sea level has a much bigger effect than Hayling as there is a much larger population density on the island, and so the floodplain size (a function of the sea level elevation) has a more pronounced effect on the estimated exposure. In Hayling there is a much smaller density of people and so the exposure is less sensitive to a slightly smaller/ bigger floodplain. In Portsea the distribution of the population has a moderate effect on exposure in the early 1800s, with an increasingly smaller effect for the more modern (better quality) census data. In Hayling there is no effect before 1920 as the low absolute exposure (less than 50 people exposed as the "best estimate") is not sensitive to changes in population distribution. As the population in Hayling started to encroach on the floodplain from 1920 onwards, the distribution has a larger relative effect. The uncertainty as a result of population size is static through time for both Portsea and Hayling as this is assumed to be 6 % for undercount and 2 % for overcount (i.e. uncertainty in census data – see ONS, 2012a).

Table A1. Modelling Assumptions and justifications.

Component	Modelling assumption	Justification
Hydrodynamic model (LISFLOOD FP)	Simplified hydraulics compared to "full" 2-D models Sea level and extremes of still water level are dominant physical drivers (waves excluded) See Bates et al. (2010)	Better than "bathtub" methods (mass conservancy and hydraulic connectivity accounted for) Widely used flood model (e.g. Wadey et al., 2012; Dawson et al., 2009; Rojas et al., 2013) Use of full models expensive (cost and computationally) and without validation improved accuracy cannot be confirmed Represents realistic storm tide inflow Waves, although important to flood events, are contentious in an inundation modelling framework (hard to validate) but recommended for inclusion in future work Model proven for coastal use (Bates et al., 2005) and with a validated model for the case study region (Wadey et al., 2012).
Residential area	Developed residential area does not change between time steps (average 20 year time step – based on availability of historic maps)	20 years is typical of long-term spatial planning time horizon (Zevenbergen et al., 2008). Constraining population to residential area improves spreading over uniformly distributing population, so best available method
Population distribution	A centroid defines a location with above average population density and is a summary point for the local area A centroid's population is distributed in the surrounding area according to some distance decay function, which has finite extent Regions may exist in the population plane in which no population is present. Assumptions from Martin (1989)	Allows for high resolution population surfaces (Martin, 1989) Method offers stability through time and ease of integration with non-population data sources (Martin et al., 2011); both are essential parts of the methodology discussed in this paper
Population change over time	The dates chosen represent a trend in population change, rather than oscillations (which do not show correlation over time). The dates chosen are representative of population change	A period of 200 years was chosen to allow for a clear trend to propagate as opposed to variation which may occur over a smaller time span The dates correspond to census years, where it is possible to get high resolution spatial population and demography data. To use other years with less sufficient data would limit the reliability of the study

Table A2. Historic maps used to create residential masks for each census year. All maps sourced from Digimap® Crown Copyright and Landmark Information Group Limited (2014). All rights reserved.

Census year	Map used to create residential layer
1801–1871	County Series Edition 1 (1870s)
1881–1891	County Series Revision 1 (1890s)
1901–1911	County Series Revision 2 (1910s)
1921–1931	County Series Revision 3 (1930s)
1941–1961	National Grid Imperial Edition 1 (1960s)
1971	National Grid Metric Edition 1 (1970s)
1981–1991	Latest National Grid (1990s)
2001–2011	MasterMap® (2012)

Acknowledgements

Andrew Stevens acknowledges a University of Southampton studentship from the EPSRC (Engineering and Physical Sciences Research Council). Thanks is extended to the University of Edinburgh's Digimap® team for their extensive work digitising census data which made this study possible. ESRI's ArcGIS was the GIS tool used.

REFERENCES

1. ASC: Adapting to climate change in the UK Measuring progress: Adaptation Sub-Committee Progress Report 2011, London, UK, 2011.
2. Bates, P. D., Dawson, R. J., Hall, J. W., Matthew, S. H. F., Nicholls, R. J., Wicks, J., and Hassan, M.: Simplified two-dimensional numerical modelling of coastal flooding and example applications, Coast. Eng., 52, 793–810, 2005.
3. Bates, P. D., Horritt, M. S., and Fewtrell, T. J.: A simple inertial formulation of the shallow water equations for efficient twodimensional flood inundation modelling, J. Hydrol., 387, 33–45, 2010.
4. Batstone, C., Lawless, M., Tawn, J., Horsburgh, K., Blackman, D., McMillan, A., Worth, D., Laeger, S., and Hunt, T.: A UK bestpractice approach for extreme sea-level analysis along complex topographic coastlines, Ocean Eng., 71, 28–39, 2013.
5. Bracken, I. and Martin, D.: The Generation of Spatial Populationdistribution from Census Centroid Data, Environ. Plann. A, 21, 537–543, 1989.

6. Church, J. A., Clark, P. U., Cazenave, A., Gregory, J. M., Jevrejeva, S., Levermann, A., Merrifield, M. A., Milne, G. A., Nerem, R. S., Nunn, P. D., Payne, A. J., Pfeffer, W. T., Stammer, D., and Unnikrishnan, A. S.: Sea Level Change, in: Climate Change 2013: The Physical Science Basis. Contribution of Working Group I to the Fifth Assessment Report of the Intergovernmental Panel on Climate Change, edited by: Stocker, T. F., Qin, D., Plattner, G.- K., Tignor, M., Allen, S. K., Boschung, J., Nauels, A., Xia, Y., Bex, V., and Midgley, P. M., Cambridge, United Kingdom and New York, NY, USA, Cambridge University Press, 2013.
7. Dawson, R. J., Dickson, M. E., Nicholls, R. J., Hall, J. W., Walkden, M. J. A., Stansby, P. K., Mokrech, M., Richards, J., Zhou, J., Milligan, J., Jordan, A., Pearson, S., Rees, J., Bates, P. D., Koukoulas, S., and Watkinson, A. R.: Integrated analysis of risks of coastal flooding and cliff erosion under scenarios of long term change, Climatic Change, 95, 249–288, 2009.
8. Dawson, R. J., Peppe, R., and Wang, M.: An agent-based model for risk-based flood incident management, Natural Hazards, 59, 167–189, 2011.
9. Dredging Today: Portsmouth Coastal Scheme About to Begin [Online], available at: http://www.dredgingtoday.com/2015/ 01/30/portsmouth-coastal-scheme-about-to-begin/ (last access: May 2015), 2015.
10. Evans, E., Ashley, R., Hall, J., Penning-Rowsell, E., Saul, A., Sayers, P., Thorne, C., and Watkinson, A.: Foresight. Future Flooding, Scientific Summary: Volume I – Future risks and their drivers, London, UK, 2004.
11. Fielding, J.: Environmental injustice or just the lie of the land: an investigation of the socio-economic class of those at risk from flooding in England and Wales, Sociological Research Online, 12, 36 pp., 2007.
12. Gouldby, B., Sayers, P., Mulet-Marti, J., Hassan, M., and Benwell, D.: A methodology for regional-scale flood risk assessment, Proceedings of the Institution of Civil Engineers – Water Management, 161, 169–182, 2008.
13. Gwilliam, J., Fedeski, M., Lindley, S., Theuray, N., and Handley, J.: Methods for assessing risk from climate hazards in urban areas, Proceedings of the Institution of Civil Engineers-Municipal Engineer, 159, 245–255, 2006.
14. Haigh, I. D., Nicholls, R., and Wells, N.: A comparison of the main methods for estimating probabilities of extreme still water levels, Coast. Eng., 57, 838–849, 2010.
15. Haigh, I. D., Nicholls, R. J., and Wells, N.: Rising sea levels in the English Channel 1900 to 2100, Proceedings of the Institution of Civil Engineers-Maritime Engineering, 164, 2011.
16. IPCC: Summary for Policymakers, in: Climate Change 2013: The Physical Science Basis. Contribution of Working Group I to the Fifth Assessment Report of the Intergovernmental Panel on Climate Change, edited by: Stocker, T. F., Qin, D., Plattner, G.-K., Tignor, M., Allen, S. K., Boschung, J., Nauels, A., Xia, Y., Bex, V., and Midgley, P. M., Cambridge, UK and New York, NY, USA: Cambridge University Press, 2013.
17. Jonkman, S. N., Vrijling, J. K., and Vrouwenvelder, A. C. W. M.: Methods for the estimation of loss of life due to floods: a literature review and a proposal for a new method, Natural Hazards, 46, 353–389, 2008.
18. Kron, W.: Flood Risk = Hazard Values Vulnerability, Water Int., 30, 58–68, 2005.
19. Kron, W.: Coasts, the Riskiest Place on Earth, edited by: Mckee Smith, J., Proceedings of the 31st International Conference of Coastal Engineering, 3–21, 2008.
20. Langford, M.: Rapid facilitation of dasymetric-based population interpolation by means of raster pixel maps, Comput. Environ. Urban, 31, 19–18, 2007.
21. Martin, D.: Mapping Population Data from Zone Centroid Locations, Transactions of the Institute of British Geographers NS, 14, 90–97, 1989.
22. Martin, D., Dorling, D., and Mitchell, R.: Linking censuses through time: problems and solutions, Area, 34, 82–91, 2002.
23. Martin, D., Lloyd, C., and Shuttleworth, I.: Evaluation of gridded population models using 2001 Northern Ireland Census data, Environ. Plann. A, 43, 1965–1980, 2011.
24. Narayan, S., Nicholls, R. J., Clarke, D., Hanson, S., Reeve, D., Horrillo-Caraballo, J., le Cozannet, G., Hissel, F., Kowalska, B., and Parda, R.: The SPR systems model as a conceptual foundation for rapid integrated risk appraisals: Lessons from Europe, Coast. Eng., 87, 15–31, 2014.
25. NFDC (New Forest District Council): North Solent Shoreline Management Plan, available at: www.northsolentsmp.co.uk (last access: May 2015), 2010
26. NFF: At risk of flooding?, National Flood Forum, available at: http://www.nationalfloodforum.org.uk/at-risk-of-flooding-2/ (last access: May 2015), 2015.
27. Nicholls, R. J.: Impacts of and responses to sea-level rise, in: Understanding Sea-Level Rise and Variability, edited by: Church, J. A., Woodworth, P. L., Aarup, T., and Wilson, W. W., WileyBlackwell, 2010.
28. Office of Population Censuses and Surveys: 1981 Census: Aggregate data (England and Wales) [computer file], available at: http://digimap.edina.ac.uk/ (last access: May 2015), 1981.
29. Office of Population Censuses and Surveys: 1991 Census: Aggregate data (England and Wales) [computer file], available at: http://digimap.edina.ac.uk/ (last access: May 2015), 1991.
30. ONS: The 2011 Census Coverage Assessment and Adjustment Process, 2011 Census: Methods and Quality Report, available at: http://www.ons.gov.uk/ons/guide-method/ census/2011/census-data/2011-census-data/2011-first-release/ first-release--quality-assurance-and-methodology-papers/ coverage-assessment-and-adjustment-process.pdf (last access: May 2015), 2012a.
31. ONS: Overcount Estimation and Adjustment. 2011 Census: Methods and Quality Report, available at: http://www.ons.gov.uk/ons/guide-method/census/ 2011/census-data/2011-census-data/2011-first-release/ first-release--quality-assurance-and-methodology-papers/overcount-estimation-and-adjustment.pdf (last access: May 2015), 2012b.

32. Penning-Rowsell, E., Johnson, C., Tunstall, S., Tapsell, S., Morris, J., Chatterton, J., and Green, C.: The Benefits of Flood and Coastal Risk Management: A Handbook of Assessment Techniques, London, UK, 2005.

33. Penning-Rowsell, E., Priest, S., Parker, D., Morris, J., Tunstall, S., Viavattene, C., Chatterton, J., and Owen, D.: Flood and Coastal Erosion Risk Management. A Manual for Economic Appraisal, Routledge, London, UK and New York, NY, USA, 2013.

34. Quinn, N., Lewis, M., Wadey, M., and Haigh, I.: Assessing the temporal variability in extreme storm-tide time series for coastal flood risk assessment, J. Geophys. Res.-Oceans, 119, 4983– 4998, 2014.

35. Registrar General for England and Wales: 1971 Census: Aggregate data (Great Britain) [computer file], Digitised Census data, available at: http://digimap.edina.ac.uk/ (last access: May 2015), 1971.

36. RIBA and ICE: Facing up to rising sea-levels: retreat? defend? attack? The future of our coastal and estuarine cities, available at: www.buildingfutures.org.uk (last access: May 2015), 2008.

37. Rojas, R., Feyen, L., and Watkiss, P.: Climate change and river floods in the European Union: Socio-economic consequences and the costs and benefits of adaptation, Global Environ. Chang., 23, 1737–1751, 2013.

38. Ruocco, A. C., Nicholls, R. J., Haigh, I. D., and Wadey, M. P.: Reconstructing coastal flood occurrence combining sea level and media sources: a case study of the Solent, UK since 1935, Natural Hazards, 59, 1773–1796, 2011.

39. Samuels, P. and Gouldby, B.: Language of Risk Project Definitions, 2nd Edn., FLOODsite Consortium, available at: http://www.floodsite.net/html/partner_area/project_docs/T32_ 04_01_FLOODsite_Language_of_Risk_D32_2_v5_2_P1.pdf (last access: May 2015), 2009.

40. Sayers, P., Hall, J., Rosu, C., Chatterton, J., and Deakin, R.: Risk assessment of flood and coastal defences for strategic planning (RASP) – a high level methodology, Environment Agency, London, UK, 2002.

41. Thrush, D., Burningham, K., Fielding, J., and Agency, E.: Flood Warning for Vulnerable Groups: Measuring & Mapping Vulnerability R&D Technical Report W5C-018/4, Bristol, UK, 2005.

42. United Nations and Birkmann, J.: Measuring Vulnerability to Natural Hazards: Towards Disaster Resilient Societies, United Nations University Press, New York, NY, USA, 2006.

43. USACE, The Japanese Ministry of Land Infrastructure Transport and Tourism, Environment Agency & Rijkswaterstaat: Flood Risk Management Approaches As being practiced in Japan, Netherlands, United Kingdom, and United States, 2011.

44. Wadey, M. P., Nicholls, R. J., Hutton, C.: Coastal Flooding in the Solent: An Integrated Analysis of Defences and Inundation, Water, 4, 430–459, 2012.

45. Wadey, M. P., Cope, S. N., Nicholls, R. J., Mchugh, K., Grewcock, G., and Mason, T.: Extreme Sea Levels and Flood Events in a Coastal Town: Community Participatory Exercise and Inundation Analysis, Environ. Model. Softw., 31, 1–22, 2015.

46. Wahl, T., Haigh, I. D., Woodworth, P. L., Albrecht, F., Dillingh, D., Jensen, J., Nicholls, R. J., Weisse, R., and Wöppelmann, G.: Observed mean sea level changes around the North Sea coastline from 1800 to present, Earth-Sci. Rev., 124, 51–67, 2013.

47. Wolf, J., Brown, J. M., and Howarth, M. J.: The wave climate of Liverpool Bay – observations and modelling, Ocean Dynam., 61, 639–655, 2011.

48. Zevenbergen, C., Veerbeek, W., Gersonius, B., Thepen, J., and van Herk, S.: Adapting to climate change: using urban renewal in managing long-term flood risk, Flood Recovery, Innovation and Response, 118, 221–233, 2008.

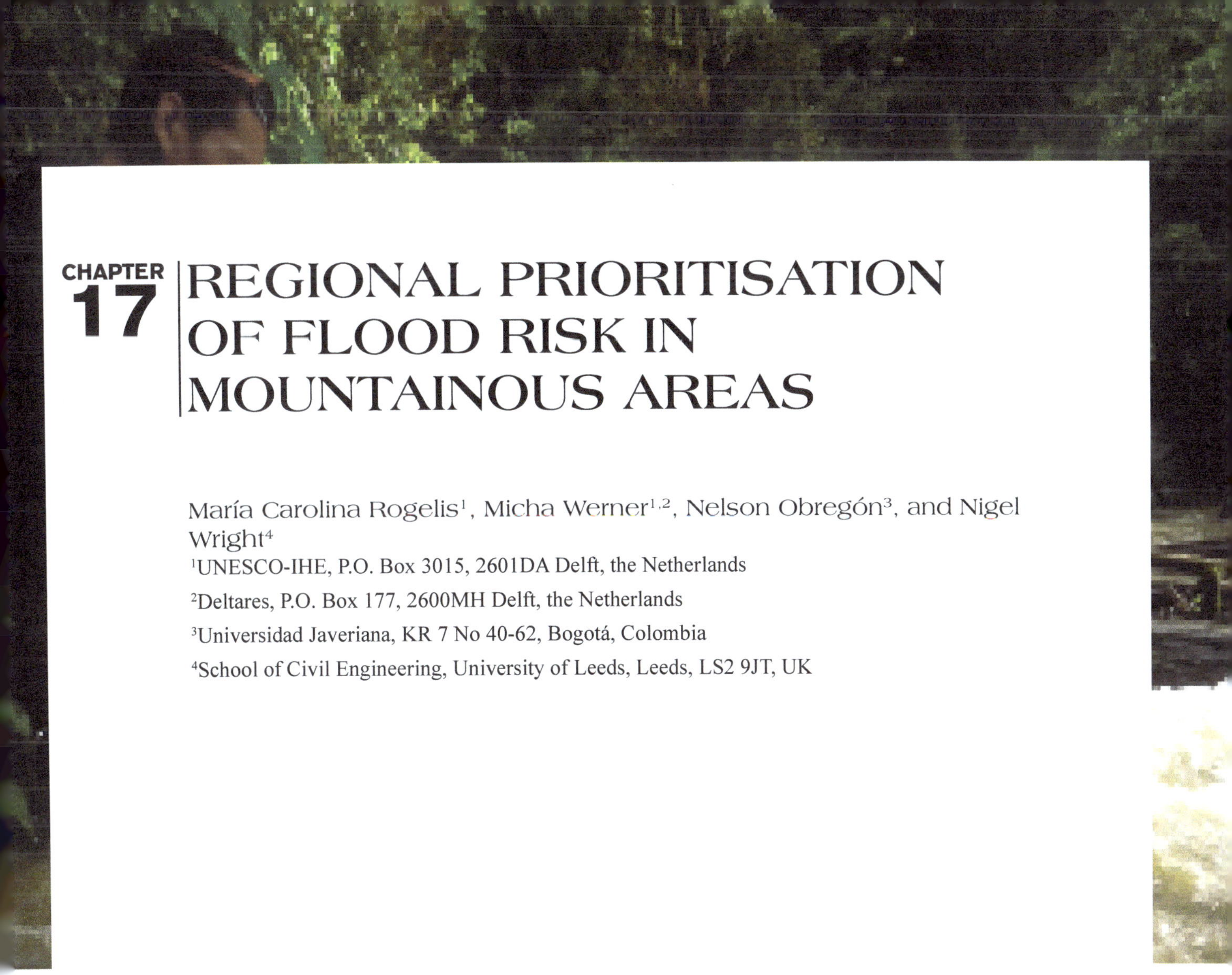

CHAPTER 17 REGIONAL PRIORITISATION OF FLOOD RISK IN MOUNTAINOUS AREAS

María Carolina Rogelis[1], Micha Werner[1,2], Nelson Obregón[3], and Nigel Wright[4]

[1]UNESCO-IHE, P.O. Box 3015, 2601DA Delft, the Netherlands

[2]Deltares, P.O. Box 177, 2600MH Delft, the Netherlands

[3]Universidad Javeriana, KR 7 No 40-62, Bogotá, Colombia

[4]School of Civil Engineering, University of Leeds, Leeds, LS2 9JT, UK

ABSTRACT

In this paper a method is proposed to identify mountainous watersheds with the highest flood risk at the regional level. Through this, the watersheds to be subjected to more detailed risk studies can be prioritised in order to establish appropriate flood risk management strategies. The prioritisation is carried out through an index composed of a qualitative indicator of vulnerability and a qualitative flash flood/debris flow susceptibility indicator. At the regional level, vulnerability was assessed on the basis of a principal component analysis carried out with variables recognised in literature to contribute to vulnerability, using watersheds as the unit of analysis. The area exposed was obtained from a simplified flood extent analysis at the regional level, which provided a mask where vulnerability variables were extracted. The vulnerability indicator obtained from the principal component analysis was combined with an existing susceptibility indicator, thus providing an index that allows the watersheds to be prioritised in support of flood risk management at regional level. Results show that the components of vulnerability can be expressed in terms of three constituent indicators: (i) socio-economic fragility, which is composed of demography and lack of well-being; (ii) lack of resilience and coping capacity, which is composed of lack of education, lack of preparedness and response capacity, lack of rescue capacity, cohesiveness of the community; and (iii) physical exposure, which is composed of exposed infrastructure and exposed population. A sensitivity analysis shows that the classification of vulnerability is robust for watersheds with low and high values of the vulnerability indicator, while some watersheds with intermediate values of the indicator are sensitive to shifting between medium and high vulnerability.

INTRODUCTION

Flood risk represents the probability of negative consequences due to floods and emerges from the convolution of flood hazard and flood vulnerability (Schanze et al., 2006). Assessing flood risk can be carried out at national, regional or local level (IWR, 2011), with the regional scale aiming to contribute to regional flood risk management policy and planning. Approaches used to assess flood risk vary widely. These include the assessment of hazard using model-based hazard analyses and combining these with damage estimations to derive a representation of risk (Liu et al., 2014; Su and Kang, 2005), as well as indicator-based analyses that focus on the assessment of vulnerability through composite indices (Chen et al., 2014; Safaripour et al.,

Citation: Rogelis, M., Werner, M., Obregón, N., & Wright, G. (2015). Regional prioritization of flood risk in mountainous areas. Natural Hazards And Earth System Sciences Discussions, 3(7), 4265-4314. http://dx.doi.org/10.5194/nhessd-3-4265-2015.

2012; Greiving, 2006). The resulting levels of risk obtained may subsequently be used to obtain grades of the risk categories (e.g. high, medium and low) that allow prioritisation, or ranking of areas for implementation of flood risk reduction measures, such as flood warning systems and guiding preparations for disaster prevention and response (Chen et al., 2014).

A risk analysis consists of an assessment of the hazard as well as an analysis of the elements at risk. These two aspects are linked via damage functions or loss models, which quantitatively describe how hazard characteristics affect specific elements at risk. This kind of damage or loss modelling typically provides an estimate of the expected monetary loss (Seifert et al., 2009; Luna et al., 2014; van Westen et al., 2014; Mazzorana et al., 2012). However, more holistic approaches go further, incorporating social, economic, cultural, institutional and educational aspects, and their interdependence (Fuchs, 2009). In most cases these are the underlying causes of the potential physical damage (Cardona, 2003; Cardona et al., 2012; Birkmann et al., 2014). A holistic approach provides crucial information that supplements flood risk assessments, informing decision makers on the particular causes of significant loss from a given vulnerable group and providing tools to improve the social capacities of flood victims (Nkwunonwo et al., 2015). The need to include social, economic and environmental factors, as well as physical factors in vulnerability assessments, is incorporated in the Hyogo Framework for Action and emphasised in the Sendai Framework for Disaster Risk Reduction 2015–2030, which establishes the need to understand disaster risks in all its dimensions as a priority (United Nations General Assembly, 2015). However, the multi-dimensional nature of vulnerability has been addressed by few studies (Papathoma-Köhle et al., 2011).

The quantification of the physical dimension of vulnerability can be carried out through empirical and analytical methods (Sterlacchini et al., 2014). However, when the multiple dimensions of vulnerability are taken into account, challenges arise in the measurement of aspects of vulnerability that can not be easily quantified. Birkmann (2006) suggests that indicators and indices can be used to measure vulnerability from a comprehensive and multidisciplinary perspective, capturing both direct physical impacts (exposure and susceptibility), and indirect impacts (socio-economic fragility and lack of resilience). The importance of indicators is rooted in their potential use for risk management since they are useful tools for (i) identifying and monitoring vulnerability over time and space, (ii) developing an improved understanding of the processes underlying vulnerability, (iii) developing and prioritising strategies to reduce vulnerability and for (iv) determining the effectiveness of those strategies (Rygel et al., 2006). However, developing, testing and implementing indicators to capture the complexity of vulnerability remains a challenge.

The use of indices for vulnerability assessment has been adopted by several authors, for example, Balica et al. (2012) describe the use of a Flood Vulnerability Index , an indicatorbased methodology that aims to identify hotspots related to flood events in different regions of the world. Müller et al. (2011) used indicators derived from geodata and census data to analyse the vulnerability to floods in a dense urban setting in Chile. A similar approach was followed by Barroca et al. (2006), organising the choice of vulnerability indicators and the integration from the point of view of various stakeholders into a software tool. Cutter et al. (2003) constructed an index of social vulnerability to environmental hazards at county level for the United States. However, several aspects of the development of these indicators continue to demand research efforts, including the selection of appropriate variables that are capable of representing the sources of vulnerability in the specific study area, the determination of the importance of each indicator, the availability of data to analyse and assess the indicators, the limitations in the scale of the analysis (geographic unit and time frame) and the validation of the results (Müller et al., 2011). Since no variable has yet been identified against which to fully validate vulnerability indicators, an alternative approach to assess the robustness of indices is to identify the sensitivity of how changes in the construction of the index may lead to changes in the outcome (Schmidtlein et al., 2008).

Vulnerability is closely tied to natural and man-made environmental degradation at urban and rural levels (Cardona, 2003; UNEP, 2003). At the same time the intensity or recurrence of flood hazard events can be partly determined by environmental degradation and human intervention in natural ecosystems (Cardona et al., 2012). This implies that human actions on the environment determine the construction of risk, influencing the exposure and vulnerability as well as enhancing or reducing hazard. For example, the construction of a bridge can increase flood hazard upstream by narrowing the width of the channel, increasing the resistance to flow and therefore resulting in higher water levels that may inundate a larger area upstream. The interaction between flood hazard and vulnerability is explored in small watersheds in a mountainous environment, where human–environment interactions that influence risk levels take place in a limited area. The hydrological response of these watersheds is sensitive to anthropogenic interventions, such as land use change (Seethapathi et al., 2008). The consequence of the interaction between hazard and vulnerability in such small watersheds is that those at risk of flooding play a crucial role themselves in the processes that enhance hazard, through modification of the natural environment. Unplanned urbanisation, characterised by a lack of adequate infrastructure and socio-economic issues (both contributors to vulnerability) may also result in environmental degradation, which increases the intensity of natural hazards (UNISDR, 2004). In the case of floods, such environmental degradation may lead to an increase in peak discharges, flood frequency and sediment load.

In this paper a method is proposed to identify montane watersheds with the highest flood damage potential at the regional level. Through this, the watersheds to be subjected to more detailed risk studies can be prioritised in order to establish appropriate flood risk management strategies. The method is demonstrated in the montane watersheds that surround the city of Bogotá (Colombia), where floods typically occur as flash floods and debris flows.

The prioritisation is carried out through an index composed of a qualitative indicator of vulnerability and a qualitative indicator of the susceptibility of the watersheds to the occurrence of flash floods/debris flows. Vulnerability is assessed through application of an indicator system that considers social, economic and physical aspects that are derived from the available data in the study area. This is subsequently combined with an indicator of flash flood/debris flow susceptibility that is based on morphometry and land cover, and was applied to the same area in a previous study (Rogelis and Werner, 2013). In the context of the flash flood/ debris flow susceptibility indicator, susceptibility is considered as the spatial component of the hazard assessment, showing the different likelihoods that flash floods and debris flow occur in the watersheds. In contrast, risk is defined as the combination of the probability of an event and its negative consequences (UNISDR, 2009). The priority index can be considered a proxy for risk, identifying potential for negative consequences but not including probability estimations. The paper is structured as follows: (i) Sect. 2 reviews the conceptual definition of vulnerability as the foundation of the paper; (ii) Sect. 3 describes the study area, and the data and methodology used; (iii) Sect. 4 presents the results of the analysis; this includes the construction of the indicators and the corresponding sensitivity analysis, as well as the prioritisation of watersheds; (iv) Sect. 5 interprets the results that lead to the final prioritisation; (v) the conclusions are summarised in Sect. 6.

CONCEPTUALISATION OF VULNERABILITY

Several concepts of vulnerability can be identified, and there is not a universal definition of this term (Thieken et al., 2006; Birkmann, 2006). Birkmann (2006) distinguishes at least six different schools of thinking regarding the conceptual and analytical frameworks on how to systematise vulnerability. In these, the concept of exposure and its relation with vulnerability, the inclusion of the coping capacity as part of vulnerability and the differentiation between hazard-dependent and hazard-independent characteristics of vulnerability play an important role. Sterlacchini et al. (2014) identifies at least two different perspectives: (i) one related to an engineering and natural science overview; and (ii) a second one related to a social science approach.

With relation to the first perspective (i), vulnerability is de- fined as the expected degree of loss for an element at risk, occurring due to the impact of a defined hazardous event (Varnes, 1984; Fuchs, 2009; Holub et al., 2012). The relationship between impact intensity and degree of loss is commonly expressed in terms of a vulnerability curve or vulnerability function (Totschnig and Fuchs, 2013), although also semi-quantitative and qualitative methods exist (Totschnig and Fuchs, 2013; Fuchs et al., 2007, Jakob et al., 2012; Kappes et al., 2012). The intensity criteria of torrent (steep stream) processes, encompassing clear water, hyperconcentrated and debris flows, has been considered in terms of impact forces (Holub et al., 2012; Quan Luna et al., 2011; Hu et al., 2012); deposit height (Mazzorana et al., 2012; Fuchs et al., 2012, 2007; Akbas et al., 2009; Totschnig et al., 2011; Lo et al., 2012; Papathoma-Köhle et al., 2012; Totschnig and Fuchs, 2013); kinematic viscosity (Quan Luna et al., 2011; Totschnig et al., 2011); flow depth (Jakob et al., 2013; Tsao et al., 2010; Totschnig and Fuchs, 2013); flow velocity times flow depth (Totschnig and Fuchs, 2013); and velocity squared times flow depth (Jakob et al., 2012). Different types of elements at risk will show different levels of damage given the same intensity of hazard (Jha et al., 2012; Albano et al., 2014; Liu et al., 2014), therefore vulnerability curves are developed for a particular type of exposed element (such as construction type, building dimensions or road access conditions). A limited number of vulnerability curves for torrent processes have been proposed, and the efforts have been mainly oriented to residential buildings (Totschnig and Fuchs, 2013). Since it can be difficult to extrapolate data gathered from place to place to different building types and contents (Papathoma-Köhle et al., 2011), different curves should be created for different geographical areas and then applied to limited and relatively homogeneous regions (Luino et al., 2009; Jonkman et al., 2008; Fuchs et al., 2007).

Regarding the second perspective (ii), social sciences de- fine vulnerability as the pre-event, inherent characteristics or qualities of social systems that create the potential for harm (Cutter et al., 2008). This definition is focused on the characteristics of a person or group and their situation that influence their capacity to anticipate, cope with, resist and recover from the impact of a hazard (Wisner et al., 2003). Social and place inequalities are recognised as influencing vulnerability (Cutter et al., 2003). The term livelihood is highlighted and used to develop models of access to resources, like money, information, cultural inheritance or social networks, influencing people's vulnerability (Hufschmidt et al., 2005). Given the different perspectives of vulnerability, it becomes apparent that only by a multidimensional approach can the overall aim of reducing natural hazards risk be achieved (Fuchs and Holub, 2012). Fuchs (2009) identifies a structural (physical) dimension of vulnerability that is complemented by economic, institutional and societal dimensions. In addition to these, Sterlacchini et al. (2014) identify a political dimension. Birkmann et al. (2014) and Birkmann et al. (2013) identify exposure, fragility and lack of resilience as key causal factors of vulnerability, as well as physical, social, ecological, economic, cultural and institutional dimensions.

In this study, physical exposure (hard risk and considered to be hazard-dependent), socioeconomic fragility (soft risk and considered to be not hazard-dependent) and lack of resilience and coping capacity (soft risk and is mainly not hazard-dependent) (Cardona, 2001) are used to group the variables that determine vulnerability in the study area. In this paper, the risk perception and the existence of a flood early warning, which are hazard-dependent, are considered as aspects influencing resilience since they influence the hazard knowledge of the communities at risk and the level of organisation to cope with floods. An analysis of physical vulnerability through vulnerability curves is not incorporated; instead, the expected degree of loss is assessed qualitatively through the consideration of physical exposure and factors that amplify the loss (socioeconomic fragility and lack of resilience). This means the expected degree of loss depends on the extent of the flash floods/debris flows, and not on the intensity of those events.

The terminology and definitions that are used in this study are as follows:

- vulnerability: propensity of exposed elements such as physical or capital assets, as well as human beings and their livelihoods, to experience harm and suffer damage and loss when impacted by a single or compound hazard events (Birkmann et al., 2014);
- exposure: people, property, systems or other elements present in hazard zones that are thereby subject to potential loss (UNISDR, 2009);

vation of 2640 m a.s.l. and is surrounded by mountains from where several creeks drain to the Tunjuelo, Fucha and Juan Amarillo rivers. These rivers flow towards the Bogotá River. Precipitation in the city is characterised by a bimodal regime with mean annual precipitation ranging from 600 to 1200 mm (Bernal et al., 2007). Despite its economic output and growing character as a global city, Bogotá suffers from social and economic inequalities, lack of affordable housing and overcrowding. Statistics indicate that there has been a significant growth in the population, which also demonstrates the process of urban immigration that the whole country is suffering not only due to industrialisation processes, but also due to violence and poverty. This disorganised urbanisation process has pushed informal settlers to build their homes in highly unstable zones and areas that can be subjected to inundation. Eighteen percent of the urban area has been occupied by informal constructions, housing almost 1 400 000 persons. This is some 22 % of the urban population of Bogotá (Pacific Disaster Center, 2006).

Between 1951 and 1982, the lower (northern) part of the Tunjuelo Basin (see Fig. 1) was the most important area for urban development in the city, where the poorest population of Bogotá were settled (Osorio, 2007). This growth has been characterised by informality and lack of planning. This change in the land use caused loss of vegetation and erosion, which enhanced flood hazard (Osorio, 2007).

The urban development of the watersheds located in the hills to the east of Bogotá (see Fig. 1) has a different characteristic to that of the Tunjuelo Basin. Not only has this taken place through both informal settlements, but also includes exclusive residential developments (Buendía, 2013). In addition, protected forests cover most of the upper watersheds. In this analysis the watersheds located in mountainous terrain that drain into the main stream of the Tunjuelo Basin, as well as the watersheds in the Eastern Hills, were considered. The remaining part of the urban area of the city covers an area that is predominantly flat, and is not considered in this study. Table 1 shows the number of watersheds in the study area, as well as the most recent and severe flood events that have been recorded.

Methodology

The prioritisation of flood risk was carried out using watersheds in the study area as units of analysis. The watershed divides were delineated up to the confluence with the Tunjuelo River, or up to the confluence with the storm water system, whichever is applicable. First a delineation of areas exposed to flooding from these watersheds using simplified approaches was carried out. Subsequently a vulnerability indicator was constructed based on a principal component analysis of variables identified in the literature as contributing to vulnerability. A sensitivity analysis was undertaken to test the robustness of the vulnerability indicator. From the vulnerability indicator, a category (high, medium and low vulnerability) was obtained that was then combined with a categorisation of flash flood/debris flow susceptibility previously generated in the study area to obtain a prioritisation category. The tool that was used to combine vulnerability and susceptibility was a matrix that relates the susceptibility levels and vulnerability levels producing a priority level as output. The combination matrix was constructed through the assessment of all possible matrices using as assessment criterion the "proportion correct". In order to obtain the proportion correct an independent classification of the watersheds was carried out on the basis of the existing damage data. A detailed explanation of the analysis is given in the following subsections.

Delineation of Exposure Areas

Flood events in the watersheds considered in this study typically occur as flash floods given their size and mountainous nature. Flash floods in such small, steep watersheds can further be conceptualised to occur as debris flows, hyperconcentrated flows or clear water flows (Hyndman and Hyndman, 2008; Jakob et al., 2004; Costa, 1988). Costa (1988) differentiates (i) clear water floods as Newtonian, turbulent fluids with non-uniform concentration profiles and sediment concentrations of less than about 20 % by volume and shear strengths less than 10 N m^{-2} ; (ii) hyperconcentrated flows as having sediment concentrations ranging from 20 to 47 % by volume and shear strengths lower than about 40 N m^{-2} ; and (iii) debris flows as being non-Newtonian visco-plastic or dilatant fluids with laminar flow and uniform concentration profiles, with sediment concentrations ranging from 47 to 77 % by volume and shear strengths greater than about 40 N m^{-2} . Debris-flow-dominated areas can be subject to hyperconcentrated flows as well as clear water floods (Larsen et al., 2001; Santo et al., 2015; Lavigne and Suwa, 2004), depending on the hydroclimatic conditions and the availability of sediments (Jakob, 2005); occurrence of all types in the same watersheds has been reported (Larsen et al., 2001; Santo et al., 2015). Therefore, the areas exposed to clear water floods and debris flows were combined. This provides a conservative delineation of the areas considered to be exposed to flooding.

Exposure areas were obtained from an analysis of the susceptibility to flooding. Areas that can potentially be affected by clear water floods and debris flows were determined using simplified methods that provide a mask where the analysis of exposed elements was carried out. The probabilities of occurrence and magnitude are not considered in the analysis, since the scope of the simplified regional assessment is limited to assessing the susceptibility of the watersheds to flooding. Areas prone to debris flows were previously identified by Rogelis and Werner (2013) through application of the Modified Single-Flow Direction model. In order to delineate the areas prone to clear water floods, or floodplains, two geomorphic-based methods were tested using a digital elevation model with a pixel size of 5 m as an input, which was obtained from contours. Floodplains are areas near stream channels shaped by the accumulated effects of floods of varying magnitudes and their associated geomorphological processes. These areas are also referred to as valley bottoms and riparian areas or buffers (Nardi et al., 2006).

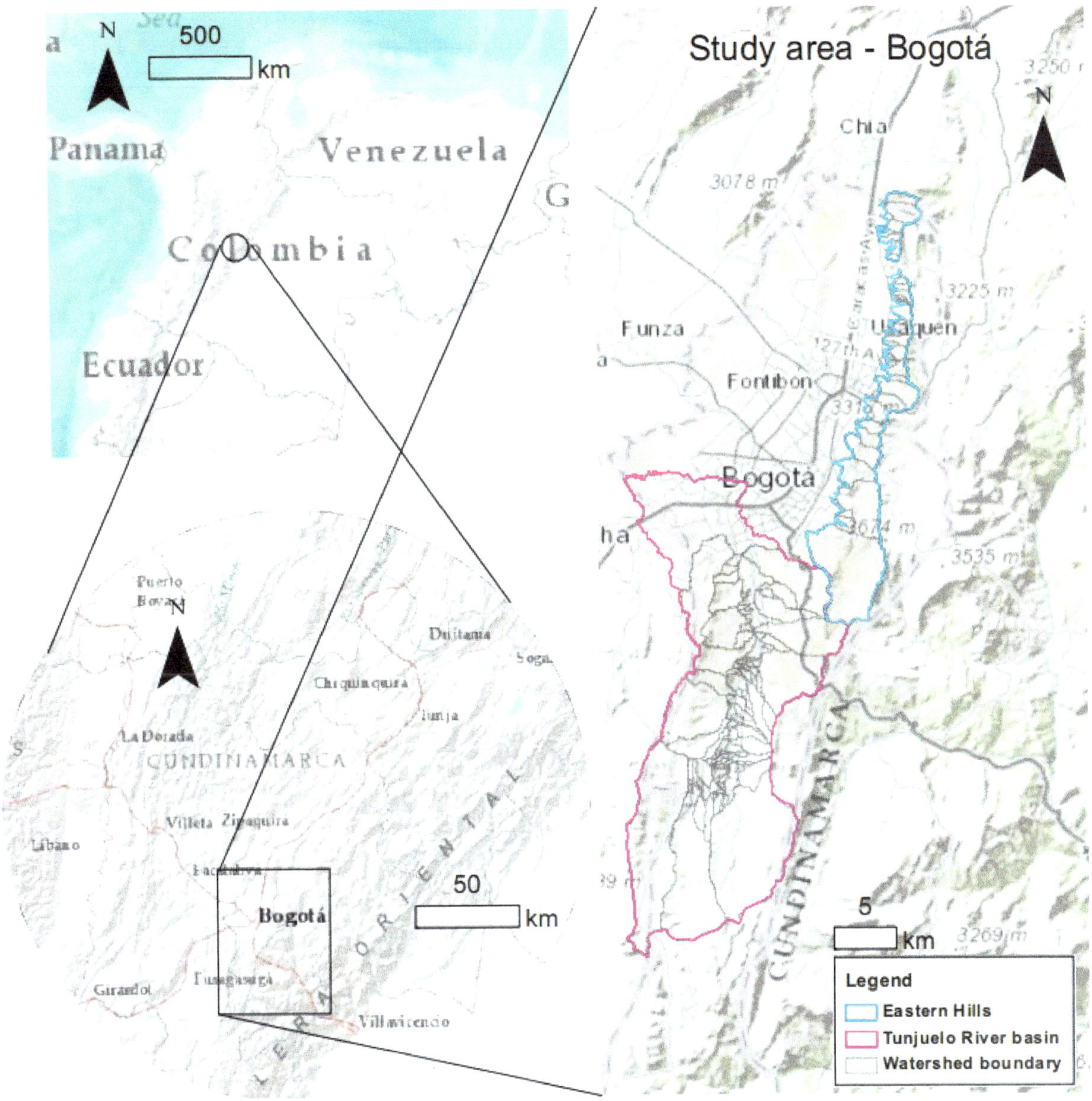

Figure 1. Location of the study areas.

Table 1. Most severe recent flooding events in the study area

	Watersheds			
Study area	Number	Average slope (%)	Area (km^2)	Recent flooding events
Tunjuelo River basin	66	12–40	0.2–57	The most severe events include the following. – In May 1994, a debris flow affected 830 people and caused the death of four people in the north-east of the basin (JICA, 2006). – In November 2003, a hyperconcentrated flow took place in the north-west of the Tunjuelo Basin. Two people were killed and 1535 were affected. A similar event occurred at the same location in November 2004 without a death toll (DPAE, 2003a, b).
Eastern Hills	40	21–59	0.2–33	The most severe events include the following. – In May 2005, a hyperconcentrated flow occurred in the central part of the area, affecting two houses (DPAE, 2005).

The first approach is the multi-resolution valley bottom flatness (MRVBF) algorithm (Gallant and Dowling, 2003). The MRVBF algorithm identifies valley bottoms using a slope classification constrained on convergent area. The classification algorithm is applied at multiple scales by progressive generalisation of the digital elevation model, combined with progressive reduction of the slope class threshold. The results at different scales are then combined into a single index. The MRVBF index utilises the flatness and lowness characteristics of valley bottoms. Flatness is measured by the inverse of slope, and lowness is measured by ranking the elevation with respect to the surrounding area. The two measures, both scaled to the range 0 to 1, are combined by multiplication and could be interpreted as membership functions of fuzzy sets. While the MRVBF is a continuous measure, it naturally divides into classes corresponding to the different resolutions and slope thresholds (Gallant and Dowling, 2003).

In the second method considered, threshold buffers are used to delineate floodplains as areas contiguous to the streams based on height above the stream level. Cells in the digital elevation model adjacent to the streams that meet height thresholds are included in the buffers (Cimmery, 2010). Thresholds for the height of 1, 2, 3, 4, 5, 7 and 10 m were tested. In order to evaluate the results of the MRVBF index and the threshold buffers, flood maps for the study area were used. These are available for only 9 of the 106 watersheds, and were developed in previous studies through hydraulic modelling for return periods up to 100 years. The delineation of the flooded area for a return period of 100 years was used in the nine watersheds to identify the suitability of the floodplain delineation methods to be used in the whole study area. With respect to areas prone to debris flows, these were validated with existing records in the study area by Rogelis and Werner (2013).

Choice of Indicators and Principal Component Analysis for Vulnerability Assessment

In this study vulnerability in the areas identified as being exposed is assessed through the use of indicators. The complexity of vulnerability requires a transformation of available data to a set of important indicators that facilitate an estimation of vulnerability (Birkmann, 2006). To this end, principal component analysis was applied to variables describing vulnerability in the study area in order to create composite indicators (Cutter et al., 2003). The variables were chosen by taking into account their usefulness according to the literature, and were calculated using the exposure areas as a mask. Table 2 shows the variables chosen to explain vulnerability in the study area. These are grouped in socio-economic fragility, lack of resilience and coping capacity and physical exposure. The variables are classified according to their social level (individual, household, community and institutional), hazard dependence and influence on vulnerability (increase or decrease). The third column specifies the spatial aggregation level of the available data. The three spatial levels considered are urban block, watershed and locality, where the locality corresponds to the 20 administrative units of the city. The data used to construct the indicators were obtained from the census and reports published by the municipality. For each variable the values were normalised between the minimum and the maximum found in the study area. In the case of variables that contribute to decreasing vulnerability, a transformation was applied so a high variable value represents high vulnerability for all variables.

In order to construct the composite indicators related to socio-economic fragility and physical exposure, principal component analysis (PCA) was applied to the corresponding variables shown in Table 2. PCA reduces the dimensionality of a data set consisting of a large number of interrelated variables, while retaining as much as possible of the variation present in the data set. This is achieved by transforming to a new set of variables, the principal components (PCs), which are uncorrelated (Jolliffe, 2002). The number of components to be retained from the PCA was chosen by considering four criteria: the Scree test acceleration factor, optimal coordinates (Cattell, 1966), the Kaiser's eigenvalue-greater than-one rule (Kaiser, 1960) and parallel analysis (Horn, 1965). Since the number of components may vary among these criteria, the interpretability was also taken into account when selecting the components to be used in further analysis, with each PC being considered an intermediate indicator. Subsequently a varimax rotation (Kaiser, 1958) was applied to minimise the number of individual indicators that have a high loading on the same principal component, thus obtaining a simpler structure with a clear pattern of loadings (Nardo et al., 2008). The intermediate indicators (PCs) were aggregated using a weight equal to the proportion of the explained variance in the data set (Nardo et al., 2008) to provide an overall indicator for socio-economic fragility and for physical exposure.

PCA has the disadvantage that correlations do not necessarily represent the real influence of the individual indicators and variables on the phenomenon being measured (Nardo et al., 2008). This can be addressed by combining PCA weights with an equal weighing scheme for those variables where PCA does not lead to interpretable results (Esty et al., 2006). In the construction of the lack of resilience and coping capacity indicator, this issue led to a separation of variables into four groups.

- Robberies and participation: these were treated separately from the rest of the variables to maintain interpretability as a measure of cohesiveness of the community. Cohesiveness of the community was identified as a factor that influences the resilience since the degradation of social networks limits the social organisation for emergency response (Ruiz-Pérez and Gelabert Grimalt, 2012). Since there are only two variables to measure this aspect of resilience, PCA was not applied, and the average of the variables was used instead.
- Risk perception and early warning: risk perception depends on the occurrence of previous floods, thus it depends on hazard exclusively. The existence of early warning is mainly an institutional and organisational issue. Therefore, an interpretation of correlation of these variables with other variables in the group of lack of resilience and coping capacity is not possible. These variables were considered separated intermediate indicators. Risk perception and early warning decrease the lack of coping capacity (Molinari et al., 2013), and therefore an equal negative weight was assigned to

these indicators, with a total of −0.2. This value was chosen so that their combined influence is less than the individual weight of the other four indicators. The sensitivity of this subjective choice was tested. The effectiveness of flood early warning is closely related to the level of preparedness as well as the available time for implementation of appropriate actions (Molinari et al., 2013). Due to the rapid hydrologic response and configuration of the watersheds in the study area, flood early warning actions are targeted at reducing exposure and vulnerability and not at hazard reduction.

- Rescue personnel: this variable was initially used in the PCA with all lack of resilience and coping capacity variables. However, it was found to increase with lack of resilience and coping capacity. This implied that the statistical behaviour of the variable did not represent its real influence on vulnerability. It was therefore treated independently.
- Level of education, illiteracy, access to information, infrastructure/accessibility, hospital beds and healthcare human resources (HR): PCA was applied to these variables, since they exhibit high correlation and are interpretable in terms of their influence on vulnerability

Table 2. Variables used to construct vulnerability indicators.

Social levels	Variable	Effect	Spatial level	Description
			Socio-economic fragility	
Individual	Age		Urban block	Percentage <10 plus percentage >65
	Disability		Urban block	% of population with any sort of disability
	Unemployment		Locality	Unemployment rate
	Income		Locality	Unsatisfied basic needs index - UBN, % of homeless, % of poor population
	Life expectancy		Locality	Life expectancy
	Household size		Locality	Average number of persons per household
	Woman-headed households		Locality	Percentage of families headed by women
Community	Illegal settlements		Urban block	Percentage of illegal settlements
	% of population of strata 1 and 2		Urban block	The socio-economic stratification system of Bogotá classifies the population into strata with similar economic characteristics on a scale from 1 to 6, with 1 as the lowest income area and 6 as the highest. Strata 1 and 2 correspond to the socio-economic classification with the lowest income.
	Life conditions		Locality	Life conditions index
	Human development index		Locality	Human development index
	Demographic pressure		Locality	Population growth rate
	Child mortality		Locality	Child mortality rate
			Lack of resilience and coping capacity	
Individual	Level of education		Locality	% of population with education level superior to high school
	Illiteracy		Locality	Illiteracy rate
	Access to information		Locality	% of homes with internet access
Household	---	---	---	---
Community	Risk perception		Watershed	Boolean indicator. A value of 1 was assigned to watersheds where floods have occurred previously and 0 if they have not.
	Robberies		Locality	Robberies per 10000 inhabitants. The robberies that occur in the locality of the watershed were used as a proxy for trust, confidence and the level at which a proper post-disaster environment could be expected, since a high probability of crime can affect the evacuation procedures and the process to recover.
	Participation		Locality	Percentage of eligible voters that voted in the most recent communal elections.
Institutional	Infrastructure/ accessibility		Locality	% of roads in good condition
	Early warning		Watershed	Boolean indicator. Existence of flood early warning systems in the watershed. Watersheds where flood early warning systems are operational were assigned a value of 1 and 0 if they do not exist.
	Hospital beds		Locality	Hospital beds per 10000 inhabitants
	Healthcare HR		Locality	Healthcare human resources per 10000 inhabitants
	Rescue personnel		Locality	Rescue personnel per 10000 inhabitants.
			Physical exposure	
Individual	Population exposed		Urban block	Number of people in flood-prone areas
	Density of population		Urban block	People per km² in flood-prone areas.
Household	Residential units		Urban block	Number of houses in flood-prone area
Community	Commercial and industrial units		Urban block	Number of commercial and industrial establishments in flood-prone area.
	Community infrastructure		Urban block	Number of community, social, cultural, healthcare infrastructure exposed
Institutional	---	---	---	---

Legend: Variable: Hazard-dependent; Hazard-independent. Effect: Increases vulnerability; Reduces vulnerability

To combine all the lack of resilience and coping capacity intermediate indicators into a composite indicator, weights with a total of 1 were assigned (see Sect. 4.3 for an explanation of the resulting intermediate indicators). The indicators corresponding to socio-economic fragility, lack of resilience and coping capacity and physical exposure were combined, assigning equal weight to the three components, to obtain an overall vulnerability indicator. The watersheds were subsequently categorised as being low, medium or high vulnerability based on the value of the vulnerability indicator and using equal intervals. This method of categorisation was chosen to avoid dependence on the distribution of the data, so monitoring of evolution in time of vulnerability can be carried out applying the same criteria.

Sensitivity of the Vulnerability Indicator

The influence of all subjective choices applied in the construction of the indicators was analysed. This included both choices made in the application of PCA, and for the weighting scheme adopted for the factors contributing to resilience and total vulnerability.

1. For the application of PCA, sensitivity to the following choices was explored.

- Four alternatives for the number of components to be retained were assessed as explained in Sect. 3.2.2.
- Five different methods in addition to the varimax rotation were considered: unrotated solution, quartimax rotation (Carroll, 1953; Neuhaus, 1954), promax rotation (Hendrickson and White, 1964), oblimin (Carroll, 1957), simplimax (Kiers, 1994) and cluster (Harris and Kaiser, 1964).

2. For the weighting scheme, the following choices were made.

Table 3. Categories of recorded damage

Category	Score	Description
Low	0	No recorded damage in the watershed
Low	1	Events that affect one house without causing injuries or fatalities and without the need of evacuation
Low	2	Events that affect one house without causing injuries or fatalities and with the need of evacuation
Low	3	Events that affect up to five houses without causing injuries or fatalities, flood depth less than 0.5 m with evacuation of families
Medium	4	Events that affect up to five houses without causing injuries or fatalities, flood depth higher than 0.5 m with evacuation of families
Medium	5	Events that affect up to 10 houses without causing injuries or fatalities, with evacuation of families
Medium	6	Events that affect 10–20 houses without causing injuries or fatalities, with evacuation of families, flood depth less than 0.5 m
High	7	Events that affect 10–20 houses without causing injuries or fatalities, with evacuation of families, flood depth higher than 0.5 m
High	8	Events that affect 20–50 houses without causing injuries or fatalities, with evacuation of families and possibility of structural damage in the houses
High	9	Events that affect more than 50 houses without causing injuries or fatalities, with evacuation of families and possibility of structural damage in the houses
High	10	Events that cause fatalities or injuries

- The weights used in the four groups of variables that describe lack of resilience and coping capacity were varied by ±10 %.
- The weights used to combine the three indicators that result in the final vulnerability composite indicator were varied by ±10 %.

All possible combinations were assessed and the results in terms of the resulting vulnerability category (high, medium and low) were compared in order to identify substantial differences as a result of the choices of subjective options.

Categories of Recorded Damage in the Study Area

A database of historical flood events compiled by the municipality was used to classify the watersheds in categories, depending on damages recorded in past flood events. For each of these events the database includes date, location, injured people, fatalities, evacuated people, number of affected houses and an indication of whether the flow depth was higher than 0.5 m or not. Unfortunately, no information on economic loss is available and as the database only covers the period from 2000 to 2012 it is not possible to carry out a frequency analysis. Complete records were only available for 14 watersheds. The event with the highest impact for each watershed was chosen from the records. Subsequently, the 14 watersheds were ordered according to their highest impact event. The criteria to sort the records and to sort the watersheds according to impact from highest to lowest were the following (in order of importance):

- fatalities
- injured people
- evacuated people
- number of affected houses.

Watersheds with similar or equal impact were grouped, resulting in 11 groups. The groups were again sorted according to damage. A score from 0 to 10 was assigned, where a score of 0 implies that no flood damage has been recorded in the watershed for a flood event, despite the occurrence of flooding, while a score of 10 corresponds to watersheds where fatalities or serious injuries have occurred (see Table 3). The 11 groups were further classified into three categories according to the emergency management organisation that was needed for the response: (i) low: the response was coordinated locally; (ii) medium: centralised coordination is needed for response with deployment of resources of mainly the emergency management agency; (iii) high: centralised coordination is needed with an inter-institutional response. This classification was made under the assumption that the more resources are needed for response, the more severe the impacts are, allowing in this way a comparison with three levels of priority classification.

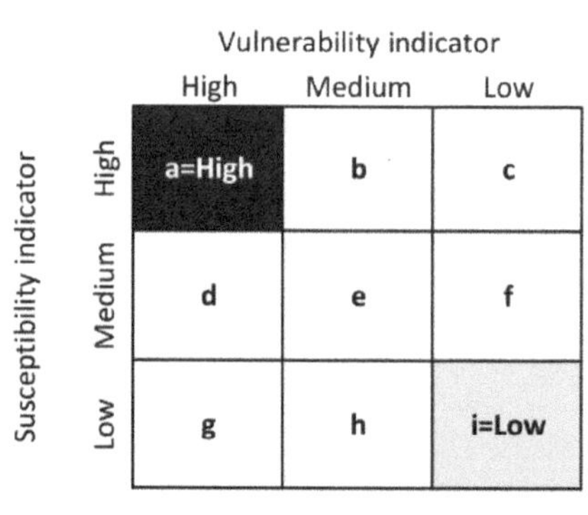

Figure 2. Initial matrix of priority

Prioritisation of Watersheds

Due to the regional character and scope of the method applied in this study, a qualitative proxy for risk was used to prioritise the watersheds in the study area. A high priority indicates watersheds where flood events will result in more severe consequences. However, the concept of probability of occurrence of these is not involved in the analysis, since the analysis of flood hazard is limited to susceptibility.

In order to combine the vulnerability and susceptibility to derive a level of risk, a classification matrix was used. This is shown in Fig. 2. The columns indicate the classification of the vulnerability indicator and the rows the classification of the susceptibility indicator. Only two priority outcomes are well defined; these are the high and low degrees assigned to the corners of the matrix corresponding to high susceptibility and high vulnerability and low susceptibility and low vulnerability (cells a and i), since they correspond to the extreme conditions in the analysis. The priority outcomes in cells from b to h were considered unknown and to potentially correspond to any category (low, medium or high priority). To define the category for these cells, the priority using all possible matrices (all possible combinations of categories of cells b to c) was assessed for the watersheds for which flood records are available. Once these watersheds are prioritised, a contingency table is constructed comparing the priority with the damage category (from Table 3) from which the proportion correct is obtained. The classification matrix that results in the highest proportion correct (best fit) was used for the prioritisation of the whole study area.

RESULTS

Exposure Areas

Figure 4 shows the results of the methods applied to identify areas susceptible to flooding through clear water floods or debris flows. Figure 4a shows the debris flow propagation extent derived for the watersheds of the Tunjuelo Basin and the Eastern Hills by Rogelis and Werner (2013). Since the method does not take into account the volume that can be deposited on the fan, this shows the maximum potential distance that the debris flow could reach according to the morphology of the area, which is in general flat to the west of the Eastern Hills watersheds. A different behaviour can be observed in the watersheds located in the Tunjuelo River basin where the marked topography and valley configuration restricts the propagation areas. Figure 4b shows the results obtained from the MRVBF index. The comparison of the index with the available flood maps in the study area shows that values of the MRVBF higher than 3 can be considered areas corresponding to valley bottoms. In areas of marked topography the index identifies areas adjacent to the creeks in most cases and the larger scale valley bottoms. However, in flat areas the index unavoidably takes high values and cannot be used to identify flood-prone areas.

Figure 4c shows the result obtained from the use of buffer thresholds. The buffers that were obtained by applying the criteria explained in Sect. 3.2.1 were compared with the available flood maps. Areas obtained for a depth criterion of 3 m were the closest to the flood delineation for a return period of 100 years, and this value was chosen as appropriate for the study area. In order to obtain the delineation of the exposure areas, the results of the debris flow propagation, the MRVBF index and the buffers were combined. The results of all three methods in flat areas do not allow for a correct identification of floodprone areas, and a criterion based on the available information and previous studies was needed to estimate a reasonable area of exposure. The resulting exposure areas are shown in Fig. 5.

Socio-Economic Fragility Indicators

The results of the principal component analysis applying a varimax rotation are shown in Table 4. Two principal components were retained as this allowed a clear interpretation to be made for each of the components. The variables included in the first principal component are related to lack of well-being (P_{LofW}), while in the second these are related to the demography (P_{demog}). The two principal components account for 79 % of the variance in the data, with the first component explaining 80 % of the variance (percentage of the variability explained, PVE) and the second 20 %.

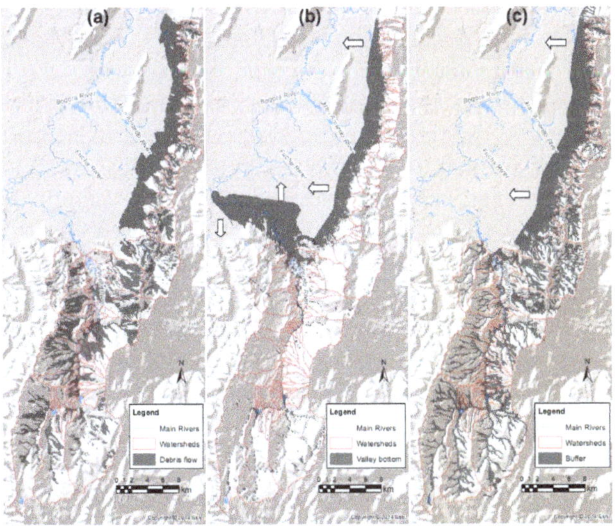

Figure 3. Clear water flood and debris flow susceptibility areas. Areas in dark grey in each map represent; (a) debris flow extent (Rogelis and Werner, 2013); (b) valley bottoms identified using the MRVBF index; (c) buffers. In the case of maps (b) and (c), the flood prone areas extend in the direction of the arrows over the flat area.

Using the factor loadings (correlation coefficients between the PCs and the variables) obtained from the analysis (see Table 4) and scaling them to unity, the coefficients of each indicator are shown in the following equations:

$$P_{\text{LofW}} = 0.10\,\text{Whh} + 0.10\,\text{UE} + 0.10\,\text{PUBNI} + 0.09\,\text{Ho} + 0.11\,\text{P} + 0.10\,\text{Pho} + 0.09\,\text{M} + 0.10\,\text{LE} + 0.08\,\text{QLI} + 0.10\,\text{HDI} + 0.04\,\text{G} \quad (1)$$

$$P_{\text{demog}} = 0.32\,\text{Age} + 0.20\,\text{D} + 0.29\,\text{PE12} + 0.19\,\text{IS}. \quad (2)$$

The impacts of the indicators imply that the higher the lack of well-being, the higher the socio-economic fragility, and equally, the higher the demography indicator, the higher the socio-economic fragility. Using the PVE of each component, the composite indicator for socio-economic fragility (P_{soc-ec}) is found to be

$$P_{soc-ec} = 0.8\,P_{LofW} + 0.2\,P_{demog}. \tag{3}$$

Lack of Resilience and Coping Capacity Indicators

The loadings of the indicators representing lack of resilience and coping capacity obtained from the PCA are shown in Table 5. Two principal components were used: the first correlated with variables related to the lack of education (P_{LEdu}) and the second correlated with variables related to lack of preparedness and response capacity ($P_{LPrRCap}$). These account for 97 % of the variance in the data with the first component explaining 53 % of the variance (PVE) and the second 47 %.

Using the factor loadings obtained from the analysis and scaling them to unity, the coefficients of each indicator are shown in the following equations:

$$P_{LEdu} = 0.33\mathrm{LEd} + 0.32\mathrm{I} + 0.35\mathrm{AI} \tag{4}$$

$$P_{LPrRCap} = 0.26\mathrm{IA} + 0.39\mathrm{Hb} + 0.35\mathrm{HRh}. \tag{5}$$

In an initial analysis, the variable rescue personnel was included in the principal component analysis. Results showed a high negative correlation of this variable with lack of education, illiteracy and access to information. This may be due to more institutional effort being allocated to deprived areas that are more often affected by emergency events, in order to strengthen the response capacity of the community. Additionally, civil protection groups rely strongly on voluntary work that seems to be more likely in areas with lower education levels.

Figure 4. Exposure areas.

Table 4. Results of the principal component analysis for socioeconomic fragility indicators.

Variable	Symbol	Loadings
Lack of well-being (PVE = 0.8)		
Women-headed households	Whh	0.94
Unemployment	UE	0.97
Poor–unsatisfied basic needs index	PUBNI	0.98
% homeless	Ho	0.92
% poor	P	0.99
Persons per home	Pho	0.94
Mortality	M	0.91
Life expectancy	LE	0.94
Quality life index	QLI	0.86
Human development index	HDI	0.97
Population growth rate	G	0.57
Demography (PVE = 0.2)		
% of children and elderly	Age	0.84
% disabled	D	0.67
% population strata 1 and 2	PE12	0.81
% illegal settlements	IS	0.64

Table 5. Results of the principal component analysis resilience indicators.

Variable	Symbol	Loadings
Lack of education (PVE = 0.53)		
Level of education	LEd	0.94
Illiteracy	I	0.96
Access to information	AI	0.93
Lack of prep. and resp. Capacity (PVE = 0.47)		
Infrastructure/accessibility	IA	0.80
Hospital beds	Hb	0.97
Healthcare HR	HRh	0.92

Since the consideration of rescue personnel changes the interpretation of the principal component that groups the lack of education and the access to information indicator together, it was decided to exclude it from the PCA and to consider this variable as an independent indicator (lack of rescue capacity). In the analysis of robberies and participation as variables describing cohesiveness of the community, it was found that the increase in crime is correlated with the lack of participation, describing the distrust of the community both of neighbours and of institutions. The corresponding composite indicator was calculated as the average of robberies and lack of participation.

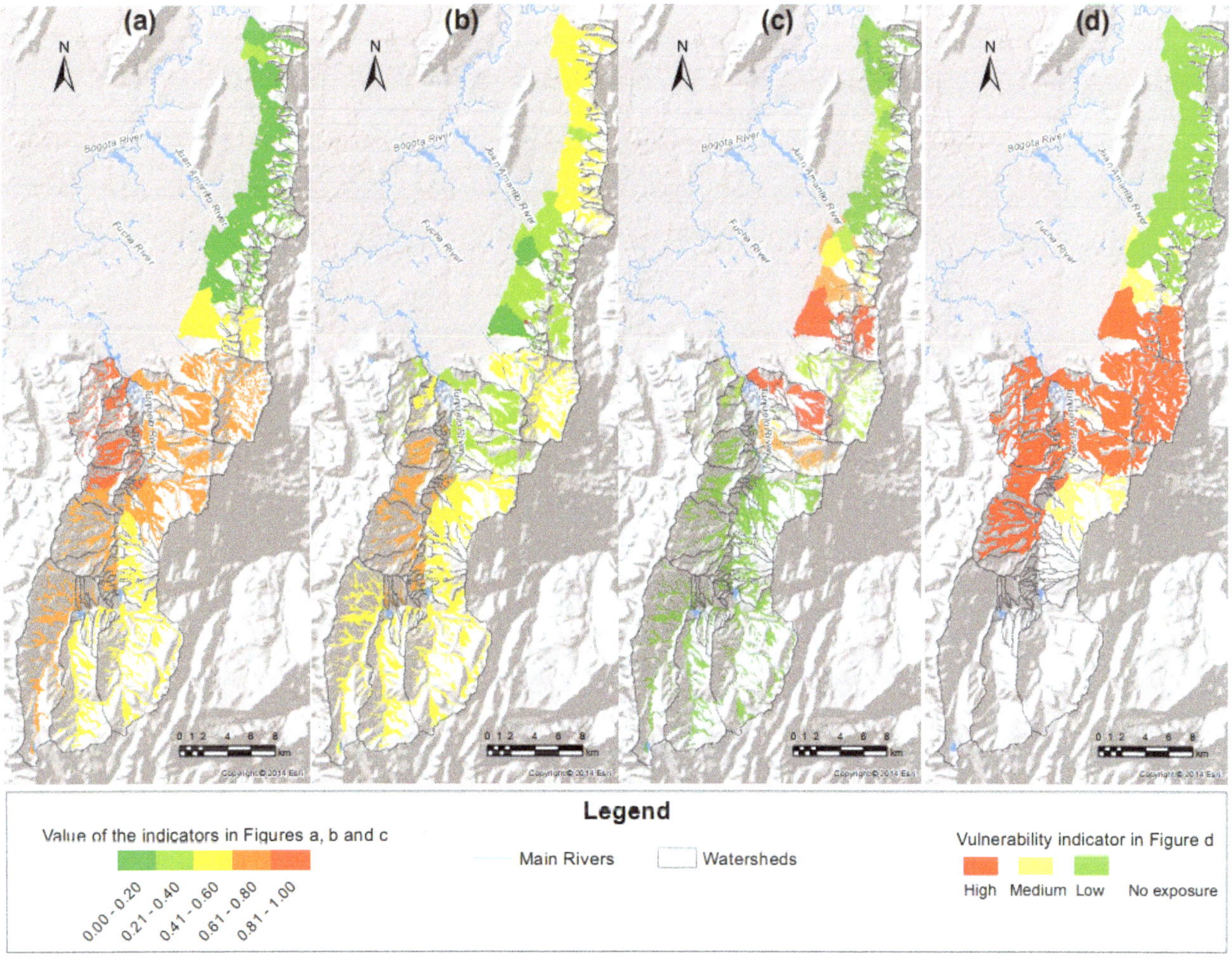

Figure 5. (a) Spatial distribution of the Socio-economic indicator; (b) Spatial distribution of the resilience indicator; (c) spatial distribution of the physical exposure indicator; (d) Spatial distribution of the total vulnerability indicator.

The equation of lack of resilience and coping capacity is shown in Eq. (6). Equal weight was assigned to the indicators reflecting lack of education, lack of preparedness and response capacity, lack of rescue capacity (P_{LRc}) and cohesiveness of the community (P_{CC}); and a weight of −0.1 to risk perception (P_{RP}) and early warning (P_{FEW}).

$$P_{LRes} = 0.25\, P_{LEdu} + 0.25\, P_{LPrRCap} + 0.25\, P_{LRc}$$
$$0.25\, P_{CC} - 0.1\, P_{RP} - 0.1\, P_{FEW} \quad (6)$$

The indicator of lack of resilience and coping capacity was rescaled between 0 and 1.

Physical Exposure Indicators

The principal component analysis of the variables selected for physical exposure shows that these can be grouped into two principal components that explain 82 % of the variability (exposed infrastructure – P_{Ei} and exposed population – P_{Ep}). The results of the analysis are shown in Table 6.

Using the factor loadings obtained from the analysis and scaling them to unity, the coefficients of each composite indicator are shown in the following equations:

$$P_{Ei} = 0.32\,\text{Ncb} + 0.37\,\text{Niu} + 0.32\,\text{Ncu} \quad (7)$$

$$P_{Ep} = 0.38\,\text{Nru} + 0.33\,\text{Pe} + 0.28\,\text{Dp}. \quad (8)$$

Using the percentage of variability explained (PVE) by each indicator, the composite indicator of physical susceptibility is found to be

$$P_{ps} = 0.52\, P_{Ei} + 0.48\, P_{Ep}. \quad (9)$$

Vulnerability Indicator

The resulting vulnerability indicator was obtained through the equal-weighted average of the indicators for socioeconomic fragility, lack of resilience and coping capacity and physical exposure. Categories of low, medium and high vulnerability for each watershed were subsequently derived based on equal bins of the indicator value. The spatial distribution is shown in Fig. 5, as well as the spatial distribution of the three constituent indicators.

Table 6. Results of the principal component analysis physical susceptibility indicators.

Variable	Symbol	Loadings
Exposed infrastructure (PVE = 0.52)		
Number of civic buildings	Ncb	0.86
Number of industrial units	Niu	0.96
Number of commercial units	Ncu	0.85
Exposed population (PVE = 0.48)		
Number of residential units	Nru	0.91
Population exposed	Pe	0.85
Density of population	Dp	0.78

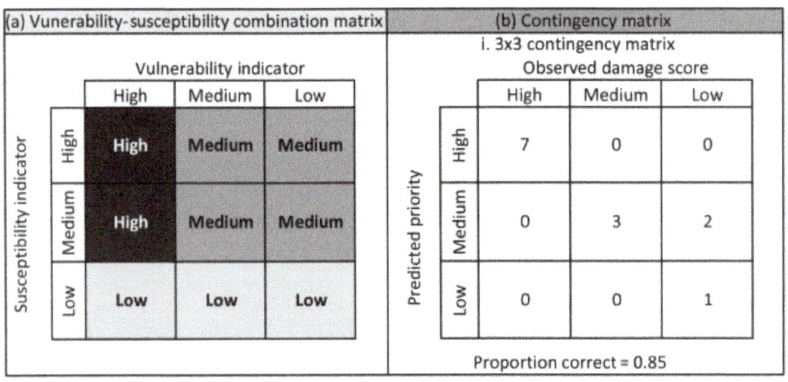

Figure 6. (a) Vulnerability–susceptibility combination matrix. (b) Contingency matrix.

Conditions of lack of well-being are shown to be concentrated in the south of the study area. The demographic conditions are more variable, showing low values (or better conditions) in the watersheds in the south, where the land use is rural. Low values also occur in the north, where the degree of urbanisation is low due to the more formal urbanisation processes (see Fig. 5a). The spatial distribution of the indicator of lack of resilience and coping capacity (Fig. 5b) shows that the highest values are concentrated in the south-west of the study area where the education levels are lower and the road and health infrastructure are poorer. The same spatial trend is exhibited by the lack of preparedness and response capacity. The area south of the study area corresponds mainly to rural use, thus the physical exposure indicator shows low values (see Fig. 5a). The highest values are concentrated in the centre of the area where the density of population is high and the economic activities are located.

The spatial distribution of the overall indicator and the derived categories show that the high vulnerability watersheds are located in the centre of the study area and in the west.

Prioritisation of Watersheds According To the Qualitative Risk Indicator and Comparison with Damage Records

The proportion correct of all possible matrices according to Sect. 3.2.5 (see Fig. 2) resulted in the optimum matrix shown in Fig. 6a, the corresponding contingency matrix is shown in Fig. 6b with a proportion correct of 0.85. The prioritisation level obtained from the application of the combination matrix to the total vulnerability indicator and the susceptibility indicator for each watershed is shown in Fig. 7a. The results were assigned to the watersheds delineated up to the discharge into the Tunjuelo River or into the storm water system, in order to facilitate the visualisation. The damage categorisation of the study area using the database with historical records according to Table 3 is shown in Fig. 7b with range categories classified as high, medium and low. This shows that the most significant damages, corresponding to the highest scores for the impact of flood events, are concentrated in the central zone of the study area. The comparison between Figs. 7a and b shows that the indicators identify a similar spatial distribution of priority levels in the central zone of the study area that is consistent with the distribution of recorded damage. This is reflected in the proportion correct of 0.85.

Sensitivity Analysis of the Vulnerability Indicator

Figure 8 shows the box plots of the values of the vulnerability indicator obtained from the sensitivity analysis in application of PCA as well as the weighting scheme as explained in Sect. 3.2.3. The values of the vulnerability indicator obtained from the proposed method were also plotted for reference. The most influential input factors correspond to the weights used both in the construction of the lack of resilience indicator and in the construction of the total vulnerability indicator. The thick vertical bars for each watershed show the interquartile range of the total vulnerability indicator, with the thin bars showing the range (min–max). While the range of the indicator for some watersheds is substantial, the sensitivity of the watersheds being classified differently in terms of low, medium or high vulnerability was evaluated through the number of watersheds for which the interquartile range intersects with the classification threshold. For seven watersheds classified as being of medium vulnerability, the interquartile range crosses the upper limits of classification of medium vulnerability, while for four watersheds classified as being of high vulnerability, the range crosses that same threshold. For the lower threshold, only two watersheds classified as being of low vulnerability are sensitive to crossing into the class of medium vulnerability.

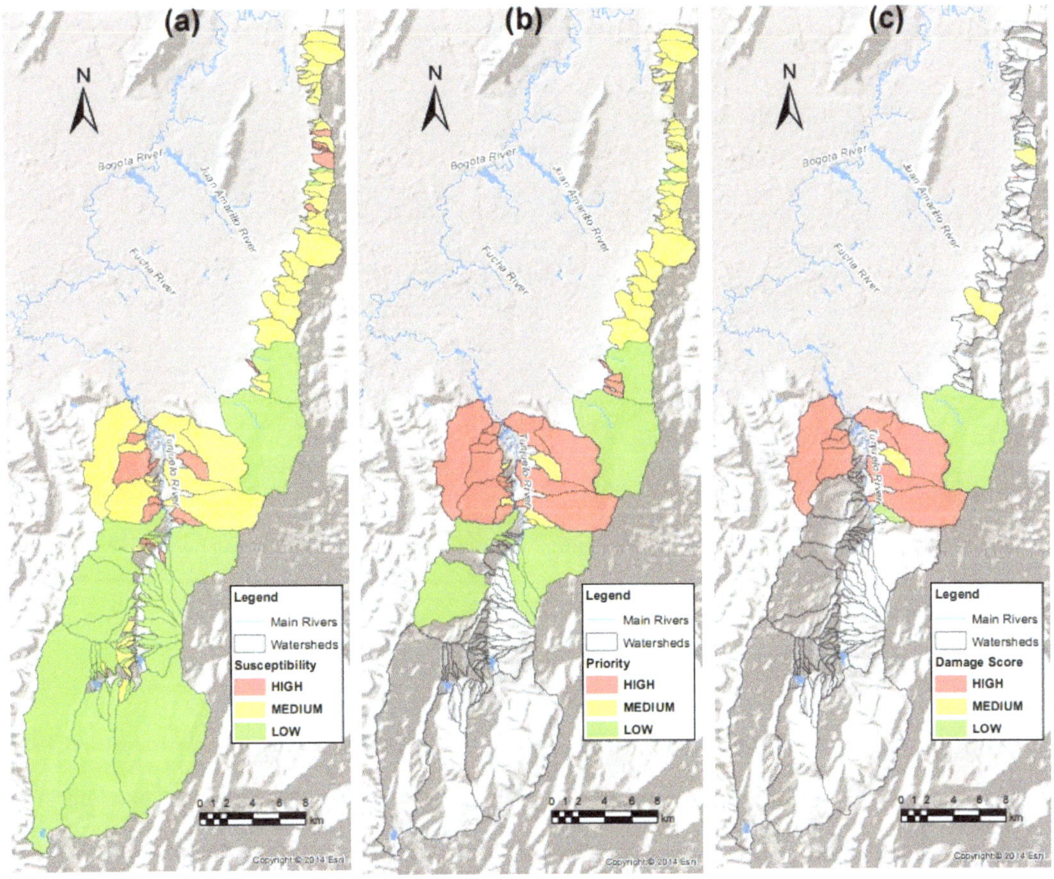

Figure 7. (a) Susceptibility classification of the study area. (b) Prioritisation according to the qualitative risk indicator. (c) Damage categorisation.

DISCUSSION

Exposure Areas

Existing flood hazard maps developed using hydraulic models that were available for a limited set of the watersheds in the study area were used to assess the suitability of the proposed simplified methods to identify flood prone areas and extend the flood exposure information over the entire study area. The areas exposed to debris flows obtained through the modified single-flow direction propagation algorithm show a good representation of the recorded events (Rogelis and Werner, 2013). However, in the Eastern Hills, where the streams flow towards a flat area, the results of the algorithms tend to overestimate the propagation areas since in these algorithms, the flood extent is dominated purely by the morphology and the flood volume is not considered, which means there is no limitation to the flood extent (see Fig. 4).

Each of the methods applied for flood plain delineation has strengths and weaknesses, while the combination of the results from these methods provides a consistent and conservative estimate of the exposure areas. The MRVBF index allows the identification of valley bottoms at several scales. In the mountainous areas, zones contiguous to the streams are identified, and in areas of marked topography the results are satisfactory, allowing a determination of a threshold of the index to define flood-prone areas. In the case of the buffers (see Fig. 4c), a depth of 3 m seems adequate to represent the general behaviour of the streams. The combination of the methods allowed the estimation of exposure areas based on the morphology (low and flat areas), elevation difference with the stream level (less than 3 m) and capacity to propagate debris flows.

Representativeness and Relative Importance of Indicators

The principal component analysis of the variables used to explain socio-economic fragility showed that the 16 variables that were chosen for the analysis could be grouped into two principal components strongly associated with the demography and the lack of well-being in the area. The latter was found to explain most of the variance in the data (80 % as shown in Table 4). The demography intermediate indicator describes the dependent population and the origin of the population. Dependent population (children, elderly and disabled) has been also identified by other authors as an important descriptor of vulnerability (Cutter et al., 2003; Fekete, 2009), associated with the limited capacity of this population to evacuate (Koks et al., 2015) and recover (Rygel et al., 2006). The origin of the population (illegal settlements and percentage of population in strata 1 and 2) shows the proportion of population resulting mainly from forced migration due to both violence and poverty (Beltrán, 2008).

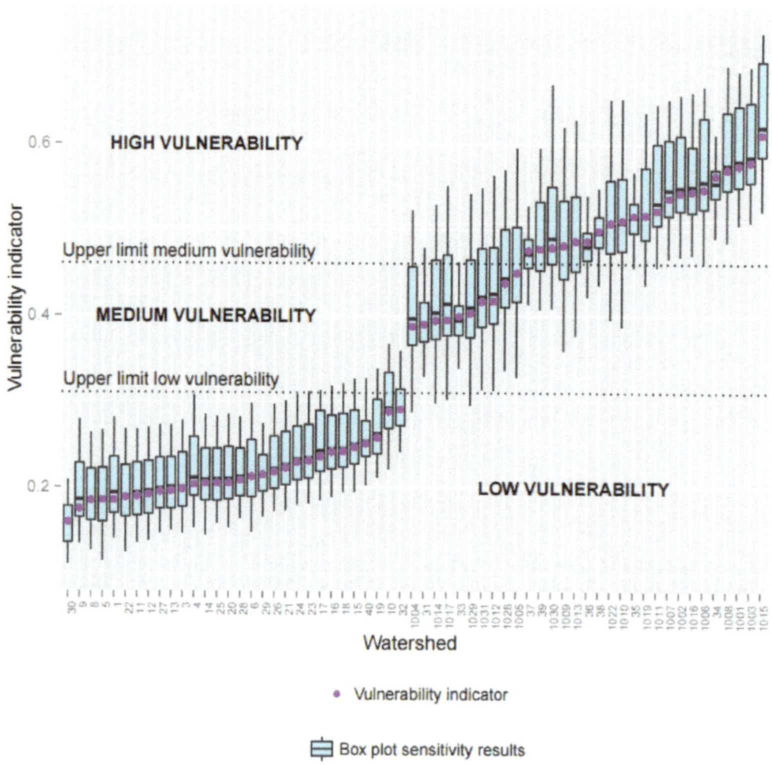

Figure 8. Sensitivity analysis of the vulnerability indicator. Note that the numbering of the watersheds in the Eastern Hills goes from 1 to 40 and in the Tunjuelo River basin from 1000 to 1066.

The lack of well-being indicator is composed of 14 strongly correlated variables that are commonly used to measure livelihood conditions. Poverty does not necessarily mean vulnerability, though the lack of economic resources is associated with the quality of construction of the houses, health and education, which are factors that influence the capability to face an adverse event (Rygel et al., 2006). The variable "women-headed households" is correlated with the principal component related to lack of well-being as identified by Barrenechea et al. (2000). Even if this condition of the families is not necessarily a criterion related to poverty, women-headed households with children are related to vulnerability conditions. The woman in charge of the family is responsible for the economic, affective and psychological well-being of other persons, especially her children and the elderly, in addition to domestic tasks and the family income. This condition suggests more assistance during emergency and recovery (Barrenechea et al., 2000).

In the case of the lack of resilience and coping capacity indicators, the PCA resulted in the intermediate indicators' lack of education and lack of preparedness and response capacity. The former captures limitations in knowledge about hazards in individuals (Müller et al., 2011) and the latter is linked to the institutional capacity for response. Risk perception and early warning are Boolean indicators. Since risk perception is based on the occurrence or non-occurrence of floods, aspects such as specific knowledge of the population about their exposure are not included. In the case of flood early warning, the effectiveness of the systems is not considered. These are aspects that can be taken into account for future research and that can help to improve the lack of resilience and coping capacity indicators.

Regarding the physical exposure, the method that was applied does not involve hazard intensity explicitly and different levels of physical fragility are not considered due to limitations in the available data. The indicators used to express physical exposure imply that the more elements exposed, the more damage, neglecting the variability in the degree of damage that the exposed elements may have. Other regional indicator-based approaches have used physical characteristics of the exposed structures to differentiate levels of damage according to structure type (Kappes et al., 2012) and economic values of the exposed elements (Liu and Lei, 2003). This is a potential area of improvement of the indicator, since the degree of damage depends on the type and intensity of the hazard and the characteristics of the exposed element. However, the development of indicators of physical characteristics and economic values is highly data-demanding, therefore future applications could be aimed at efficiently using existing information and applying innovative data collection methods at a regional level for the improvement of the physical indicators.

Sensitivity of the Vulnerability Indicator

The interquartile ranges cross the thresholds between categories of low, medium and high vulnerability only in the case of 13 watersheds (see Fig. 8). This means that only these 13 watersheds are sensitive to the criteria selected for the analysis. In 11 of these, the category changes between medium vulnerability and high vulnerability and in the remaining two, the change is from low to medium vulnerability. Watersheds with values of the vulnerability indicator out of the intermediate ranges of the thresholds are robust to the change in the modelling criteria. Clearly, these results are dependent on the number of categories. While introducing more categories may provide more information to differentiate watersheds, the identification of the category of the watersheds may become more difficult due to the sensitivity to the results. Therefore, in order to preserve identifiability of the vulnerability category of the watersheds, no more than three categories could be used. Indicator-based regional studies, which classify vulnerability into three categories, have been shown to provide useful information for flood risk management (Kappes et al., 2012; Liu and Li, 2015; Luino et al., 2012).

The impact on the proportion correct of a shift of category for the 13 watersheds mentioned above can only be assessed for the two watersheds where flood records are available. This does not result in changes in the contingency matrix shown in Fig. 6b. With respect to assigning priority to the watersheds, only 7 (7 % of the total) of the 13 watersheds that showed sensitivity to a shift of the vulnerability categories were found to be sensitive to a change in priority (high/medium), which reflects the robustness of the analysis using the considered categories.

Usefulness of the Prioritisation Indicator

The resulting vulnerability–susceptibility combination matrix, shown in Fig. 6a, shows that in the study area, high priorities are determined by high vulnerability conditions and medium and high susceptibility. This would suggest that, high vulnerability is a determinant condition of priority, since areas with high vulnerability can only be assigned a low priority if the susceptibility to flash floods/debris flows is low. This also shows that the analysis of the indicators that compose the vulnerability index allows insight to be gained into the drivers of high vulnerability conditions. Figure 5 shows that high vulnerability watersheds are the result of

- high socio-economic fragility and high lack of resilience and coping capacity (west of the lower and middle basin of the Tunjuelo River; and watershed most to the south of the Eastern Hills);
- high socio-economic fragility and high physical exposure (east of the middle basin of the Tunjuelo River);
- high physical exposure levels (south of the Eastern Hills).

This information is useful for regional allocation of resources for detailed flood risk analysis, with the advantage that the data demand is low in comparison with other indicator-based approaches (Kappes et al., 2012; Fekete, 2009). Furthermore most weights are determined from a statistical analysis with a low influence of subjective weights, which is an advantage over expert weighting, where large variations may occur depending on the expert's perspective (Müller et al., 2011). However, more detailed flood risk management decision-making cannot be informed by the level of resolution used in this study. Studies where assessments are carried out at the level of house units would be needed for planning of mitigation measures, emergency planning and vulnerability reduction (Kappes et al., 2012). Although the proposed procedure could be applied at that more detailed level, this could not be done due to the availability of information. This is a common problem in regional analyses (Kappes et al., 2012) where collecting a large amount of data at high resolution is a challenge. Nevertheless, future advances in collection of data could be incorporated in the proposed procedure, yielding results at finer resolutions. The challenge not only lies in collecting data of good quality at high resolution that can be transformed into indicators, but also in producing data at the same pace as significant changes in variables that contribute to vulnerability take place in the study area. In this research, vulnerability was assessed statically, however, there is an increasing need for analyses that take into account the dynamic characteristics of vulnerability (Hufschmidt et al., 2005). Methods such as the one applied in this study can provide a tool to explore these dynamics since it can be adapted to different resolutions according to the data available.

CONCLUSIONS

In this paper a method to identify mountainous watersheds with the highest flood risk at the regional level is proposed. Through this, the watersheds to be subjected to more detailed risk studies can be prioritised in order to establish appropriate flood risk management strategies. The method is demonstrated in the steep, mountainous watersheds that surround the city of Bogotá (Colombia), where floods typically occur as flash floods and debris flows. The prioritisation of the watersheds is obtained through the combination of vulnerability with susceptibility to flash floods/debris flows. The combination is carried out through a matrix that relates levels of vulnerability and susceptibility with priority levels.

The analysis shows the interactions between drivers of vulnerability, and how the understanding of these drivers can be used to gain insight into the conditions that determine vulnerability to floods in mountainous watersheds. Vulnerability is expressed in terms of composite indicators: socio-economic fragility, lack of resilience and coping capacity and physical exposure. Each of these composite indicators is formed by an underlying set of constituent indicators that reflect the behaviour of highly correlated variables, and that represent characteristics of the exposed elements. The combination of these three component indicators allowed the calculation of a vulnerability indicator, from which a classification into high, medium and low vulnerability was obtained for the watersheds of the study area. Tracing back the composite indicators that generate high vulnerability provided an understanding of the conditions of watersheds that are more critical, allowing these to be targeted for more detailed flood risk studies. In the study area it is shown that those watersheds with high vulnerability are categorised to be of high priority, unless the susceptibility is low, indicating that the vulnerability is the main contributor to risk. Furthermore, the contributing components that determine high vulnerability could be identified spatially in the study area. The developed methodology can be applied to other areas, although adaptation of the variables considered may be required depending on the setting and the available data. The proposed method is flexible to the availability of data, which is an advantage for assessments in mountainous developing cities and when the evolution in time of variables that contribute to vulnerability is taken into account.

The results also demonstrate the need for a comprehensive documentation of damage records, as well as the potential for improvement of the method. Accordingly, further research should be focused on (i) the use of smaller units of analysis than the watershed scale, which was used in this study; (ii) improvement of physical exposure indicators, incorporating type of structures and economic loss; and (iii) incorporation of more detailed information about risk perception and flood early warning.

ACKNOWLEDGEMENTS

This work was funded by the UNESCO-IHE Partnership Research Fund – UPaRF – in the framework of the FORESEE project. We wish to express our gratitude to the Fondo de Prevención y Atención de Emergencias de Bogotá for providing the flood event data for this analysis. We also would like to thank Sven Fuchs and another anonymous reviewer, whose comments helped improve and clarify this paper.

REFERENCES

1. Akbas, S., Blahut, J., and Sterlacchini, S.: Critical assessment of existing physical vulnerability estimation approaches for debris flows, in: Landslide processes: from geomorphological mapping to dynamic modelling, 229–233, 2009.
2. Albano, R., Sole, A., Adamowski, J., and Mancusi, L.: A GIS-based model to estimate flood consequences and the degree of accessibility and operability of strategic emergency response structures in urban areas, Nat. Hazards Earth Syst. Sci., 14, 2847–2865, doi:10.5194/nhess-14-2847-2014, 2014.
3. Balica, S. F., Wright, N. G., and van der Meulen, F.: A flood vulnerability index for coastal cities and its use in assessing climate change impacts, Nat. Hazards, 64, 73–105, doi:10.1007/s11069- 012-0234-1, 2012.
4. Barrenechea, J., Gentile, E., González, S., and Natenson, C.: Una propuesta metodológica para el estudio de la vulnerabilidad social en el marco de la teoría social del riesgo, in: IV Jornadas de Sociología, edited by: Facultad de Ciencias Sociales Facultad de Ciencias Sociales, UBA, Buenos Aires, Argentina, 1–13, 2000.
5. Barroca, B., Bernardara, P., Mouchel, J. M., and Hubert, G.: Indicators for identification of urban flooding vulnerability, Nat. Hazards Earth Syst. Sci., 6, 553–561, doi:10.5194/nhess-6-553- 2006, 2006.
6. Bernal, G., Rosero, M., Cadena, M., Montealegre, J., and Sanabria, F.: Estudio de la Caracterización Climática de Bogotá y cuenca alta del Río Tunjuelo, Tech. rep., Instituto de Hidrología, Meteorología y Estudios Ambientales IDEAM – Fondo de Prevención y Atención de Emergencias FOPAE, Bogotá, Colombia, 2007.
7. Birkmann, J., Cardona, O. D., Carreño, M. L., Barbat, A. H., Pelling, M., Schneiderbauer, S., Kienberger, S., Keiler, M., Alexander, D., Zeil, P., and Welle, T.: Framing vulnerability, risk and societal responses: the MOVE framework, Nat. Hazards, 67, 193–211, 2013.
8. Birkmann, J., Cardona, O. D., Carreño, M. L., Barbat, A. H., Pelling, M., Schneiderbauer, S., Zeil, P., and Welle, T.: Theoretical and Conceptual Framework for the Assessment of Vulnerability to Natural Hazards and Climate Change in Europe, in: Assessment of Vulnerability to Natural Hazards: A European Perspective, edited by: Birkmann, J., Kienberger, S., and Alexander, D., Elsevier, San Diego, California, USA, chap. 1, 1–19, 2014.
9. Cardona, O.: The need for rethinking the concepts of vulnerability and risk from a holistic perspective: a necessary review and criticism for effective risk management, in: Mapping Vulnerability: Disasters, Development and People, edited by: Bankoff, G., Frerks, G., and Hilhorst, D., chap. 3, Earthscan Publishers, London, UK, 37–51, 2003.
10. Cardona, O. D., Van Aalts, M. K., Birkmann, J., Fordham, M., Glenn, M., Perez, R., Pulwarty, R. S., Schipper, L. F., and Sinh, B. T.: Determinants of Risk : Exposure and Vulnerability, in: Managing the Risks of Extreme Events and Disasters to Advance Climate Change Adaptation. A Special Report of Working Groups I and II of the Intergovernmental Panel on Climate Change (IPCC), Cambridge University Press, Cambridge, UK, and New York, NY, USA, chap. Determinan, 65–108, 2012.
11. Carroll, J.: An analytical solution for approximating simple structure in factor analysis, Psychometrika, 18, 23–38, doi:10.1007/BF02289025, 1953.
12. Costa, J.: Rheologic, geomorphic, and sedimentologic differentiation of water floods, hyperconcentrated flows, and debris flows, in: Flood Geomorphology, edited by: Baker, V. R., Kochel, R. C., and Patton, P. C., Wiley, New York, USA, 113–122, 1988.
13. Cutter, S., Boruff, B., and Shirley, W.: Social vulnerability to environmental hazards, Soc. Sci. Quart., 84, 242–261, doi:10.1111/1540-6237.8402002, 2003.
14. Cutter, S. L., Barnes, L., Berry, M., Burton, C., Evans, E., Tate, E., and Webb, J.: A place-based model for understanding community resilience to natural disasters, Global Environ. Chang., 18, 598–606, 2008.
15. DPAE: Diagnóstico Técnico 1836, Tech. rep., Direccion de Prevención y Atención de Emergencias de Bogotá, Bogotá, Colombia, 2003a
16. DPAE: Diagnóstico Técnico 1891, Tech. rep., Direccion de Prevención y Atención de Emergencias de Bogotá, Bogotá, Colombia, 2003b.
17. Esty, D., Srebotnjak, T., Kim, C., Levy, M., Sherbinin, A., and Anderson, B.: Pilot 2006 Environmental Performance Index, Tech. rep., Yale Center for Environmental Law & Policy, New Haven, USA, 2006.
18. Fekete, A.: Validation of a social vulnerability index in

context to river-floods in Germany, Nat. Hazards Earth Syst. Sci., 9, 393– 403, doi:10.5194/nhess-9-393-2009, 2009.

19. Fuchs, S. and Holub, M.: Reducing Physical Vulnerability to Mountain Hazards, in: 12th Congress INTERPRAEVENT 2012, Grenoble, France, 675–686, 2012.
20. Fuchs, S., Heiss, K., and Hübl, J.: Towards an empirical vulnerability function for use in debris flow risk assessment, Nat. Hazards Earth Syst. Sci., 7, 495–506, doi:10.5194/nhess-7-495- 2007, 2007.
21. Fuchs, S., Tsao, T.-C., and Keiler, M.: Quantitative Vulnerability Functions for use in Mountain Hazard Risk Management, in: 12th Congress INTERPRAEVENT 2012, Grenoble, France, 885–896, 2012.
22. Gallant, J. and Dowling, T.: A multiresolution index of valley bottom flatness for mapping depositional areas, Water Resour. Res., 39, 1347–1360, 2003.
23. Greiving, S.: Multi-risk assessment of Europe's regions, Measuring Vulnerability to Natural Hazards: Toward Disaster Resilient Societies, United Nations University, New York, USA, 2006.
24. Harris, C. and Kaiser, H.: Oblique factor analytic solutions by orthogonal transformations, Psychometrika, 29, 347–362, doi:10.1007/BF02289601, 1964.
25. Horn, J.: A rationale and test for the number of factors in factor analysis, Psychometrika, 30, 179–185, doi:10.1007/BF02289447, 1965.
26. Hu, K. H., Cui, P., and Zhang, J. Q.: Characteristics of damage to buildings by debris flows on 7 August 2010 in Zhouqu, Western China, Nat. Hazards Earth Syst. Sci., 12, 2209–2217, doi:10.5194/nhess-12-2209-2012, 2012.
27. Hufschmidt, G., Crozier, M., and Glade, T.: Evolution of natural risk: research framework and perspectives, Nat. Hazards Earth Syst. Sci., 5, 375–387, doi:10.5194/nhess-5-375-2005, 2005
28. IWR: Flood Risk Management Approaches. As being practiced in Japan, the Netherlands, United Kingdom and United States, United States Army Corps of Engineers, Washington, DC, USA, 2011.
29. Jakob, M.: Debris-flow hazard analysis, in: Debris-flow hazards and related phenomena, edited by: Jakob, M. and Hungr, O., Springer-Verlag, Berlin, Germany, chap. 17, 411–443, 2005.
30. Jakob, M., Porter, M., Savigny, K. W., and Yaremko, E.: A geomorphic approach to the design of pipeline crossings of mountain streams, Proceedings of IPC 2004: International Pipeline Conference, 4–8 October 2004, Calgary, Alberta, Canada, 1–8, 2004.
31. Jha, A., Bloch, R., and Lamond, J.: Cities and Flooding A guide to Integrated Urban Flood Risk Management for the 21st Century, The World Bank Location: Washington, DC, USA, 2012.
32. Kaiser, H.: The varimax criterion for analytic rotation in factor analysis, Psychometrika, 23, 187–200, doi:10.1007/BF02289233,1958.
33. Kappes, M., Papathoma-Köhle, M., and Keiler, M.: Assessing physical vulnerability for multi-hazards using an indicator-based methodology, Appl. Geogr., 32, 577–590, http://linkinghub. elsevier.com/retrieve/pii/S0143622811001378, 2012.
34. Kiers, H.: SIMPLIMAX: Oblique rotation to an optimal target with simple structure, Psychometrika, 59, 567–579, doi:10.1007/BF02294392, 1994.
35. Larsen, M., Wieczorek, G., Eaton, L., and Torres-Sierra, H.: Natural Hazards on Aluvial Fans: The Debris Flow and Flash flood disaster of December 1999, Vargas State, Venezuela, in: Proceedings of the Sixth Caribbean Islands Water Resources Congress, edited by: Sylva, W., Mayagüez, Puerto Rico, 00965, 1–7, 2001.
36. Lavigne, F. and Suwa, H.: Contrasts between debris flows, hyperconcentrated flows and stream flows at a channel of Mount Semeru, East Java, Indonesia, Geomorphology, 61, 41–58, 2004.
37. Liu, Y., Zhou, J., Song, L., Zou, Q., Guo, J., and Wang, Y.: Efficient GIS-based model-driven method for flood risk management and its application in central China, Nat. Hazards Earth Syst. Sci., 14, 331–346, doi:10.5194/nhess-14-331-2014, 2014.
38. Lo, W.-C., Tsao, T.-C., and Hsu, C.-H.: Building vulnerability to debris flows in Taiwan: a preliminary study, Nat. Hazards, 64, 2107–2128, http://www.springerlink.com/index/10. 1007/s11069-012-0124-6, 2012.
39. Luino, F., Cirio, C. G., Biddoccu, M., Agangi, A., Giulietto, W., Godone, F., and Nigrelli, G.: Application of a model to the evaluation of flood damage, Geoinformatica, 13, 339–353, 2009.
40. Luino, F., Turconi, L., Petrea, C., and Nigrelli, G.: Uncorrected land-use planning highlighted by flooding: the Alba case study (Piedmont, Italy), Nat. Hazards Earth Syst. Sci., 12, 2329–2346, doi:10.5194/nhess-12-2329-2012, 2012.
41. Mazzorana, B., Levaggi, L., Keiler, M., and Fuchs, S.: Towards dynamics in flood risk assessment, Nat. Hazards Earth Syst. Sci., 12, 3571–3587, doi:10.5194/nhess-12-3571-2012, 2012.
42. Molinari, D., Molini, S., and Ballio, F.: Flood Early Warning Systems: Knowledge and Tools for Their Critical Assessment, WIT Press, Southampton, UK, 2013.
43. Müller, A., Reiter, J., and Weiland, U.: Assessment of urban vulnerability towards floods using an indicator-based approach – a case study for Santiago de Chile, Nat. Hazards Earth Syst. Sci., 11, 2107–2123, doi:10.5194/nhess-11-2107-2011, 2011.
44. Nardi, F., Vivoni, E. R., and Grimaldi, S.: Investigating a floodplain scaling relation using a hydrogeomorphic delineation method, Water Resour. Res., 42, 1–15, 2006.
45. Nkwunonwo, U., Whitworth, M., and Baily, B.: Relevance of Social Vulnerability Assessment to Flood Risk Reduction in the Lagos Metropolis of Nigeria, British Journal of Applied Science & Technology, 8, 366–382, http://www.sciencedomain.org/ abstract.php?iid=1072&id=5&aid=8949, 2015.
46. Osorio, J. A.: El río Tunjuelo en la historia de Bogotá,

1900–1990, Alcaldia Mayor de Bogotá, Bogotá, Colombia, 2007.

47. Papathoma-Köhle, M., Kappes, M., Keiler, M., and Glade, T.: Physical vulnerability assessment for alpine hazards: state of the art and future needs, Nat. hazards, 58, 645–680, http://link.springer. com/10.1007/s11069-010-9632-4, 2011.

48. Papathoma-Köhle, M., Keiler, M., Totschnig, R., and Glade, T.: Improvement of vulnerability curves using data from extreme events: debris flow event in South Tyrol, Natural Hazards, 64, 2083–2105, http://link.springer.com/10.1007/ s11069-012-0105-9, 2012.

49. Quan Luna, B., Blahut, J., van Westen, C. J., Sterlacchini, S., van Asch, T. W. J., and Akbas, S. O.: The application of numerical debris flow modelling for the generation of physical vulnerability curves, Nat. Hazards Earth Syst. Sci., 11, 2047–2060, doi:10.5194/nhess-11-2047-2011, 2011.

50. Rogelis, M. C. and Werner, M.: Regional debris flow susceptibility analysis in mountainous peri-urban areas through morphometric and land cover indicators, Nat. Hazards Earth Syst. Sci., 14, 3043–3064, doi:10.5194/ nhess-14-3043-2014, 2014.

51. Rygel, L., O'sullivan, D., and Yarnal, B.: A method for constructing a social vulnerability index: an application to hurricane storm surges in a developed country, Mitigation and Adaptation Strategies for Global Change, 11, 741–764, doi:10.1007/s11027-006- 0265-6 , 2006.

52. Safaripour, M., Monavari, M., and Zare, M.: Flood risk assessment using GIS (case study: Golestan province, Iran), Pol. J. Environ. Stud., 21, 1817–1824, 2012.

53. Santo, A., Santangelo, N., Di Crescenzo, G., Scorpio, V., De Falco, M., and Chirico, G. B.: Flash flood occurrence and magnitude assessment in an alluvial fan context: the October 2011 event in the Southern Apennines, Nat. Hazards, http://link.springer.com/ 10.1007/s11069-015-1728-4, 2015.

54. Schmidtlein, M. C., Deutsch, R., Piegorsch, W. W., and Cutter, S. L.: Building indexes of Vulnerability : a sensitivity analysis of the Social Vulnerability Index, Risk Anal., 28, 1099–1114, 2008

55. Sterlacchini, S., Akbas, S. O., Blahut, J., Mavrouli, O.-C., Garcia, C., Luna, B. Q., and Corominas Dulcet, J.: Methods for the characterization of the vulnerability of elements at risk, in: Mountain risks: from prediction to management and governance, Springer Netherlands, Dordrecht, the Netherlands, 33–273, http: //hdl.handle.net/2117/20764, 2014.

56. Thieken, A., Merz, B., Kreibich, H., and Apel, H.: Methods for flood risk assessment: Concepts and challenges, International Workshop on Flash Floods in Urban Areas Muscat, 4–6 September 2006, Oman, 1–12 2006.

57. Totschnig, R. and Fuchs, S.: Mountain torrents: Quantifying vulnerability and assessing uncertainties, Eng. Geol., 155, 31–44, doi:10.1016/j.enggeo.2012.12.019, 2013.

58. Totschnig, R., Sedlacek, W., and Fuchs, S.: A quantitative vulnerability function for fluvial sediment transport, Nat. Hazards, 58, 681–703, http://link.springer.com/10.1007/ s11069-010-9623-5, 2011.

59. Tsao, T.-C., Hsu, W.-K., Cheng, C.-T., Lo, W.-C., and Chen, C.- Y.: A preliminary study of debris flow risk estimation and management in Taiwan, International symposium interpraevent in the Pacific Rim, 26–30 April 2010, Taipei, 930–939, 2010.

60. United Nations General Assembly: Resolution adopted by the General Assembly on 3 June 2015, Sendai Framework for Disaster Risk Reduction 2015–2030, 08955, 1–24, http://www. preventionweb.net/files/ resolutions/N1516716.pdf, 2015.

61. van Westen, C., Kappes, M., Luna, B., Frigerio, S., Glade, T., and Malet, J.-P.: Medium-Scale Multi-hazard Risk Assessment of Gravitational Processes, in: Mountain Risks: From Prediction to Management and Governance SE – 7, 34, 201–231, http://dx.doi.org/10.1007/978-94-007-6769-0_7, 2014.

62. Varnes, D. J.: Landslide hazard zonation: a review of principles and practice, Tech. rep., Commission on landslides of the IAEG, UNESCO, 1984.

63. Wisner, B., Blaikie, P., and Cannon, T.: At Risk: natural hazards, people's vulnerability and disasters, Routledge, second edn., 2003.

INDEX